Immunologic Defects in Laboratory Animals 2

Immunologic Defects in Laboratory Animals 2

Edited by

M. Eric Gershwin
University of California School of Medicine
Davis, California

and

Bruce Merchant
Food and Drug Administration
Bethesda, Maryland

PLENUM PRESS • NEW YORK AND LONDON

Library of Congress Cataloging in Publication Data

Main entry under title:

Immunologic defects in laboratory animals.

Includes index.

1. Immunological deficiency syndromes. 2. Laboratory animals. 3. Mice as laboratory animals. I. Gershwin, M. Eric, 1946– . II. Merchant, Bruce.

QR188.35.I47 616.07'9 80-28192

ISBN 978-1-4684-8654-4 ISBN 978-1-4684-8652-0 (eBook)

DOI 10.1007/978-1-4684-8652-0

Softcover reprint of the hardcover 1st edition 1981

A Division of Plenum Publishing Corporation
233 Spring Strreet, New York, N. Y. 10013

Contributors

PATRICIA M. BEALMEAR, Departments of Biological Resources and Dermatology, Roswell Park Memorial Institute, Buffalo, New York 14263

H. C. DUNG, Department of Anatomy, The University of Texas Health Science Center at San Antonio, San Antonio, Texas 78284

MICHAEL M. FRANK, Laboratory of Clinical Investigation, National Institue of Allergy and Infectious Diseases, National Institutes of Health, Bethesda, Maryland 20205

THELMA GAITHER, Laboratory of Clinical Investigation, National Institue of Allergy and Infectious Diseases, National Institutes of Health, Bethesda, Maryland 20205

M. ERIC GERSHWIN, Section of Rheumatology–Clinical Immunology, School of Medicine, University of California, Davis, California 95616

CARL H. HAMMER, Laboratory of Clinical Investigation, National Institue of Allergy and Infectious Diseases, National Institutes of Health, Bethesda, Maryland 20205

PAUL W. KINCADE, Sloan–Kettering Institute for Cancer Research, Donald S. Walker Laboratory, Rye, New York 10580

S. P. LEIBO, Biology Division, Oak Ridge National Laboratory, Oak Ridge, Tennessee 37830

DONALD L. LODMELL, Laboratory of Persistent Viral Diseases, Rocky Mountain Laboratories, National Institute of Allergy and Infectious Diseases, National Institutes of Health, Hamilton, Montana 59840

TRAVIS C. MCGUIRE, Department of Microbiology and Pathology, College of Veterinary Medicine, Washington State University, Pullman, Washington 99164

DAVID R. MILICH, Section of Rheumatology–Clinical Immunology, School of Medicine, University of California, Davis, California 95616

EDWIN D. MURPHY, The Jackson Laboratory, Bar Harbor, Maine 04609

LANCE E. PERRYMAN, Department of Microbiology and Pathology, College of Veterinary Medicine, Washington State University, Pullman, Washington 99164

JOHN L. PORTIS, Laboratory of Persistent Viral Diseases, Rocky Mountain Laboratories, National Institute of Allergy and Infectious Diseases, National Institutes of Health, Hamilton, Montana 59840

FRED WILLIAM QUIMBY, Division of Laboratory Animal Medicine and Department of Pathology, Cornell University Medical College, New York, New York 10021

NOEL R. ROSE, Department of Immunology and Microbiology, Wayne State University School of Medicine, Detroit, Michigan 48201

ROBERT J. SCIBIENSKI, Department of Medical Microbiology, School of Medicine, University of California, Davis, California 95616

ROY S. SUNDICK, Department of Immunology and Microbiology, Wayne State University School of Medicine, Detroit, Michigan 48201

Preface

Tiselius demonstrated that the immunologically active components of immune sera migrated electrophoretically in the gamma globulin region. His findings illuminated the classic observations of Jenner regarding development of resistance to infection, and those of von Pirquet, Pasteur, and Arthus regarding the transfer and specificity of resistance. Conceptual integration of these observations provided the impetus for the present modern era of immunology. Subsequent to Tiselius's work, multiple, rapid advances have occurred in the study of congenital and acquired immune deficiency states in mice, chickens, and humans. These studies have readily demonstrated that the immunologic ability of an organism to protect itself from environmental influences is a prerequisite for survival. Indeed, this necessity for protection from microenvironmental influences has promoted the evolutionary development of immunologic diversification, namely, host dependence upon a sophisticated, multifaceted network of cells and effector mechanisms responsible for the clearance and neutralization of toxins and potentially harmful pathogens. The obligate dependence of animals upon the functional integrity of their immunologic systems is illustrated by the ready invasion of ubiquitous organisms when the host is in a state of immune defense derangement. Nevertheless, derangements in immune function can range from partial to complete and can be compatible with survival. The consequences of such derangements run the gamut from subclinical disease to inevitable mortality.

To further our understanding of such events, investigators have concentrated on the study and manipulation of a number of spontaneously occurring animal models that have genetically associated immune defects. Although direct extrapolation of data from animal to man must proceed with caution, the use of such immunodeficient animal systems, both *in vitro* and *in vivo,* has been an enormous help in the development of immunologic concepts.

Our intention in preparing these books was to place between covers a functionally rather than taxonomically organized exposition of the major immunologic defects as they occur in laboratory animals. We hope that this effort will be of service to the scientific community and that it will stimulate further investigations of immunologic defects. This work was not intended to be exhaustive, but rather illustrative of the better-studied immune defects and of the diverse and interrelated categories that they manifest.

There has been a rapidly accelerating recognition of immune defects in recent years.

This accelerating recognition of immune defects, especially in laboratory mice, leads us to believe that we have only begun to identify the many forms of immune defects that may take place. Given sufficient time, every possible immune defect should ultimately occur. Eventually, immunologists may well identify and characterize a myriad of immune defects representative of each successive biochemical step in each separate component of the complex immune system.

Despite our limited current knowledge of the extent and variation of immune deficiencies, defects in species as diverse as horses, mink, dogs, guinea pigs, rats, and chickens are represented within this work. In addition, pigs, sheep, and quokkas have been given additional attention for pecularities expressed by their immune systems either for reasons of unique environment or because of unique ontologic development. Nonetheless these books deal predominantly with immune defects as recognized in various strains of laboratory mice. They generally concentrate on those defects already subjected to some systematic studies and thus reflect the vagaries of contemporary inquiry. Immune defects in mice have received much attention, but many other species have been relatively neglected. If dogs, for example, were all tiny animals with very short gestation periods and were available as highly inbred strains, it is quite possible that they might represent the most convenient and preferred laboratory animal and thus be the species in which we would have our greatest knowledge of immune defects.

Any functional treatment of immune defects in laboratory animals might be expected to concentrate not only on reasonably distinct T-cell, B-cell, macrophage, and hematologic defects, but also on complex, less well-defined defects which, for example, could be manifested in the form of autoimmune diseases. In this respect, our work is no exception. We have also provided chapters, however, devoted to examination of the immune system under unique environmental circumstances as, for example, with the evaluation of immunity in germfree animals and with the analysis of immune potential during very early ontogenetic development.

Generally, the various sections and chapters of this book are devoted to comparisons between animals bearing genetically determined immune defects and normal animals of the same species. Except for an effort to compare animal immune defects with their relatively obvious counterparts in human clinical syndromes, comparison between species has been largely neglected. This has been intentional partly because of the recognized dangers of such extrapolation. The lack of comparison between species has also been due to the conceptual difficulties inherent in such comparisons. Nonetheless, one day we would hope to comprehend why whole species may lack some generally recognized immune capability that is ordinarily accepted as essential to survival in other species. For instance, the capacity to mount active humoral immune responses to soluble polysaccharide antigens is recognized as having distinct survival advantages both in humans and in mice, yet no one has acquired sufficient immunologic insight to explain why the total failure to perform this "essential" immune function does not seem to place rabbits, as a species, at any great evolutionary disadvantage.

Immune defects have often been referred to as experiments of nature, and it is apparent that these experiments often diminish the survival capacity of the individual. However, those complex immune defects that promote autoimmune disease with aging seem to exert little evolutionary pressure if most defective animals have already reached reproductive maturity prior to clinical expression of their autoimmune syndrome. On the other hand, some immune defects may represent experiments of nature that actually succeed. Mutative loss of an immune function that is no longer relevant to the environment of a given species may actually represent a merciful excision of excess genetic baggage. Some such immune defects may be critical experiments of nature that actually function to enhance alternative immune adaptations.

Although a primary objective of this work was to promote unification of our concepts of the immune system, on reflection one of its more essential messages may well be that the immune system still escapes full definition. Rather than representing a monolithic system imposed on many related species it probably represents a polylithic system, components of which have been selected and amplified in those species that found them most useful. There may well be numerous different immunologic solutions to similar evolutionary problems. As an example of such thinking, evidence is presented herein that suggests the existence, beyond B cells and T cells, of a third specific immune recognition system—possibly subserved by nonlymphoid cells—which has evolved in marsupial species such as the quokka. This evidence certainly invokes the question of whether such a system has been retained in most other mammals. If such a third specific recognition system exists in humans, it might well account for some examples of chronic graft rejection that appear refractory to all known efforts at immunosuppression.

When one considers that defects of any and all features of the immune system are ultimately probable, it is rather astounding that certain types of mutations have led to immune defects that have been so constrained in their expression. At least five separate mutations in three separate rodent species have resulted in thymic agenesis with coexistent lack of hair follicle development. Although minor phenotypic differences certainly exist between these mutations, they are each characterized by the loss of multiple parameters of T-cell function, such that their similarities are much more notable than their differences. When compared with osteopetrotic mutations, which are markedly diverse in character, or with the multiple beigelike mutations of rodents, the various mutations that produce athymia are notably constrained in their overall phenotypic expression. Does this indicate that certain stretches of genomic information that code for thymic development are especially prone to mutative change? And does it indicate that thymic development is invariably associated with development of the integument and its adnexae? Might the converse be true? To the best of our knowledge there has been no concerted or systematic effort aimed at identifying rodents with excessive hair development or in testing them for possible robust or excessive T-cell function.

The immune system has evolved almost certainly as a host adaptation aimed at meeting the demands of self recognition for hosts that must exist in a sea of microorganisms. But defense against microorganisms has not been the only consequence of immune system evolution. A unique commensal association has developed between the immune system and these microorganisms, since the immune system depends to a great extent for its own development upon the microbial flora of its natural environment. The immune systems of animals raised in the absence of microbial flora or their products are incompletely developed and, for all practical purposes, defective in expression of their genotypic capacities. T-cell functions appear to be much less essential in axenic than in conventional animals. Indeed, if thymic ablation experiments had been conducted only under germfree conditions, our current concepts of T-cell function would, in all likelihood, not have been deduced. Thus our appreciation of immune defects may often be relative and dependent upon the environment in which investigations of such possible defects are conducted.

Although these books concentrate heavily on genetic, anatomic, pathologic, and functional definitions of immune defects and on their conceptual and theoretical relationships, thcy also contain information aimed at coping practically with the study of immune disorders. Notable in this practical emphasis is the chapter on freezing of mammalian embryos. This sophisticated approach to genetic conservation should allow development of repositories of mouse embryos harboring one or more immunologic defects. The logistical aid that can be provided by this new approach is considerable for both immunologic and genetic studies. In the future, it seems likely that many rare, diverse, and exotic immune defects of laboratory animals may well be retained and protected from inadvertent loss by the use of this powerful technology. If we have indeed only scratched the surface in our detection, recognition, and characterization of immune defects, then capabilities such as mammalian embryo freezing will in all probability be instrumental to our future capability to categorize and characterize the many potential defects of the immune system.

Finally, this preface would not be complete without significant appreciation expressed to Jody Wall and Jeanie Redding for their untiring assistance.

M. E. G.
B. M.

Contents

4 The Pathogenesis of Autoimmunity in New Zealand Mice

DAVID R. MILICH AND M. ERIC GERSHWIN

5 Hemopoietic Abnormalities in New Zealand Black and Motheaten Mice

PAUL W. KINCADE

6 Lymphoproliferation (*lpr*) and Other Single-Locus Models for Murine Lupus

EDWIN D. MURPHY

Contents of Volume 1

PART IV. B-CELL DEFECTS

PART V. T-CELL DEFECTS

PART VI. MACROPHAGE DEFECTS

Part VII
Complex Defects

1
Autoimmune Thyroiditis in Obese-Strain Chickens

ROY S. SUNDICK AND NOEL R. ROSE

1. Introduction

The autoimmune thyroiditis spontaneously occurring in obese-strain chickens is remarkably similar to the human autoimmune disease, Hashimoto's thyroiditis. This fact is frequently surprising to both physicians and the lay public. However, upon closer examination the great degree of conservation of both the thyroid gland and the immune system during the evolutionary process is obvious. The histology, function, and hormonal regulation of the thyroid gland are practically identical among all vertebrates. Basic aspects of the immune system are also well conserved: T, B, and null cells, complement, and the structure of IgG, IgA, and IgM. Some characteristics of the immune system do, of course, differ between avian and mammalian species, most notably the presence of a bursa of Fabricius in the former.

To understand this spontaneous form of autoimmune thyroid disease it is important to review the procedure employed by R. K. Cole to develop the obese strain. It dates back to 1935 when a breeding program was initiated among a random-bred flock of White Leghorns (Cole and Hutt, 1973). Breeders were selected for the following characteristics: resistance to disease, especially neoplasms, maximum egg production, and optimal fertility and hatchability. By the late 1960s dramatic gains were recorded in all areas. Mortality over a 500-day period dropped from 52% to 13%, egg production doubled, and fertility and hatchability were high despite the limited inbreeding that occurred. This strain is called the Cornell C strain (CS).

Among the birds of the Cornell flock Cole noticed a few females (0.6%) that had the clinical symptoms of hypothyroidism (Cole, 1966). They were mated with CS males. Some of the offspring had similar symptoms and they in turn were selected. After several gen-

ROY S. SUNDICK AND NOEL R. ROSE • Department of Immunology and Microbiology, Wayne State University School of Medicine, Detroit, Michigan 48201.

FIGURE 1. Adult female White Leghorn chickens. The chicken on the right has severe autoimmune thyroiditis and has the phenotypic symptoms of hypothyroidism. These include much reduced wattle size, rumpled feathers, and a short "plump" stature. The chicken on the left is normal. [Reproduced with permission from Rose *et al.* (1978).]

erations of selection for hypothyroidism not only the females, but the males as well, developed this disease. At the present time most of the birds in this flock develop hypothyroidism. Due to one of the prominent symptoms, subcutaneous fat deposition, they were named the obese strain (OS) (Fig. 1).

Many of the details of this strain have already been described in earlier review articles. Readers primarily interested in the pathology and serology of OS chickens are directed to the review by Wick *et al.* (1974). Individuals interested in genetic factors predisposing toward thyroid autoimmunity are encouraged to read the review by Rose *et al.* (1976). This chapter will briefly summarize the information already reviewed plus more recent information. An attempt will be made to describe the etiology of this disease in the context of contemporary immunologic and endocrinologic theory.

To acquaint the reader with our perspective and that of others studying OS chickens, a series of questions follows. These comprise some of the more important ones that have been raised, and in a few cases answered. Why do B lymphocytes produce thyroglobulin (Tg) antibody? Why do lymphocytes infiltrate the thyroid gland? Are there decreased numbers of Tg-specific T-suppressor cells? Is the immune system hyperresponsive to all antigens (foreign and self antigens) or just to Tg? Are thyroid abnormalities present in OS chickens that allow lymphocyte sensitization and/or infiltration? How can the disease be prevented? What are the mechanisms for damage? How do genes at the major histocompatibility complex affect thyroiditis? Answers to the above questions would significantly enhance our understanding of autoimmune disease and of the normal functioning of the immune system. Similar questions have been posed in the study of human autoimmune thyroiditis and other autoimmune disorders.

2. Serology and Thyroid Pathology

Virtually all chickens of the OS flock spontaneously develop antibodies to Tg, but both the incidence and the mean titer are influenced by age (Table I). They first appear in birds of about 2 weeks of age and reach a maximum at about 17 weeks, declining somewhat thereafter. It is noteworthy that during the 9-year period from 1971 to 1980, the OS birds developed higher Tg antibody titers at an earlier age. Heterogeneous in character, they may be found in either IgG or IgM classes, and exhibit no evidence of oligoclonality in gel electrophoresis. In contrast to human Hashimoto's disease, OS antibodies to Tg fix complement. On the other hand, antibodies to thyroid microsomes are not a feature of OS disease. Although antibodies that bind thyroxine could be demonstrated, reactions with nonthyroidal antigens are absent, save for the low-titered reactions that occur with virtually all normal chicken sera. An exceptional finding is the antibody reactive with chicken erythrocyte nuclei described by Wick and his associates, which will be discussed later. In addition to circulating antibody, delayed hypersensitivity to Tg develops in OS chickens, as evidenced by a positive wattle test. However, this reaction occurs relatively late in the course of the disease.

Lymphoid cells first infiltrate the thyroid gland at about 1 week of age. Electron microscopic studies by Wick and Graf (1972) suggested that these early cells were plasma cells, although confirmatory studies using other techniques (e.g., anti-T-cell or anti-B-cell reagents) were not performed. The disease progresses rapidly with large numbers of lymphocytes infiltrating the gland. Some areas of the thyroid gland are diffusely infiltrated and other areas are so densely packed with lymphoid cells and germinal centers that they are indistinguishable from peripheral lymphoid tissues. By 3 or 4 weeks of age the thyroid glands of most OS $B^{13}B^{13}$ chickens are completely replaced by infiltrating cells (Figs. 2 and 3). The thyroid epithelial cells no longer incorporate $^{131}I^-$ and serum levels of thyroid hormones steadily fall (R. S. Sundick, personal observation). T and B lymphocytes from OS and CS chickens migrate into the already infiltrated gland, but not into a healthy "uninfiltrated" CS thyroid gland, or even into the lymphocyte-free areas of a young OS thyroid gland (Moorhead *et al.*, 1974). It is, therefore, apparent that when a critical number or

TABLE I
Antibody to Thyroglobulin in Obese-Strain Chickens

Age (weeks)	1970[a]		Age (weeks)	1980[b]	
	n	$\bar{X}$ $\log_2$ Titer ± S.D.		n	$\bar{X}$ $\log_2$ Titer ± S.D.
2–4	68	1.4 ± 1.6			
5–7	82	2.2 ± 2.4	3	42	7.4 ± 2.5
8–10	70	4.1 ± 2.7	4	43	7.7 ± 2.2
11–13	68	4.8 ± 3.4	5	38	9.5 ± 2.2
14–16	51	5.6 ± 2.8	6	11	9.2 ± 2.4
20–22	22	6.2 ± 2.6			
23–30	22	3.3 ± 3.2			

[a]Data reported by Rose *et al.* (1971) of randomly bred OS chickens in the late 1960s (with permission).
[b]Unpublished data of R. S. Sanker on OS chickens homozygous for the B^{13} allele in 1979–1980.

density of lymphoid cells is reached, a heterogeneous population of lymphoid cells with various antigenic specificities migrates freely into the gland. This concept has been verified by Schauenstein and Wick (1974). They demonstrated by immunofluorescent techniques that lymphocytes in the thyroid gland of OS chickens immunized with BSA produce antibodies not only to Tg, but to BSA as well.

It is perhaps appropriate at this time to emphasize one intriguing point. It is not yet known why lymphoid cells first migrate into the thyroid gland. Neither the antigenic specificity nor the type of lymphocyte (T, B, or other) has been identified. The following indirect evidence suggests that B cells are not necessary for infiltration to begin. When newly hatched OS birds were treated with cyclophosphamide at a dosage high enough to (1) suppress a humoral response to two injections of sheep erythrocytes and (2) prevent detectable antibody to Tg, they still showed a significant, albeit lower, amount of thyroidal infiltration (Polley and Bacon, 1976; R. S. Sundick, personal observation). It is also unlikely that antibody-dependent cytotoxic (K) cells are the earliest cells on the scene, for lymphocytes appear in the thyroid gland by 1 week of age, and antibody is not usually detectable in the serum until 2 weeks of age. A second experimental approach also suggests that K cells are not the initial cells infiltrating the thyroid gland. CS chickens given multiple injections of

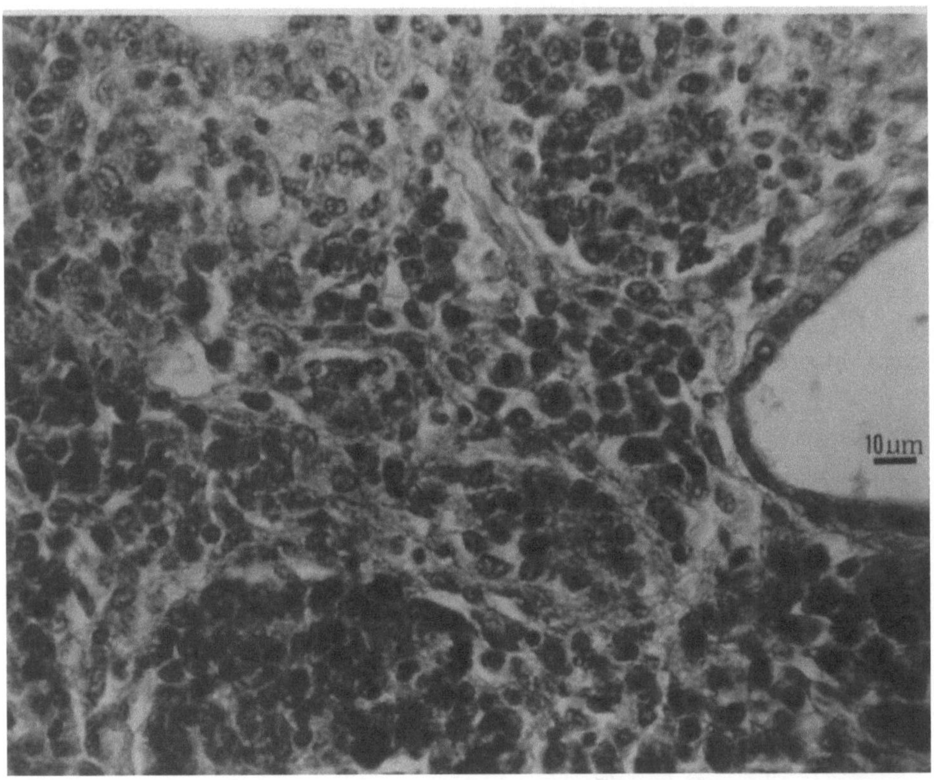

FIGURE 2. Thyroid gland of a 6-week-old OS chicken revealing complete destruction by infiltrating mononuclear cells. Plasma cells and mitotic figures are readily observable. One damaged thyroid follicle is visible on the right side (hematoxylin and eosin).

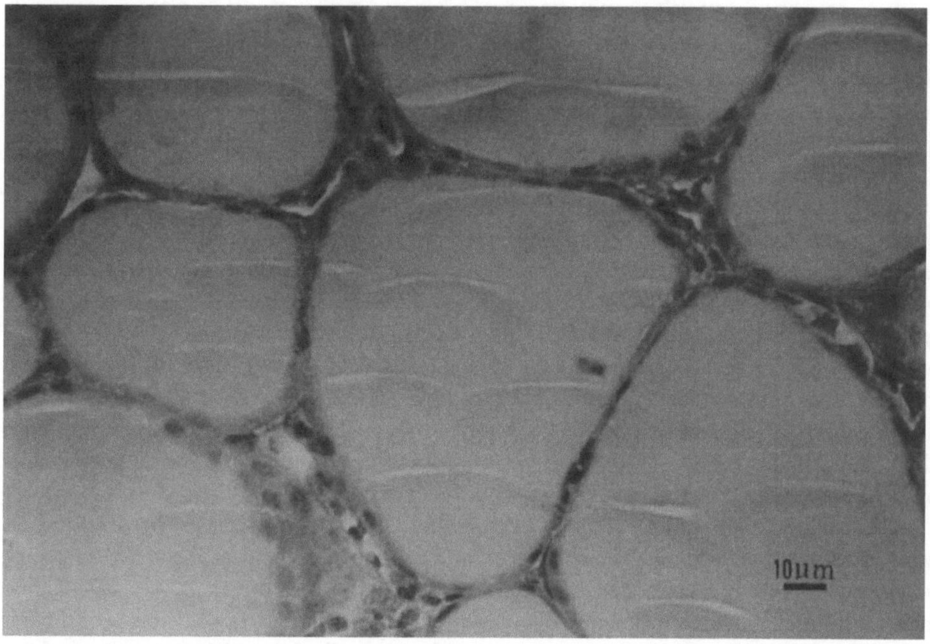

FIGURE 3. Thyroid gland of a 9-week-old CS chicken. Each of the follicles is filled with colloid. No lymphoid cells are present.

OS serum containing antibodies to Tg do not have significant numbers of lymphoid cells in their thyroid glands (Jaroszewski *et al.*, 1978), even though it is likely that the antibody has fixed to the thyroid gland (Sundick and Wick, 1974).

While the nature of the earliest-infiltrating cells is unknown, some of the later pathological events have been more thoroughly investigated. Owing to the presence of many germinal centers in the OS thyroid gland and the local production of Tg antibody it is obvious that B cells, and almost certainly helper T cells, specifically committed to Tg, are playing an active role in pathogenesis. This is analagous to the situation of experimentally induced disease in rabbits (Weigle, 1977) and human autoimmune thyroid diseases (Tötterman, 1978). Immune complexes have been detected in the thyroid glands of OS chickens, and they may also contribute to pathology (Kalderon *et al.*, 1977; Albini and Kite, 1980).

3. Defects in the Thyroid Gland May Be a Predisposing Factor in the Autoimmune Thyroiditis of OS Chickens

The thyroid gland of neonatal OS chickens appears normal by histologic examination. No lymphoid cells are present. By 1 week of age, lymphocytes and plasma cells are present and some of the thyroid epithelial cells show degenerative changes (Wick and Graf, 1972).

Several unusual functional characteristics have been observed in the thyroid glands of OS chicks. At hatching the thyroid incorporates more $^{131}I^-$ than outbred normal and CS chickens (Sundick and Wick, 1974). The hyperactivity of newly hatched OS thyroid glands is also manifested in another way, an elevated ratio of monoiodotyrosine to diiodotyrosine (Rose *et al.,* 1976). However, this hyperactivity of the early steps in thyroid hormone biosynthesis does not cause thyrotoxicosis. A detailed analysis of the neonatal serum levels of the thyroid hormones, thyroxine (T_4) and triiodothyronine (T_3), revealed no significant differences among OS, CS, and normal chicken strains (Sundick *et al.,* 1980b). The finding of high thyroidal uptake of $^{131}I^-$ in the face of normal serum levels of T_4 and T_3 was puzzling, but not without precedent. Patients with the syndrome "euthyroid Graves' disease" behave similarly. A test for pituitary regulation of the thyroid gland was therefore performed to clarify this situation (Sundick *et al.,* 1979). The pituitary glands of OS, CS, and normal chicken strains were suppressed by feeding high levels of T_4 for the first 2 weeks of age. Twenty-hour thyroidal $^{131}I^-$ uptakes were then measured. The thyroid glands of four normal chicken strains behaved as expected; they were suppressed by 98%. The OS thyroid was not as completely suppressible, and the CS thyroid was even more abnormal. Further tests revealed that both OS and CS thyroid glands were poorly suppressible with regard to the coupling reaction that produces T_3 and T_4.

Two likely explanations for this presumably "TSH-independent" thyroid function were investigated. OS and CS chicks were immunosuppressed and at the same time suppressed with T_4 and then tested for thyroidal $^{131}I^-$ uptake (Sundick *et al.,* 1979). The immunosuppression had no effect, strongly implying that the poor suppressibility of the OS and CS thyroids was not due to a thyroid stimulatory antibody (which is the cause of the hyperactivity and TSH-independence of the thyroid in human Graves' disease). Another experiment indicated that the hyperactivity of OS and CS thyroids was due to an intrinsic property of the gland genetically determined, and probably in a codominant fashion (Livezey and Sundick, 1980). This gene might determine an increased sensitivity of the thyroid to low levels of TSH or even enable the gland to function in the absence of TSH.

We do not know what role, if any, this unusual thyroid dysfunction has on the subsequent development of autoimmune thyroiditis. Several possibilities are being considered; serum levels of Tg could be affected or even the immunogenicity of Tg. Studies of human thyroid physiology and pathology (Van Herle *et al.,* 1979) have indicated that both TSH and thyroid stimulatory antibodies have significant effects on serum concentrations of Tg. Neoplastic changes in the human thyroid gland often result in increased serum levels of Tg and, in addition, TSH-independent thyroid function (DeGroot and Stanbury, 1979). The immunogenicity of Tg might also be affected by TSH-independent thyroid function and high iodide uptake, as the degree of iodination of Tg significantly affects its chemical stability (Velenta, 1974). The neonatal injection of OS chicks with thyroid extract (Sundick *et al.,* 1978) or with 1 mg of purified Tg (Sunick *et al.,* 1980a) significantly reduces the severity of their autoimmune thyroiditis. This suggests that OS chickens might develop autoantibodies to Tg because their serum levels of Tg during the neonatal period are insufficient to maintain T-cell tolerance, a concept stressed by Weigle (1977).

4. Evidence for an Abnormal Immunologic System

There now exist several types of data indicating the presence of a hyperactive B-cell system in OS chickens. First, Albini and Wick (1974) demonstrated a proportional increase in the B-/T-cell ratio in the spleen as early as the 19-day embryo stage. This splenic ratio was elevated in OS chicks at all ages tested. A surprisingly high percentage of B cells was detected in the 1.5-week-old thymus (28%) and the percentage was still quite high at 4 weeks (25%). Normal chickens had lower B-/T-cell ratios in the thymus at 1.5 weeks (11%) and at 4 weeks (3%). Second, Nowak *et al.* (1978) determined the percentage of Fc(IgG) receptor-bearing lymphocytes in the lymphoid tissues of OS and CS chickens. The largest difference between the two strains was detected in the spleens of newly hatched chicks; 29% of OS and 17% of CS spleen cells were Fc(IgG) positive. Third, Albini and Wick (1971) detected significantly elevated antibody titers to avian red cell nuclei in the OS chickens compared to a normal chicken strain. Finally, Richter *et al.* (1975) detected a significantly higher incidence and titer of naturally occurring heteroagglutinins in 10- to 20-week-old, but not in 4-week-old, OS chickens, compared to a normal chicken strain maintained under identical conditions.

All of the above studies support the concept of a hyperactive *peripheral* B-cell system in OS chickens. No significant abnormalities were detected in the bursa of Fabricius of OS chickens. In fact, experiments by Polley and Bacon (1976) and Polley *et al.* (1977) indicated that bursal cells obtained from neonatal CS chickens were just as effective as those of OS chickens in causing thyroiditis when transferred to chemically bursectomized OS chicks (Figs. 4 and 5). When these bursal cell preparations were transferred to bursectomized CS chicks, thyroiditis did not occur.

Several unusual properties of the OS thymus and/or peripheral T cells have been observed. Jakobisiak *et al.* (1976) reported data suggesting that more OS than CS effector T cells emigrated from the thymus prior to hatching. Luster *et al.* (1977) found that OS chickens had significantly reduced serum concentrations of IgA and elevated amounts of

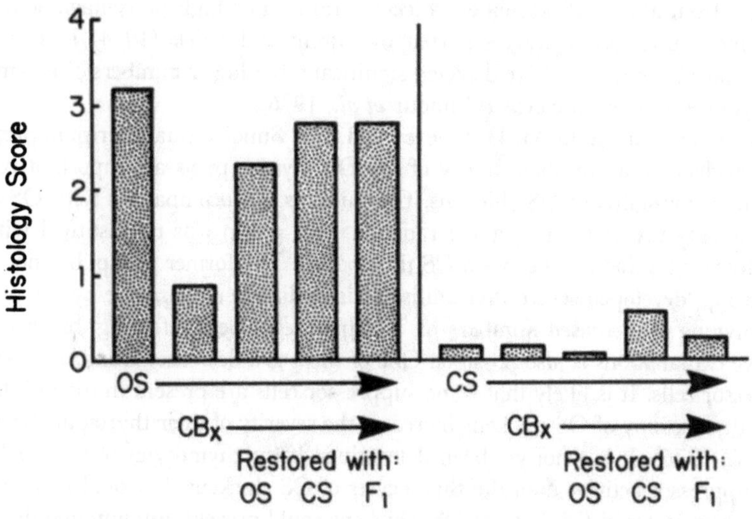

FIGURE 4. Thyroid pathology. Bursal cells obtained from CS chickens are just as effective as those from OS chickens or (OS × CS)F_1 chicks in causing thyroiditis in chemically bursectomized OS chicks. CS chickens chemically bursectomized and reconstituted with OS, CS, or F_1 bursal cells do not develop thyroiditis. The thyroid pathology score ranges from 0 (no infiltration) to 4.0 (100% infiltration). [Based on the data of Polley *et al.* (1977).]

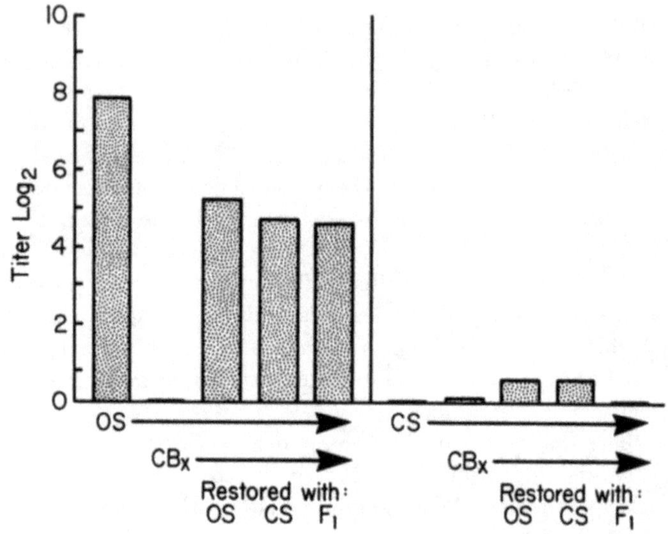

FIGURE 5. Six-week antibody response to thyroglobulin. Bursal cells obtained from CS chickens are just as effective as those from OS chickens or (OS × CS)F_1 chickens in producing thyroglobulin antibody when transferred to chemically bursectomized OS chicks. No thyroglobulin antibody is produced in bursa-reconstituted CS chicks. Antibody titers are determined by passive hemagglutination of thyroglobulin-coated chicken red blood cells. [Based on the data of Polley *et al.* (1977).]

IgM, which is most likely explainable by abnormal T-cell function. Waytes *et al.* (1980) detected a thymic alloantigen in 90% of OS chickens (and in only 45% of CS chickens) that was absent from almost all unrelated chicken strains. The high percentage of B lymphocytes in the young OS thymus reported by Albini and Wick (1974) is also unusual, although other investigators have detected significant, but lower numbers of B lymphocytes in the thymuses of normal chicks (Gilmour *et al.*, 1976).

A recent experiment by M. D. Livezey and R. S. Sundick (manuscript in preparation) provides evidence that an abnormality of the OS thymus plays an important role in the autoimmune thyroiditis of OS chickens. CS chickens, histocompatible with OS chickens, were neonatally thymectomized and irradiated. One group was reconstitued with 6-day-old OS thymocytes and another with CS thymocytes. The former group, but not the latter control group, developed severe thyroiditis. This finding is explainable by the presence in the OS thymus of increased numbers of T-helper cells specific for Tg, but a number of alternative explanations is also possible. One of them is a decreased number of Tg-specific T-suppressor cells. It is likely that some suppressor cells are present in the OS thymus, as neonatal thymectomy of OS chickens increases the severity of their thyroiditis (reviewed by Rose *et al.*, 1976). It has not yet been determined if the thymocytes of CS chickens have greater suppressor activity than the thymocytes of OS chickens. It would be interesting to see if the thymocytes of CS, but not OS, chickens could prevent autoimmune thyroiditis in neonatally thymectomized, irradiated, and T-cell-reconstitued OS chicks.

Several conclusions can be drawn concerning the immune system of OS chickens. First, there appears to be a generalized hyperresponsiveness of the peripheral B-cell system.

However, no functional defect has been detected in the bursa of Fabricius of OS chicks, particularly with regard to its ability to produce Tg-specific B cells. It is, therefore, likely that B cells emigrating from the bursa of Fabricius of an OS chick come in contact with Tg and are activated under the influence of an "abnormal" thymus and/or mature T cell. The thymus might have increased numbers of Tg-specific helper T cells, cytotoxic T cells, and/or decreased numbers of Tg-specific suppressor T cells.

One point needs to be emphasized. The OS chickens were genetically selected for, and developed, severe thyroiditis. They produce high titers of antibody to Tg, increased numbers of Tg-specific B cells (Richter and Wick, 1975), and a delayed hypersensitivity response to Tg (Welch and Kite, 1971). They also produce other autoantibodies—to red blood cell nuclei and to liver extracts. However, no other autoimmune diseases have been detected in OS chickens (reviewed by Wick *et al.,* 1974). If OS chickens have a generalized defect in T-cell regulation, it is clearly not as profound with respect to other autoantigens and foreign antigens as it is to autologous Tg.

5. Relationship between the Major Histocompatibility Complex *(B)* and Autoimmune Thyroiditis

There are many reports in the literature concerning significant associations between alleles within the *HLA* complex and the incidence of human autoimmune thyroid disease (Terasaki, 1978; Volpé 1978). These associations are statistically significant, but the *HLA* alleles are probably not the only, or even the major, genetic factors in susceptibility to autoimmune thyroid disease. This is evidenced by the generally low-risk factors associated with *HLA* alleles, which is in marked contrast to the major influence of all genes combined in susceptibility to autoimmune thyroid disease (Volpé, 1978).

The association between alleles at the major histocompatibility complex *(B)* and susceptibility to autoimmune thyroiditis has also been studied in OS chickens (Rose *et al.,* 1976; Bacon and Rose, 1979). The primary conclusion is surprisingly similar to the data in humans; alleles within the *B* locus have a significant effect on autoimmune thyroiditis. A significant effect of the *B*-locus alleles on thyroiditis was demonstrated in the closed flock of OS chickens maintained at Cornell University (Rose et al., 1976). That flock was not highly inbred and when investigated by Bacon *et al.* (1974), it was found to be segregating for three different alleles at the *B* locus. Starting in 1972, three families of OS chickens were selected on the basis of good hatchability and segregation of two alleles at the *B* locus, B^{13} (formerly called B^{1}) and B^{5} (formerly called B^{4}). Sibling matings within the three families were set up using $B^{13}B^{5}$ heterozygote breeders. After several generations these three families were partially inbred and designated as the OSA, OSB, and OSC sublines. Both *B* alleles had been purposely preserved in all three sublines. When the effects of the *B* alleles were examined in these sublines, some interesting differences emerged (Bacon et al., 1974; Bacon and Rose, 1979). Within the OSA subline, birds homozygous or heterozygous for the B^{13} allele had more severe thyroiditis and higher Tg antibody titers than homozygous B^{5} chicks. In the other two sublines, the effects of the B^{13} and B^{5} alleles were not as dramatic. Birds homozygous or heterozgous for B^{13} had higher Tg antibody titers

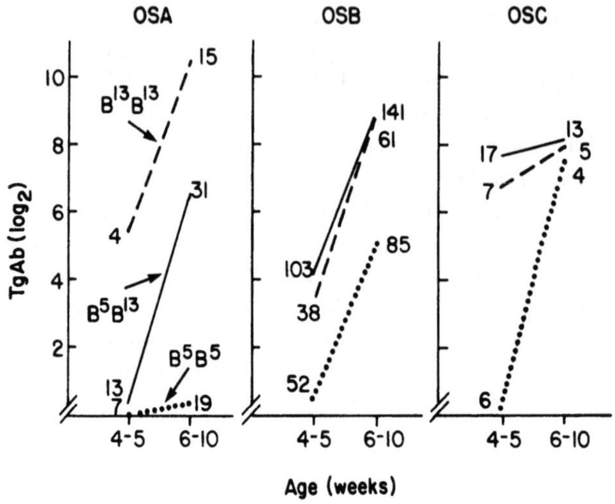

FIGURE 6. Median TgAb in OS sublines. The offspring of selected matings within three inbred sublines of OS chickens were tested for thyroglobulin antibody and *B*-locus alleles. The *B* locus had the most significant effect on the OSA subline, but the other sublines were also affected. Numerals indicate number of animals. [From Bacon *et al.* (1978) with permission.]

(at least at 4 or 5 weeks of age, but not at 6 to 10 weeks) than homozygous B^5 birds (Fig.6). However, the *B* alleles had almost no effect on the degree of thyroid infiltration in OSB and OSC chicks sacrificed at 6 weeks of age. It was concluded that genes at the *B* locus were only one of several factors influencing autoimmune thyroiditis, and that under certain (unknown) conditions genes at other loci play a predominant role.

The mechanism by which genes within the *B* locus influence disease susceptibility in OS chickens is unknown. Two experimental approaches suggested that OS chickens with the *B*-locus allele coding for mild disease (B^5) have elevated T-suppressor activity, but alternative explanations for the data are also possible. Neonatal thymectomy of OSA B^5B^5 chickens caused a great increase in the severity of thyroiditis and in the level of Tg antibody production (Rose *et al.*, 1976). Injection of outbred OS B^5B^5 chicks with antibodies to Tg caused severe thyroiditis (Jaroszewski *et al.*, 1978). These two treatments perhaps eliminated suppressor T cells or circumvented the need for helper T cells and lymphocytes then infiltrated the thyroid gland. There is good evidence from the experimentally induced thyroiditis in mice that genes within the major histocompatibility complex influence autoimmune thyroiditis by their effects on T cells (Vladutiu and Rose, 1975; Christadoss *et al.*, 1978), but this has not yet been examined in OS chickens.

6. Summary

Obese-strain chickens were developed by selecting breeders afflicted with hypothyroidism. It was subsequently learned the hypothroidism was due to an early and severe immunologically mediated destruction of the thyroid gland. Backcross experiments indicated the polygenic nature of the disease. Immunogenetic studies revealed that genes at the major histocompatibility complex influenced disease severity, but genes at other loci also played an important role.

The B-cell system of OS chickens appears hyperactive. However, this is most likely due to abnormal regulation by T cells, and not a defective bursal stem cell. The B cells of OS chickens are regulated to some extent because they respond to only a limited number of autoantigens by producing autoantibodies, and OS chickens do not develop other autoimmune diseases. OS thymocytes and/or T cells functioned abnormally in a number of experiments and it is likely that these abonormalities played a major role in this disease. Experiments to quantitate Tg-specific helper, cytotoxic, and suppressor T cells in OS chickens would likely clarify many of these points.

The initial triggering event in lymphoid infiltration of the thyroid is unknown. Subacute thyroiditis in man may be triggered by virus infection of the thyroid (Stamcek *et al.*, 1975) and this is also possible in OS chickens (Wick and Graf, 1972). Abnormal thyroid function in young OS chicks could perhaps be a result of virus infection, but it is more likely independent of it. The TSH-independent thyroid function of OS and CS chickens has some precedence in the autonomous thyroid nodules of humans (Miller and Black, 1970). It is not known how this unusual property of the thyroid gland might predispose it to lymphoid infiltration, but several possibilities have been suggested, including its possible effect on the serum concentration of Tg during the neonatal period.

ACKNOWLEDGMENTS

The excellent typing of Mrs. Barbara Paton is gratefully acknowledged. Much of our research on the OS chickens was supported by Research Grant AM20028 from the National Institutes of Health.

References

Albini, B., and Kite, J., 1980, Immune complex deposition in the thyroid and kidney of obese (OS) chickens, *Fed. Proc.* **39**:471.

Albini, B., and Wick, G., 1971, Erythrocyte-specific antinuclear factors in chickens of the obese strain (OS), *Immunology* **21**:957.

Albini, B., and Wick, G., 1974, Proportional increase of bursa-derived cells in chickens of the obese strain, *Nature (London)* **249**:653.

Bacon, L. D., and Rose, N. R., 1979, Influence of major histocompatibility haplotype on autoimmune disease varies in different inbred families of chickens, *Proc. Natl. Acad. Sci. USA* **76**:1435.

Bacon, L. D., Kite, J. H., Jr., and Rose, N. R., 1974, Relation between the major histocompatibility *(B)* locus and thyroiditis in obese chickens, *Science* **186**:274.

Bacon, L. D., Cole, R. D., Polley, C. R., and Rose, N. R., 1978, Genetic aspects of autoimmune thyroiditis in OS chickens, in: *Genetic Control of Autoimmune Disease* (N. R. Rose, P. E. Bigazzi, and N. L. Warner, eds.), p. 259, Elsevier/North-Holland, Amsterdam.

Christadoss, P., Kong, Y. M., Elrehewy, M., Rose, N. R., and David, C. S., 1978, Genetic control of T-lymphocyte proliferative autoimmune response to thyroglobulin in mice, in: *Genetic Control of Autoimmune Disease* (N. R. Rose, P. E. Bigazzi, and N. L. Warner, eds.), p. 445, Elsevier/North-Holland, Amsterdam.

Cole, R. K., 1966, Hereditary hypothyroidism in the domestic fowl, *Genetics* **53**:1021.

Cole, R. K., and Hutt, F. B., 1973, Selection and heterosis in Cornell white leghorns: A review with special consideration of interstrain hybrids, *Animal Breeding Abstracts* **41**:103.

DeGroot, L. J., and Stanbury, J. B. (eds.), 1975, Thyroid neoplasia, in: *The Thyroid and Its Diseases*, pp. 666–733, Wiley, New York.

Gilmour, D. G., Brand, A., Donnelly, N., and Stone, H. A., 1976, Bu-1 and Th-1, two loci determining surface antigens of B or T lymphocytes in the chicken, *Immunogenetics* **3**:549.

Jakobisiak, M., Sundick, R. S., Bacon, L. D., and Rose, N. R., 1976, Abnormal response to minor histocompatibility antigens in obese strain chickens, *Proc. Natl. Acad. Sci. USA* **73**:2877.

Jaroszewski, J., Sundick. R. S., and Rose, N. R., 1978, Effects of antiserum containing thyroglobulin antibody on the chicken thyroid gland, *Clin. Immunol. Immunopathol.* **10**:95.

Kalderon, A. E., Bogaars, H. A., Jolly, G., and Diamond, I., 1977, Electrondense deposits in the follicular basal lamina of obese strain chickens with spontaneous autoimmune thyroiditis: An electron microscope study, *Lab. Invest.* **37**:487.

Livezey, M. D., and Sundick, R. S., 1980, Intrinsic thyroid hyperactivity in avian strains susceptible to autoimmune thyroiditis, *Gen. Comp. Endocrinol.* **41**:243.

Luster, M. I., Bacon, L. D., Rose, N. R., and Leslie, G. A., 1977, Immunogenetic and ontogenetic studies of chickens with selective IgA deficiency and autoimmune thyroiditis, *Cell. Immunol.* **32**:417.

Miller, J. M., and Black, M. A., 1970, Functional autonomy in multinodular goiter, *J. Am. Med. Assoc.* **214**:535.

Moorhead, J. W., Kite, J. H., McCluskey, R. T., Werdelin, O., and Wick, G., 1974, Migration patterns of thymus and bursa lymphocytes in normal chickens and obese strain chickens with spontaneous thyroiditis, *Clin. Immunol. Immunopathol.* **2**:160.

Nowak, J. S., Bacon, L. D., and Rose, N. R., 1978, Fc receptor-bearing lymphoid cells in the chicken. II. Relative increase of Fc (IgG) receptor bearing cells in obese strain chickens, *Immunol. Commun.* **7**:621.

Polley, C. R., and Bacon, L. D., 1976, Restoration of bursal function in chemically bursectomized non-inbred chicks, *Poult. Sci.* **55**:2081 (abstract).

Polley, C. R., Bacon, L. D., and Rose, N. R., 1977, Failure of functional bursa cells transplanted between susceptible OS and resistant CS strains of chickens to influence spontaneous autoimmune thyroiditis, *Fed. Proc.* **36**:1207 (abstract).

Richter, E., and Wick, G., 1975, Thyroglobulin binding lymphoid cells in obese strain (OS) chickens, *J. Immunol.* **114**:757.

Richter, E., Sundick, R. S., and Wick, G., 1975, Heteroagglutinins and spontaneous rosette-forming cells in obese strain (OS) chickens, *Z. Immunitaetsforsch.* **149**:61.

Rose, N. R., Kite, J. H., Jr. Flanagan, T. D., and Witebsky, E., 1971, Humoral and cellular immune factors in spontaneous autoimmune disease, in: *Celluar Interactions in the Immune Response* (S. Cohen, G. Cudkowicz, and R. McCluskey, eds.), pp. 264–281, Karger, Basel.

Rose, N. R., Bacon, L. D., and Sundick, R. S., 1976, Genetic determinants of thyroiditis in the OS chicken, *Transplant. Rev.* **31**:264.

Rose, N. R., Bacon, L. D., Sundick, R. S., and Kong, Y. M., 1978, The role of genetic factors in autoimmunity in: *The Menarini Series on Immunopathology* (P. A. Miescher, ed.) Vol. 1, pp. 225–238, Schwabe, Basel.

Schauenstein, K., and Wick, G., 1974, Local production of immunoglobulin in the thyroid gland of obese strain (OS) chickens, *Clin. Exp. Immunol.* **17**:637.

Stamcek, D., Stancekova-Gressnervoa, M., and Janotka, M., 1975, Isolation and some serological and epidemiological data on the viruses recovered from patients with subacute thyroiditis de Quervain, *Med. Microbiol. Immunol.* **16**:133.

Sundick, R. S., and Wick, G., 1974, Increased I uptake by the thyroid gland of obese strain (OS) chickens derived from non-protamone-supplemented hens, *Clin. Exp. Immunol.* **3**:272.

Sundick, R. S., Accavitti, M. A., and Rose, N. R., 1978, Reduction in severity of spontaneous thyroiditis by the neonatal injection of thyroid extract, *Fed. Proc.* **37**:1374 (abstract).

Sundick, S., Bagchi, N., Livezey, M. D., Brown, T. R., and Mack, R. E., 1979, Abnormal thyroid regulation in chickens with autoimmune thyroiditis, *Endocrinology* **105**:493.

Sundick, R. S., Sanker, A. J., and Rose, N. R., 1980a, Etiology of spontaneous thyroiditis in chickens, in: *Immunoregulation and Autoimmunity* (R. Krakauer and M. Cathcart, eds.), pp. 80–90, Elsevier/North-Holland, Amsterdam.

Sundick, R. S., Livezey, M. D., and Rose, N. R., 1980b, Spontaneous autoimmune thyroiditis in chickens, in: *Thyroid Research VIII* (J. R. Stockigt and S. Nagataki, eds.) pp.773–776, Australian Academy of Sciences, Camberra.

Terasaki, P. I., 1978, HLA frequencies in patients with Graves' disease and Hashimoto's thyroiditis: A second study, *Ann. Intern. Med.* **88**:384.

Tötterman, T. H., 1978, Distribution of T-, B-, and thyroglobulin-binding lymphocytes infiltrating the gland in Graves' disease, Hashimoto's thyroiditis, and de Quervain's thyroiditis, *Clin. Immunol. Immunopathol.* **10**:270.

Valenta, L. J., 1974, Differential stability of iodine poor and iodine rich rat thyroglobulin, *Acta Endocrinol.* **75**:33.

Van Herle, A. J., Vassart, G., and Dumont, J. E., 1979, Control of thyroglobulin synthesis and secretion, *N. Engl. J. Med.* **301**:307.

Valdutiu, A. O., and Rose, N. R., 1975, Cellular basis of the genetic control of immune responsiveness to murine thyroglobulin in mice, *Cell. Immunol.* **17**:106.

Volpé, R., 1978, The genetics and immunology of Graves' and Hashimoto's diseases, in: *Genetic Control of Autoimmune Disease* (N. R. Rose, P. E. Bigazzi, and N. L. Warner, eds.), pp. 43–56, Elsevier/North-Holland, Amsterdam.

Waytes, T., Bacon, L. D., and Rose, N. R., 1980, A thymus alloantigen in obese strain chickens, *Fed. Proc.* **39**:471.

Weigle, W. O., 1977, Cellular events in experimental autoimmune thyroiditis, allergic encephalomyelitis, and tolerance to self, in: *Autoimmunity: Genetic, Immunologic, Virologic and Clinical Aspects* (N. Talal, ed.), pp. 141–170, Academic Press, New York.

Welch, P. C., and Kite, J. H., 1971, Delayed hypersensitivity in spontaneous and experimentally induced autoimmune thyroiditis in chickens, *Fed. Proc.* **30**:306.

Wick, G., and Graf, J., 1972, Electron microscope studies in chickens of the obese strain with spontaneous hereditary autoimmune thyroiditis, *Lab. Invest.* **27**:400.

Wick, G., Sundick, R. S., and Albini, B., 1974, The obese strain (OS) chickens: An animal model with spontaneous autoimmune thyroiditis, *Clin. Immunol. Immunopathol.* **3**:272.

2
Lethargic Mice

H. C. DUNG

1. Description of the Mutant and Its Neurologic Symptoms

A mutation appeared spontaneously in 1962 in an inbred BALB/cGn strain of the mouse colony maintained at The Jackson Laboratory, Bar Harbor, Maine. It was reported as a single recessive gene mutation, and the affected mice were called lethargic (*lh*) because of their inactivity and sluggish behavior (Dickie, 1964). Such abnormal neurological symptoms led to the classification of the affected mice as a neurological mutation of the mouse by Sidman *et al.* (1965).

The mice carrying the mutant gene (*lh*) have been maintained as a highly inbred strain in The Jackson Laboratory as well as in ours. However, after many generations of inbreeding, the viability of these mice becomes diminished and their reproductivity decreases. Therefore, outcrossed breeding has been instituted both in The Jackson Laboratory and in our own. Such outcrossed breeding has caused no visible alterations in the phenotypic characteristics of the affected mice in terms of their neurologic symptoms.

Affected lethargic mice first become recognizable at 15 days of age. Prior to this age, no phenotypic differences can be noted between normal littermates and lethargic mice. After this age, the mutant mice have several physical and behavioral characteristics that can be used to distinguish them from normal animals.

1.1. Physical Appearance

The body size of the affected mice is invariably smaller than that of phenotypically normal animals. The mice shown in Fig. 1 illustrate the typical appearance of lethargic and normal mice of the same age. Besides the difference in size, one can also note that the lethargic mouse has a ruffled coat and a hunched posture. This appearance resembles that

H. C. DUNG • Department of Anatomy, The University of Texas Health Science Center at San Antonio, San Antonio, Texas 78284.

FIGURE 1. One-month-old male lethargic (left) and phenotypically normal (right) mice. At this age, body weights were 5.6 g for the lethargic (which weighed 6.4 g at 15 days of age) and 15.2 g for the normal mouse.

of mice with wasting disease caused by neonatal thymectomy (Miller, 1962; Parrott and East, 1962) and that of mice with the runting syndrome caused by the graft-versus-host reaction (Billingham, 1958).

1.2. Growth and Body Weight

The smaller body size of lethargic mice is reflected in their body weight. A study was done to determine the changes occurring in the body weight of normal and lethargic mice between 15 and 95 days of age. The results of this study are summarized in Fig. 2. It appears that lethargic mice weigh the same as their normal littermates at 15 days of age. Immediately after the onset of their neurologic symptoms, the mutant mice show a loss in body weight, and they soon weigh significantly less than their normal littermates. This loss of body weight is observed in all lethargic mice, although the degree of wasting varies. Those mutant mice that survive this period regain some weight but, nevertheless, they stay significantly lighter than normal mice of the same sex and age, and very few of the mutants ever had a weight equal to that of normal mice.

1.3. Mortality

Most lethargic mice die during the period in which they show a weight loss. Approximately 68–80% of the mutants die before the age of 45 days. The cause of their death is possibly infectious, because a large number of them have diarrhea. If they survive for more than 2 months, the mortality rate decreases. A few live a normal life span (longer than 20 months). The average life span of the BALB-strain mouse is 16 months (Green *et al.*,

1966). Use of Terramycin in the drinking water (50 mg/250 ml) and restriction of litter size were found to improve the survival of lethargic mice and allowed more of them to pass through the critical wasting period.

1.4. Body Temperature and Its Relation to Cold Exposure

Young lethargic mice less than 45 days of age are hypothermic. Their rectal temperature is at least 1.5°C lower than that of normal mice kept in the same environment. Many young lethargic mice have a rectal temperature as low as 30°C. However, lethargic mice older than 2 months have a rectal temperature as high as that of normal mice (36 ± 1°C).

Hypothermia in lethargic mice can be enhanced by placing the animal in a 4°C cold room. No lethargic mice of any age can maintain a rectal temperature above 20°C for more than 4 hr at 4°C, yet at least 85% of normal mice as young as 18 days of age are able to maintain a rectal temperature above 30°C for 4 hr or longer.

1.5. Stress Susceptibility

Lethargic mice are susceptible to stress as shown by their reduced capacity to resist cold. In an effort to see how they would respond to another type of stress, a swimming test

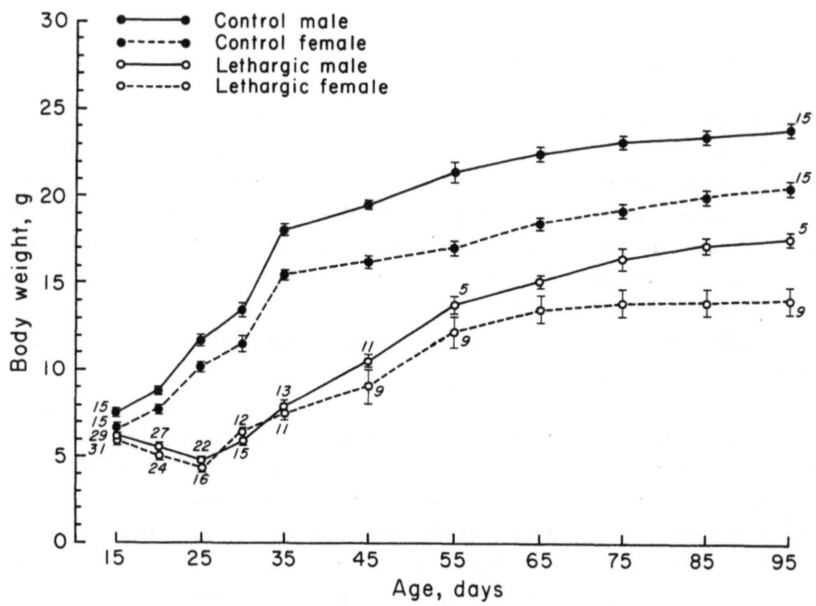

FIGURE 2. Growth curves for normal and lethargic mice of both sexes from 15 to 95 days of age. The numbers near the data points represent the number of mice from which the mean values were determined. Due to a high mortality in affected mice, the number of lethargic mice decreased with time.

was performed. Both normal and mutant mice swam individually either in 35°C or in 23°C water. None of 135 lethargic mice of different ages swam for longer than 3 hr, and more than 50% of them swam less than 1 hr in water at either temperature. For normal mice older than 18 days of age, at least 42% swam for as long as 4 hr in water at either temperature. In general, the swimming capacity of lethargic mice improved as their age increased. However, they lagged far behind normal mice of the same age.

1.6. Fertility and Sexual Maturity

Most lethargic mice are not able to reproduce; only about 25% of females can produce a litter, and fertility is delayed for about 1 month. The average age at which normal females give birth to their first litter is 81 days. Fertile lethargic females give birth to their first litter at about 115 days. Only about 10% of lethargic males are able to sire young.

The period during which reproduction is possible is much shorter for lethargic than for normal mice. Normal females produce an average of seven litters within 12 consecutive months, whereas 7 of the 30 fertile lethargic females we have studied produced only one litter, and none produced more than three litters. Furthermore, lethargic females are poor mothers. All of the young mice produced by lethargic mothers either die of neglect or are cannibalized shortly after birth (Dung and Swigart, 1971).

1.7. Neurologic Abnormalities of the Mutant

The most striking difference between normal and mutant mice after 15 days of age is the neurologic abnormality manifested in their ambulatory behavior. According to neuroontogenetic studies reported by Fox (1965), a normal mouse becomes adultlike in its ambulatory behavior at about 15 days of age, and it should have stable locomotive activities. Young mice between 15 and 30 days of age are hyperactive (Fox, 1965). They run with great rapidity in all directions when they are disturbed. In contrast, lethargic mice at this age do not run about as quickly as their normal littermates, and when they run they show instability of gait. After running for a very short distance, they fall on their sides, sometimes in a seizure. Thereafter, they move with great difficulty.

At first (15 days of age), these behavioral abnormalities are mild. They then become more severe during the next few days. The affected mice become weaker and less active and thus may be trampled by normal littermates. By 25 days of age, many lethargic mice can hardly move. At times they rear back and sit motionless on their haunches with their backs arched. Seizures are a common occurrence and they increase in severity when the mice are subjected to stress such as that which occurs when they are handled. Sidman *et al.* (1965) stated that seizure observed in lethargic mice sometimes show a typical "Jacksonian march" pattern beginning on one side of the face, moving to the ipsilateral forelimb, then to the ipsilateral hind limb, the tail, the contralateral hind limb, the contralateral forelimb, and finally the contralateral face. Such a complete "march" has been rarely observed in our mutant mice.

2. Light Microscopic Studies

Because of the nerologic abnormality, our early studies on the lethargic mutant were focused on a pathological examination of the central nervous system. Comparisons of the normal and lethargic tissues showed that: (1) the relative proportions of gray and white matter were normal, (2) fiber tracts appeared to be normally myelinated, (3) chromatolysis and gliosis were not observed, (4) roots of the spinal nerves and dorsal root ganglia were normal, (5) the six layers of the cerebal neocortex were clearly distinguishable and normally arranged, and (6) the folia of the cerebellum were typical as were the three layers of the cerebellar cortex, including the Purkinje cells. Thus, the central nervous system of the mutant studied at the light microscopic level revealed no lesions, no degenerative changes, and no other pathologic abnormalities (Dung and Swigart, 1972).

In addition to the central nervous system, many other tissues including skeletal muscles, submaxillary glands, thymuses, spleens, lymph nodes, Peyer's patches in the intestine, lungs, adrenal glands, kidneys, testes, ovaries, seminal vesicles, livers, and skin were examined by light microscopy. Except for the lymphoid tissue (spleen, thymus, lymph nodes, and Peyer's patches), which will be described subsequently, histopathological observations will be briefly summarized only for those remaining organs that showed abnormalities. Unless specified, all of the histological preparations were stained with hematoxylin–eosin or periodic acid–Schiff (Pas)–hematoxylin methods.

2.1. The Submaxillary Gland

A sexual dimorphism is known to exist in the morphology of the mouse submaxillary gland (Shackelford and Klapper, 1962). Secretory tubules in glands of the mature male mouse were described as larger than those of the female (Caramia, 1966). However, the submaxillary gland of many lethargic males remained infantile (Fig. 3). Secretory tubules in the gland were small and poorly developed. Between 5 and 12 months of age the gland weighed 132.5 $\pm$ 4.8 g for normal and 108.3 $\pm$ 6.7 g for mutant mice. This difference in weight is statistically significant ($p < 0.005$).

2.2. The Adrenal Gland

Numerous morphological changes were found in the adrenal cortex of the mutant mice. Of these changes, a decrease in the number of lipid granules in cortical cells demonstrated by osmium tetroxide, Sudan black B, or oil red O stains was the most consistent observation (Fig. 4). Such a decrease in lipid granules was seen in the adrenal cortex of lethargic mice as young as 22 days and as old as 8 months, although it was more severe during the third and fourth weeks after birth than in older mice (Dung, 1976). Another noticeable change occurred only in the adrenal gland of older mutant mice. Adrenals from lethargic mice at 4 months of age or older showed groups of spindle-shaped cells radiating from the gland capsule into the cortex (Fig. 5). These spindle-shaped cells were seen in all adrenals of lethargic mice surviving longer than 8 months. They were rarely seen in the adrenal glands of normal mice younger than 10 months of age. "Wear and tear" has been

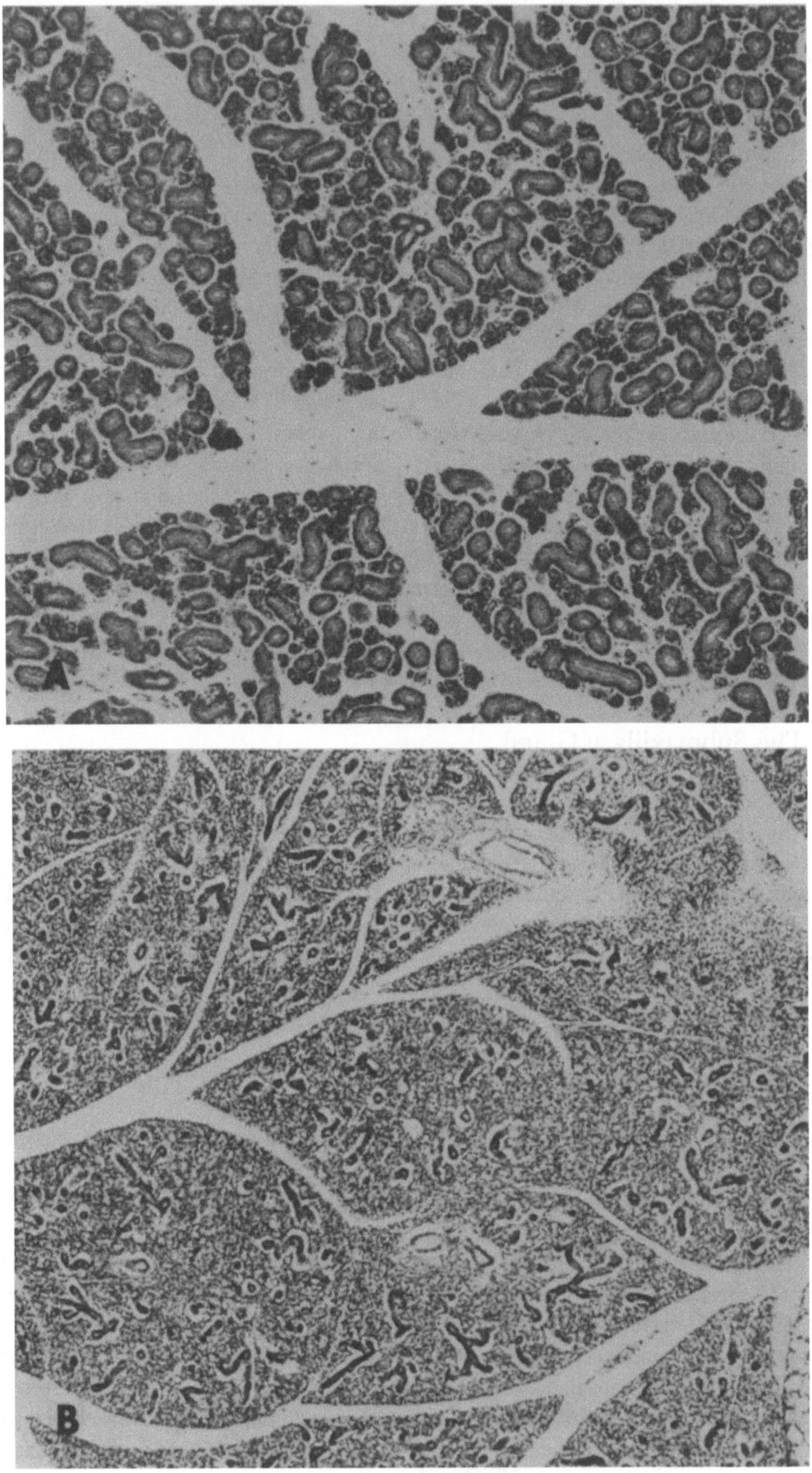

FIGURE 3. Submaxillary glands of 3-month-old normal (A) and lethargic (B) male mice. Note atrophic secretory tubules in (B). Both 100×.

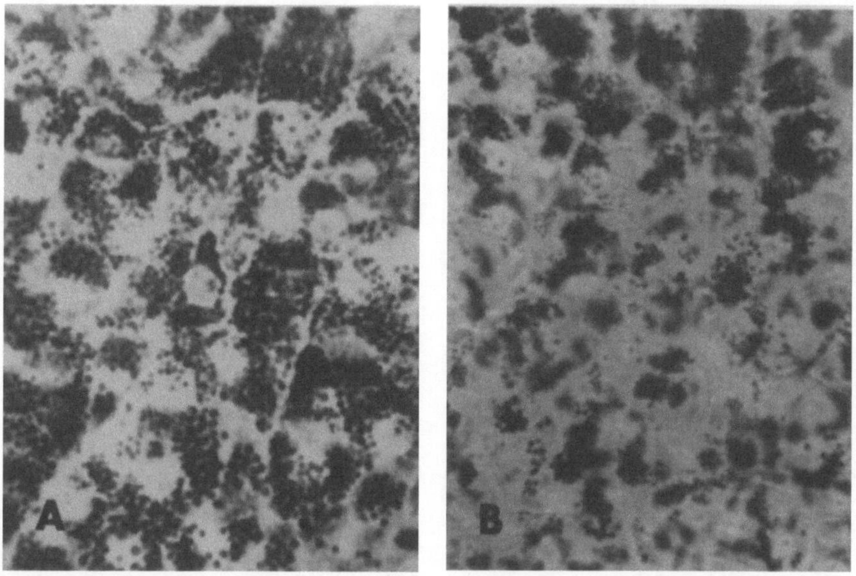

FIGURE 4. Zona fasciculata of the adrenal cortex of 24-day-old normal (A) and lethargic (B) mice. Note fewer lipid granules (stained with osmium tetroxide) in (B). Both 480×.

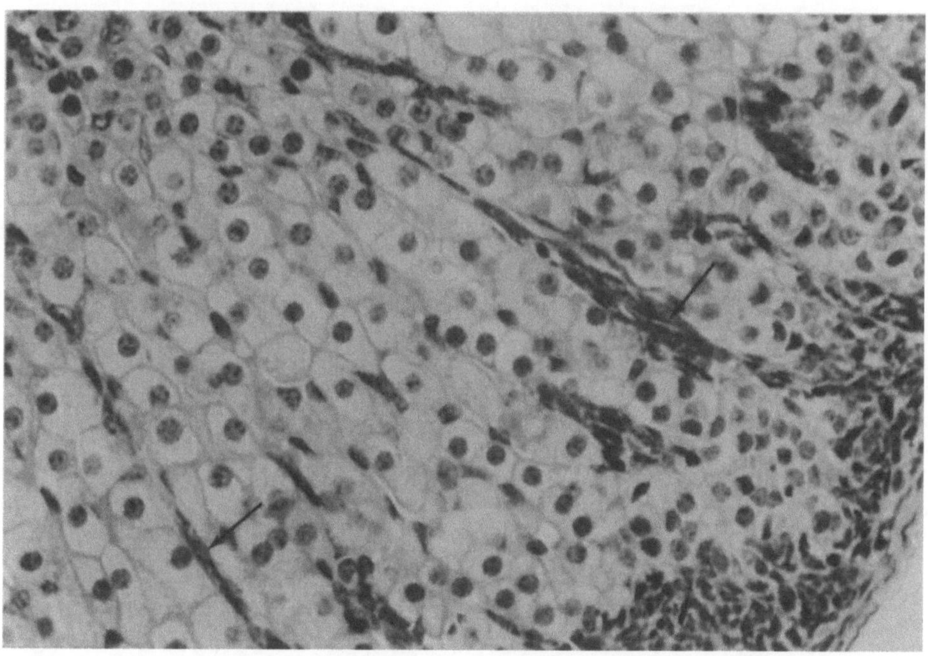

FIGURE 5. Adrenal cortex of 4-month-old lethargic mouse showing groups of spindle-shaped cells (arrows) radiating from the capsule into the zona glomerulosa and fasciculata. 480×.

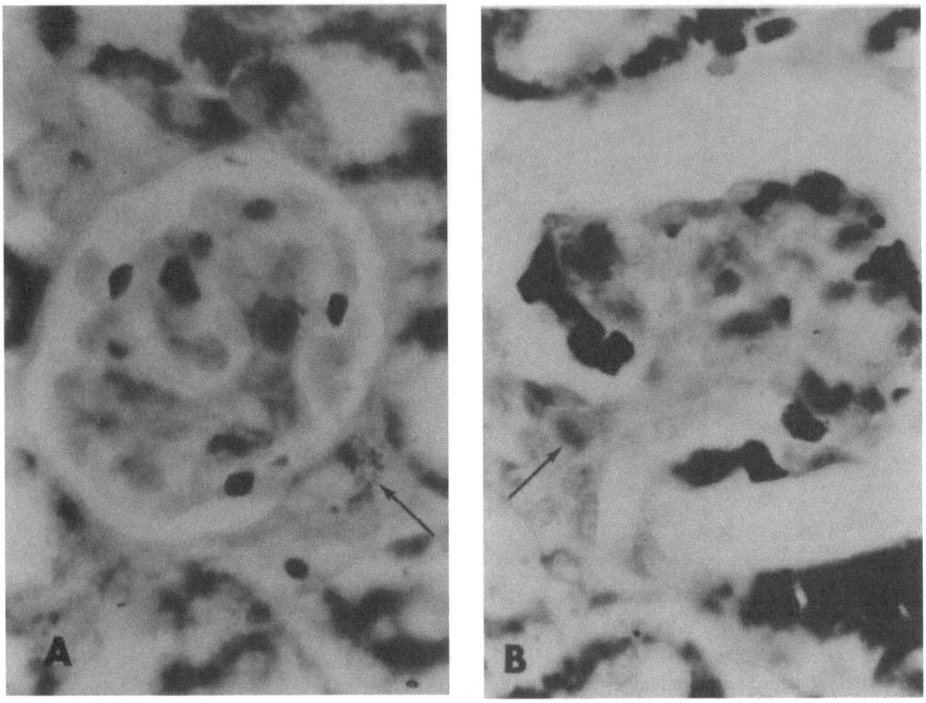

FIGURE 6. Renal corpuscles of normal (A) and lethargic (B) mice. Note barely visible juxtaglomerular granules (arrow) in (B). Both 480×.

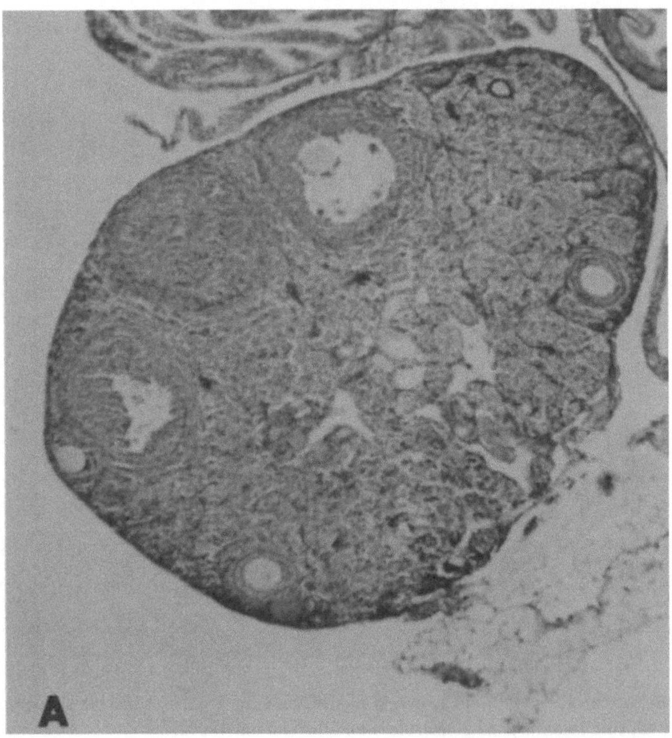

FIGURE 7. (A) Section of an ovary of a 4-month-old lethargic female mouse. 80×.

suggested as a basis for the spindle cells in the adrenal gland and they could be the "scar" cells associated with aging (Whitehead, 1932).

2.3. The Urinary System

No remarkable changes were found in the kidneys of the mutant mice, using hematoxylin–eosin staining methods. Unilateral or bilateral polycystic kidneys were found infrequently in lethargic mice at autopsy. Most of the cases occurred in younger affected mice.

Using Bowie's stain on paraffin sections of kidneys (Smith, 1966), juxtaglomerular cells in the kidney of the mutant invariably contained fewer stained granules than those of normal mice (Fig. 6). Very often it was difficult to identify a juxtaglomerular appartus in the lethargic kidneys because of poor granulation. Juxtaglomerular granules in lethargic mice were barely visible in most cases.

The urinary bladder of lethargic mice was often enlarged and appeared to be substantially larger than that of normal mice.

2.4. The Reproductive Organs

No marked abnormality was observed in the testes or seminal vesicles. Indications of spermatogenesis were obvious in all testis sections examined. The spermatozoa of mutant

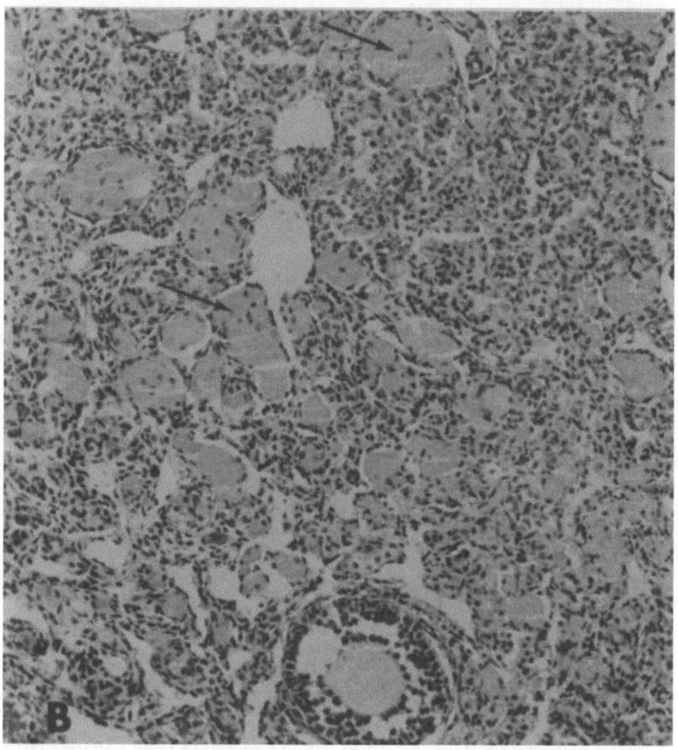

(B) Ovary of a 7-month-old lethargic female showing accumulation of pigment (arrow). 140×.

mice were as motile as those from normal mice when suspended in saline. However, in the female, the ovaries of mature lethargic mice showed morphologic abnormalities characterized by the absence of corpora lutea and the accumulation of yellow pigment (Fig. 7). Such ovarian dysgenesis was also observed in mice with wasting disease induced by neonatal thymectomy (Nishizuka and Sakakura, 1969, 1971). Accumulations of pigment are known to occur in the ovaries of mice older than 12 months of age (Deane and Fawcett, 1952). However, in lethargic mice, pigment accumulation was commonly seen in the ovaries of animals as young as 4 months of age.

2.5. The Liver and Gallbladder

A high incidence of hepatic degeneration localized around enlarged central veins was commonly found in lethargic mice (Fig. 8). Similar hepatic necrosis was seen in about 60%

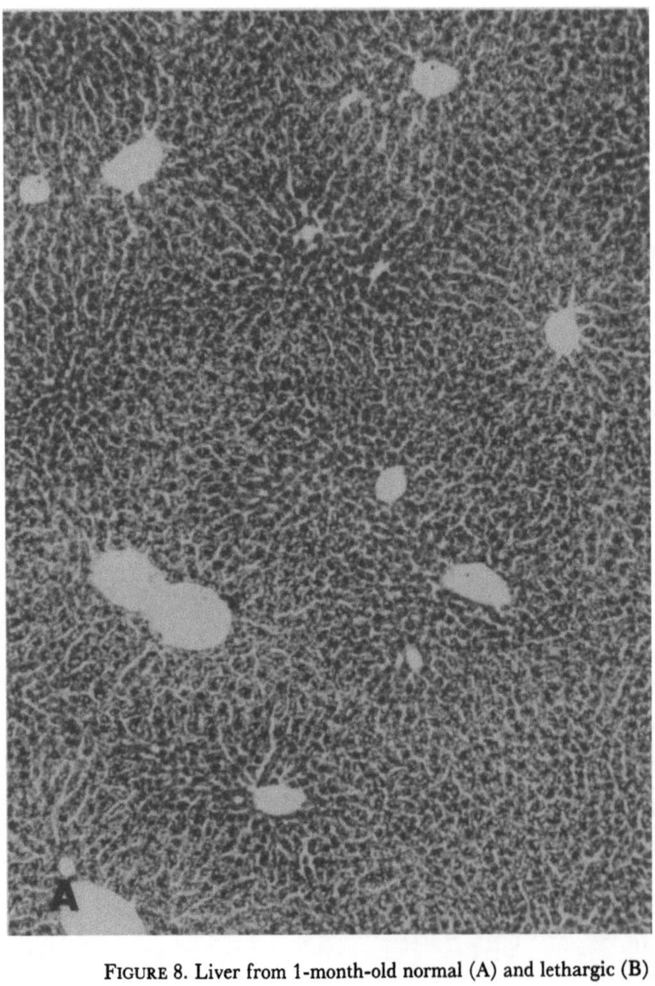

FIGURE 8. Liver from 1-month-old normal (A) and lethargic (B) mice. Note hepatic

of neonatally thymectomized mice (Parrott and East, 1962). Livers of lethargic mice younger than 5 months of age weighed significantly less (1101.4 ± 34.5 mg/100 g body wt.) than those of normal mice (1527.3 ± 36.7 mg/100 g body wt.). The difference was statistically significant ($p < 0.001$). For weight determinations, the tissues were lyophilized. At autopsy, the gallbladders of young lethargic mice were invariably distended. The gallbladders in lethargic mice were at least 3 times larger than in normal animals.

2.6. The Skin

Figure 9 shows cross sections of the skin of normal and lethargic mice. As seen in mice with wasting disease, the skin of the mutant was thin and there was a lack of subcutaneous fat (Parrott and East, 1962). This lack of subcutaneous fat in the skin was noted in every lethargic mouse sacrificed before 3 months of age.

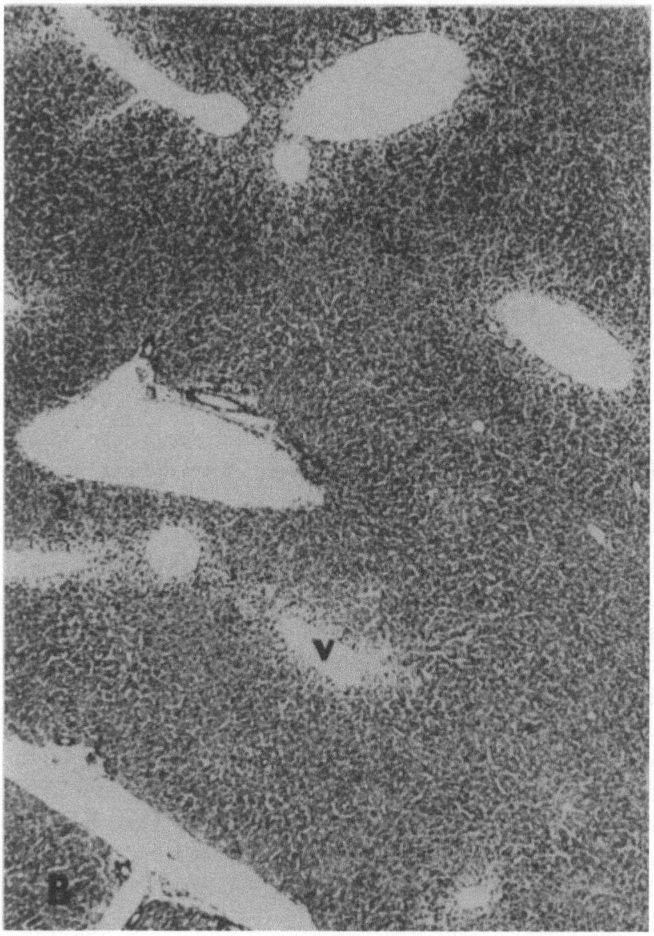

degeneration and enlargement of central veins (V) in (B). Hematoylin–eosin; 140×.

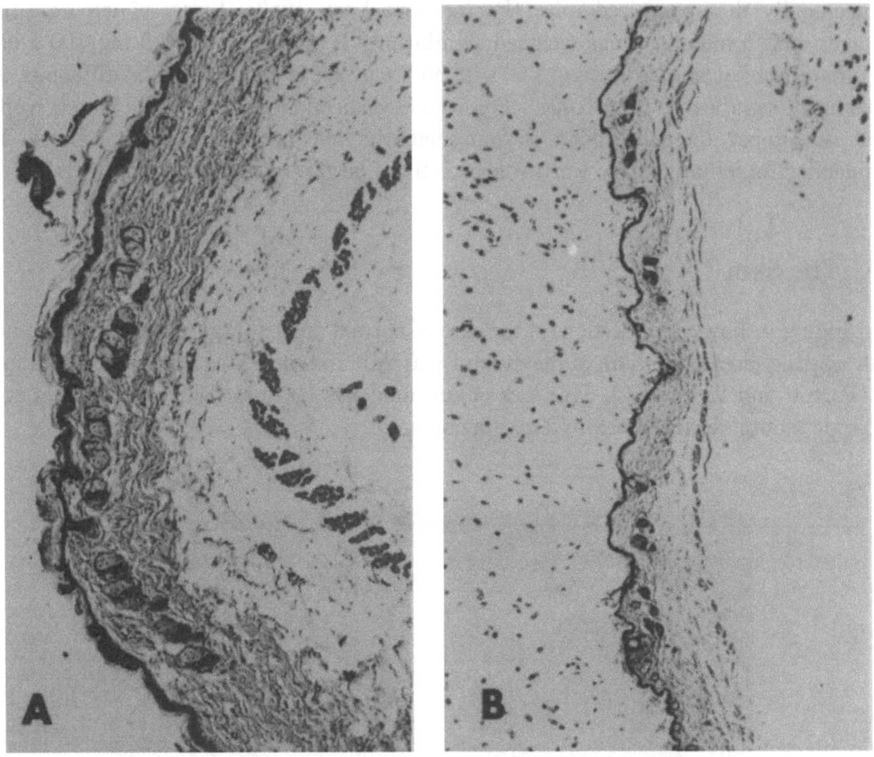

FIGURE 9. The skin of normal (A) and lethargic (B) mice at 1 month of age. Tissues were taken from the abdominal region. Note the difference in the thickness of the subcutaneous fat. Hematoxylin–eosin.

3. Electrophysiologic Studies on Peripheral Nerves

Histological examination at the light microscopic level did not produce clear evidence to indicate that the neurologic symptoms of lethargic mutant mice were a result of lesions in the central nervous system. To assess possible peripheral nerve involvement, electrophysiologic studies to measure motor nerve conduction velocity (MNCV), evoked muscle action potential (EMAP), and distal latency were done using the tail of the mouse (Herring *et al.,* 1979). The method used was a modification of the standard technique employed clinically to measure MNCV in human peripheral nerves, the principal difference being the special surface electrodes required because of the small size of the animal.

The electrophysiologic studies clearly showed that lethargic mutants have certain peripheral neuropathies. They had a much slower MNCV (20.1 $\pm$ 4.0 m/sec at a 35°C rectal temperature) than that of normal mice (30.8 $\pm$ 5.3 m/sec at a 35°C rectal temperature). The difference was highly significant ($p < 0.001$). Their EMAP amplitude (1.5 $\pm$ 0.6 mV) was significantly reduced, and their distal motor latency (2.0 $\pm$ 0.5 msec) was significantly prolonged. EMAP amplitude and distal latency for normal mice at the same rectal temperature were 2.9 $\pm$ 0.6 mV and 1.6 $\pm$ 0.4 msec, respectively. These differences were also highly significant ($p < 0.001$).

To determine how soon the neuropathy manifests itself electrophysiologically and how the parameters of nerve function change with age, the MNCV, EMAP amplitude, and distal latency were measured in normal and lethargic mice at monthly intervals beginning at 1 month of age and continuing for 10 months. The chronological studies showed that there were highly significant differences in the MNCV, EMAP amplitude, and distal latency between normal and lethargic mice beginning at 2 months of age and thereafter (Herring *et al.,* 1981). The neuropathy exhibited by the lethargic mouse was not improved by a parabiotic procedure that established communication between the blood circulation of a normal mouse and an affected mutant (Dung *et al.,* 1981). Thus, electrophysiologic studies led to the speculation that the mutation may create an inheritable dysmyelination in the affected mice due to a mild to moderate dysfunction of myelinogenesis (Herring *et al.,* 1981).

4. Pathologic Changes in Lymphoid Organs

When intracardiac perfusion was performed to fix and preserve the brain and spinal cord for neuropathological examination, it was noted that every lethargic mouse had a small thymus. Frequently, the thymus was so small that it was difficult to visualize and dissect for removal.

4.1. Organ Weight Studies

In general, the lymphoid organs, including thymus, spleen, lymph nodes, and Peyer's patches, of lethargic mice are smaller than those of their normal littermates. Peyer's patches are fewer in number in lethargic mice than in normal mice. The results of organ weight studies (Dung and Swigart, 1971) showed that dry weights of both the thymus and the spleen are significantly less in the mutant than in normal littermates soon after the onset of the neurological symptoms. For instance, the spleens of 16- and 17-day-old lethargic mice weighed 63.2 $\pm$ 1.2 mg/100 g body wt, whereas the spleens of normal littermates weighed 127.2 $\pm$ 3.8 mg on the same body weight basis. The thymuses of 24- to 32-day-old mutants weighed 22.4 $\pm$ 1.7 mg/100 g body wt, and those of normal littermates weighed 122.0 $\pm$ 5.4 mg/100 g body wt. This significantly reduced weight of the thymus persists throughout the entire life of the mutant, whereas the spleen weight of the mutant increases to that of normal mice after about 5 months of age.

4.2. Hematologic Studies

The average hematocrits of lethargic mice (36.5%) at 1 month of age are lower than those of their normal littermates (41.1%). Total and differential white cell counts reveal that lethargic mice have severe leukocytopenia. Total white cell counts for normal mice between 15 and 31 days of age range from 5000 to 6000 per mm^3 of blood. Total white cell counts for lethargic mice in the same age group range from 1000 to 2000 per mm^3 of blood. Most of the decrease in total leukocytes is due to a decrease in lymphocytes (Dung and Swigart, 1971).

4.3. Morphologic Studies

Histologic sections of the thymus, spleen, inguinal lymph nodes, and Peyer's patches in the ileum have been studied by light microscopy. Morphologic changes observed in the thymuses of lethargic mice were very similar to those described for the thymuses of mice that had received injections of adrenocortical hormones (Ito and Hoshino, 1962; Haelst, 1967).

The rate of thymic involution in lethargic mice varied from animal to animal. The rapidity of the thymic involution was proportional to the severity of the symptoms in the affected mice. The first indication of thymic involution was characterized by the appearance of large numbers of pycnotic lymphocytes. Most of the pycnotic lymphocytes were found in the peripheral cortical zone of the thymus. Pycnosis occurred infrequently among medullary thymocytes. A result of the massive thymocyte pycnosis was a substantial loss of cells in the thymic cortex of lethargic mice. Thymocyte loss in the cortex initially created a thymic condition in which no visible boundary existed between the cortex and the medulla. As the loss of cortical thymocytes continued, the thymus became smaller and smaller, and eventually thymocyte depletion in the cortex was so complete that the medulla stained darker than the cortex (Fig. 10). At this stage of involution, the medulla contained a number of "giant bodies" that resemble the Hassall's corpuscles described for the human thymus. The mouse thymus rarely contains Hassall's corpuscles (Clark, 1963). When examined under electron microscope, the thymus in lethargic mice has a lamellar structure formed by layers of densely packed epithelial cells.

When thymic tissues were stained by the PAS method, numerous PAS-positive cells were seen in the cortical areas. The number of PAS-positive cells increased in lethargic mice older than 5 months of age.

Morphologic changes observed in the spleen, lymph nodes, and Peyer's patches were similar to those described in mice with wasting disease caused by neonatal thymectomy (Archer *et al.,* 1962). In the spleen, lymphocyte depletion of the white pulp was the most remarkable change, beginning at 17 days and continuing through 35 days of age. Loss of lymphocytes, first observed in the marginal zones, was followed by diminution in the entire white pulp area. The white pulp areas of most 30-day-old lethargic mice were completely depleted of lymphocytes so that only reticular tissue remained. Erythrocytes frequently infiltrated into such white pulp areas.

Morphologic changes in the lymph nodes and Peyer's patches of lethargic mice were not as remarkable as those in the thymus and spleen, though both types of lymphoid tissues in the mutant contained fewer lymphatic follicles or nodules than those of normal animals (Dung and Swigart, 1972).

Electron microscopic studies (Dung, 1973) reconfirmed thymic involution in lethargic mice. The cells participating in the involution were cortical thymocytes, macrophages, and epithelial cells. The pycnotic nuclei of degenerating thymocytes appeared to be engulfed quickly by cortical macrophages. The macrophages were also characterized by a variety of

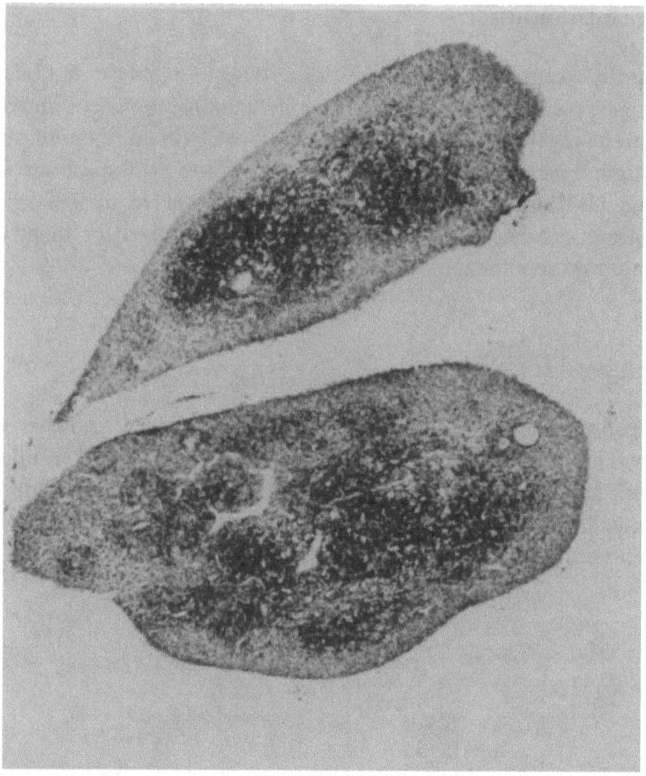

FIGURE 10. The thymus from a 28-day-old lethargic mouse. Note that the medullary area is darker than the cortical area. Hematoxylin–eosin.

cytoplasmic inclusions. These cytoplasmic inclusions were of two different types: (1) tonofibrils in markedly increased numbers as compared to a noninvoluted thymus, and (2) large vacuoles with dense bodies and/or myelin figures. Similar cytoplasmic inclusions were also found in the epithelial cells. In addition, the involuting thymuses in lethargic mice contained a large number of lipid-laden cells that were not seen in the thymuses of normal mice.

5. Deficiencies in Cellular Immunity

The pathologic changes observed in the lymphoid tissues of lethargic mice strongly suggested that they would have deficiencies in cell-mediated immunity. Experimental studies designed to test the integrity of the cellular immunity in this mutant mouse were performed (Dung, 1977).

5.1. Skin Transplantation

Skin allografts were made from DBA donor mice to lethargic mutants. It was found that lethargic mice younger than 30 days of age were unable to reject skin allografts. However, mutant mice over 45 days of age appeared to have a normal rejection capacity. Sixteen of twenty lethargic mice receiving skin transplants before 30 days of age died within 21 days of grafting. No indication of skin rejection was observed in lethargic recipients at autopsy. Skin grafts on those four lethargic mice that survived for longer than 21 days remained intact until the animals' death.

5.2. Graft-versus-Host Reaction

Spleen cell suspensions from normal and lethargic mice were transferred into F_1 newborn recipients. The recipients were produced by crossing BALB/cGn mice carrying the mutant gene with DBA mice. Spleen cell donors were all 24 ± 1 days old. Approximately 75% of recipients died by the end of the third week after inoculation with 4.5 to 5 $\times$ 10^6 spleen cells from normal mice.

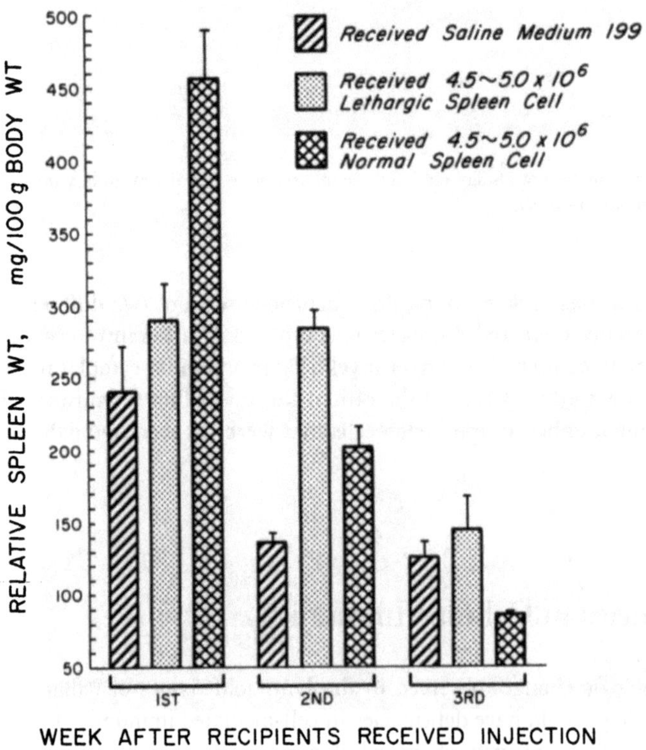

FIGURE 11. Spleen weight changes of F_1 recipients after the inoculation of a parental spleen cell suspension in a graft-versus-host test.

Graft-versus-host reactions were also measured by their effects on body weight, total white cell count, and organ weight. The spleen cells from normal mice were more effective in suppressing weight gain in F_1 recipients than were mutant spleen cells. The leukocytopenic effect of mutant spleen cells on the recipient was not as prompt as that of spleen cells from normal mice. Spleen cells from normal mice were also more potent in producing splenomegaly in the recipient (Fig. 11).

Although lethargic mice have a definite deficiency in cell-mediated immunity between 3 and 4 weeks of age (Dung, 1977), this deficiency appears to be transitional. For the minority of lethargic mice that survive longer than 2 months, the deficiency can no longer be shown. It is not clear whether the deficiency in cellular immunity among young mutants is a primary defect directly due to gene mutation or whether it is secondary to some acquired disease process.

6. Humoral Immunity

6.1. Serum Immunoglobulins

Sera obtained from 23- to 25-day-old lethargic mice were assayed on agarose immunodiffusion plates for IgG1, IgG2a, IgG2b, IgA, and IgM levels. In both normal and lethargic mice of this age, only IgG1 and IgG2a levels were high enough to be measurable. Serum IgG1 and Ig2a levels were significantly higher in the mutant mice than in their normal littermates. The mean for all IgG1 readings in the mutant mice was 148.0 ± 28.9 mg/100 ml of serum compared to 89.2 ± 43.0 mg/100 ml of serum for the normal mice (Dung *et al.,* 1977).

6.2. Serum Electrophoresis

Lethargic mice showed a lower percentage (7.45 ± 0.26) of α_1-protein following serum electrophoresis on cellulose acetate than normal mice (8.65 ± 0.28). The difference ($p < 0.01$) was significant (Dung *et al.,* 1977).

Serum samples from mice between 15 days and 18 months of age were analyzed electrophoretically on polyacrylamide gel. The number of discrete protein bands resolved varied from young to old animals. However, 11 bands were discernible at all ages except in 15-day-old and a few 17-day-old mice. Thus, the densitometric tracings for each gel were divided into 11 different bands (using albumin as Band 1) as shown in Fig. 12 for subsequent peak analysis of protein fractions.

Results of peak analysis showed differences in serum proteins between normal and lethargic mice seen primarily at young ages (17 to 27 days of age). During that interval splenic atrophy and thymic involution were occurring in the mutant, and the most striking and consistent difference was seen in Band 2 (H. C. Dung *et al.,* unpublished data). A remarkable observation was that the protein concentration in Band 2 was inversely proportional to lymphoid organ weight. Proteins in Band 2 quantitatively increase when the lymphoid organs of the mouse rapidly decrease in weight. When lymphoid organs were

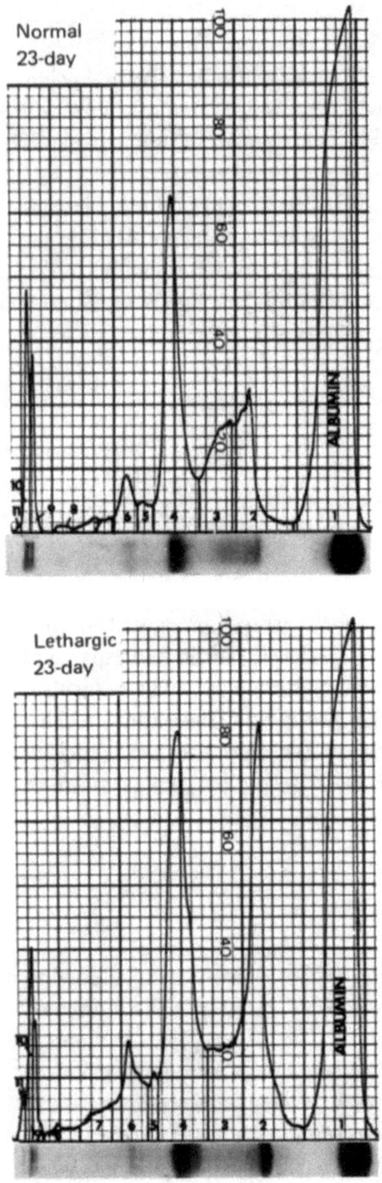

FIGURE 12. Electrophoretograms of sera from 23-day-old normal (upper) and lethargic (lower) mice. Note quantitative difference in Band .

rapidly gaining weight, the protein concentration in Band 2 was low. It has been speculated that Band 2 may contain protein or proteins that are capable of inhibiting cell proliferation in lymphoid tissue (Dung and Yu, 1975).

7. Lymphocyte Kinetics by Autoradiography

Autoradiographic analyses were performed to determine if spontaneous thymic involution and splenic atrophy in the mutant were the result of altered cell kinetics. Lethargic and normal littermates 23 days old were used for the study. Each mouse was given 2 Ci of [^{3}H]thymidine per gram of body weight as a single intraperitoneal injection. One hour later the mice were sacrificed to obtain the thymus and spleen for autoradiographic examination. Tissue slides were dipped in Kodak NTB2 emulsion, exposed for 4 weeks at 4°C, and later stained with hematoxylin and eosin.

Autoradiographic analysis was performed by counting labeled cells. Only cells with four or more grains were considered to be positive in determining the labeling index (Dung and Adrian, 1978).

One hour after the injection, labeled cells in the thymic cortex of normal mice (12.1 $\pm$ 2.2%) were three times more numerous than in lethargic mutants (4.4 $\pm$ 1.4%). The difference was significant ($p < 0.01$). In the thymic medulla, the percentage of labeled cells did not differ significantly between normal and mutant mice. Labeled cells were primarily large and medium-sized thymocytes in the cortical area. Small thymocytes were rarely labeled (Fig. 13).

The thymus is known to have very high proliferative activity (Andreasen and Christensen, 1949). It has been estimated that it can replace its entire lymphocyte population within 4 days (Hess, 1968). The very low percentage of labeled cells in the thymuses of lethargic mice strongly suggested that the cells in this organ may have nearly ceased to proliferate. The existence of mitotic inhibitors has been reported (Attallah *et al.*, 1975). Because we have found that the protein concentration in Band 2 obtained by serum elec-

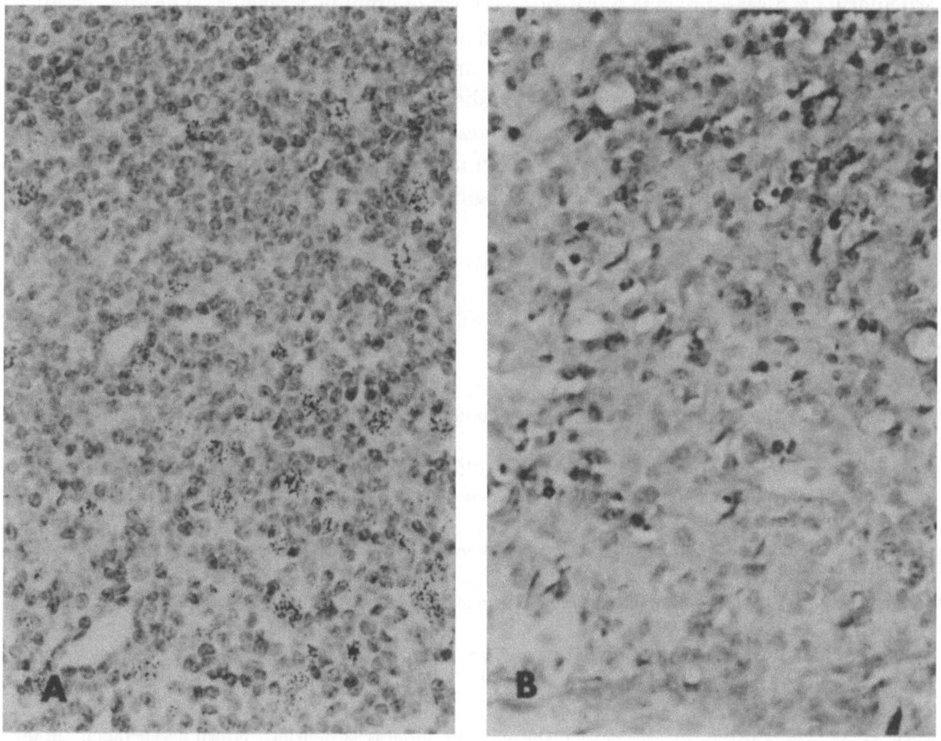

FIGURE 13. Autoradiographic picture of thymic cortex from 23-day-old normal (A) and lethargic (B) mice. Note very few labeled cells in (B).

trophoresis was inversely proportional to that of the lymphoid organ weight of the mouse, we have speculated that Band 2 may contain antimitotic substances capable of inhibiting lymphocyte proliferation in the thymus. The lymphoid organ weight decrease could then be the result of an increase in this endogenous mitotic inhibitor in lethargic mice.

In the spleen, labeled cells were also significantly fewer in the red pulp of lethargic micc (12.8 ± 2.8%) than in normal mice (26.8 ± 3.7%). When mice were challenged by immunization with sheep red blood cells, labeled cells in the white pulp of normal mice increased almost threefold. No such increase was seen in the white pulp of lethargic mice after similar antigenic stimulation. Thus, cell kinetics in the spleens of lethargic mice also appeared to be altered.

8. Conclusions

Lethargic, classified as a neurological mutation of the mouse, is associated with obvious phenotypic behavioral abnormalities. The etiology of the neurologic abnormalities is not known. A morphologic basis for the neurologic disorder has not been clearly established. It is questionable if the mutation is a useful model for experimental studies of neurologic diseases in man. However, it is believed that the lethargic mutant will be valuable animal for studies leading to a better understanding of diseases with immunologic defects, particularly for diseases having a transitional form of immunodeficiency. The affected mice have unquestionable defects in cell-mediated immunity, which are invariably expressed from 3 to 4 weeks after birth. The humoral immunity of this mutant has not been thoroughly investigated, although indications of aberrant humoral immunity are also detected in young mutant mice. Recent reports have shown that many neurologic diseases are related to defects in immunologic functions. Myasthenia is one of the best examples. One wonders if the neurologic disorders and immunologic deficiencies in lethargic mice are only coincidental.

References

Andreasen, E., and Christensen, S., 1949, The rate of mitotic activity in the lymphoid organs of the rat, *Anat. Rec.* **103**:401.

Archer, O. K., Papermaster, B. W., and Good, R. A., 1962, Thymectomy in rabbit and mouse: Consideration of time of lymphoid peripheralization, in: *The Thymus in Immunobiology* (R. A. Good and A. E. Gabrielsen, eds.), p. 414, Harper & Row, New York.

Attallah, A. M., Sunshine, G. H., Hunt, C. V., and Houck, J. C., 1975, The specific and endogenous mitotic inhibitor of lymphocytes, *Exp. Cell Res.* **93**:283.

Billingham, R. E., 1958, Studies on the reaction of injected homologous lymphoid tissue cells against the host, *Ann. N.Y. Acad. Sci.* **73**:782.

Caramia, F., 1966, Ultrastructure of mouse submaxillary gland, *J. Ultrastruct. Res.* **16**:505.

Clark, S. L., Jr., 1963, The thymus in mice of strain 129/J, studied with the electron microscope, *Am. J. Anat.* **112**:1.

Deane, H. W., and Fawcett, D. W., 1952, Pigmented interstitial cells showing "brown degeneration" in the ovaries of old mice, *Anat. Rec.* **113**:239.

Dickie, M. M., 1964, Lethargic (lh) mutant mouse, *Mouse News Letter* **30**:31.
Dung, H. C., 1973, Electron microscopic study of involuting thymus of lethargic mutant mice, *Anat. Rec.* **177**:585.
Dung, H. C., 1976, Relationship between the adrenal cortex and thymic involution in lethargic mutant mice, *Am. J. Anat.* **147**:255.
Dung, H. C., 1977, Deficiency in the thymus-dependent immunity in lethargic mutant mice, *Transplantation* **23**:39.
Dung, H. C., and Adrian, E. K., Jr., 1978, Lymphoid tissues of lethargic mice, *Dev. Comp. Immunol.* **2**:327.
Dung, H. C., and Swigart, R. H., 1971, Experimental studies of lethargic mutant mice, *Texas Rep. Biol. Med.* **29**:273.
Dung, H. C., and Swigart, R. H. 1972, Histo-pathologic observations of the nervous and lymphoid tissues of lethargic mutant mice, *Texas Rep. Biol. Med.* **30**:23.
Dung. H. C., and Yu, J., 1979, Serum protein of mice with splenomegaly, *Experientia* **35**:1108.
Dung, H. C., Lawson, R. L., and Stevens, M., 1977, A study of the increased serum level of IgG1 in lethargic mice combined with a depressed thymus-dependent lymphoid system, *J. Immunogenet.* **4**:287.
Dung. H. C., Herring, J. M., and Yu, J., 1981, Peripheral motor nerve conduction in parabiotic pairs of lethargic and normal mice, *Electromyogr. Clin. Neurophysiol.*, in press.
Fox, M. W., 1965, The neuro-ontogeny of neuromuscular mutant mice, *J. Hered.* **56**:55.
Green, E. L., and Staff of The Jackson Laboratory, 1966, *Biology of the Laboratory Mouse*, 2nd ed., McGraw-Hill, New York.
Haelst, U. Von, 1967, Light and electron microscopic study of the normal and pathological thymus of the rat. I. The normal thymus, *Z. Zellforsch. Mikrosk. Anat.* **77**:534.
Herring, J. M., Dung, H. C., and Yu, J., 1979, Peripheral motor nerve conduction studies in lethargic mice, *Electromyogr. Clin. Neurophysiol.* **19**:147.
Herring, J. M., Dung, H. C., Yoo, J. H., and Yu, J., 1981, Chronological studies of peripheral motor nerve condition in lethargic mice, *Electromygr. Clin. Neurophysiol.*, in press.
Hess, M. W., 1968, *Experimental Thymectomy*, pp. 15–23, Springer-Verlag, New York.
Ito, T., and Hoshino, T., 1962, Histological changes of the mouse thymus during involution and regeneration following administration of hydrocortisone, *Z. Zellforsch. Mikrosk. Anat.* **56**:445.
Miller, J. F. A. P., 1962, Effect of neonatal thymectomy on the immunological responsiveness of the mouse, *Proc. R. Soc. London Ser. B* **156**:415.
Nishizuka, Y., and Sakakura, T., 1969, Thymus and reproduction: Sex-linked dysgenesis of the gonad after neonatal thymectomy in mice, *Science* **166**:753.
Nishizuka, Y., and Sakakura, T., 1971, Effect of combined removal of thymus and pituitary on post-natal ovarian follicular development in the mouse, *Endocrinology* **89**:902.
Parrott, D. M. V., and East, J., 1962, Studies on fatal wasting syndrome in mice thymectomized at birth, in: *The Thymus in Immunobiology* (R. A. Good and A. E. Gabrielsen, eds.), Harper & Row, New York.
Shackelford, J. M., and Klapper, C. E., 1962, A sexual dimorphism of hamster submaxillary mucin, *Anat. Rec.* **142**:495.
Sidman, R. L., Green, M. C., and Appel, S. H., 1965, *Catalog of Neurological Mutants of the Mouse*, Harvard University Press, Cambridge, Mass.
Smith, C. L., 1966, Rapid demonstration of juxtaglomerular granules in mammals and birds, *Stain Technol.* **4**:291.
Whitehead, R., 1932, Abnormalities of the mouse suprarenal, *J. Pathol. Bacteriol.* **35**:415.

3

Immunologic and Genetic Aspects of Aleutian Disease

DONALD L. LODMELL AND JOHN L. PORTIS

1. Introduction

Aleutian disease (AD) is a common infection of ranch-bred mink *(Mustela vison)* caused by a temperature-sensitive, nondefective, xenotropic parvovirus (ADV). Hartsough and Gorham (1956) first described AD in mutant mink homozygous for the recessive Aleutian coat color gene which arose on an Oregon mink ranch in 1941. The mutant mink were called Aleutian because of the similarity of their blue-gray coat (Fig. 1) to that of the Aleutian blue fox. It subsequently has been shown that ADV affects Aleutian and non-Aleutian genotypes (Porter and Larsen, 1964) with persistent progressive, persistent nonprogressive, or nonpersistent nonprogressive infections that are determined by virus strain and mink genotype. AD has been classified in the past as a "slow virus disease" akin to kuru, Creutzfelt–Jacob, visna, and mink encephalopathy. However, this term is a misnomer (Oldstone, 1971) as shown by demonstration of rapid virus replication shortly after inoculation (Porter *et al.,* 1969).

Mink with clinical disease manifest rapid virus replication, persistent viremia, polyclonal hypergammaglobulinemia, high titers of anti-ADV antibody, and immune-complex-mediated lesions. Animals usually die from 2 months to 2 years postinoculation of virus. The cause of death is usually uremia due to kidney disease and/or hemorrhage due to vasculitis.

AD can be characterized generally as a disease of the immune system mediated by a persistent virus. Its study has been fostered in the past because of its use as a potential model for immune-complex disease of man, such as systemic lupus erythematosus. However, the exaggerated immune response to the virus resembles in many respects persistent

DONALD L. LODMELL AND JOHN L. PORTIS • Laboratory of Persistent Viral Diseases, Rocky Mountain Laboratories, National Institute of Allergy and Infectious Diseases, National Institutes of Health, Hamilton, Montana 59840.

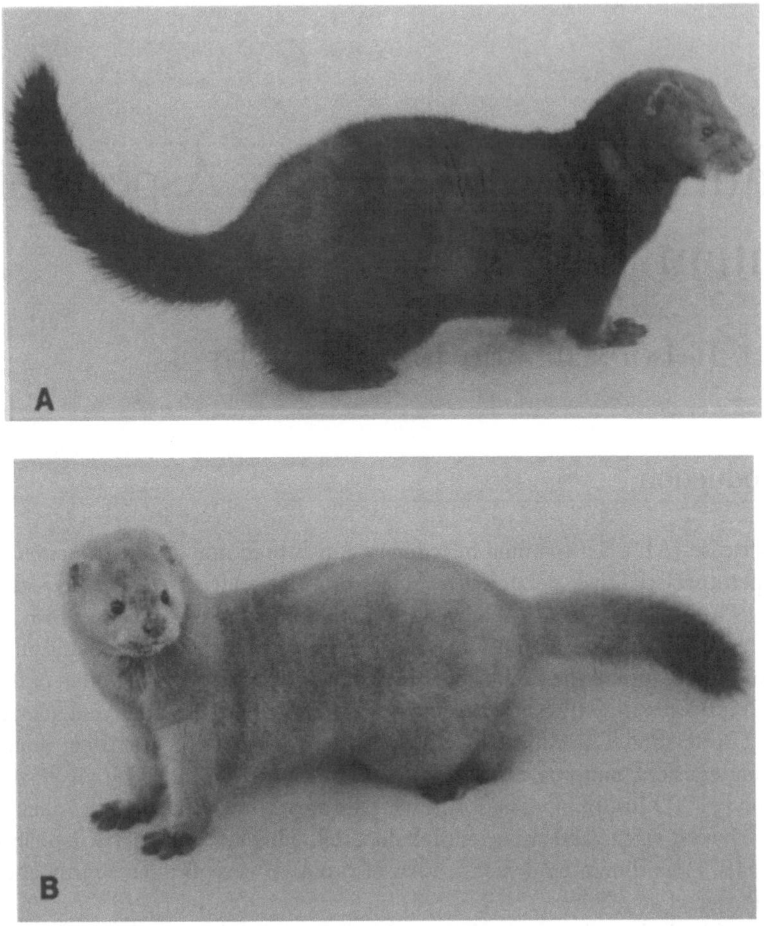

FIGURE 1. Two color phases of mink used in the study of AD. (A) non-Aleutian (pastel) and (B) Aleutian (sapphire) (pictures selected by J. Lodmell and N. Portis).

viral diseases in humans such as the rubella syndrome and perinatal cytomegalovirus infection (Williams *et al.,* 1968). Both of these diesases are manifested by persistent viral infection and polyclonal hypergammaglobulinemia.

In this review we have attempted to synthesize relevant information on viral persistence, possible effects of host and viral genetics on the disease course, and effect of persistent virus infection on the immune system.

1.1. Genetics

The genetics of coat color in mink has been exhaustively studied for economic reasons. However, little is known of the gene(s) controlling susceptibility to AD. Mink have 30 chromosomes (Fredga, 1961) and are highly inbred for various genes determining coat color, but are outbred for the rest of the genome. At least 19 genes are known to be involved in determining coat color (Shackelford, 1973). Mink of the Aleutian color phase (*aa*), a very desirable homozygous recessive genotype, were observed to be more susceptible to fatal AD than were other color phases. It was found that outbreeding this gene into unrelated stocks tended to prolong the life span. However, all mink homozygous for *aa* are susceptible to AD. Mink that are used extensively in AD research include sapphire and violet (both *aa*), pastel (*Aa* or *AA*), and standard dark (*AA*). The former two color phases henceforth will be referred to as Aleutian, the latter two as non-Aleutian.

The only known allotypic system, other than a rather complex array of blood groups (Schackelford, 1973), is the *Lpm* system described by Baranov and Yermalaev (1978). There are to date eight serologically defined allotypic markers on the very-high-density α lipoproteins in mink, but no attempt has been made to determine any possible linkage to the *aa* genotype.

We have found no published studies on the detection of, or response to, cellular alloantigens in mink. However, R. E. Race (personal communication) has studied a sarcoma that arose in a pastel mink. This tumor is transplantable by intraperitoneal passage into either neonatal pastel or sapphire mink, but is rejected by adult mink of either color phase. Thus, alloantigenic differences in mink appear to be substantial. Because of the limited breeding season for mink (one litter per year) and the economic constraints of a breeding program, the genetic influences on AD probably will be restricted to comparing mink with and without the Aleutian genotype.

1.2. Clinical Manifestations

The duration of preclinical disease in experimental infection with ADV is rarely less than 6 weeks and is a function of viral strain, concentration of virus inoculated, route of administration, and recipient genotype. The clinical disease in experimentally and spontaneously infected mink is similar. Aleutian mink develop a more rapid clinical course and die within a shorter time after experimental infection than do non-Aleutian mink (Henson *et al.,* 1976).

The earliest manifestations of clinical involvement are partial to total anorexia that, once established, persists until the animals' death. Coincident with anorexia is progressive weight loss that, in animals that exhibit prolonged survival, eventually results in severe emaciation. During the interim, the animals' coats become progressively more unkept and dull in appearance but are rarely soiled (R. E. Race, personal communication). Dehydration and anemia are generally apparent. Polydipsia is common, but sometimes difficult to appreciate. Impending death is marked by a transition in fecal appearance from normal to a black tarlike consistency, and other hemorrhagic episodes are common including bleeding from the nose, mouth, and mucosal surfaces. Animals often are comatose during the few hours preceding death or may die in convulsive seizures (R. E. Race, personal communication).

Clinical disease of several months' duration is not unusual and occasionally animals will survive a year or more. However, the course may be abbreviated especially in mink housed outdoors in severely cold weather. The cause of death is usually hemorrhage (often cerebral) or uremia (R. E. Race, personal communication).

1.3. Diagnosis

The first method used for detecting AD was the simple iodine agglutination test (IAT) as described by Mallén *et al.* (1950). AD also can be diagnosed on the basis of gross and macroscopic lesions in mink tissues, and by increased levels of serum gamma globulins as determined by cellulose acetate electrophoresis (Larsen, 1965). Recent advances for the detection and quantitation of ADV antigen and antibody have been developed. Porter *et al.* (1969, 1972) described an indirect immunofluorescence (IF) assay for the quantitation of anti-ADV antibody and a direct fluorescence assay for the detection of viral antigen. With the use of a complement-fixation (CF) test, McGuire *et al.* (1971) detected ADV antibody, and Cho and Ingram (1972) demonstrated antibody against ADV antigen by a precipitin reaction using counterimmunoelectrophoresis (CIE) in agar gels. More recently, Shahrabadi and Cho (1977) have detected and localized ADV by electron microscopy with an immunoferritin technique, and Burger *et al.* (1978) have detected antibody and antigen in AD-affected mink with an enzyme-linked immunoabsorbent assay (ELISA).

The specificity, reproducibility, and sensitivity of these various assays recently have been determined (Crawford *et al.*, 1977). Sera were collected from four mink from 12 days to $1\frac{1}{2}$ years after experimental inoculation with the virulent Utah-1 strain of ADV. The IF, CF, and CIE assays were specific for anti-ADV antibody and no false positive reactions were reported. IF was less reproducible than the other systems, and the standard CIE test was the least sensitive. IF and CF were more sensitive than CIE, but were limited by background staining and anticomplementary activity, respectively, when used with undiluted sera. Because CIE is the simplest and most rapid test and is reliably sensitive and specific, it was determined to be the assay of choice for routine detection and quantitation of mink antibody. Nonetheless, until more data are available on the antigens and antibody classes that are detected by these tests, caution should be observed when comparing data (Crawford *et al.*, 1977).

2. Pathology

The reader is referred to several excellent reports on the pathology of AD for a more detailed description (Helmboldt and Jungherr, 1958; Obel, 1959; Trautwein, 1964; Henson *et al.*, 1966, 1976). Briefly, AD is characterized by multisystem infiltrations of mononuclear cells (predominantly plasma cells) (Fig. 2). In fact, the disease has been termed "infectious plasmacytosis" of mink. The plasma cell infiltrates are accompanied by hypergammaglobulinemia and occasionally "monoclonal" spikes in cellulose acetate electrophoresis of serum. In mink with monoclonal gamma-globulin spikes, urinary excretion of Bence Jones protein (Porter and Dixon, 1965b) has been seen. However, osteolytic lesions have never been demonstrated.

Other notable manifestations of AD include a proliferative glomerulopathy (Pan *et al.*, 1970a) (Fig. 3) and acute to subacute arteritis (Porter *et al.*, 1973). In both of these lesions host Ig, C3, and viral antigen have been demonstrated by IF (Henson *et al.*, 1969;

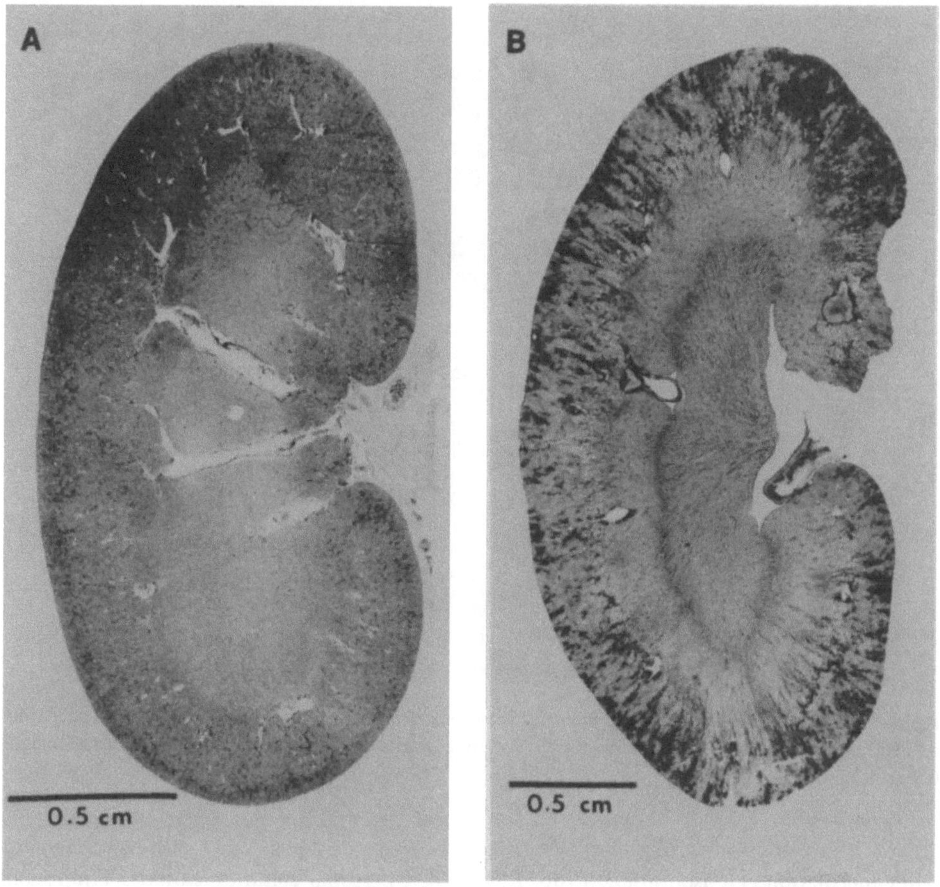

FIGURE 2. Low-power photomicrographs of mink kidneys. (A) Normal kidney and (B) kidney from a mink with AD. Note the difference in scale of the two plates. The AD kidney is about 1½ times the normal size. Also observe the cellular infiltrate (dark areas of predominantly plasma cells) in the cortex of the AD kidney.

Porter *et al.*, 1969; Trautwein and Seidler, 1972; Johnson *et al.*, 1975) (Fig. 4). The glomerulopathy examined by electron microscopy reveals electron-dense deposits in both subendothelial and subepithelial locations (Henson *et al.*, 1967; Kindig *et al.*, 1967). This agrees with findings of granular deposits of Ig, C3, and viral antigen along the glomerular capillary basement membrane and in the mesangium. The arteritis involves small to medium-sized arteries and resembles that of experimental serum sickness. Bile duct proliferation (Drommer and Trautwein, 1975) is noted in the liver and also may be related to deposition of immune complexes (Porter and Larsen, 1967).

The pathologic lesions of AD do not appear to be the direct result of viral infection. Although virus titers in serum and organs peak 10 days to 2 weeks after experimental infection (Porter *et al.*, 1969), glomerulitis and arteritis are not seen until 1 to 2 months

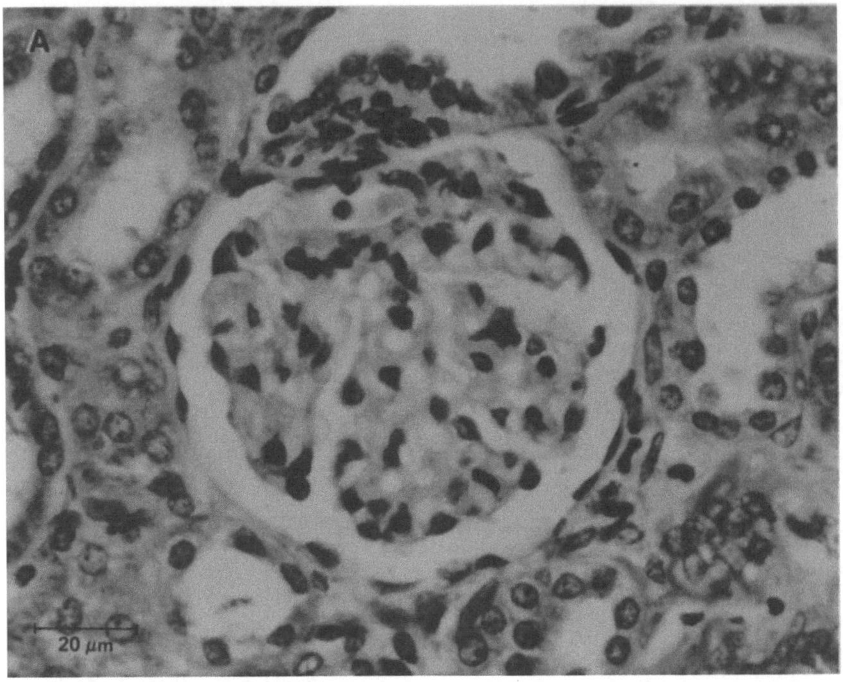

FIGURE 3. Photomicrographs of normal glomerulus (A) and two glomeruli from mink with AD (B, C). Note cellular proliferation (B), adhesion (arrow in C), and accumulation of homogeneous material in the mesangium (C).

after inoculation (Henson *et al.*, 1966; R. E. Race, personal communication). The lesions do appear to correlate temporally with the rise in gamma-globulin concentration and the appearance of anti-ADV antibody (see Section 6). Although ADV antigen has been detected in the kidney, there is no evidence that it replicates there. Likewise, electron microscopy has not revealed difinitive evidence of ADV virions in glomeruli.

Mink of the Aleutian color phase manifest the Chediak–Higashi syndrome (CHS), an autosomal recessive genetically transmitted disease (Leader *et al.*, 1963; Davis *et al.*, 1971). Features of this syndrome include partial occulocutaneous albinism, photophobia, frequent pyogenic infections, and large lysosomal granules in leukocytes and other cell types (Blume and Wolff, 1972). This disease has been identified in man, mink, cattle, and mice and is discussed in detail in Chapter 5 of Volume 1. Of particular note is the accelerated form of CHS that is characterized by systemic lymphomatosis infiltrates and hepatosplenomegaly. A similar lymphoreticular disorder has been observed in Aleutian mink, but is not related to exposure to ADV (Hadlow *et al.*, 1972).

Thus, the histopathology of AD resembles, on the one hand, a myelomalike disease, and on the other, an immune-complex disease akin to experimental serum sickness of rabbits and systemic lupus erythematosus of man.

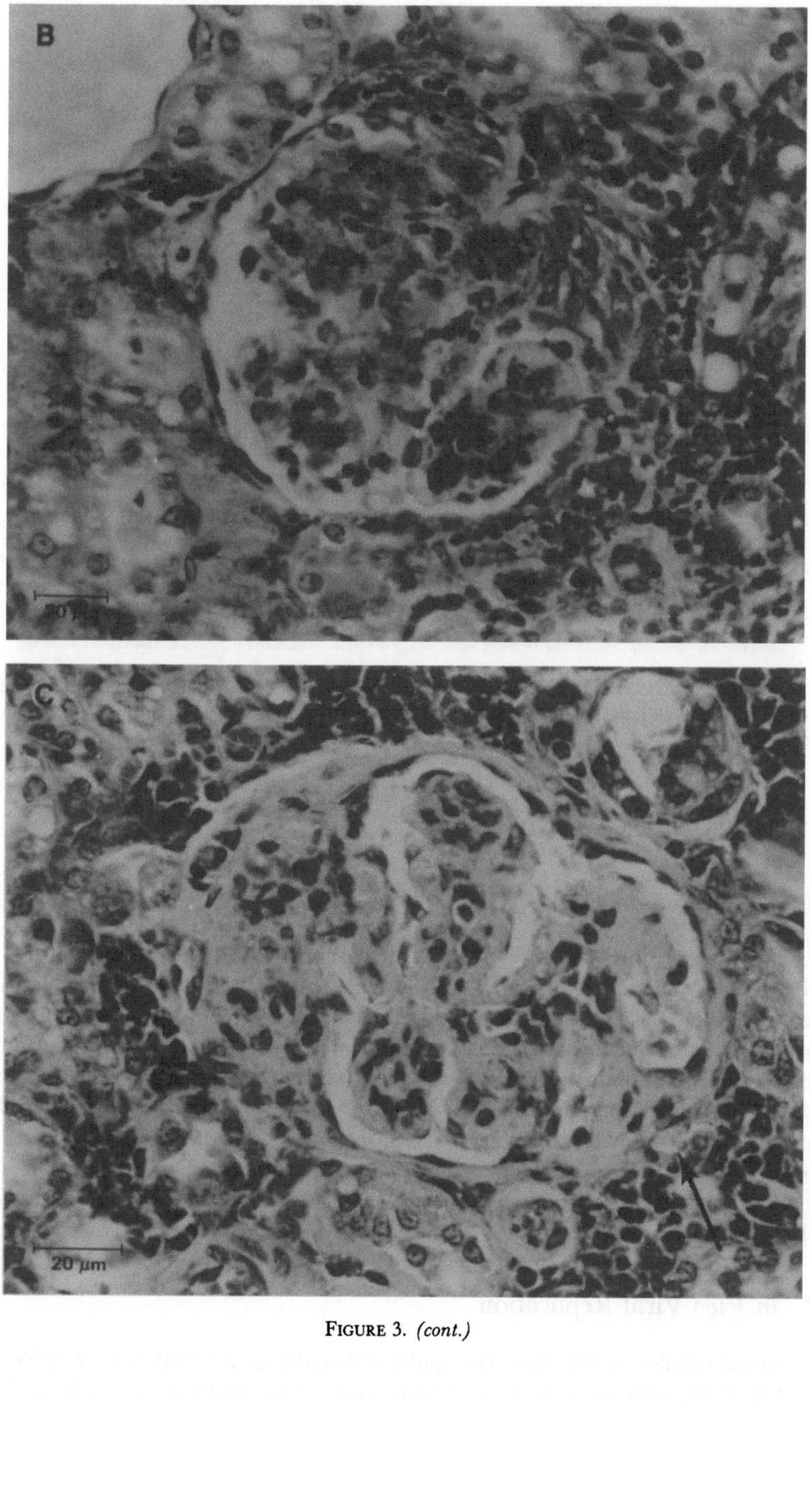

FIGURE 3. *(cont.)*

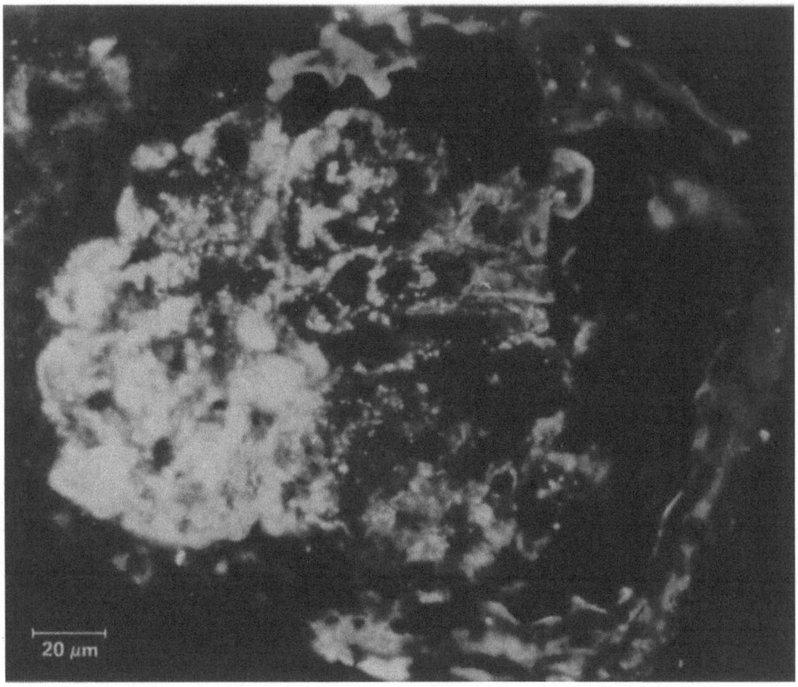

FIGURE 4. Immunofluorescence of a glomerulus from an AD-affected sapphire mink stained with fluorescein-labeled rabbit anti-mink IgA. Note the fine granular deposits along capillary walls and dense accumulations in the mesangium.

3. Aleutian Disease Virus

3.1. Etiology

The transmissible nature of AD was first described in 1962 (Henson *et al.,* 1962; Karstad and Pridham, 1962; Trautwein and Helmboldt, 1962) by demonstrating that the disease could be reproduced by inoculating mink with infected tissue suspensions or cell-free filtrates. Subsequent characterization of the infectious agent has revealed that ADV is a temperature-sensitive, nondefective, xenotropic parvovirus (Porter *et al.,* 1969; McGuire *et al.,* 1971; Cho and Ingram, 1972, 1973a; Crawford, 1973; Chesebro *et al.,* 1975; Porter *et al.,* 1977a; Shahrabadi and Cho, 1977; Shahrabadi *et al.,* 1977; Burger *et al.,* 1978; Bloom *et al.,* 1980). Studies by others, however, suggested that ADV has characteristics of the picornavirus group (Kenyon *et al.,* 1973; Yoon *et al.,* 1973, 1975). These discrepancies will be discussed in Section 3.3.

3.2. *In Vivo* Viral Replication

Several excellent articles have been published on the *in vivo* replication of ADV (Kenyon *et al.,* 1963; Gorham *et al.,* 1964; Eklund *et al.,* 1968; Porter *et al.,* 1969; Gorham *et *al.*, 1976; Ingram, 1978; Porter, 1978; Porter and Cho, 1980; Porter *et al.*, 1980). We will mention their salient features. Infectious virus has been isolated from urine, feces, saliva, whole blood, serum, and tissues of AD-affected mink. Following intraperitoneal inoculation of the highly virulent Utah-1 isolate of ADV, maximal viral titers of 10^8 to 10^9 ID_{50}/g of spleen, liver, and lymph nodes were recovered 10 days after infection and were coincident with the development of anti-ADV antibody (Porter *et al.*, 1969). The viral titers then slowly decrease and eventually plateau 2 or more months after infection at 10^5 ID_{50}/g of spleen and 10^4 ID_{50}/ml of serum (Porter and Larsen, 1974).

Interestingly, conflicting data have resulted when ADV isolates of different virulence were compared for their degree of replication in Aleutian and non-Aleutian mink. Porter *et al.* (1969) found no substantial difference in viral replication in the early phase of viral growth in mink of different genotypes inoculated with the highly virulent Utah-1 isolate, whereas Eklund *et al.* (1968) reported that the less virulent Pullman isolate replicated to much higher titers in mink homozygous for the Aleutian gene as compared to non-Aleutian mink.

At present, IF studies indicate that ADV replicates *in vivo* in the cytoplasm and nuclei of macrophages (Porter *et al.*, 1969; D. D. Porter, unpublished observations). These studies have been confirmed by electron microscopic studies (Shahrabadi and Cho, 1977).

3.3. *In Vitro* Viral Replication

The *in vitro* replication of ADV has been a baffling enigma for many years. Basrur *et al.* (1963) reported a cytopathic effect of an ADV isolate in mink testes and kidney cell cultures, and ability to reproduce AD in mink with *in vitro* passages of the agent. To date others have been unable to confirm these findings (D. D. Porter, unpublished observations; D. L. Lodmell, unpublished observations; M. E. Bloom, unpublished observations). The virus isolated by Kenyon *et al.* (1973) replicated in mouse L cells and mink kidney cells. These results also have not been confirmed by various laboratories (unpublished reports). Hahn *et al.* (1977), however, did show that the Connecticut strain of ADV produced viral antigens without infectivity in feline cells.

Because of its importance to the future of ADV research, the isolation of ADV in cell culture by Porter *et al.* (1977a) will be discussed in this review. The virulent Utah-1 strain of ADV recovered from mink tissue replicated in a continuous line of Crandall feline renal cells (CRFK) or in primary feline kidney cells when cultures were maintained at 31.8°C, but not at 37 or 39°C in primary isolation attempts. The cells had to be rapidly dividing to permit viral replication, no plaques were produced, and ADV was found to be strongly cell-associated. Furthermore, cultured virus had equal infectivity for cell cultures and mink, and could be reisolated from mink 10–180 days after inoculation. At 31.8°C virus could be passed continuously, but cultures lost their virus after one or two passages at 37°C (D. D. Porter, unpublished observations).

The tropism of ADV for feline renal cells was shown when the ninth *in vitro* passage of Utah-1 induced antigen and produced infectious virus in primary feline kidney cells and CRFK cells at 31.8 and 37°C, but no infectious virus at either temperature in primary ferret spleen, lung, or kidney cells, primary mink kidney or spleen cells, three mink embryo cell lines, or two spontaneously transformed mink embryo cell lines. In each of these cells

ADV nuclear antigen was induced at 31.8 but not 37°C in the same number of cells as in CRFK cultures (Porter *et al.,* 1977a), indicating viral infection but lack of replication.

In contrast to the results of Porter *et al* (1977a), in which cultured virus had equal infectivity for cell cultures and mink, Bloom *et al.* (1980) have shown that a spleen-derived isolate of Utah-1 became progressively attenuated during cell culture passage. By the fifth CRFK passage at 31.8°C, the ADV resembled the less virulent Pullman strain with respect to its behavior in mink; it had approximately equal infectivity for both Aleutian and non-Aleutian mink, but produced disease only in Aleutian mink. Further alteration occurred by the tenth CRFK passage in that inoculation of $< 10^7$ focus-forming units of cultured virus did not induce antibody or cause disease in Aleutian mink. The loss of infectivity could be restored by *in vivo* blind passage, indicating that loss of virulence did not represent irreversible attenuation. This attenuation did not appear to be a peculiarity of the culture conditions, as another phenotypic variant of Utah-1 retained the virulence of the parent strain following *in vitro* passage.

In contrast to the studies of Porter *et al.* (1977a), Yoon *et al.* (1973, 1975) identifed an ADV isolate in cell culture that produced a marked cytopathic effect in mink kidney cells and mouse L cells, neither of which has been shown to be permissive for ADV by Porter or other investigators. The 24-nm particle size, a major-density peak of 1.334 g/ml and a minor-density peak of 1.295 g/ml in CsCl, and incorporation of [^{3}H]uridine into virions prompted these investigators to suggest that their isolate was a picornavirus. Because the isolate bound to ADV antibody immunoadsorbent columns and caused hypergammaglobulinemia and characteristic tissue lesions of AD, Yoon *et al.* (1975) determined their isolate to be ADV. Porter *et al.* (1977a) believe the isolate of Yoon *et al.* may be a picornavirus that was contaminated with a small amount of ADV, possibly obtained from the IgG used to prepare the immunoadsorbent column, especially because infectious ADV–antibody complexes are well known to be present in AD-affected mink serum (Porter and Larsen, 1967).

3.4. Physicochemical Characteristics

Virus purification techniques and the physicochemical characteristics of ADV have been reviewed in detail (Burger *et al.,* 1965; Porter *et al.,* 1969; Cho and Ingram, 1973a, 1974; Chesebro *et al.,* 1975; Brummerstedt, 1976; Cho, 1976, 1977; Porter *et al.,* 1977a; Shahrabadi and Cho, 1977; Shahrabadi *et al.,* 1977; Porter and Cho, 1980; Porter *et al.,* 1980). Briefly, the virions are nonenveloped icosahedra, 23–25 nm in diameter with a CsCl boyant density of 1.405–1.430 g/ml. The virion appears to have 32 capsomers which are hollow tubes with a 1- to 2-nm central hole. Three visible bands at buoyant densities of 1.295, 1.332, and 1.405–1.416 g/ml in CsCl gradients are detected with concentrated ADV prepared from infected mink tissues. The lightest band contains mainly empty particles whereas the other two bands have both complete and empty particles. The virion contains single-stranded DNA of large, intermediate, and small size with approximate molecular masses of 1.2×10^6, 0.5×10^6, and 0.23×10^6 daltons, respectively, and three or four structural polypeptides. All of the above-mentioned characteristics with the exception of the two smaller molecules of DNA, in addition to the known resistance of ADV to 56°C, lipid solvents and other chemicals, are typical of other nondefective parvoviruses.

Recent results of Bloom *et al.* (1980) are in agreement with previously published results and confirm that ADV is a nondefective parvovirus. Analysis of ADV DNA on alkaline agarose gels reveals that it has the same molecular mass as adeno-associated virus DNA, approximately 1.4×10^6 daltons, which is similar to the largest molecule detected by Shahrabadi *et al.* (1977). Furthermore, ADV DNA is 85% susceptible to digestion with the single-strand specific mung bean nuclease, implying the presence of a largely single-stranded genome containing a stable duplex region (Bloom *et al.*, 1980).

Exhaustive immunological studies have shown that there is no antigenic relationship between ADV isolated from feline renal cell culture and other parvoviruses including mink enteritis virus (Porter *et al.*, 1977a).

4. Immune Response of Normal and Aleutian Disease-Affected Mink

4.1. Immunoglobulins of Normal Mink

Three Ig classes have been defined in mink (6–7 S IgG, 12–13 S IgA, and 19 S IgM) (Coe and Hadlow, 1972). Serum Ig concentrations were found to be comparable to those of other species (IgG 4–8 mg/ml, IgA 0.2–0.3 mg/ml, and IgM 0.5–1.5 mg/ml) (Porter *et al.*, 1977c; Coe, personal communication). Immunization of mink subcutaneously (s.c.) with Keyhole limpet hemocyanin (KLH) in complete Freund's adjuvant (CFA) induced serum antibody in all three classes. In addition, IgA antibody was detected in the succus entericus after s.c. antigen administration.

Mink appear to have only one L-chain type (λ) (Hood *et al.*, 1967). Coe (1972) reported two coexisting distinct L-chain populations, one being antigenically deficient with respect to the other. These two L-chain types may prove useful in antigenic analysis of oligoclonal immunoglobulins seen in mink with AD (Coe, 1972).

IgG and IgM are synthesized shortly after birth (Coe and Race, 1978). IgG also is transferred from dams to kits in the colostrum (Porter, 1966; Coe and Race, 1978). Serum IgA, first detected at 39 days of age, is likely acquired from the dam for synthesis of IgA in lymphoid tissues of kits was not seen until 2½ months of age (Coe and Race, 1978). This delayed synthesis of IgA may relate to the pathogenesis of lesion production in AD (see Section 6).

It should be noted that five 7 S Ig classes have been reported by Tabel and Ingram (1972). These Ig subpopulations were detected by spurs in immunolectrophoresis; attempts to isolate these subpopulations have been uniformly unsuccessful (J. E. Coe, personal communication).

4.2. Immune Response of Normal Mink

4.2.1. Antibody Response to Non-ADV Antigens

Porter *et al.* (1965a) reported that mink administered a single injection of bovine gamma globulin (BGG) or KLH in CFA elicited significant responses to KLH and a minimal response to BGG. A response to egg albumin was not detected. Lodmell *et al.*

(1970) have shown that healthy Aleutian mink made slightly less hemagglutinating antibody to KLH than non-Aleutian mink after primary and secondary immunization. *Brucella abortus* (Kenyon, 1966), horseradish peroxidase (Trautwein *et al.,* 1974), and goat erythrocytes (GRBC) (Lodmell *et al.,* 1971) also have been shown to be good antigens in mink. In contrast, bovine insulin (Munoz *et al.,* 1974), bovine serum albumin, and poliovirus induce little if any antibody in mink (Porter *et al.,* 1980).

The different responses of Aleutian and non-Aleutian mink to the less virulent Pullman strain of ADV suggested that the antibody-forming potential of these two color phases may be different. It was determined that on the fourth day after primary i.p. inoculation of GRBC, 5- to 18-fold more IgM plaque-forming cells (PFC) were detected in non-Aleutian than Aleutian mink lymphoid tissues (Lodmell *et al.,* 1971). In addition, there was a delay in the formation of IgG PFC in Aleutian mink. Following a secondary i.p. inoculation, the PFC were markedly different in the two color phases. Insignificant numbers of IgM and IgG PFC were detected in peripheral lymph nodes (PLN) of Aleutian mink. In contrast, there was a prominent anamnestic IgM and IgG PFC response in non-Aleutian mink PLN. Similar, but less marked differences were detected in spleens of the two genotypes. In corroboration with the different PFC responses detected between the genotypes, IgM and IgG hemolysin titers of non-Aleutian mink exceeded those of Aleutian mink subsequent to primary and secondary GRBC inoculations. The results from these and other studies (Lodmell *et al.,* 1970) indicate that Aleutian mink are not immunological cripples, nor immunologically hyperactive, but that they appear to be, in some instances, strikingly less responsive to secondary stimulation than non-Aleutian mink. The explanation for this is not apparent at present, but may be related to the total number of cells available for antigenic stimulation, or possibly there may be a deficiency in macrophage function of Aleutian mink (possibly related to the Chediak–Higashi syndrome) that renders them less able to process antigens for subsequent stimulation of appropriate antibody-forming cells.

Russel and Percy (1974) have reported that mink are the first species to be described in which antipolynucleotide antibodies have been detected in the absence of disease. The sera of normal healthy mink were shown to contain IgG antibodies that reacted with single-stranded DNA, polyinosinic acid, and polyinosinic-polycytidylic acid, but not with native DNA, UV-treated DNA, poly-A, poly-U, or poly-C. This antibody activity was unrelated to the presence of the Aleutian gene.

4.2.2. Cell-Mediated Response to Non-ADV Antigens

There are limited studies of immediate and delayed-type hypersensitivity reactions of healthy mink. In our experience (Lodmell *et al.,* 1970), weak skin test reactions to KLH were observed at 4 and 24 to 72 hr, whereas others have failed to demonstrate immediate or delayed skin reactions (Munoz *et al.,* 1974). These discrepancies may have been due to the mink genotype, or the concentration of KLH used to elicit the reactions. In fact, mycobacteria-sensitized Aleutian mink appear to be more reactive than non-Aleutian mink to skin tests with purified protein derivative (Munoz *et al.,* 1974).

Polyclonal lymphocyte stimulation by mitogens and antigen-specific proliferative responses have been studied by Perryman *et al.* (1975). Both Aleutian and non-Aleutian mink responded equally to PHA and Con A. Lymphocytes from healthy mink sensitized with KLH also responded to the specific antigen. Race *et al.* (1980) showed that normal mink sensitized to KLH or human gamma globulin failed to respond to ADV antigen, but did respond to the respective sensitizing antigen. Studies to determine if there is a difference in antigen-specific blastogenesis between healthy mink of the Aleutian and non-Aleutian genotypes are in progress.

4.3. Immunoglobulins of AD-Affected Mink

The hypergammaglobulinemia seen in AD is composed of predominantly IgG (Porter *et al.,* 1977c; J. E. Coe, personal communication). In the study by Porter *et al.* (1977c), progressive elevation of serum IgG concentration was seen beginning 2 to 3 weeks postinoculation, reaching 50 mg/ml (normal 5 to 10 mg/ml) by 90 days postinoculation. Serum IgM concentrations showed an initial minimal elevation of 2 to 3 times normal concentration and then returned to normal levels by day 60 postinoculation (Porter *et al.,* 1977c; J. E. Coe, personal communication). The concentrations of IgA also were elevated during the course of AD (Porter *et al.,* 1977c). J. E. Coe (personal communication) found markedly elevated levels of IgA after inoculation of Aleutian mink with the Pullman strain of ADV. These mink had a mean IgA concentration about 8 times normal levels in the late stages of disease.

Using an indirect fluorescence assay, anti-ADV antibody has been detected in all three Ig classes of AD-affected mink. Low levels of IgM and IgA antibody and high titers of IgG antibody (1:10,000) were observed (Porter *et al.,* 1977c).

4.4. Immune Response of AD-Affected Mink

4.4.1. Antibody Response to ADV Antigen

After experimental infection, anti-ADV antibody can be detected occasionally at 4 days, and consistently at 7 days with the CIE test (Cho and Ingram, 1973b) and at 9 or 10 days by immunofluorescence (Porter *et al.,* 1969). In all assays extremely high specific anti-ADV antibody titers have been observed: 100,000 by immunofluoresence (Porter *et al.,* 1969), 8000–260,000 by complement fixation (McGuire *et al.,* 1971), and 5000–8000 by CIE (Cho and Ingram, 1973b). Once the viral specific antibody titers attain maximal levels, they usually plateau. Exceptions to this rule occur in non-Aleutian mink without clinical disease, in which a marked decrease in (Bloom *et al.,* 1975; Larsen and Porter, 1975) or prolonged low titers of anti-ADV antibody are seen (An and Ingram, 1977).

Evidence suggests that a significant proportion of the IgG in hypergammaglobulinemic mink is viral specific. (1) There is a statistical correlation between the temporal course of hypergammaglobulinemia and the titer of anti-ADV antibody (Porter *et al.,* 1969; Larsen and Porter, 1975). (2) Extremely high titers of anti-ADV antibody have been observed. (3) Virus-infected cells are stained with as little as 1 μg/ml of IgG from persistently infected mink (Porter and Larsen, 1974). Nonetheless, evidence contrary to this suggestion has been reported. Bloom *et al.* (1975) showed that anti-ADV antibody titers plateau while gamma-globulin levels are still increasing (Fig. 5). This could be explained by ADV in the serum binding free antibody and preventing its detection. Kenyon *et al.* (1976)

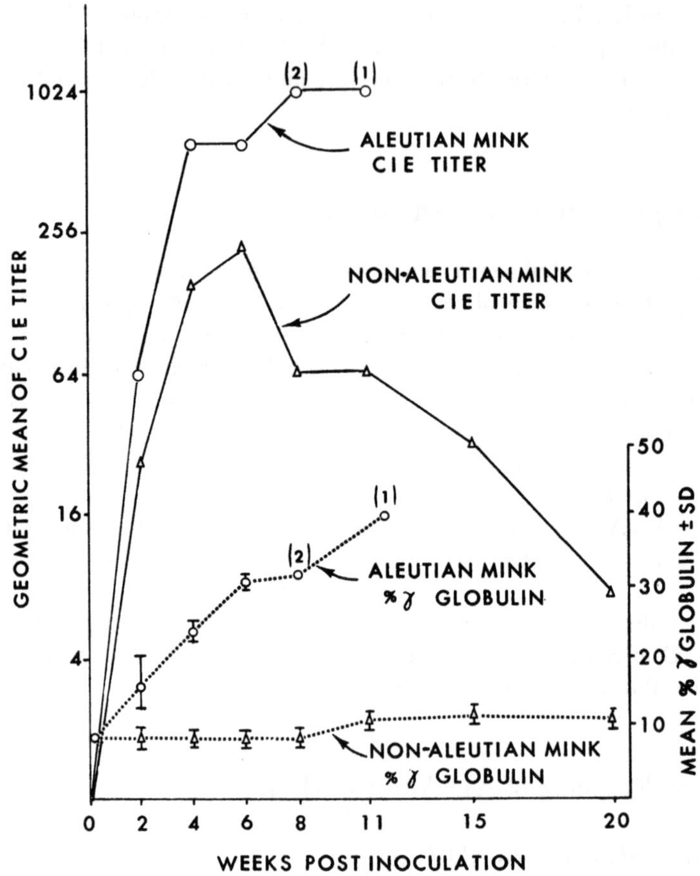

FIGURE 5. Antibody and serum gamma-globulin responses of groups of three Aleutian and non-Aleutian mink inoculated intraperitoneally with 300,000 LD_{50} Pullman-strain ADV. Antibody is expressed as group geometric mean of counterimmunoelectrophoresis titers (CIE). Serum gamma globulin is expressed as group mean percent gamma globulin ± standard deviation. Numbers in parentheses represent number of mink remaining in group (data provided by M. E. Bloom).

reported that Levamisole, a phenylimidothiazole, which has immunomodulating activity, reduced the antibody titer to ADV antigens but did not affect the development of gammapathy associated with AD. Nonetheless, it is clear that AD-affected mink produce extremely high titers of anti-ADV antibody. The mechanism by which this response occurs is unclear (see Section 6.2).

4.4.2. Cell-Mediated Response to ADV Antigen

To our knowledge only one study has been done on the T-lymphocyte response of AD-affected mink to specific viral antigen. Nylon-wool-nonadherent lymphocytes (T lymphocytes) from lymph nodes of AD-affected mink responded to ADV-associated antigen in a lymphoproliferative assay (Race *et al.,* 1980). ADV responsiveness was detected within

2 weeks of inoculation, subsequently increased in magnitude, and was maintained throughout the disease course. If cells were incubated in the presence of terminal autologous serum, a depressed response was detected. Results of this study indicate that T-cell responsiveness to ADV antigens is preserved throughout disease.

4.4.3. Antibody Response to Non-ADV Antigens

In general, it appears that AD-affected mink are immunodepressed. AD-affected mink made approximately one-fourth the amount of antibody to KLH as did normal controls (Porter *et al.,* 1965a). Lodmell *et al.* (1970) also showed that primary antibody responses to KLH in Aleutian mink with clinical AD were extremely poor. Furthermore, the most severely affected mink had the lowest primary titers and the poorest secondary response. Thus, the ability to respond to non-ADV antigens appeared to be inversely related to the severity of the disease. These studies have been confirmed using the Jerne plaque assay (Lodmell *et al.,* 1973). The above results are in contrast to those of Kenyon (1966), who showed that mink that survived AD for more than 6 months had a nearly normal response when sensitized with *Brucella abortus.* The mechanism of the immunodepression seen in AD appears to be quite complex as evidenced by the immune response observed in non-Aleutian mink inoculated with the less virulent Pullman strain of ADV (Lodmell *et al.,* 1973). In many of these animals, immunodepression was not seen even in those animals with clinical disease. These contradictory data will have to be reevaluated before firm conclusions can be made.

Studies of healthy mink administered antigens before or concomitantly with ADV indicate that the titer to the non-ADV antigens falls during development of hypergammaglobulinemia (Porter *et al.,* 1965a; Porter and Larsen, 1968; Lodmell *et al.,* 1970). These decreasing titers may be due to the increased portein catabolism of infected mink (Porter *et al.,* 1965a). However, in similarly inoculated mink, there appears to be no deficiency in the anamnestic response to KLH compared to normal controls (Lodmell *et al.,* 1970) (Fig. 6). It is clear that the immunodepression that has been reported in mink with clinical AD is not a constant finding. Varying the immunization schedule or genotype of mink and virus strain appears to affect the degree of immunodepression to non-ADV antigens. A proposal of a unifying mechanism is not appropriate at this time.

4.4.4. Cell-Mediated Response to Non-ADV Antigens

Delayed-hypersensitivity reactions to PPD (Munoz *et al.,* 1974) and KLH (Lodmell *et al.,* 1970) have been shown to be reduced in AD-affected as compared to normal Aleutian mink.

Perryman *et al.* (1975) have reported that mink inoculated with the Utah-1 strain of ADV had a significantly lower response than healthy controls to PHA and Con A, which are T-cell mitogens in mice, and that the difference in the response increased progressively with time after infection. Similarly, the proliferative response to KLH in specifically sensitized mink was decreased after appearance of clinical disease. Recent studies of Race *et al.* (1980) confirm these observations.

These studies suggest some T-cell functions are depressed in AD-infected mink. A similar decreased lymphocyte response to the T-cell mitogens PHA and Con A has been

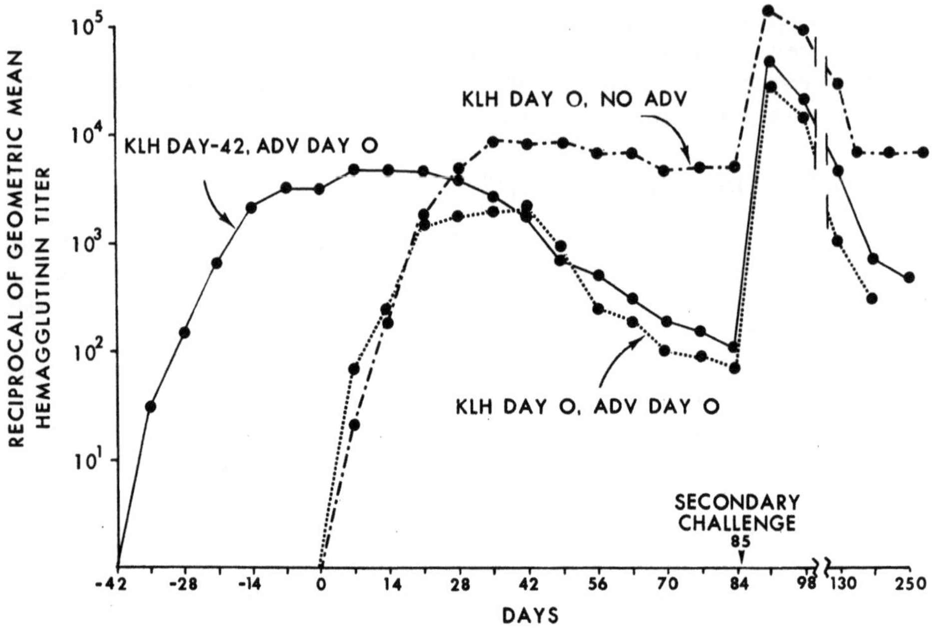

FIGURE 6. Hemagglutinin antibody response of groups of four Aleutian mink inoculated subcutaneously with 100 μg of keyhole limpet hemocyanin (KLH) emulsified with Freund's complete adjuvant, or KLH similarly inoculated and 10,000 LD_{50} of Pullman-strain ADV inoculated intraperitoneally. The secondary challenge consisted of 100 μg of KLH inoculated subcutaneously with incomplete Freund's adjuvant. Antibody titers are expressed as group reciprocal of geometric mean hemagglutinin antibody titers.

shown in adult cats infected with the parvovirus of feline panleukopenia (Schultz *et al.*, 1976). These results are the antithesis of the T-cell responsiveness of AD-affected mink to ADV antigens as shown by Race *et al.* (1980). Possible mechanisms for these observations will be discussed in Section 6.2.

4.5. Unanswered Questions

One of the more perplexing questions of AD is why high titers of anti-ADV antibody do not neutralize virus, and infectious virus usually persists throughout disease. Porter and Larsen (1967) demonstrated that the majority, if not all of the circulating infectious virus was in the form of virus–antibody complexes. Furthermore, anti-ADV antibody from mink persistently infected with ADV, but without clinical disease, does not neutralize virus (An and Ingram, 1978). To understand how this might occur, it is necessary to consider several possibilities:

1. Current evidence indicates that attachment of antibody to virions does not cause an irreversible alteration in viral surface structures, and that early low-affinity antibody can be dissociated from virus with reestablishment of infectivity (Daniels, 1975).

2. Certain surface viral proteins called "critical" sites are needed to manifest infectivity and other "noncritical" sites are important to viral structural integrity. Thus, if antibody to critical sites is sterically hindered by previously bound immunoglobulin to noncritical sites, the virus would remain infectious (Daniels, 1975).
3. The aggregation of ADV by anti-ADV antibody could account for persistence of infectious virus in the center of such aggregates. This is a distinct possibility, as large ADV–antibody aggregates have been detected in AD-affected mink (Porter and Larsen, 1967).
4. Neutralization may be complement dependent, and it is known that complement components are consumed in AD-affected mink (Ingram, 1978). Lysis of ADV by complement in the absence of antibody (virolysis) is doubtful, for only enveloped viruses are susceptible to lytic damage by this mechanism. However, nonenveloped viruses, such as parvoviruses, can interact with antibody and early reacting factors of the complement system with resulting neutralization (Cooper and Welsh, 1979). Evidence that addition of complement to virus–antibody aggregates could significantly reduce their size has been provided in the HSV system by Wallis and Melnick (1971). The significance of this possibility *in vivo,* however, is not known.
5. The state of thymic function in AD-affected mink could effect the production of neutralizing antibody. Burns *et al.* (1975) have shown that sustained antibody responses in nude mice to 12 viruses from nine major groups, including parvoviruses, were markedly thymus dependent. The modulation of T-cell function, therefore, may result in deficient antibody production as well as a diminution of effector cells involved in T-cell-mediated immunity to viral infections. As it is known that there is thymic atrophy in AD-affected mink (J. L. Portis, personal communication), and that T-cell blastogenic responses and thymus-dependent antibody responses are decreased, the possibility exists that these altered responses may be a reflection of the inability of AD-affected mink to produce neutralizing antibody.
6. ADV may undergo antigenic drift, as has been shown with visna virus infection of sheep (Narayan *et al.,* 1977, 1978) and equine infectious anemia virus infection of horses (Kono *et al.,* 1973). In these studies, animals inoculated with a single genotype of virus responded with antibody that neutralized the original inoculum strain of virus, but not other antigenic variants that appeared during the course of infection. Furthermore, Narayan *et al.* (1978) showed that the mutant viruses were antigenically stable, did not revert to parental antigenicity, replicated in cell culture as efficiently as the parental virus, and were virulent for sheep.
7. Antibody may be stripped from phagocytosed immune complexes by macrophages thereby releasing infectious virus, as suggested by Porter *et al.* (1980).

It still is not clear that anti-ADV antibody has no neutralizing capacity. Animals infected transplacentally develop lower antibody titers and higher virus titers than mink infected as adolescents (Porter *et al.,* 1977b). Furthermore, preliminary evidence by M. E. Bloom *et al.* (in preparation) indicates that ADV propagated *in vitro* is neutralized >2 $\log_{10}$ with dialyzed AD-affected mink serum as demonstrated by inhibition of fluorescent foci in tissue culture. Studies to determine if *in vivo* infectivity also is neutralized are in progress. Obviously, the apparent nonneutralization of ADV is an important question that now can be studied in detail with the advent of virus purification procedures and the use of *in vitro* systems.

Depression of the immune response to non-ADV antigens in AD-affected mink is an additional unexplained problem. Several possible mechanisms are offered:

1. Macrophage function may be altered, as it is known that ADV replicates in macrophages and that these cells are important in the inductive phase of immune reactivity. Unfortunately, *in vitro* studies to determine the effect of ADV on macrophage function have not been done.
2. ADV could suppress immune reactivity through destruction of T and B lymphocytes. There are at least three possible mechanisms by which this could occur: (a) viral replication within lymphocytes, (b) stimulation through viral replication of corticosteroid hormones which are capable of destroying lymphocytes, and (c) viral interaction with lymphocytes could induce changes in lymphocyte membranes which might lead to their removal or destruction through phagocytosis by macrophages (Woodruff and Woodruff, 1975). The removal or destruction of B lymphocytes seems unlikely, for Perryman *et al.* (1975) have shown that there is a highly significant increase in the percentage of B cells in the peripheral blood in AD-affected mink. However, this same study suggested that the absolute number of T cells is decreased. T-lymphocyte markers in mink have not been defined, and the possibility of T-lymphocyte depletion has not been studied. Obviously, identification of such markers would greatly enhance the understanding of AD.
3. Antigenic competition may cause immunodepression if the increased number of B lymphocytes is committed to the ADV virion or its antigenic components (Porter *et al.*, 1980).
4. Inhibition of T-helper cells or activation of T-suppressor cells also may be important. These possibilities are further discussed in Section 6.3.

5. Genetic Effects on Aleutian Disease

5.1. Host Effects

It is generally agreed that ADV of high virulence, such as Utah-1 and Guelph strains, are those that cause progressive disease and death in non-Aleutian mink, while low-virulence virus, such as Pullman strain, usualy produces disease and death only in Aleutian mink (Porter, 1978). Lethal disease has been detected, however, in non-Aleutian mink inoculated with high-titered tissue suspenions of Pullman ADV (Padgett *et al.*, 1967a; Eklund *et al.*, 1968). In the classic study by Eklund *et al.* (1968), it was shown that the number of non-Aleutian mink that became infected varied from 25 to 50% depending upon the concentration of virus inoculated. Furthermore, in non-Aleutian mink that did respond to ADV, the level of serum gamma globulin became elevated at the same time as Aleutian mink, but it rose more slowly over a longer period. Similar results were shown by Padgett *et al.* (1968) in that no statistical difference was detected in the time at which either genotype first exhibited hypergammaglobulinemia. However, the rate of progression of glomerulonephritis (Padgett *et al.,* 1967b; Johnson *et al.,* 1975) and arteritis (Porter *et al.,* 1973) was much faster in Aleutian than non-Aleutian mink and a more severe and rapid disease course occurred (Padgett *et al.,* 1968). Additional evidence that Aleutian mink are more susceptible to ADV has come from mink ranchers who have consistently noted a greater prevalence of AD in Aleutian as compared to non-Aleutian mink (Padgett *et al.,* 1968).

Aleutian mink also have a greater tendency toward inflammatory diseases such as periodontitis (Gustafson, 1969), a lymphoreticular proliferative disease (Hadlow *et al.,* 1972), avian tuberculosis (W. J. Hadlow, personal communication), and infection with bacterial pathogens such as *Pasteurella multocida* and a *Corynebacterium* sp. (Padgett *et al.,* 1968) and staphylococci (Helgebostad, 1963). The increased susceptibility to pyogenic infections is probably due to the Chediak–Higashi syndrome noted in the Aleutian genotype (Leader *et al.,* 1963).

5.2. Persistent and Nonpersistent Infections

Studies with non-Aleutian mink infected with ADV have indicated that at least three types of infection occur: (1) persistent progressive infection with hypergammaglobulinemia, high levels of anti-ADV antibody, histopathological lesions of typical AD, and eventual death; (2) nonpersistent nonprogressive infection with a transient elevation of serum gamma globulin, specific anti-ADV antibody titers that decrease to very low levels, and absence of typical AD lesions (Larsen and Porter, 1975); and (3) persistent nonprogressive infection with no development of hypergammaglobulinemia, low specific anti-ADV antibody titers for prolonged periods, absence of typical AD lesions, and persistence of infectious virus in tissues and blood (An and Ingram, 1977).

Persistent nonprogressive infections initially were detected by Eklund *et al.* (1968), who isolated ADV from mesenteric lymph nodes of non-Aleutian mink that remained healthy for 22 months after intraperitoneal inoculation of Pullman-strain ADV. An extensive documentation of this form of disease in naturally infected non-Aleutian mink has been reported by An and Ingram (1977; An *et al.,* 1978) [see (3) above for characteristics]. In these studies, ADV was detected in blood, mesenteric lymph nodes, kidney, liver, and spleen. Though titrations were not done, it was concluded that the amount of virus in the mink was small. Additional studies by An *et al.* (1978) showed that stress did not initiate progressive disease, mink were not resistant as they developed progressive AD after challenge with the virus, and no neutralizing antibody was detected. Interestingly, it was shown that the mink transmitted AD horizontally by direct or indirect contact, and vertically from dam to kits (An and Ingram, 1978).

The lack of pathologic changes in these mink was postulated to be due to the limited amounts of virus and moderate concentrations of antibody in the circulation that formed insufficient complexes to precipitate tissue damage (An and Ingram, 1977). It also was considered that infection may have been controlled by interferon, as Russell *et al.* (1975) have demonstrated that some normal mink treated with the interferon inducer poly(I·C) after ADV inoculation had normal target organ histology. Furthermore, poly(I·C) will induce interferon in mink (D. L. Lodmell, in preparation). It is important to note that the studies of An and Ingram were done on naturally infected mink and that this form of disease has not been duplicated by experimental infection (An and Ingram, 1977).

5.3. Unanswered Questions

5.3.1. Host Effects

The previous discussion indicates the importance of host genetics in control of ADV infection. In pondering explanations for this occurrence we must consider the nature of the virus, the ability of the host's defense system to respond to it, and the marked genetic heterogeneity of mink. The observation that some non-Aleutian mink die after inoculation with the less virulent Pullman strain of ADV would suggest that a small percentage of these mink are as susceptible as the Aleutian genotype. Explosive occurrences of natural disease in non-Aleutian mink also would indicate that extremely susceptible mink of this genotype already exist (Henson *et al.,* 1976) or, alternatively, that they had been infected with a highly virulent strain of ADV. Presumably, by inbreeding, genetic lines of non-Aleutian mink could be obtained that are uniformly susceptible or resistant to ADV. This does not appear to be the case, for attempts to develop genetically resistant mink have met with uniform failure (Larsen and Porter, 1975; Henson *et al.,* 1976). Furthermore, the genetic control of ADV infection appears to be complex and not simply due to a single gene (Larsen and Porter, 1975). This is not surprising as it is known that the host response to many viruses, including murine oncornaviruses (Lilly and Pincus, 1973), is mediated by a multitude of genes. Even though there exists precedence for modulation of virus infection through host genetic composition in many species, the importance of host genetics in resistance to AD will remain unknown until exhaustive inbreeding experiments are done and inbred mink subsequently are inoculated with freshly isolated strains of ADV.

5.3.2. Virulence of ADV

At present, no data are available on the physicochemical or antigenic properties of high- or low-virulence strains of ADV. Perhaps the advent of today's more sophisticated biochemical, virological, and immunological techniques will provide this information. However, one study has indicated that there may be genetic variation, or at least a change in *in vivo* virulence of one preparation of Utah-1 ADV after *in vitro* passage (Bloom *et al.,* 1980) (see Section 3.3). The loss of virulence of this strain was unstable in that virulence could be restored by *in vivo* passage. In light of these results, the possibility exists that a genetic change in ADV virulence also may occur in nature or in artificial passage of virus from mink to mink in the laboratory; thus the more virulent Utah-1 and less virulent Pullman strain may be manifestations of these genetic variations.

Differences in virulence between Utah-1 and Pullman strains of ADV may be explained by their ability to gain access to susceptible cells, to replicate in these sites, or to promote pathogenicity. Possibilities that need to be addressed are as follows: (1) Does sensitivity to interferon participate in the degree of virulence? It is known that more-virulent viruses are in general less sensitive to interferon and stimulate less interferon from cells. (2) Are there differences in replication rate between various virus strains? Parvoviruses are known to infect cells in the S phase of the cell cycle (Tattersall and Ward, 1978). Thus, Utah-1 may have been selected for its ability to stimulate the S phase. (3) Is there a difference in susceptibility to neutralization by anti-ADV antibody? (4) The more virulent strains may initiate a greater activation of complement components that would enhance generation of chemotactic factors, attracting more leukocytes to sites of immune-complex deposition. (5) Infection may be limited by the inability of less virulent ADV to bind to specific virus receptors or, alternatively, the receptors may already be bound with defective-interfering (DI) particles that may be produced by less virulent ADV strains. (6) The less virulent Pullman strain may be a temperature-sensitive mutant that is unable to replicate efficiently *in vivo*.

5.3.3. Persistence of ADV

The persistence of ADV is more than likely related to its apparent inability to be neutralized (see Section 4.5). Another possibility for persistence could be the temperature-sensitive nature of ADV replication as shown *in vitro*. Viral persistence has been associated with a temperature-sensitive lesion in the viral genome (Preble and Youngner, 1975). In addition, DI particles have been demonstrated in other nondefective parvovirus infections such as minute virus of mice, Kilham rat virus, and feline panleukopenia virus (Muller *et al.*, 1978), and some evidence suggests their occurrence in ADV (Cho, 1977). DI particles may compete with infectious ADV for anti-ADV antibody allowing for persistent productive infection.

6. Pathogenesis

The general thrust of research on AD in the past 20 years has revolved around two basic problems: (1) the transmissible agent that initiates the disease process and (2) the pathogenesis of lesions mediated by immune-complex deposition that lead, ultimately, to the host's demise. Much of this information has already been summarized in several excellent reviews (Porter and Larsen, 1974; Henson *et al.*, 1976; Porter and Cho, 1980; Porter *et al.*, 1980). A third area, which has not been studied in any detail, is the mechanism of the extraordinary hypergammaglobulinemia seen in this disease. We will discuss the second and third problems in this section.

6.1. Vasculitis and Glomerulitis

The pathology of AD appears not to be related to virus replication, but to the host response to the virus, or to antigens therefrom. Lesions are not seen until 1 to 2 months after experimental inoculation, and yet peak virus titers are demonstrated by days 8 to 10. Thereafter, virus titers in the serum tend to fall somewhat but remain at detectable levels until death. The host immune response to the virus or viral antigens appears to be the critical factor in determining the disease course. This is supported by several pieces of evidence. (1) Immunization of mink with formalin-killed ADV or administration of anti-ADV antisera prior to inoculation with viable ADV augments the development of lesions and hastens death (Porter *et al.*, 1972). (2) Immunosuppression with cyclophosphamide completely prevents lesion development, yet has little effect on replication of ADV *in vivo* (Cheema *et al.*, 1972). (3) The kinetics of replication of the Utah-1 strain of ADV is comparable in Aleutian and non-Aleutian mink, yet Aleutian mink get a severe form and non-Aleutian mink a milder form of disease (Porter *et al.*, 1969). (4) The severity of clinical and pathologic signs of AD are directly related to the degree of hypergammaglobulinemia (Porter and Larsen, 1964; Eklund *et al.*, 1968; Porter *et al.*, 1969, 1973). (5) Mink of the Aleutian genotype infected *in utero* with ADV develop a milder form of disease than those infected as adolescents. Yet, transplacentally infected mink have higher virus titers than those infected as adolescents (Porter *et al.*, 1977b).

Porter *et al.* (1965a) reported the first evidence that the lesions of AD are caused by deposition of antigen–antibody complexes in the tissues. The deposition of C3 and Ig has been observed by IF techniques in a granular pattern along the capillary basement membranes of glomeruli, in the glomerular mesangium (Porter and Larsen, 1967; Henson *et al.*, 1969; Porter *et al.*, 1969; Pan *et al.*, 1970a; Trautwein and Seidler, 1972; Johnson *et al.*, 1975), and in the media of inflammed vessels (Porter *et al.*, 1973). The presence of viral antigen in these tissues is suggested by the following. (1) After elution of tissue-bound Ig by acid treatment of frozen sections, fluoresceinated serum from AD-affected mink stained material localized in a distribution similar to deposited C3 (Porter *et al.*, 1973; Porter and Larsen, 1974). However, although AD mink serum contains abundant anti-ADV antibody, antibody with specificity to other antigens could be present. (2) Elution of kidney fragments and isolated glomeruli with acid yields material that stains Kupffer cells in liver sections of AD-affected mink (Porter *et al.*, 1969) and produces a precipitin line in CIE with Freon-extracted ADV (Cho and Ingram, 1973c). These data must be interpreted with some caution. AD is unique among the immune-complex diseases with respect to the extraordinary hypergammaglobulinemia and extremely high titers of serum anti-ADV antibody. In order to rule out serum contamination in the kidney eluates, the quantitation of Ig classes in the serum and eluates or comparison of the quantity of anti-ADV antibody/total Ig for eluates and serum should be done (Oldstone and Dixon, 1975).

Porter demonstrated by analytical ultracentrifugation that in serum of mink with severe disease there were Ig-containing complexes that sedimented in two peaks, 9–14 S and 22–30 S (Porter *et al.*, 1965a). In experimental immune-complex disease in rabbits (Cochrane and Hawkins, 1968; Cochrane, 1971; Mannik and Arend, 1971) and in natural immune-complex disease of mice (Izui *et al.*, 1979), circulating complexes of similar size have been demonstrated. Thus, based on the histopathology of the lesions and immunologic studies thereon, the glomerular and arterial disease of AD is initiated by the deposition of circulating immune complexes. The nature of the antigen(s) involved in these complexes is unknown. Shahrabadi *et al.* (1977) have demonstrated four viral proteins from Freon-extracted preparations. Studies on the immune-complex disease induced by lymphocytic choriomeningitis virus in mice (Buchmeier and Oldstone, 1978; Oldstone *et al*, 1978) and that of NZB/W (Izui *et al.*, 1979) and SL/Ni mice (Yoshika *et al.*, 1979) have revealed that complexes contain antibody to viral structural proteins. Further identification of ADV structural proteins is being carried out in the LPVD at the Rocky Mountain Laboratories. This work should help answer the question of which viral antigens are involved in the formation of pathogenic complexes.

Another problem that may bear on the mechanism of the immunopathology of AD is the Ig class(es) involved in the pathogenic complexes. IF techniques have revealed the deposition of Ig in the glomerular and arterial lesions of AD, but identification of the Ig class(es) involved has been more difficult. Many workers have presented evidence that IgG (see above), and in one report IgM (Johnson *et al.,* 1975), is deposited in the tissues. However, the anti-IgG reagents have not been appropriately absorbed to remove cross-reacting light-chain reactivity, and the anti-IgM reagents depend on a cross-reaction between human and mink IgM (Johnson *et al.,* 1975). Recent studies from our laboratory using class-specific rabbit anti-mink Ig antisera indicate that in Aleutian mink infected with the Pullman strain of ADV, IgA was the predominant Ig class deposited in affected glomeruli (Portis and Coe, 1979).

IgA-mediated nephropathy has been reported in a number of diseases of humans including Schönlein–Henoch purpura (Urizar *et al.,* 1968) and Bergers disease (Berger and Hinglais, 1968). IgA is reported to fix complement only via the alternate pathway (Götze and Müller-Eberhard, 1971; Spiegelberg, 1974). Ingram (1978) noted that no variations in the serum properdin levels were seen during the course of AD, suggesting that the alternate pathway was not involved in pathogenesis. Recent evidence, however, indicates that IgA can fix complement via the classical as well as the alternate pathway (Burritt *et al.,* 1977). Experimental immune-complex nephritis also has been induced in mice using preformed IgA complexes (Rifai *et al.,* 1979).

In general, experimental immune-complex glomerulonephritis can be induced most readily by antigen–antibody complexes made in antigen excess (Weigle, 1958; Cochrane and Koffler, 1973), although other physical properties of the complexes are certainly important (Kijlstra *et al.,* 1979). The extremely high titers of anti-ADV antibody in animals with clinical AD would suggest that the complexes must be in antibody excess. However, a minor subpopulation of Ig with specificity for a major viral structural protein might lead to complexes of appropriate size and antigen/antibody ratio to deposit in the tissues.

The reticuloendothelial system (RES) appears to be of primary importance in removing circulating complexes (Benacerraf *et al.,* 1959), probably through binding to Fc receptors (Arend and Mannik, 1971; Mannik *et al.,* 1971; Haakenstad and Mannik, 1976). Fc receptors in most species have greatest binding avidity for IgG and particularly IgG immune complexes (Spiegelberg, 1974). Thus, it is appealing to view the immune-complex disease of AD as a manifestation of Fc receptor blockade (Haakenstad and Mannik, 1974) due to high concentrations of circulating IgG–antigen complexes, allowing complexes in antigen excess (perhaps containing IgA) to localize in the tissues.

The ontogeny of the mink Ig classes may provide further insight into this problem. Coe and Race (1978) reported that IgA synthesis by mink lymphoid tissues was not seen until about 2.5 months of age, whereas both IgG and IgM were produced shortly after birth or perhaps *in utero* in the late stages of gestation. When mink were inoculated *in utero* with ADV, the course of clinical disease was found to be slower than in animals inoculated as adolescents (Porter *et al.,* 1977b). Thus, the slower tempo of disease in these transplacentally infected mink appears to correlate with the delayed appearance of IgA synthesis. Indeed, when hypergammaglobulinemic AD mink sera were quantified for Ig classes, IgG showed the greatest absolute increase over normal serum, but IgA showed variable increases up to 5- to 10-fold over normal levels (Porter *et al.,* 1977c; J. E. Coe, personal communication).

It is of interest to note that particularly high titers of ADV are observed in mesenteric nodes, and in non-Aleutian mink with persistent nonprogressive infection induced by Pullman virus, ADV was found exclusively in the mesenteric lymph node.

It has been suggested that the difference in susceptibility of Aleutian and non-Aleutian mink may be related to the Chediak–Higashi syndrome seen in the former genotype. This disease, mentioned previously, is characterized by large lysosomal granules in leukocytes and impaired degradation of exogenous protein by reticuloendothelial cells (Prieur *et al.*, 1972; Lyons and Pitot, 1978). It is thought that the glomerular mesangium may serve an important function in removing and disposing of immune complexes deposited in glomerular capillaries (Schreiner *et al.*, 1978; Hunsicker *et al.*, 1979; Striker *et al.*, 1979). Thus, impaired degradation of immune complexes both within the glomerulus and in the rest of the RES could, in part, account for the exaggerated disease seen in Aleutian mink.

Human glomeruli express a receptor for C3 (Gelfand *et al.*, 1975) that appears to be localized in the membranes of glomerular epithelial cells (Shin *et al.*, 1977). Binding of C3 to this receptor may contribute to localization of complexes in a subepithelial site (Shin *et al.*, 1977). Subepithelial localization of immune complexes in AD is suggested by electron microscopic studies (Henson *et al.*, 1968; Pan *et al.*, 1970b). It would be of interest to compare Aleutian and non-Aleutian mink for the expression of this receptor.

6.2. Hypergammaglobulinemia

Several findings strongly suggest that ADV infection of susceptible mink promotes a massive expansion of the B-lymphocyte compartment of the immune system. (1) Using the overlay technique of Shevach *et al.* (1973), a systemic infiltration of B lymphocytes was demonstrated by adherence of sheep erythrocytes coated with IgM and complement to lymphoid infiltrates in frozen sections (J. L. Portis and J. E. Coe, unpublished observations). (2) An increase in the proportion of circulating B lymphocytes in the blood has been reported (Perryman *et al.*, 1975). (3) A systemic plasmacytosis is seen that, at least in the kidneys resembles a leukemic infiltrate. (4) The hypergammaglobulinemia can account for 40 to 70% of the total serum proteins. The question we wish to address in this section is what are the possible mechanisms responsible for this enormous B-cell proliferative response?

One possibility that must be raised is whether AD is in reality a neoplasm of the B-cell lineage. Several pieces of evidence speak against this hypothesis. (1) The plasma cell infiltrates do not resemble those of myeloma in mice or man, and no osteolytic lesions are seen in AD (Porter *et al.*, 1965b). (2) The hypergammaglobulinemia, at least initially, is polyclonal with respect to its electrophoretic mobility (Porter *et al.*, 1980).

Much, if not all, of the hypergammaglobulinemia of AD appears to have specificity for ADV (see Section 4.4.1). It is clear, therefore, that ADV infection induces an extreme stimulus of the immune system to synthesize Ig, a large proportion of which has specificity to ADV. This stimulus affects a large number of B-cell clones as evidenced by the broad electrophoretic pattern seen in the gamma region (Henson *et al.*, 1961). However, with time after infection, animals may show restricted gamma-globulin heterogeneity, and in approximately 10% of mink infected with the Utah-1 strain of ADV for more than 1 year, a monoclonal and rarely a biclonal pattern can be seen (Porter *et al.*, 1965b). The monoclonal spikes for each individual mink have different electrophoretic mobility. Occasionally a monoclonal pattern has been observed to convert to a polyclonal pattern (D. D. Porter, personal communication). In our experience with the Pullman strain, however, oligoclonal patterns are very rare.

Several possible mechanisms have been proposed for the exaggerated immune response in AD. (1) ADV induces a generalized state of immunostimulation. Several laboratories have shown that the primary and secondary antibody responses to antigens other than ADV are usually depressed (see Section 4.4.3). Likewise, AD depresses the proliferative response of peripheral blood lymphocytes to Con A and PHA (Perryman *et al.,* 1975; Race *et al.,* 1980). Both of these lectins are specific T-cell stimulators in mice. Thus, after animals have developed clinical AD, the T- and B-cell responsiveness to other anitgens appears to be compromised. (2) The persistence of ADV antigen in the host produces a constant state of antigen-specific immunostimulation. Examination of non-Aleutian mink has shed some light on this possibility in that a proportion of these mink infected with ADV develop persistent viremia without hypergammaglobulinemia (An and Ingram, 1977; An *et al.,* 1978). Thus, persistent viremia is not by itself a prerequisite to development of hypergammaglobulinemia, but may be a contributing factor. In addition, exaggerated antibody responses to persisting antigen are not a general finding in immunology. Thus, repeated immunization does not, as a rule, result in the production of progressively more antibody. Serum antibody concentrations increase to a certain level, then plateau or even decline (Mota, 1964) despite continued immunization. Furthermore, antigens that are not readily degraded and persist in the animal, such as pneumococcal polysaccharide (Howard *et al.,* 1970) and synthetic polypeptides of D-amino acids (Janeway and Humphrey, 1968), do not induce an exaggerated immune response. (3) A third possible mechanism for the hypergammaglobulinemia of AD is an aberration in immunoregulation, perhaps mediated by virus infection or an exaggeration of a preexisting regulatory imbalance. With the discovery of suppressor T lymphocytes in the early 1970s (Gershon and Kondo, 1971; Katz and Benacerraf, 1972) and the proposal of the network theory of immunologic homeostasis in 1974 (Jerne, 1974), it has been a popular notion that various immunologically mediated diseases may be a manifestation of altered immunoregulation (Allison *et al.,* 1971). For example, the autoimmune disease of aged NZB/W mice has been correlated with a generalized decrease in both antigen-specific and -nonspecific T-cell-mediated immunosuppression (Gelfand and Steinberg, 1973; Barthold *et al.,* 1974; Gerber *et al.,* 1974; Krakawer *et al.,* 1976). Yet a clear picture of a single defect in immunoregulation has not emerged, as alterations in both T- and B- cell functions have been described in these mice (Primi *et al.*, 1978). In addition, several reports indicate either no defect in suppressor T-cell function in aged NZB/W mice (Roder *et al.,* 1977; Creighton *et al.,* 1979) or increased suppressor T-cell activity (Ranney and Steinberg, 1976). Because essentially no studies on immunoregulation in mink have been done, it is difficult to discuss the relevance of immunoregulation to AD. There are several aspects of the disease, however, that suggest that this would be a fruitful avenue to pursue.

As mentioned previously, the augmented B-cell response in AD is apparently selective with respect to antigen specificity. That is, B-cell subpopulations committed to anti-ADV antibody production expand at the expense of other populations committed to non-ADV antigens. Porter has suggested that this may be the result of antigenic competition (Porter *et al.,* 1980). The concept of antigenic competition is purely operational and may have many mechanisms (Pross and Eidinger, 1974; Liacopoulos and Ben-Efraim, 1975). The finding that a decreased antibody response to non-ADV antigens correlates temporally with the appearance of hypergammaglobulinemia and not with the appearance of peak virus titers gives credence to the idea of preemption of the immune response due to an overwhelming response to ADV. Three inconsistent findings, however, suggest that the mechanism may be more complicated. (1) Kenyon reported that the serum antibody response to *Brucella abortus* (a T-independent antigen in mice; Mond *et al.,* 1978) 5 days after immunization was depressed in AD-affected mink up to at least 3 months after ADV infection. However, the response returned to normal if animals lived for 6 months, even though their serum gamma-globulin level was 40% of the total serum proteins (Kenyon, 1966). (2) D. L. Lodmell and co-workers noted that animals primed to KLH before infection with ADV made a near-normal secondary response on day 84 when clinical disease was apparent and serum gamma globulins were between 50 and 60 mg/ml (Lodmell *et al.,* 1970). (3) Immunization of mink with formalin-inactivated ADV induced anti-ADV antibody in only 1 of 26 animals, indicating that, at least, the inactivated preparation was not particularly immunogenic (Porter *et al.,* 1972). Thus, it appears that hyperresponsiveness to ADV requires viral replication.

Immunodepression has been observed with other persistent viruses such as the oncogenic viruses MuLV, FeLV, and MTV (Stutman, 1975) and nononcogenic viruses such as measles, Junin, LCM, and cytomegalovirus (Notkins *et al.,* 1970; Salaman, 1970). It has been suggested that direct infection of T and/or B lymphocytes may exert a direct negative influence on their capacity to respond to antigen. Active mechanisms of suppression (i.e., via the induction of suppressor T lymphocytes) have also been proposed (Kumar *et al.,* 1976). There are several exceptions such as LDV (Notkins *et al.,* 1966) and VEE (Craig *et al.,* 1969), which appear to enhance nonspecifically the immune response of the host.

AD appears to be unique among the persistent viral diseases with respect to its effect on the immune system. Thus, immunodepression of the response to nonviral antigens is seen concomitant with hyperresponsiveness to viral antigens. There are factors in this disease that might have a positive (i.e., enhancing) effect on the immune response. The constant presence of circulating immune complexes could have a direct enhancing effect on B-cell responsiveness through binding to Fc or C3 receptors on B lymphocytes (Oppenheim, 1972; Dukor and Hartman, 1973; Möller and Coutinho, 1975; Shearer *et al.,* 1975). However, one would expect that the enhancement would be generalized, effecting a wide variety of B-cell clones. Yet in AD the B-cell hyperactivity is selective. The intimate association of ADV with macrophages may generate a "superantigen" as has been demonstrated both for soluble proteins (Mitchison, 1969; Unanue and Cerottini, 1970) and for synthetic polypeptides (Pierce *et al.,* 1974). Such a mechanism would provide specificity to the hyperresponsiveness.

ADV may infect a subpopulation of suppressor T cells, altering their function. Again, this scheme could theoretically lead to selective enhancement of the response to ADV antigens. A wide variety of RNA viruses selectively infect activated lymphocytes (Berg, 1961; Duc-Nuguyen and Henle, 1966; Edelman and Wheelock, 1966; Willems *et al.,* 1969). These cells have been activated nonspecifically using mitogens (Edelman and Wheelock, 1966) or specifically using antigens (Bloom *et al.,* 1970). Recently, it was reported that VSV under certain *in vitro* conditions inactivates a nylon-adherent subpopulation of suppressor T lymphocytes (Minato and Katsura, 1978a,b). The effect of VSV infection *in vitro* was an enhancement of the secondary PFC response to SRBC. Because VSV produces a lytic infection, this effect was regarded as a result of suppressor cell death. However, selective effects of viruses on lymphocyte function have been reported in the apparent absence of cell death. Thus, measles virus in the mouse appears to inactivate a population of T-helper cells (McFarland, 1974), and Friend and Moloney leukemia viruses selectively suppress memory B-cell function (Cerny and Waner, 1975). The parvovirus group, of which ADV is a member, is characterized by the predilection for infecting actively proliferating cells both *in vivo* and *in vitro* (Tennant *et al.,* 1969; Tattersall and Ward, 1978). Two viruses in this group, minute virus of mouse (MVM) and Kilham rat virus (KRV), are common contaminants of tumor-cell lines. It is thought that their tropism for neoplasms is due to their requirement for replication in dividing cells. These two viruses are of particular interest because of their suppressive effects on the immune response. KRV can replicate in rat lymphocytes stimulated with Con A and inhibits their capacity to divide (Campbell *et al.,* 1977). MVM infects EL-4, a T-lymphocytic murine leukemia-cell line, and markedly inhibits the mixed lymphocyte response *in vitro* (Bonnard *et al.,* 1976). Thus, there is certain precedent for believing that parvoviruses as a group can selectively infect activated lymphocytes and alter cellular function. Although ADV antigens have been detected exclusively within macrophages, the techniques used (i.e., direct IF) may not be of sufficient sensitivity to reveal viral infection of other cell types. This has been found to be the case with several latent viral infections including herpes simplex in humans (Stevens and Cook, 1971; Puga *et al.,* 1978) and MuLV in mice (Rowe *et al.,* 1971). The possibility that ADV infects lymphocytes will have to await the development of appropriate probes for detecting the viral genome. The functional effects of ADV on various subpopulations of lymphocytes have not been studied, but the techniques required for such studies are available.

6.3. Unanswered Questions

6.3.1. Immune-Complex Disease

With respect to the immune-complex disease, several problems need to be studied:

1. Which antigens are involved in formation of pathogenic complexes? In LCM, kidney eluates contain antibody to virtually all viral structural proteins (Oldstone *et al.,* 1978). Similar studies need to be done with AD. As mentioned, AD presents a particular problem in this respect due to the extraordinarily high levels of circulating anti-ADV antibody. Thus, techniques for isolating complexes from the circulation such as the Clq (Zubler *et al.,* 1976) or the Raji cell techniques (Theofilopoulos *et al.,* 1976) will probably have to be used. More definitive identification of the viral structural proteins of ADV will be required in this regard.
2. Which Ig classes are involved in the formation of pathogenic complexes? This will require the use of well-defined class-specific reagents.
3. Are autoreactive elements important in the pathogenesis of AD? Although rheumatoidlike substances have been reported in AD (Porter *et al.,* 1980), this appears to be a rare finding. The presence of anti-DNA and antinuclear antibody has been reported (Barnett *et al.,* 1969; Russell and Percey, 1974). However, this also appears to be a rare occurrence (Porter *et al.,* 1980) and caution must be exercised due to the presence of anti-DNA substances other than Ig in normal mink sera (Thoburn *et al.,* 1972). Although hemolytic anemia is not a feature of AD, a pos-

itive Coombs test has been reported. The titer of the antiglobulin test appeared to correlate with the severity of the histologic lesions (Saison *et al.,* 1966) and the Ig eluted from erythrocytes reacted with ADV in the CIE test (Cho and Ingram, 1973c). Porter *et al.* (1980) have suggested that ADV may, like other parvoviruses, express a hemagglutinin determinant, thus explaining this apparent autoimmune phenomenon. Autoimmunity, thus, does not appear to be a prominent feature in AD. Yet this aspect of the disease should be reexamined with the more sensitive techniques that are now available.

4. What is the role of the cell-mediated response in this disease? This is more a general problem in the pathogenesis of immune-complex disease, as there is recent evidence that T-cell recognition of *in situ* antigen–antibody complexes may contribute, in part, to tissue injury (Bhan *et al.,* 1978).
5. Does the Chediak–Higashi syndrome, in mink homozygous for the Aleutian gene, contribute to the accelerated AD seen in these color phases? It is likely that RES function in these mink may be compromised such that pathogenic immune complexes are not as readily cleared from the circulation. In addition, the large lysosomal granules seen within leukocytes in CHS mink may, as suggested by Gustafson (1969), contain large amounts of proteolytic enzymes that further contribute to tissue injury.

6.3.2. Hypergammaglobulinemia

The effect of ADV on the immune system is one of the most interesting aspects of this disease. Clearly, most of the cellular work will have to be done in autologous systems as all mink are outbred. Many of the questions outlined below will best be answered by the use of cell separation and immunization techniques developed in other species to assess T- and B-lymphocyte responses.

1. Evidence has been reported that mink with AD show decreased nonspecific responsiveness to T-cell mitogens and lower specific responses to protein antigens than uninfected controls. Whether the decreased responsiveness is the result of active suppression is not known.
2. Is the enhanced B-cell response to ADV the result of diminished immunosuppression or enhanced B-cell responsiveness? This question can best be approached by developing *in vitro* techniques for assessing T- and B-cell responses to ADV antigens and studying the requirements for macrophages and the effects of added sera, cells, and fractions thereof.
3. Is the hypergammaglobulinemia of AD T-cell-dependent? What is the effect of T-cell depletion *in vivo* on the course of AD?
4. What is the avidity of anti-ADV antibody as a function of time after inoculation? Information on this question may shed light not only on the nature of the pathogenic immune complexes, but also the possibility that autoantiidiotypic antibody may be produced during disease. The latter point relates to a recent report raising the possibility that low-avidity antibody may, under certain conditions, be autoantiidiotypic antibody (i.e., a response not to antigen but to the antibody induced by the antigen) (Oudin and Cazenave, 1971; Wikler *et al.,* 1979).
5. Does ADV infect lymphocytes and, if so, is there a subset of lymphocytes prefer-

entially susceptible to infection? This problem can be solved by development of a viral-DNA-specific probe that can be used to localize, within the cell, viral DNA sequences.

6. Lodmell *et al.* (1971) reported a difference between normal non-Aleutian and Aleutian mink with respect to their secondary PFC response to GRBC. Aleutian mink made essentially no IgG PFC in their peripheral lymph nodes after an antigen boost. This study suggests that Aleutian mink have a more active immunosuppressive system than do non-Aleutian mink, a finding that may provide insight into the clear difference in susceptibility of these two color phases to the Pullman strain of ADV, and the difference in the tempo of the disease induced by the Utah-1 strain. This study, therefore, needs to be confirmed and amplified.

The ferret may be a more suitable model than mink for studying the hypergammaglobulinemia of AD. Ranch ferrets develop, on occasion, a disease similar to AD but without a component of immune-complex lesions (Kenyon *et al.*, 1966, 1967; Ohshima *et al.*, 1978). These animals have systemic plasmacytosis, hypergammaglobulinemia, and serum antibody that cross-reacts with ADV in CIE tests. The disease appears to be caused by a virus similar to or identical with ADV. This model offers the advantages of studying the affect of ADV on the immune response *in vivo* without the problems of intercurrent complications caused by immune-complex deposition in the tissues.

7. Summary

Aleutian disease of mink is a progressive lethal disease induced by a nondefective parvovirus. It is manifested by rapid viral replication, persistent virus infection, extreme hypergammaglobulinemia, and immune-complex-mediated lesions. The tempo of disease appears to be under genetic control in that mink of the Aleutian color phase have a rapidly progressive course, and non-Aleutian mink a slowly progressive course. Animals generally die in uremia due to a glomerulopathy often complicated by hemorrhage.

In this report we have addressed several aspects of AD including viral persistence, the exaggerated antibody response to the virus, and the pathogenesis of the histologic lesions. In examining the genetic aspects of this disease, we have discussed differences between the immune response of Aleutian and non-Aleutian mink to ADV and non-ADV antigens.

The exaggerated antiviral antibody response seen in AD may be a manifestation of an imbalance in the immunoregulatory system of AD-affected mink. Possible mechanisms for this phenomenon are discussed.

ACKNOWLEDGMENTS

The authors thank Drs. Bloom, Coe, Hadlow, and Race for use of data from studies that have been submitted for publication or are in progress. Special thanks are offerred to Mrs. Helen Blahnik for typing the manuscript and to Joni Lodmell and Nina Portis for their unrelenting patience. Last, we acknowledge Dr. David Porter for his support and encouragement.

References

Allison, A. C., Denman, A. M., and Barnes, R. D., 1971, Cooperating and controlling functions of thymus-derived lymphocytes in relation to autoimmunity, *Lancet* **2**:135.

An, S. H., and Ingram, D. G., 1977, Detection of inapparent Aleutian disease virus infection in mink, *Am. J. Vet. Res.* **38**:1619.

An, S. H., and Ingram, D. G., 1978, Transmission of Aleutian disease from mink with inapparent infections, *Am. J. Vet. Res.* **39**:309.

An, S. H., DePauli, F. J., Wright, P., and Ingram, D. G., 1978, Characteristics of inapparent Aleutian disease virus infection in mink, *Res. Vet. Sci.* **24**:200.

Arend, W. P., and Mannik, M., 1971, Studies of antigen–antibody complexes. II. Quantification of tissue uptake of soluble complexes in normal and complement depleted rabbits, *J. Immunol.* **107**:63.

Baranov, O. K., and Yermalaev, V. I., 1978, Molecular distribution of allotypic determinants LpM 7 and LpM 8 of mink very high density α_2-lipoprotein, *Immunochemistry* **15**:629.

Barnett, E. V., Williams, R. C., Kenyon, A. J., and Henson, J. E., 1969, "Nuclear" antigens and antinuclear antibodies in mink sera, *Immunology* **16**:241.

Barthold, D. R., Kysela, S., and Steinberg, A. D., 1974, Decline in suppressor T cell function with age in female NZB mice, *J. Immunol.* **112**:9.

Basrur, P. K., Gran, D. P., and Karstad, L., 1963, Aleutian disease (plasmacytosis) of mink. III. Propagation of the virus in mink tissue cultures, *Can. J. Comp. Med. Vet. Sci.* **27**:301.

Benacerraf, B., Sebestyen, M., and Cooper, N. S., 1959, The clearance of antigen–antibody complexes from the blood by the reticuloendothelial system, *J. Immunol.* **82**:131.

Berg, R. B., 1961, Multiplication of ECHO 9 virus in suspensions of human leukocytes, *Proc. Soc. Exp. Biol. Med.* **108**:742.

Berger, J., and Hinglais, N., 1968, Les depots intercapillaires d'IgA-IgG, *J. Urol. Nephrol. (Paris)* **74**:694.

Bhan, A. K., Schneeberger, E. E., Collins, A. B., and McCloskey, R. T., 1978, Evidence for a pathogenic role of a cell-mediated immune mechanism in experimental glomerulonephritis, *J. Exp. Med.* **148**:246.

Bloom, B. R., Jimenez, L., and Marcus, P. I., 1970, A plaque assay for enumerating antigen-sensitive cells in delayed type hypersensitivity, *J. Exp. Med.* **132**:16.

Bloom, M. E., Race, R. E., Hadlow, W. J., and Chesebro, B., 1975, Aleutian disease of mink: The antibody response of sapphire and pastel mink to Aleutian disease virus, *J. Immunol.* **115**:1034.

Bloom, M. E., Race, R. E., and Wolfinbarger, J. B., 1980, Characterization of Aleutian disease virus as a parvovirus, *J. Virol.* **35**:836.

Blume, R. S., and Wolff, S. M., 1972, The Chediak–Higashi syndrome. Studies in four patients and a review of the literature, *Medicine* **51**:247.

Bonnard, G. D., Manders, E. K., Campbell, D. A., Herberman, R. B., and Collins, M. J., 1976, Immunosuppressive activity of a subline of the mouse EL-4 lymphoma. Evidence for minute virus of mouse causing the inhibition, *J. Exp. Med.* **143**:187.

Brummerstedt, E., 1976, Preparation of antigen for the counterimmunoelectrophoretic test for plasmacytosis in mink, *Acta Vet. Scand.* **17**:395.

Buchmeier, M. J., and Oldstone, M. B. A., 1978, Virus-induced immune complex disease: Identification of specific viral antigens and antibodies deposited in complexes during chronic lymphocytic choriomeningitis virus infection, *J. Immunol.* **120**:1297.

Burger, D., Gorham, J. R., and Leader, R. W., 1965, Some physical and chemical characteristics of partially purified Aleutian disease virus, in: *Slow, Latent and Temperate Virus Infections* (D. C. Gadjuske, C. J. Gibbs, and M. Alpers, eds.), pp. 307–313, NINDB Monograph No. 2, U.S. Gov. Printing Office, Washington, D.C.

Burger, D., Srironganathan, N., and Gorham, J. R., 1978, Detection of antibody and antigen in Aleutian disease virus infected mink with enzyme-linked immunosorbent assay (ELISA), Abstr. Annu. Meet. Am. Soc. Microbiol., p. 65.

Burns, W. H., Billups, L. C., and Notkins, A. L., 1975, Thymus dependence of viral antigens, *Nature (London)* **256**:654.

Burritt, M. F., Caloanico, N. J., Mehta, S., and Tomasi, T. B., 1977, Activation of the classical complement pathway by Fc fragments of human IgA, *J. Immunol.* **118**:723.

Campbell, D. A., Staal, S. P., Manders, E. K., Bonnard, G. D., Oldham, R. K., Salzman, L. A., and Herberman, R. B., 1977, Inhibition of *in vitro* lymphoproliferative responses by *in vivo* passaged rat 13762 mammary adenocarcinoma cells. II. Evidence that Kilham rat virus is responsible for the inhibitory effect, *Cell. Immunol.* **33**:378.

Cerny, J., and Waner, E. B., 1975, Specific susceptibility of sensitized (memory) B cells to suppression and antigenic alteration by murine leukemia virus, *J. Immunol.* **114**:571.

Cheema, A., Henson, J. B., and Gorham, J. R., 1972, Aleutian disease of mink. Prevention of lesions by immunosuppression, *Am. J. Pathol.* **66**:543.

Chesebro, B., Bloom, M., Hadlow, W., and Race, R., 1975, Purification and ultrastructure of Aleutian disease virus of mink, *Nature (London)* **254**:456.

Cho, H. J., 1976, Purification and structure of Aleutian disease virus, *Front. Biol.* **44**:159.

Cho, H. J., 1977, Demonstration of heavy and light density populations of Aleutian disease virus, *Can. J. Comp. Med.* **41**:215.

Cho, H. J., and Ingram, D. G., 1972, Antigen and antibody in Aleutian disease in mink. I. Precipitation reaction by agar-gel electrophoresis, *J. Immunol.* **108**:555.

Cho, H. J., and Ingram, D. G., 1973a, Isolation, purification and structure of Aleutian disease virus by immunological techniques, *Nature New Biol.* **243**:174.

Cho, H. J., and Ingram, D. G., 1973b, Antigen and antibody in Aleutian disease in mink. II. The reaction of antibody with Aleutian disease agent using immunodiffusion and immunoelectroosmophoresis, *Can. J. Comp. Med.* **37**:217.

Cho, H. J., and Ingram, D. G., 1973c, Pathogenesis of Aleutian disease of mink: Nature of the antiglobulin reaction and elution of antibody from erythrocytes and glomeruli of infected mink, *Infect. Immun.* **8**:264.

Cho, H. J., and Ingram, D. G., 1974, The antigen and virus of Aleutian disease in mink, *J. Immunol. Methods* **4**:217.

Cochrane, C. G., 1971, Mechanisms involved in the deposition of immune complexes in tissues, *J. Exp. Med.* **134**:755.

Cochrane, C. G., and Hawkins, D., 1968, Studies on circulating immune complexes. III. Factors governing the ability of circulating complexes to localize in blood vessels, *J. Exp. Med.* **127**:137.

Cochrane, C. G., and Koffler, D., 1973, Immune complex disease in experimental animals and man, *Adv. Immunol.* **16**:185.

Coe, J. E., 1972, Studies on immunoglobulins of mink: Definition of two populations of light chains, *Immunochemistry* **9**:147.

Coe, J. E., and Hadlow, W. J., 1972, Studies on immunoglobulins of mink: Definition of IgG, IgA and IgM, *J. Immunol.* **108**:530.

Coe, J. E., and Race, R. E., 1978, Ontogeny of mink IgG, IgA and IgM, *Proc. Soc. Exp. Biol. Med.* **157**:289.

Cooper, N. R., and Welsh, R. M., Jr., 1979, Antibody and complement-dependent viral neutralization, *Springer Sem. Immunopathol.* **2**:285.

Craig, C. P., Reynolds, S. L., Airhart, J. W., and Staab, E. V., 1969, Alterations in immune responses by attenuated Venezuelan equine encephalitis vaccine. I. Adjuvant effect of VEE virus infection in guinea pigs, *J. Immunol.* **102**:1220.

Crawford, T. B., 1973, Purification of the Aleutian disease agent by isodensity ultracentrifugation, *Fed. Proc.* **32**:842.

Crawford, T. B., McGuire, T. C., Porter, D. D., and Cho, H. J., 1977, A comparative study of detection methods for Aleutian disease viral antibody, *J. Immunol.* **118**:1249.

Creighton, W. D., Katz, D. H., and Dixon, F. J., 1979, Antigen-specific immunocompetency B cell function, and regulatory helper and suppressor T cell activities in spontaneous autoimmune mice, *J. Immunol.* **123**:2627.

Daniels, C. A., 1975, Mechanisms of viral neutralization, in: *Viral Immunology and Immunopathology* (A. L. Notkins, ed.), pp. 79–97, Academic Press, New York.

Davis, W. C., Spicer, S. S., Green, W. B., and Padgett, G. A., 1971, Ultrastructure of bone marrow granulocytes in normal mink and mink with the homolog of the Chediak–Higashi trait of humans. I. Origin of the abnormal granules present in the neutrophils of mink with the C-HS trait, *Lab. Invest.* **24**:303.

Drommer, W., and Trautwein, G., 1975, Pathogenesis of Aleutian mink disease. VII. Chronic hepatitis with bile duct proliferation, *Vet. Pathol.* **12**:77.

Duc-Nguyen, H., and Henle, W., 1966, Replication of mumps virus in human leukocyte cultures, *J. Bacteriol.* **92:**285.

Dukor, P., and Hartman, K. U., 1973, Hypothesis: Bound C3 as a second signal for B cell activation, *Cell. Immunol.* **7:**349.

Edelman, R., and Wheelock, E. F., 1966, Vesicular stomatitis virus replication in human leukocyte cultures: Enhancement by phytohemagglutinin, *Science* **154:**1053.

Eklund, C. M., Hadlow, W. J., Kennedy, R. C., Boyle, C. C., and Jackson, T. A., 1968, Aleutian disease of mink: Properties of the etiologic agent and the host responses, *J. Infect. Dis.* **118:**510.

Fredga, K., 1961, The chromosomes of mink, *J. Hered.* **52:**90.

Gelfand, M. C., and Steinberg, A. D., 1973, Mechanism of allograft rejection in New Zealand mice. I. Cell synergy and its age-dependent loss, *J. Immunol.* **110:**:1652.

Gelfand, M. C., Frank, M. M., and Green, I., 1975, A receptor for the third component of complement in the human renal glomerulus, *J. Exp. Med.* **142:**1029.

Gerber, N. L., Hardin, J. A., Chused, T. M., and Steinberg, A. D., 1974, Loss with age in NZB/W mice of thymic suppressor cells in the graft-vs-host reaction, *J. Immunol.* **113:**1618.

Gershon, R. K., and Kondo, K., 1971, Infectious immunological tolerance, *Immunology* **21:**903.

Gorham, J. R., Leader, R. W., and Henson, J. B., 1964, The experimental transmission of a virus causing hypergammaglobulinemia in mink: Sources and modes of infection, *J. Infect. Dis.* **114:**341.

Gorham, J. R., Henson, J. B., Crawford, T. B., and Padgett, G. A., 1976, The epizootiology of Aleutian disease, in: *Slow Virus Diseases of Animals and Man* (R. H. Kimberlin, ed.), pp. 135–158, North-Holland, Amsterdam.

Götze, O., and Müller-Eberhard, H. J., 1971, The C3 activator system: An alternate pathway of complement activation, *J. Exp. Med.* **134:**905.

Gustafson, G. T., 1969, Increased susceptibility to periodontitis in mink affected by a lysosomal disease, *J. Periodontol. Res.* **4:**259.

Haakenstad, A. O., and Mannik, M., 1974, Saturation of the reticuloendothelial system with soluble immune complexes, *J. Immunol.* **112:**1939.

Haakenstad, A. O., and Mannik, M., 1976, The disappearance kinetics of soluble immune complexes prepared with reduced and alkylated antibodies and with intact antibodies in mice, *Lab. Invest.* **35:**283.

Hadlow, W. J., Race, R. E., and Jackson, T. A., 1972, Lymphoreticular proliferative disease in mink homozygous for the Aleutian gene, *J. Natl. Cancer Inst.* **49:**1455.

Hahn, E. C., Ramos, L., and Kenyon, A. J., 1977, Expression of Aleutian mink disease antigen in cell culture, *Infect. Immun.* **15:**204.

Hartsough, G. R., and Gorham, J. R., 1956, Aleutian disease in mink, *Nat. Fur News* **28:**10.

Helgebostad, A., 1963, The Aleutian disease, *Fur Trade J. Can.* **40:**10.

Helmboldt, C. F., and Jungherr, E. L., 1958, The pathology of Aleutian disease in mink, *Am. J. Vet. Res.* **119:**212.

Henson, J. B., Leader, R. W., and Gorham, J. R., 1961, Hypergammaglobulinemia of mink, *Proc. Soc. Exp. Biol. Med.* **107:**919.

Henson, J. B., Gorham, J. R., Leader, R. W., and Wagner, B. M., 1962, Experimental hypergammaglobulinemia in mink, *J. Exp. Med.* **116:**357.

Henson, J. B., Leader, R. W., Gorham, J. R., and Padgett, G. A., 1966, The sequential development of lesions in spontaneous Aleutian disease of mink, *Pathol. Vet.* **3:**289.

Henson, J. B., Gorham, J. R., and Tanaka, Y., 1967, Renal glomerular ultrastructure in mink affected by Aleutian disease, *Lab. Invest.* **17:**123.

Henson, J. B., Gorham, J. R., Tanaka, Y., and Padgett, G. A., 1968, The sequential development of ultrastructural lesions in the glomeruli of mink with experimental Aleutian disease, *Lab. Invest.* **19:**153.

Henson, J. B., Gorham, J. R., Padgett, G. A., and Davis, W. C., 1969, Pathogenesis of the glomerular lesions in Aleutian disease. Immunofluorescent studies, *Arch. Pathol.* **87:**21.

Henson, J. B., Gorham, J. R., McGuire, T. C., and Crawford, T. B., 1976, Pathology and pathogenesis of Aleutian disease, in: *Slow Virus Diseases of Animals and Man* (R. H. Kimberlin, ed.), pp. 175–205, North-Holland, Amsterdam.

Hood, L., Gray, W. R., Sanders, B. G., and Dreyer, W. J., 1967, Light chain evolution, *Cold Spring Harbor Symp. Quant. Biol.* **32:**133.

Howard, J. G., Christie, G. H., Jacob, M. J., and Elson, J., 1970, Studies on immunological paralysis. III.

Recirculation and antibody-neutralizing activity of ^{14}C-labeled type III pneumococcal polysaccharide in paralyzed mice, *Clin. Exp. Immunol.* 7:583.

Hunsicker, L. G., Shearer, T. P., Plattner, S. B., and Weisenburger, D., 1979, The role of monocytes in serum sickness nephritis, *J. Exp. Med.* **150**:413.

Ingram, D. G., 1978, Aleutian disease of mink. An immunological disease caused by a persistent virus infection, in: *Animal Models of Comparative and Developmental Aspects of Immunity and Disease* (M. E. Gershwin and E. L. Cooper, eds.) pp. 308–321, Pergamon Press, New York.

Izui, S., McConahey, P. J., Theofilopoulos, A. N., and Dixon, F. J., 1979, Association of circulating retroviral gp70–anti-gp70 immune complexes with murine systemic lupus erythematosus, *J. Exp. Med.* **149**:1099.

Janeway, C. A., and Humphrey, J. H., 1968, Synthetic antigens composed exclusively of L- or D-amino acids. II. Effect of optical configuration on the metabolism and fate of synthetic polypeptide antigens in mice, *Immunology* **14**:225.

Jerne, N. K., 1974, Toward a network theory of the immune system, *Ann. Immunol. (Inst. Pasteur)* **125c**:373.

Johnson, M. I., Henson, J. B., and Gorham, J. R., 1975, The influence of genotype on the development of glomerular lesions in mink with Aleutian disease virus. A correlated light, fluorescent, and electron microscopic study, *Am. J. Pathol.* **81**:321.

Karstad, L., and Pridham, T. J., 1962, Aleutian disease of mink. I. Evidence of its viral etiology, *Can. J. Comp. Med.* **26**:97.

Katz, D. H., and Benacerraf, B., 1972, The regulatory influence of activated T cells on B cell responses to antigen, *Adv. Immunol.* **15**:1.

Kenyon, A. J., 1966, Immunologic deficiency in Aleutian disease of mink, *Am. J. Vet. Res.* **27**:1780.

Kenyon, A. J., Helmboldt, C. F., and Nielson, S. W., 1963, Experimental transmission of Aleutian disease with urine, *Am. J. Vet. Res.* **24**:1066.

Kenyon, A. J., Magnamo, T. S., Helmboldt, C. F., and Buko, L., 1966, Aleutian disease in the ferret, *J. Am. Vet. Med. Assoc.* **149**:920.

Kenyon, A. J., Howard, E., and Buko, L., 1967, Hypergammaglobulinemia in ferrets with lymphoproliferative lesions (Aleutian disease), *Am. J. Vet. Res.* **28**:1167.

Kenyon, A. J., Gander, J. E., Lopez, C., and Good, R. A., 1973, Isolation of Aleutian mink disease virus by affinity chromatography, *Science* **179**:187.

Kenyon, A. J., Kassel, R., Notani, G., and Hahn, E. C., 1976, Separation of Aleutian disease antibody response from induction of gammapathy with levamisole, *Fed. Proc.* **35**:569.

Kijlstra, A., Knutson, D. W., Daha, M. R., and Van Es, L. A., 1979, Clearance and glomerular localization of preformed DNP anti-DNP immune complexes, *Scand. J. Immunol.* **10**:421.

Kindig, D., Sporgo, B., and Kirsten, W. H., 1967, Glomerular response in Aleutian disease of mink, *Lab. Invest.* **16**:436.

Kono, Y., Kobayashi, K., and Fukunaga, Y., 1973, Antigenic drift of equine infectious anemia virus in chronically infected horses, *Arch Gesamte Virusforsch.* **41**:1.

Krakawer, R. S., Waldmann, T. A., and Stober, W., 1976, Loss of suppressor T cells in adult NZB/NZW mice, *J. Exp. Med.* **144**:662.

Kumar, V., Coruso, T., and Bennett, M., 1976, Mechanisms of genetic resistence to Friend virus leukemia. III. Susceptibility of mitogen-responsive lymphocytes mediated by T cells, *J. Exp. Med.* **143**:728.

Larsen, A. E., 1965, Electrophoresis, IAT and the elusive Aleutian disease, *Nat. Fur News* **37**:18.

Larsen, A. E., and Porter, D. D., 1975, Pathogenesis of Aleutian disease of mink: Identification of nonpersistent infections, *Infect. Immun.* **11**:92.

Leader, R. W., Padgett, G. A., and Gorham, J. R., 1963, Studies of abnormal lymphocyte bodies in the mink, *Blood* **22**:477.

Liacopoulos, P., and Ben-Efraim, S., 1975, Antigenic competition, *Progr. Allergy* **18**:97.

Lilly, F., and Pincus, T., 1973, Genetic control of murine viral leukemogenesis, *Adv. Cancer Res.* **17**:231.

Lodmell, D. L., Hadlow, W. J., Munoz, J. J., and Whitford, H. W., 1970, Hemagglutinin antibody response of normal and Aleutian disease-affected mink to keyhole limpet hemocyanin, *J. Immunol.* **104**:878.

Lodmell, D. L., Bergman, R. K., Hadlow, W. J., and Munoz, J. J., 1971, Cellular and humoral antibody responses of normal pastel and sapphire mink to goat erythrocytes, *Infect. Immun.* **3**:221.

Lodmell, D. L., Bergman, R. K., and Hadlow, W. J., 1973, Antibody-forming cells and serum hemolysin responses of pastel and sapphire mink inoculated with Aleutian disease virus, *Infect. Immun.* **8**:769.

Lyons, R. T., and Pitot, H. C., 1978, Protein degradation in normal and beige (Chediak-Higashi) mice, *J. Clin. Invest.* **61**:260.

McFarland, H. F., 1974, The effect of measles virus infection on T and B lymphocytes in the mouse. I. Suppression of helper cell activity, *J. Immunol.* **113**:1978.

McGuire, T. C., Crawford, T. B., Henson, J. B., and Gorham, J. R., 1971, Aleutian disease of mink: Detection of large quantities of complement-fixing antibody to viral antigen, *J. Immunol.* **107**:1481.

Mallén, M. S., Ugalde, E. L., Balcazar, M. R., Bolívar, J. I., and Meyran, S., 1950, Precipitation of abnormal serums by Lugol's solution, *Am. J. Clin. Pathol.* **20**:39.

Mannik, M., and Arend, W. P., 1971, Fate of preformed immune complexes in rabbits and rhesus monkeys, *J. Exp. Med.* **134**:195.

Mannik, M., Arend, W. P., Hall, A. P., and Gilliland, B. C., 1971, Studies on antigen-antibody complexes. I. Elimination of soluble complexes from rabbit circulation, *J. Exp. Med.* **133**:713.

Minato, N., and Katsura, Y., 1978a, Virus-replicating T cells in the immune response of mice. II. Characterization of T cells capable of replicating vesicular stomatitis virus, *J. Exp. Med.* **148**:837.

Minato, N., and Katsura, Y., 1978b, Virus-replicating T cells in the immune response of mice. III. Role of vesicular stomatitis virus-replicating T cells in the antibody response, *J. Exp. Med.* **148**:850.

Mitchison, N. A., 1969, The immunogenic capacity of antigen taken up by peritoneal exudate cells, *Immunology* **16**:1.

Möller, G., and Coutinho, A., 1975, Role of C'3 and Fc receptors in B-lymphocyte activation, *J. Exp. Med.* **141**:647.

Mond, J. J., Sher, I., Mosier, D. E., Blaese, M., and Paul, W. E., 1978, T-independent responses in B cell-defective CBA/N mice to *Brucella abortus* and to trinitrophenyl (TNP) conjugates of *Brucella abortus*, *Eur. J. Immunol.* **8**:459.

Mota, I., 1964, The mechanism of anaphylaxis. I. Production and biological properties of mast cell sensitizing antibodies, *Immunology* **7**:681.

Muller, H. P., Gautschi, M., and Siegl, G., 1978, Defective particles of parvovirus Lu III, in: *Replication of Mammalian Parvoviruses* (D. C. Ward and P. Tattersall, eds.), pp. 231–240, Cold Spring Harbor, Long Island, N.Y.

Munoz, J., Lodmell, D. L., Race, R. E., Jackson, T. A., and Hadlow, W. J., 1974, Delayed hypersensitivity in normal and Aleutian disease-affected mink, *Z. Immunitaetsforsch.* **147**:53.

Narayan, O., Griffin, D. E., and Chase, J., 1977, Antigenic shift of visna virus in persistently infected sheep, *Science* **197**:376.

Narayan, O., Griffin, D. E., and Clements, J. E., 1978, Progressive antigenic drift of visna virus in persistently infected sheep, in: *Persistent Viruses,* ICN–UCLA Symposia on Molecular and Cellular Biology (J. G. Stevens, G. J. Todaro, and C. F. Fox, eds.), pp. 663–669, Academic Press, New York.

Notkins, A. L., Mergenhagen, S. E., Rizzo, A. A., Scheele, C., and Waldmann, T. A., 1966, Elevated gamma globulin and increased antibody production in mice infected with lactic dehydrogenase virus, *J. Exp. Med.* **123**:347.

Notkins, A. L., Mergenhagen, S. E., and Howard, R. J., 1970, Effect of virus infections on the function of the immune system, *Annu. Rev. Microbiol.* **24**:525.

Obel, A.-L., 1959, Studies on a disease in mink with systemic proliferation of the plasma cells, *Am. J. Vet. Res.* **20**:384.

Ohshima, K., Shen, D. T., Henson, J. B., and Gorham, J. R., 1978, Comparison of the lesions of Aleutian disease in mink and hypergammaglobulinemia in ferrets, *Am. J. Vet. Res.* **39**:653.

Oldstone, M. B. A., 1971, Autoimmunity and viruses—fact or fiction: Persistent LCM viral infection, anti-LCM viral immune response, and tissue injury, *Am. J. Clin. Pathol.* **56**:299.

Oldstone, M. B. A., and Dixon, F. J., 1975, Immune complex diseases associated with viral infections, in: *Viral Immunology and Immunopathology* (A. L. Notkins, ed.), pp. 341–356, Academic Press, New York.

Oldstone, M. B. A., Buchmeier, M. J., and Doyle, M. V., 1978, Immunobiology of persistent virus infection, in: *Persistent Viruses,* ICN–UCLA Symposia on Molecular and Cellular Biology (J. G. Stevens, G. J. Todaro, and C. F. Fox, eds.), pp. 43–56, Academic Press, New York.

Oppenheim, J. J., 1972, Modulation of *in vitro* lymphocyte transformation by antibodies: Enhancement by antigen-antibody complexes and inhibition by antibody excess, *Cell. Immunol.* **3**:341.

Oudin, J., and Cazenave, P. A., 1971, Similar idiotypic specificities in immunoglobulin fractions with different antibody functions or even without detectable antibody function, *Proc. Natl. Acad. Sci. USA* **68**:2616.

Padgett, G. A., Gorham, J. R., and Henson, J. B., 1967a, Epizootiologic studies of Aleutian disease. I. Transplacental transmission of the virus, *J. Infect. Dis.* **117**:35.

Padgett, G. A., Reiquam, C. W., Gorham, J. R., Henson, J. B., and O'Mary, C. C., 1967b, Comparative studies of the Chediak–Higashi syndrome, *Am. J. Pathol.* **51**:553.

Padgett, G. A., Reiquam, C. W., Henson, J. B., and Gorham, J. R., 1968, Comparative studies of susceptibility to infection in the Chediak–Higashi syndrome, *J. Pathol. Bacteriol.* **95**:509.

Pan, I. C., Tsai, K. S., and Karstad, L., 1970a, Glomerulonephritis in Aleutian disease of mink. Histological and immunofluorescence studies, *J. Pathol.* **101**:119.

Pan, I. C., Tsai, K. S., Gringer, I., and Karstad, L., 1970b, Glomerulonephritis in Aleutian disease of mink: Ultrastructural studies, *J. Pathol.* **102**:33.

Perryman, L. E., Banks, K. L., and McGuire, T. C., 1975, Lymphocyte abnormalities in Aleutian disease virus infection of mink: Decreased T lymphocyte responses and increased B lymphocyte levels in persistent viral infection, *J. Immunol.* **115**:22.

Pierce, C. W., Kapp, J. A., Wood, D. D., and Benacerraf, B., 1974, Immune response *in vitro*. X. Functions of macrophages, *J. Immunol.* **112**:1181.

Porter, D. D., 1965, Transfer of gamma globulin from mother to offspring in mink, *Proc. Soc. Exp. Biol. Med.* **119**:131.

Porter, D. D., 1978, Genetic aspects of Aleutian disease of mink, in: *Genetic Control of Autoimmune Disease* (N. R. Rose, P. E. Bigazzi, and N. L. Warner, eds.), pp. 309–318, Elsevier/North-Holland, Amsterdam.

Porter, D. D., and Cho, H. J., 1980, Aleutian disease of mink: A model for persistent infection, in: *Comprehensive Virology* (H. Fraenkel-Conrat and R. R. Wagner, eds.), Vol. 16, pp. 233–256, Plenum Press, New York.

Porter, D. D., and Dixon, F. J., 1966, Electrophoretic and immunoelectrophoretic characterization of normal mink serum proteins, *Am. J. Vet. Res.* **27**:335.

Porter, D. D., and Larsen, A. E., 1964, Statistical survey of Aleutian disease in ranch mink, *Am. J. Vet. Res.* **25**:1226.

Porter, D. D., and Larsen, A. E., 1967, Aleutian disease of mink: Infectious virus–antibody complexes in the serum, *Proc. Soc. Exp. Biol. Med.* **126**:680.

Porter, D. D., and Larsen, A. E., 1968, Virus–host interactions in Aleutian disease of mink, *Perspect. Virol.* **6**:173.

Porter, D. D., and Larsen, A. E., 1974, Aleutian disease of mink, *Progr. Med. Virol.* **18**:32.

Porter, D. D., Dixon, F. J., and Larsen, A. E., 1965a, Metabolism and function of gamma globulin in Aleutian disease of mink, *J. Exp. Med.* **121**:889.

Porter, D. D., Dixon, F. J., and Larsen, A. E., 1965b, The development of a myeloma-like condition in mink with Aleutian disease, *Blood* **25**:736.

Porter, D. D., Larsen, A. E., and Porter, H. G., 1969, The pathogenesis of Aleutian disease of mink. I. *In vivo* viral replication and the host antibody response to viral antigen, *J. Exp. Med.* **130**:575.

Porter, D. D., Larsen, A. E., and Porter, H. G., 1972, The pathogenesis of Aleutian disease of mink. II. Enhancement of tissue lesions following the administration of a killed virus vaccine or passive antibody, *J. Immunol.* **109**:1.

Porter, D. D., Larsen, A. E., and Porter, H. G., 1973, The pathogenesis of Aleutian disease of mink. 3. Immune complex arteritis, *Am. J. Pathol.* **71**:331.

Porter, D. D., Larsen, A. E., Cox, N. A., Porter, H. G., and Suffin, S. C., 1977a, Isolation of Aleutian disease virus of mink in cell culture, *Intervirology* **8**:129.

Porter, D. D., Larsen, A. E., and Porter, H. G., 1977b, Reduced severity of lesions in mink infected transplacentally with Aleutian disease virus, *J. Immunol.* **119**:872.

Porter, D. D., Porter, H. G., and Larsen, A. E., 1977c, Immunoglobulin class of Aleutian disease viral antibody, *Fed. Proc.* **36**:1268.

Porter, D. D., Larsen, A. E., and Porter, H. G., 1980, Aleutian disease of mink, *Adv. Immunol.*, **29**:261.

Portis, J. L., and Coe, J. E., 1979, Deposition of IgA in renal glomeruli of mink affected with Aleutian disease, *Am. J. Pathol.* **96**:227.

Preble, O. T., and Youngner, J. S., 1975, Temperature-sensitive viruses and the etiology of chronic and inapparent infections, *J. Infect. Dis.* **131**:467.

Prieur, D. J., Davis, W. C., and Padgett, G. A., 1972, Defective function of renal lysosomes in mice with the Chediak–Higashi syndrome, *Am. J. Pathol.* **67**:227.

Primi, D., Hamarstrom, L., and Smith, C. I. E., 1978, Genetic control of lymphocyte suppression. I. Lack of suppression in aged NZB mice is due to a B cell defect, *J. Immunol.* **121**:2241.

Pross, H. F., and Eidinger, D., 1974, Antigenic competition: A review of nonspecific antigen-induced suppression, *Adv. Immunol.* **18**:133.

Puga, A., Rosenthal, J. D., Openshaw, H., and Notkins, A. L., 1978, Herpes simplex virus DNA and mRNA sequences in acutely and chronically infected trigeminal ganglia of mice, *Virology* **89**:102.

Race, R. E., Coe, J. E., and Bloom, M. E., 1980, Blastogenesis of lymphocytes from mink infected with Aleutian disease virus (ADV) *Fed. Proc. Fed. Ann. Soc. Exp. Biol.* **39**:356 (abstract).

Ranney, D. F., and Steinberg, A. D., 1976, Differences in the age-dependent release of a low molecular weight suppressor (LMWS) and stimulators by normal and NZB/W lymphoid organs, *J. Immunol.* **117**:1219.

Rifai, A., Small, P. A., Teague, P. O., and Ayoub, E. M., 1979, Experimental IgA nephropathy, *J. Exp. Med.* **150**:1161.

Roder, J. C., Bell, O. A., and Singhal, S. K., 1977, Regulation of the immune response in autoimmne NZB/NZW F_1 mice. I. The spontaneous generation of splenic suppressor cells, *Cell. Immunol.* **29**:272.

Rowe, W. P., Hartley, J. W., Lander, M. R., Pugh, W. E., and Teich, N., 1971, Noninfectious AKR mouse embryo cell lines in which each cell has the capacity to be activated to produce infectious murine leukemia virus, *Virology* **46**:866.

Russell, A. S., and Percy, J. S., 1974, The antigenicity of polynucleotides in mink, rabbit and man, *J. Rheumatol.* **1**:66.

Russell, A. S., Percy, J. S., and Cho, H. J., 1975, Poly IC therapy in Aleutian disease of mink, *Can. J. Comp. Med.* **39**:240.

Saison, R., Karstad, L., and Pridham, T. J., 1966, Viral plasmacytosis (Aleutian disease) in mink. VI. The development of positive Coombs tests in experimental infections, *Can. J. Comp. Med.* **30**:151.

Salaman, M. H., 1970, Immunosuppressive effects in infection, *Proc. R. Soc. Med.* **63**:11.

Schreiner, G. F., Cotran, R. S., Pardo, U., and Unanue, E. R., 1978, A mononuclear cell component in experimental immunologic glomerulonephritis, *J. Exp. Med.* **147**:369.

Schultz, R. D., Mendel, H., and Scott, F. W., 1976, Effect of feline panleukopenia virus infection on development of humoral and cellular immunity, *Cornell Vet.* **66**:324.

Shackelford, R. M., 1973, Genetics of mink, in: *Bluebook of Fur Farming*, p. 7.

Shahrabadi, M. S., and Cho, H. J., 1977, Detection and localization of Aleutian disease virus and its antigens *in vivo* by immunoferritin technique, *Can. J. Comp. Med.* **41**:435.

Shahrabadi, M. S., Cho, H. J., and Marusyk, R. G., 1977, Characterization of the protein and nucleic acid of Aleutian disease virus, *J. Virol.* **23**:353.

Shearer, W. T., Atkinson, J. P., Frank, M. M., and Parker, C. W., 1975, Humoral immunostimulation. IV. Role of complement, *J. Exp. Med.* **141**:736.

Shevach, E. M., Jaffe, E. S., and Green, I., 1973, Receptors for complement and immunoglobulin on human and animal lymphoid cells, *Transplant. Rev.* **16**:3.

Shin, M. L., Gelfand, M. C., Nagle, R. B., Carlo, J. R., Green, I., and Frank, M. M., 1977, Localization of receptors for activated complement on visceral epithelial cells of the human renal glomerulus, *J. Immunol.* **118**:869.

Spiegelberg, H. L., 1974, Biological activities of immunoglobulins of different classes and subclasses, *Adv. Immunol.* **19**:259.

Stevens, J. G., and Cook, M. L., 1971, Latent herpes simplex virus in spinal ganglia of mice, *Science* **173**:843.

Striker, G. E., Mannik, M., and Tung, M. Y., 1979, Role of marrow derived monocytes and mesangeal cells in removal of immune complexes from renal glomeruli, *J. Exp. Med.* **149**:127.

Stutman, O., 1975, Immunodepression and malignancy, *Adv. Cancer Res.* **22**:261.

Tabel, H., and Ingram, D. G., 1972, Immunoglobulins of mink. Evidence for five immunoglobulin classes of 7S type, *Immunology* **22**:933.

Tattersall, P., and Ward, D., (eds.), 1978, The parvoviruses—An introduction, in: *Replication of Mammalian Parvoviruses*, pp. 3–12, Cold Spring Harbor, Long Island, N.Y.

Tennant, R. W., Layman, K. R., and Hand, R. E., 1969, Effect of cell physiological state on infection by rat virus, *J. Virol.* **4**:872.

Theofilopoulos, A. N., Wilson, C. B., and Dixon, F. J., 1976, The Raji cell radioimmune assay for detecting immune complexes in human sera, *J. Clin. Invest.* **57**:169.

Thoburn, R., Hurwitz, A. I., and Kunkel, H. G., 1972, A DNA-binding protein in the serum of certain mammalian species, *Proc. Natl. Acad. Sci. USA* **69**:3327.

Trautwein, G., 1964, Experimentelle Untersuchungen uber die Aleuten krankheit (Aleutian disease) der Nerze, *Arch. Exp. Veterinaermed.* **18:**287.

Trautwein, G. M., and Helmboldt, C. F., 1962, Aleutian disease of mink. I. Experimental transmission of the disease, *Am. J. Vet. Res.* **23:**1280.

Trautwein, G., and Seidler, D., 1972, Studies on the pathogenesis of Aleutian mink disease. V. Immunofluorescent microscopical findings in the kidney, *Zentralbl. Veterinaermed.* **19:**144.

Trautwein, G., Schneider, P., and Ernst, E., 1974, Pathogenesis of the Aleutian mink disease. 8. Depression of antibodies against peroxidase from horseradish, *Zentralbl. Veterinaermed. Reihe B* **21:**467.

Unanue, E. R., and Cerottini, J., 1970, The immunogenicity of antigen bound to the plasma membrane of macrophages, *J. Exp. Med.* **131:**711.

Urizar, R. E., Michael, A., Sisson, S., and Vernier, R. L., 1968, Anaphylactoid purpura. II. Immunofluorescent and electron microscopic studies of the glomerular lesions, *Lab. Invest.* **19:**437.

Wallis, C., and Melnick, J. L., 1971, Herpesvirus neutralization: The role of complement, *J. Immunol.* **107:**1235.

Weigle, W. O., 1958, Elimination of antigen–antibody complexes from sera of rabbits, *J. Immunol.* **81:**204.

Wikler, M., Franssen, J.-D., Collignon, C., Leo, O., Mariame, B., van Dewalle, P., De Groote, D., and Urbain, J., 1979, Idiotypic regulation of the immune system. Common idiotypic specificities between idiotypes and antibodies raised against anti-idiotypic antibodies in rabbits, *J. Exp. Med.* **150:**184.

Willems, F. T. C., Melnick, J. L., and Rawls, W. E., 1969, Replication of polio-virus in phytohemagglutinin-stimulated human lymphocytes, *J. Virol.* **3:**451.

Williams, R. C., Jr., Kenyon, A. J., and Huntley, C. C., 1968, Immunoglobulins, viruses, and speculation on their interrelationship in certain human and animal disease states, *Blood* **31:**522.

Woodruff, J. F., and Woodruff, J. J., 1975, The effect of viral infections on the function of the immune system, in: *Viral Immunology and Immunopathology* (A. L. Notkins, ed.), pp. 393–418, Academic Press, New York.

Yoon, J. W., Kenyon, A. J., and Good, R. A., 1973, Demonstration of Aleutian mink disease virus in cell culture, *Nature New Biol.* **245:**205.

Yoon, J. W., Dunker, A. K., and Kenyon, A. J., 1975, Characterization of Aleutian mink disease virus, *Virology* **64:**575.

Yoshika, T., Hayasaka, T., Fukatsu, R., Shirai, T., Itoh, T., Ikeda, H., and Katagiri, M., 1979, The structural proteins of murine leukemia virus and the pathogenesis of necrotizing arteritis and glomerulonephritis in SL/Ni mice, *J. Immunol.* **122:**1812.

Zubler, R. H., Lange, G., Lambert, P. H., and Meischer, P. A., 1976, Detection of immune complexes in unheated sera by a modified ^{125}I-Clq binding test, *J. Immunol.* **116:**232.

4

The Pathogenesis of Autoimmunity in New Zealand Mice

DAVID R. MILICH AND M. ERIC GERSHWIN

1. Introduction

New Zealand Black (NZB) mice and the F_1 hybrid produced by the mating of NZB with New Zealand White (NZW) mice [(NZB × NZW)F_1] were the first described models of spontaneous autoimmune disease. The NZB strain was generated by Dr. Marianne Bielschowsky at the Otago University Medical School, New Zealand, by inbreeding for coat color for use in cancer research. During their development it was recognized that NZB mice died prematurely with severe anemia and hepatosplenomegaly. These symptoms were shown to result from an autoimmune hemolytic anemia (Bielschowsky *et al.*, 1959). Subsequently, in a series of hybrid experiments it was shown that the autoimmune trait was inherited as a dominant characteristic, with incomplete and differing expressions dependent upon the non-NZB parent. When NZW mice, developed independently by Mr. W. H. Hall, were crossed with NZB, the F_1 hybrid expressed a markedly different autoimmune process characterized by the formation of antinuclear antibodies, lupus erythematosus (LE) cell, and a renal disease resembling lupus nephritis (Helyer and Howie, 1963a,b) Indeed, the spontaneous disease that NZB and (NZB × NZW)F_1 (B/W) mice develop are so similar to the human counterparts of autoimmune hemolytic anemia and systemic lupus erythematosus (SLE) that they have become the most versatile animal model in clinical immunobiology. The diseases of New Zealand (NZ) mice provide an opportunity to study the etiology and pathogenesis of a variety of autoimmune phenomena. Furthermore, they have provided insight into basic imunologic processes involved in self recognition and the interrelationships between viral and genetic factors in autoimmunity.

This paper will review pertinent information regarding the natural history and immunopathology of NZ mice with specific emphasis on recent observations. The histo-

DAVID R. MILICH AND M. ERIC GERSHWIN • Section of Rheumatology–Clinical Immunology, School of Medicine, University of California, Davis, California 95616.

pathology of NZ mice has been described in great detail elsewhere (East, 1970; Howie and Helyer, 1968; Mellors, 1966a; Helyer and Howie, 1963a) and will be only briefly summarized herein. However, the antigenic specificity of autoantibodies produced by NZ mice and their immunopathologic significance, including genetic and viral factors, will be discussed in considerable detail. Specifically, description of recently elucidated cellular defects in the T-cell, B-cell, macrophage, and stem-cell compartments and their relative contribution to the pathogenesis of the autoimmune disease, including hormonal influences and numerous therapeutic agents, will be reviewed and an attempt will be made to present this information from a historical perspective.

The numerous and varied studies of this inbred strain and the plethora of theories of pathogenesis emerging from them have reflected our rapidly changing understanding of the cellular basis of the immune response to self and non-self antigens. The dicotomy between T and B cells, the existence of immunoregulatory cells (Gershon, 1974; Katz *et al.,* 1973), the identification of subpopulations of T and B cells and the complex interaction between these subpopulations necessary for effector function have been relatively recent findings (Cantor and Boyse, 1975). Furthermore, that this cell–cell interaction is mediated through recognition of various self markers (i.e., anti-id) and that autoimmunity may result from the breakdown of such communication (Cantor, 1978) are far removed from the earlier view that deletion at birth of all lymphocytes bearing receptors for self antigens prevented autoimmune disease (Jerne, 1971; Nossal, 1965; Burnet, 1959). Therefore, our understanding and interpretation of observations have constantly changed. It is appropriate that some twenty years after the discovery of the NZ inbred strains we place the current thoughts in perspective.

2. The Natural Life History and Pathology of NZ Mice

2.1. The NZB Strain

The mean survival time for NZB mice varies with sex and reproductive history. Males live a mean of 467 days, females 431 days, and breeder females 418 days (Howie and Simpson, 1975). The most consistent abnormality occurring in NZB mice is the development of an autoimmune hemolytic anemia early in life. Their anemia is produced by an antierythrocyte autoantibody appearing at 3 months with an incidence of 100% by 12 to 15 months of age. The prevalence of this process is similar in male and virgin females but is delayed in breeder females. Initially, the antierythrocyte autoantibody is of the incomplete warm antibody type and is detected by the direct antiglobulin test (DAT). The antibody class has been characterized as IgG (Norins and Holmes, 1964), in contrast to an IgM antierythrocyte antibody occurring later in life (Mellors, 1965). The antigenic specificities of these and other erythrocyte autoantibodies found in NZB mice are discussed in a subsequent section. The hemolytic anemia develops 1–5 months after the appearance of autoantibody and is associated with a significant reticulocytosis, which in the majority of animals reaches 25% (Howie and Helyer, 1968). Erythrocyte survival time is greatly reduced and both liver and spleen participate in the sequestration of antibody-coated erythrocytes, although in severe hemolysis the liver is more active (Lindsey *et al.,* 1966). The incidence of severe anemia is 87% in virgin females, 77% in males, and 56% in breeder females (Howie and Helyer, 1968). Splenomegaly is a feature of the hemolytic disease but is not invariably present. The splenic enlargement is due mainly to expansion of splenic pulp by both extramedullary hematopoietic tissue and lymphoid hyperplasia. It has been suggested that the spleen is the main site of autoantibody formation; however, it is also an important hematopoietic organ as shown by the rapid drop in hematocrit following splenectomy (Mellors, 1965; Helyer and Howie, 1963b). Intracellular and extracellular hemosiderin increases in abundance with age.

Hepatomegaly is less commonly seen than splenomegaly and is most common in virgins. The most consistent histological change is the accumulation of hemosiderin in Kupffer cells. Kupffer cell hyperplasia and local lymphoid infiltration of centrilobular and portal regions have been reported (De Vries and Hijmans, 1967) along with liver necrosis in older animals (Helyer and Howie, 1963b).

Renal disease in the NZB strain develops insidiously and remains clinically latent for the greater part of the life span, although a few animals develop a nephrotic syndrome (Bielschowsky and D'Ath, 1971; Hicks and Burnet, 1966; Mellors, 1966b; Helyer and Howie, 1963a). A more active form of renal insufficiency, resembling the renal disorder seen in NZB/NZW females, develops toward the end of the life span particularly in virgin females and has been characterized as severe proliferative glomerulonephritis (Mellors, 1966b, 1965). The nephropathy commences with the accumulation of hyaline material in the mesangial and capillary basement membranes. The hyaline material is strongly PAS positive and shows an irregular lumpy structure with antimouse IgG by fluorescence microscopy (Mellors, 1965). In addition to the glomerular pathology, aggregations of lymphoid cells are present around the small renal vessels; some arteries show a necrotizing arteritis (Hicks, 1966). Estimations of the prevalence of antinuclear antibody (ANA) in NZB mice have been extremely variable. However, it appears that positive LE tests and the occurence of ANA in high titer are found in a minority of NZB mice (Talal and Steinberg, 1974). There is evidence to suggest that the anti-DNA component of ANA correlates best with the incidence of nephritis (Steinberg *et al.,* 1971).

Hyperplasia of lymphoid tissue involving the spleen, lymph nodes, lungs, kidneys, liver, bone marrow, salivary glands, and thymus are consistent features of the disorders of NZB mice. Two phases of lymphoproliferative changes have been described (Denman and Denman, 1970; East, 1970; East *et al.,* 1965). In the first phase, 3–11 months of age, there is extensive development of large lymphoid follicles containing multiple germinal centers in the white pulp of the spleen and cortices of lymph nodes. The second phase, occurring later in life, consists of extreme hyperplasia of plasma cells and reticulum cells involving the medullary areas of spleen and lymphoid tissue throughout the body. This may result in frank lymphoid malignancy in as many as 20% of the animals, although the incidence of lymphoid neoplasia in NZB mice is variably reported in the literature (De Vries and Hijmans, 1966; Helyer and Howie, 1963b; Holmes and Burnet, 1963a; Bielschowsky and Bielschowsky, 1962; Casey, 1968a). This may be due to the difficulty in distinguishing between nonmalignant lymphoid hyperplasia and neoplasia.

Histological lesions of the thymus have been described. These include lymphoid hyperplasia and follicular aggregations of lymphocytes in the medulla (De Vries and Hijmans 1967; East *et al.,* 1965), numerous mast cells (Viklicky and Polackova, 1969; Burnet, 1965), premature thymic involution with a predominant degeneration and vacuolization of epithelial cells (De Vries and Hijmans, 1966). Recently Gershwin *et al.* (1978) have shown an age-dependent loss of both functional and morphological characteristics of thymic epithelial cells cultured *in vitro*. These changes could be of considerable significance in relation to the established deficiency of thymus-dependent immune function; however, their relation to autoimmune phenomena remains uncertain.

There is a pulmonary abnormality in NZB mice that consists of peribronchial aggregations of lymphoid cells. This increases with age, is present in 80% of mice over 1 year of age, and may represent either a specific immune reaction against components of pulmonary tissue or alternatively the general immunoproliferative changes in this strain (Howie and Helyer, 1968). NZB mice also develop chronic peptic ulceration distal to the pyloric ring that shows a striking histological resemblance to chronic peptic ulcers in humans (Howie and Simpson, 1976). The sera of both conventional and specific-pathogen-free NZB mice contain increased concentrations of IgM. These increased levels develop at an early age before expression of autoimmune disease and do not correlate with age, sex, DAT positivity, or clinical status (Warner and Wistar, 1968; Kamm *et al.,* 1966). Considerable abnormalities occur in plasma proteins as the mice age and develop proteinuria (Mellors, 1965). Finally, NZB mice have been found to be particularly susceptible to bacterial infection (Kaye and Hook, 1964); this may be the cause of death in some animals. However, the majority die of their autoimmune processes, i.e., hemolytic anemia and renal disease. Lymphoid malignancy contributes to a minor proportion of the mortality.

2.2. (NZB $\times$ NZW)F_1 (B/W) Hybrids

B/W mice develop a high titer of antinuclear antibodies associated with development of rapidly progressive and lethal glomerulonephritis (Lambert and Dixon, 1970; Dubois *et al.,* 1966; Dubois and Arterberry, 1966; Mellors,1966a; Burnet and Holmes, 1965a; Helyer and Howie, 1963b). Virtually 100% of B/W mice develop this syndrome, although there are significant sex differences in the disease pattern. The mean survival time for males is 438.8 days and for virgin and breeder females 279.8 days (Howie and Helyer, 1968). Hemolytic anemia is much less significant in B/W than in NZB mice and anti-nuclear antibodies rather than antierythrocyte antibodies are of prime pathological importance. Antinuclear antibody can first be detected at 2 months in females and reaches 100% prevalence by 12 months of age (McGiven and Ghose, 1968). In contrast, such antibodies appear at 4 months of age in male B/W mice and are present in all animals by 16 months (Howie and Helyer, 1968). The antinuclear antibodies are heterogeneous with respect to specificity and include reactivity to single- and double-stranded DNA and RNA, nucleic acid–protein complexes, nuclear proteins, and native transfer RNA (Talal and Steinberg, 1974). Additionally, viral antigens and antibody have been demonstrated in the kidneys of NZ mice. However, anti-double-stranded DNA appears to be the antibody with nephritogenic properties and it most closely correlates with renal disease (Steinberg *et al.,* 1971; Lambert and Dixon, 1968). The renal changes have the morphological features of an immune-complex type of nephritis and are associated with the deposition of IgG in irregular lumpy masses in the mesangium and along the glomerular capillary basement membrane. This process occurs between 3 and 6 months of age (Lambert and Dixon, 1968). The anti-DNA antibody has no specific activity for glomerular basement membrane, is predominantly of the IgG2a subclass, and is associated with complement. DNA is present in association with the immunoglobulin and anti-DNA can be eluted from the glomerular lesions (Lambert and Dixon,1968; McGiven and Ironside, 1968). The earliest change, as seen by light microscopy, is a glomerulitis appearing in females between 5 and 6 months and characterized by areas of endothelial and mesangial hypercellularity particularly in the peripheral capillary loops. After an interval of 1–2 months the glomerulonephritis develops with focal and irregular hyaline thickening and deformity of the capillary basement membrane. The glomerulonephritis becomes more florid and clearly fibrinoid and necrotic over the next 2 months (Howie and Helyer, 1968). The incidence of proteinuria increases after 6–7 months of age and death from renal failure generally follows the onset of proteinuria by 3–4 weeks (Lambert and Dixon, 1968).

Positive DAT occurs in less than 50% of females and in an even smaller proportion of males. Anemia parallels the development and degree of renal insufficiency and is uremic in type with the presence of burr cells and a mild reticulocytosis (Howie and Helyer, 1968). Splenomegaly and generalized lymph node enlargement are uncommon and lymphoid neoplasia occurs with a lower frequency in B/W than in NZB mice. Malignancy does occur, especially thymoma, with a frequency of between 1 and 5% (Talal and Steinberg, 1974). Conspicuous medullary lesions of the thymus comparable to those seen in NZB mice are present in B/W mice. In addition, hyperplasia of the large epithelial cells of the thymic medulla, increased numbers of Hassall's corpuscles, and invasion of these structures by lymphoid cells are seen in the neonatal period (DeVries and Hijmans, 1967).

There are numerous pathologic alterations of parenchymal organs in B/W mice. Female B/W mice develop focal necrosis of the liver parenchyma, hyperplasia of Kupffer cells, and aggregation of plasma cells around bile ducts in the portal tracts (Chou and Herdson, 1969). Both NZB and B/W mice develop lesions of their lacrimal and salivary glands characterized by mononuclear infiltration around intralobular ducts and small vessels. In fact, B/W female mice provide an excellent model of Sjögrens syndrome (Kessler, 1968). NZB and B/W mice both spontaneously develop carditis characterized by an inflammatory process including pancarditis, valvulitis, and arteritis. Neither antibody nor complement has been demonstrated in any of the cardiac lesions described (Pansky and Freimer, 1974). The possibility that this carditis is a result of cell-mediated injury has been strengthened by recent observations that syngeneic heart and kidney (but not liver cells) induce stimulation of DNA synthesis in old and young B/W spleen cells (Senitzer *et al.*, 1979). Aging B/W mice also develop irregular immunoglobulin deposits along the dermal epidermal junction similar in appearance to those seen in the skin of SLE patients. These deposits apparently represent nDNA–anti-DNA complexes and it has been suggested that the nDNA released locally from epidermal nuclear breakdown precipitates with serum anti-DNA (Sontheimer and Gilliam, 1979; Pertschuk *et al.*, 1976; Gilliam, 1975). There has been at least one report of immune-complex deposition in the choroid plexus of B/W mice (Lambert and Oldstone, 1973). This may be significant in view of the recent observation of a naturally occurring antibody in NZ mice specific for autologous cerebellar cells

TABLE I
Natural History and Histopathology of New Zealand Mice

Parameter	NZB	NZB/NZW
Primary disease	Autoimmune hemolytic anemia	Immune-complex glomerulonephritis
Pathological autoantibody	Antierythrocyte	Antinuclear
Lymphoid infiltration	Intense, all organs	Fewer organs involved
Lymphoid malignancy	10–20%	1–5%
Sex predominance	Slight, female	Extreme, female
Mean survival time (days)	Female: 430	Female: 280
	Male: 467	Male: 438
Cause of death	1. Hemolytic anemia	Uremia
	2. Renal disease	
	3. Malignancy	

(Harbeck *et al.,* 1978; Hoffman *et al.,* 1978). Table I represents a comparison between the natural life history and histopathology of NZB and B/W mice.

3. Autoantibody Production

NZB mice and their B/W F_1 hybrids spontaneously produce numerous autoantibodies with varying specificities and pathogenetic significance. Antibodies with specificity for erythrocytes, nucleic acid, thymocytes, brain cells, liver cells, and fibroblasts will be discussed.

3.1. Antierythrocyte Autoantibodies

The antierythrocyte responses of NZB mice have been well characterized by DeHeer and Edgington and co-workers. There are at least four independent and genetically unrelated antierythrocyte responses expressed by NZB mice, anti-X, anti-HB, anti-HOL, and anti-I (Linder and Edgington, 1972a; Costea *et al.,* 1970; Holborow *et al.,* 1965). The antibody of primary pathogenetic significance is an IgG immunoglobin with anti-X specificity. The autoantigen is exposed on all murine erythrocytes (Linder and Edgington, 1972a) and is detected by the direct antiglobulin test. The anti-X antibody displays heterogeneity with respect to subclass distribution and binding affinities. In these respects the response has features of a normal polyclonal immune response and does not appear to result from an expansion of a single autonomous clone of cells (DeHeer and Edgington, 1974). Alternatively, Warner (1974) has suggested that the response is monoclonal or at least of restricted heterogeneity based on immunoglobulin allotype studies.

The other autoantibodies are directed toward cryptic erythrocyte antigens not exposed *in vivo.* The anti-HB, IgM response is directed to the cryptic HB autoantigen, which can be exposed *in vitro* by bromelin treatment. Exposure of this autoantigen may occur *in vivo* following the interaction of anti-X with intact erythrocytes either through complement activation or reticuloendothelial sequestration (Linder and Edgington, 1972a; Abramson *et al.*, 1970). The HOL autoantigen is exposed in dried red cell smears and on erythrocyte stromata. Anti-HOL antibody appears to react with fibril-forming antigen located at the inner aspect of the erythrocyte membrane (Linder, 1977). The anti-I response is an IgM autoantibody with specificity for the I antigen on human erythrocytes and it can be detected on murine stromata. Anti-HB and anti-I occur in many mouse strains, anti-HOL in NZB and NZB hybrids predominantly, and anti-X in NZB and NZB hybrids exclusively (DeHeer and Edgington, 1976a). Soluble analogs of erythrocyte-bound X and HB antigens exist in mouse plasma (Linder and Edgington, 1973). The possibility that these substances may generate autologous immune complexes that contribute to the glomerulonephritis in these animals has been suggested (Linder and Edgington, 1972b).

The cellular correlates of anti-X and anti-HB antibody production have also been examined (DeHeer and Edgington, 1976b). Antigen-binding lymphocytes (ABL) of T and B classification were reactive with the HB antigen in all mouse strains, although elevated in NZB mice. Elevated ABL and plaque-forming cells (PFC) specific for the HB antigen have also been observed in nonautoimmune strains when an immune hemolytic anemia is induced by rat erythrocyte immunization (Milich and Gershwin, 1979b). These data suggest the absence of tolerance to the HB cryptic antigen in all mouse strains. ABL of the B-lymphocyte class with receptors for the pathogenetic X autoantigen were also observed in all mouse strains; however, ABL of the T-lymphocyte class were present only in NZB and NZB hybrid mice. These results suggest that B cells in all mouse strains are immunocompetent to bind the X autoantigen and that NZB mice lack tolerance to the X autoantigen at the T-cell level. Aberrant maturation of the anti-X and anti-HB responses in NZB mice has also been described (DeHeer and Edgington, 1975).

3.2. Antinuclear Autoantibodies

Antibodies to DNA and RNA develop progressively in NZ mice and occur with greatest frequency in B/W mice (Schur *et al.*, 1971; Talal, 1970; Steinberg *et al.*, 1969). High-titer antibody to double-stranded DNA is found virtually only in NZ mice and in patients with SLE or SLE-like syndromes. Naturally occurring antinuclear antibodies with specificity to single- and double-stranded DNA and RNA as well as antibodies to nucleic acid–protein complexes, nuclear proteins (Talal and Steinberg, 1974), and native transfer RNA (Eilat *et al.*, 1977) have been detected in B/W mice. However, immunofluorescent and elution studies have indicated that DNA, as well as its antibody, is concentrated in renal deposits, and makes up 50% of the eluted antibody, thus implicating this particular immune complex in the pathogenesis of nephritis (Lambert and Dixon, 1968). As double-stranded RNA (dsRNA) is relatively rare in mammalian tissues, unless infected with virus, naturally occurring antibodies to RNA may have arisen because of immunization to viral nucleic acids. This view is supported by the fact that these antibodies have greatest specificity for viral dsRNA, next for synthetic dsRNA [poly(I·C)] and least for mammalian ribosomal or transfer RNA (Talal, 1970).

Antibodies to nucleic acids have been induced in experimental animals by a variety of immunization procedures. Antibody to native and synthetic nucleic acid can be produced by coupling them to an immunogenic protein carrier [i.e., methylated BSA (mBSA)].

Antibodies to dsRNA and ssDNA have been elicited in this manner; however, antibody to native DNA has not (Stollar, 1970). It was originally assumed that nucleic acids were haptens and required attachment to protein carriers for expression of immunogenicity (Plescia and Braun, 1968; Yachnin, 1962). In studies conducted in NZ mice given multiple injections of poly(I·C), it was observed that they exhibited an increased production of antibodies to RNA (Steinberg *et al.*, 1969). Subsequently, studies in nonautoimmune mice confirmed that poly(I·C) is immunogenic without a protein carrier (Steinberg *et al.*, 1971, 1970). Thymus-deprived animals are capable of responding to poly(I·C) in aqueous solution, and antithymocyte sera plus poly(I·C) leads to acceleration and not depression of antibody formation (Chused *et al.*, 1973). These studies suggest that the immune response to nucleic acids is relatively T independent. The predominant anti-dsDNA antibody in young B/W mice is of the IgM class, but as the animal ages, there is an increase in the level of the IgG2a and IgG2b subclasses. This IgM–IgG switch is thymic dependent, occurs in females before males, and coincides with the age of onset of immune-complex disease (Steward and Hay, 1976). Immunization of young B/W mice with denatured DNA-mBSA or poly(I·C) leads to an accelerated production of antinuclear antibodies and a corresponding acceleration in the onset of renal disease (Carpenter *et al.*, 1970; Lambert and Dixon, 1968). Both B/W and NZW mice respond quite well to immunization with denatured DNA-mBSA, whereas the NZB parent has a low response. It has been proposed that the NZW genotype contributes to the F_1 and ability for high responsiveness to DNA, whereas the NZB contributes features that result in autoimmunity (Lambert and Dixon, 1970). There is a striking difference in disease activity between male and female B/W mice, the female mice being affected more severely. Castration and sex hormone modification studies have demonstrated that androgen has a suppressive effect on the production of antinuclear antibodies and mortality (Roubinian *et al.*, 1977a) and that it delays the switch from IgM to IgG antibodies to both DNA and RNA so that the immune response of castrated males resembles the response of females (Papoian *et al.*, 1977).

3.3. Natural Thymocytotoxic Autoantibody

New Zealand mice spontaneously develop an antibody with specificity for T cells (Shirai and Mellors, 1971). This naturally occurring thymocytotoxic autoantibody (NTA) was detected by a cytotoxicity test of sera from young NZB mice; it was first detectable by 1 month of age and reached 100% incidence by 3 months of age. Some B/W mice are NTA positive by 2 months and the incidence reaches 50% by 3–6 months of age (Shirai *et al.*, 1973; Shirai and Mellors, 1972). However, (NZB × C57BL/6)F_1 mice with much milder autoimmune disease than B/W mice have an equivalent prevalence of NTA. Similarly, nonautoimmune mice demonstrate NTA, but at an older age and with lower incidence and titer than NZB mice. Gel filtration, mercaptoethanol sensitivity, and thermal characteristics have demonstrated NTA to be of the IgM class. Further, NTA is cytotoxic for 98% of thymocytes, 62% of thoracic duct cells, 64% of peripheral blood leukocytes, 55% of lymph node cells, 33% of spleen cells, 3% of bone marrow cells, and less than 1% of fetal liver cells. These findings suggested that the antigenic specificity of NTA is very similar to that of the antitheta antibodies obtained by immunization (Raff and Cantor, 1971). Like antitheta, NTA was absorbed by mouse brain, but differed in that it was cytotoxic for thymocytes bearing either the Thy-1.1 (AKR) or the Thy-1.2 (C3H) allotypic form of the theta antigen. It was concluded that NTA reacted with a single antigen common to mouse thymocytes that was different from the theta antigen (Shirai and Mellors, 1972). Alternatively, based on co-migration patterns, it has been suggested that NTA reacts with a portion of the theta antigen that is common to both allotypes (Parker *et al.*, 1974).

NTA has been implicated as one of the causative factors for the age-associated decline in suppressor T-cell function (Krakauer *et al.*, 1976; Barthold *et al.*, 1974; Allison *et al.*, 1971) and helper T-cell function (Gelfand *et al.*, 1974; Shirai *et al.*, 1972) in NZ mice. Recently, experiments were designed to ascertain the effect of NTA on suppressor and helper T cells in an *in vitro* secondary antibody response in NZB mice. It was found that suppressor T cells were more sensitive to NTA than helper T cells (Shirai *et al.*, 1978). Similarly, neonatal B/W mice injected with NTA had a significant increase in ability to induce a graft-versus-host reaction, an inability to generate suppressor cells after concanavalin A treatment, and an increased or enhanced ability to either reject skin allografts or respond to SRBC immunization. Such data likewise suggest that NTA causes a selective loss of suppressor cells in B/W mice (Klassen *et al.*, 1977a).

3.4. Other Autoantibodies

An autoantibody present in the sera of diseased NZB mice has been demonstrated to react with autochthonous and syngeneic fibroblasts, but not with either H-2-identical or nonidentical allogeneic cells (Liburd *et al.*, 1974). A spontaneously occurring antibody cytotoxic to dissociated cells of the neonatal mouse cerebellum was reported in the sera of NZB and B/W mice. Moreover, no such activity was found in the sera of six nonautoimmune murine strains (Harbeck *et al.*, 1978). The degree of cytotoxicity of this antibody was similar against syngeneic and allogeneic target cells and was absorbed with brain and kidney homogenates. Liver absorbed less reactivity, and absorption with thymocytes was ineffective in reducing antibody activity. Finally, this anticerebellum antibody was of the IgM class as determined by gel filtration and was not found in cerebrospinal fluid. NZB mice have also been reported to produce an IgM antiliver autoantibody. The appearance of this autoantibody increases with age but the significance is unclear (Elson and Naysmith, 1969; Flad *et al.*, 1969). Table II provides a summary of the autoantibodies found in NZ mice.

4. Genetic Factors

The genetic analysis of the autoimmune disease of NZ mice is accompanied by numerous problems. First, the pathogenesis of autoimmune disease is clearly a complex process influenced by hormonal as well as viral factors. Second, the phenotypic markers primarily used in genetic studies, DAT-positive hemolytic anemia, antinuclear antibodies with immune-complex nephritis, and NTA positivity, are not subject to absolute analysis. These phenotypic expressions may vary according to the methodologies used for detection, the age at which they are determined, associated environmental factors (i.e., viral infec-

TABLE II
Autoantibodies in New Zealand Mice

Specificity	Incidence NZB	NZB/NZW	Nonautoimmune	Class	Properties
Murine erythrocytes					
X	++	+	0	IgG	Panagglutinin, exposed antigen
HB	+	+	+	IgM	Cryptic antigen
HOL	++	+	±	ND[a]	Cryptic antigen
I	+	+	+	IgM	Cryptic antigen
Nucleic acids	+	++	0	IgM (initially), IgG switch	Female predominance
Thymocytes (NTA)[b]	++	+	±	IgM	Possibly selective for Ts[c]
Cerebellar cells	+	+	0	IgM	Cytotoxic for syngeneic, allogeneic targets
Fibroblasts	+	+	0	ND	Cytotoxic for syngeneic targets
Liver cells	+	ND	0	IgM	

[a]ND, not determined.
[b]NTA, natural thymocytotoxic autoantibodies.
[c]Ts, T suppressor.

tions), the influence of unrelated pathological processes of a nonautoimmune parental strain, and varying degrees of suppression or exaggeration in hybrids according to characterisitcs of the nonautoimmune parent. In earlier genetic studies, these complexities were not fully appreciated; however, in recent studies they have been considered.

Large numbers of backcross mating studies have been performed in NZB mice. The hemolytic anemia observed in homozygous NZB mice was modified to glomerulonephritis in (NZB × NZY)F_1 hybrids (Helyer and Howie, 1961) and (NZB × NZW)F_1 hybrids (Helyer and Howie, 1963b). Hybridization with nonautoimmune strains and backcross to the parental strain led to animals with phenotypic expressions of autoimmunity characterized by variation in incidence of antierythrocyte autoantibodies, variation in age of onset, and variation in survival time (Burnet and Holmes, 1965b; Howie and Helyer, 1965; Holmes and Burnet, 1964). These studies suggested that several genes were involved and could be accounted for by, for example, a three- to five-gene complex in NZB mice (Burnet and Holmes, 1965b). In contrast, Bielschowsky and Bielschowsky (1964) observed a 100% incidence of hemolytic anemia in F_1 mice derived from NZB × NZC matings and a 74% incidence in F_2 mice and suggested that a single dominant gene was operative. Nevertheless, studies of the NZB × NZC hybrid led Warner (1973) to postulate a domnant and a recessive gene for antierythrocyte antibody production in NZB mice, with only the recessive gene present in NZC mice. From the analysis of F_1, F_2, and backcross NZB × NZW mice, Braverman (1968a) concluded that there is a dominant gene in NZB mice that determines DAT positivity and an additional modifying gene that allows the expression of antinuclear antibody. The examination of (NZB × NZW) and (NZB × BALB/c)F_1, F_2, and backcross progeny by Ghaffar and Playfair (1971) suggested one or possibly two recessive genes underlying antierythrocyte autoantibody production. The autoimmune disorder is not sex linked based on similar disease expression in NZB × NZW and NZW × NZB mice (Ghaffar and Playfair, 1971; Burnet and Holmes, 1965b). In other studies, it was observed that transplantation of ova from NZB into CFW mice did not alter the disease process compared to naturally derived NZB. Similarly, allophenic (tetraparental) NZB–CFW chimeras behaved as naturally derived F_1 hybrids with respect to autoimmune expression (Barnes *et al.*,1972; Barnes and Tuffrey, 1969). It was concluded that the disorder was not determined by passage of maternal factors either *in utero* or via colostrum and that the stimulus to develop disease is present prior to ovum implantation, thus supporting a genetic basis for the autoantibody response.

More recently, studies of (NZB × DBA/2)F_1 reciprocal hybrids, and backcross mice indicated that the ability to respond to ssDNA-mBSA is inherited as an autosomal dominant trait regulated by sex hormones (Raveche *et al.*, 1978) and that it is not sex linked as previously suggested (Moses and Fuchs, 1974). Similar sex hormonal regulation for the sponataneous production of antinuclear antibodies in B/W mice has been reported (Papoian *et al.*, 1977). In addition, the production of NTA by NZB hybrid mice has been shown to be regulated by sex hormones, with androgens producing a suppressive effect. Female (NZB × DBA)F_1 hybrids produced NTA, male (NZB × DBA)F_1 hybrids did not; however, castration abolished this difference (Raveche *et al.*, 1976). These studies indicate that no meaningful genetic analysis can be made of antinuclear or NTA production unless only female or castrated offspring are studied. To determine the mode of inheritance of NTA and anti-ssDNA in NZB mice, Raveche and co-workers (1978) used castrated progeny. The results indicated that NTA production was controlled by a single codominant gene and that an independent dominant gene controlled the production of autoantibodies to ssDNA. The quantity of each autoantibody was determined either by gene dosage or by an independent regulatory gene (probable in anti-ssDNA production). While the genetic control of levels of ssDNA and NTA antibodies appeared to be distinct, the level of anti-ssDNA and the quantity of serum IgM in individual mice were found to be correlated. This suggests that the gene that regulates anti-ssDNA levels may be the same as or linked to the gene controlling the quantity of serum IgM. Additionally, the locus controlling production of either autoantibody appeared not to be *H-2* linked. Testing the possibility that production of these autoantibodies is linked to immunoglobulin heavy-chain allotypes awaits the development of congenic lines differing only at these loci.

In a study of the occurrence of lupus nephritis in female (B/W)F_1 mice, it was concluded that at least two dominant or codominant genes were involved. A 50% incidence of nephritis in the (NZB × NZW)F_1 × NZW backcross mice indicated that the NZB strain contributed only one gene (or cluster of closely linked genes), which must be dominant or codominant in that the effect was expressed in the heterozygous state (Knight *et al.*, 1977). The NZB-contributed gene(s) was loosely linked to the D end of the *H-2* complex, with a crossover frequency of $32.6 \pm 3.1\%$. Because the NZB gene(s) contributing to the lupus nephritis was not closely linked to the *H-2* complex, it was concluded that it cannot be a histocompatibility-linked *Ir* gene and it was suggested that an immunoglobulin V gene(s) was the most likely candidate.

Recent analysis of the hemolytic patterns in the F_1, F_2, and backcross NZB × NZC mice confirmed earlier results of Bielschowsky and Bielschowsky (1964) and indicated that a single dominant gene waas involved in the development of DAT positivity in this hybrid (Warner, 1978). However, the incidence of IgG2 immunoglobulin subclass autoantibodies was found to be considerably lower in the (NZB × NZC)F_1 than in the inbred NZB. This suggested a gene-dosage phenomenon or the presence of an operative modifying locus. In addition, (NZB × NZW)F_1 hybrid autoantibody expression is less pronounced than in the NZB × NZC hybrid and in nonautoimmune (NZB)F_1 hybrids the autoimmune trait is least expressed. This indicated that several genetic loci were involved and that there was something different about the NZC genetic contribution. Warner (1978) has developed a congenic mouse in which the chocolate coat color was introduced on an NZB background (NZB.ch). The NZB.ch had an incidence of hemolytic anemia of 38% instead of the expected 90% incidence observed in NZB or (NZB × NZC)F_1 hybrids. The introduction of genes determining coat color or closely linked genes suppressed the expression of the dominant NZB gene. This is compatible with the view that few genes are essential for the expression of autoimmune hemolytic anemia. Warner has suggested an interesting hypothesis of three different loci involved in autoimmunity of the NZ mouse (Table III). It is proposed that at least one dominant gene (A) is involved in autoimmunity of a generalized nature, because this influences both the lupus syndrome and hemolytic anemia. The effect of this locus could be modified by another locus (M); this modifying locus may control various hormone or steroid levels. The wild-type M gene would depress autoimmune expression, the NZB allele (m) would be inactive, and the NZC (m^1) allele would be active in depressing autoimmune expression only in the homozygous state as in the NZB.ch mouse. The G gene represents a specific *Ir* gene for the ANA response. In support of the concept of a G gene, NZB mice are low responders and NZW and (NZB × NZW)F_1 mice are high responders to immunization with ssDNA-mBSA (Lambert and Dixon, 1970). Moreover, similar results were obtained in a T-cell proliferative response to ssDNA-mBSA (Papoian and Talal, 1980). It was shown that B/W mice recognize and respond to ssDNA-mBSA presented by macrophages of B/W or NZW but not of NZB haplotypes. It has also been suggested that only one gene involved in lupus nephritis is *H-*

TABLE III
Proposed Gene Loci Involved in NZ Autoimmunity[a]

Strain	Autoimmunity (generalized)	Predisposition to antinuclear antibodies	Modifying locus
NZB	*AA*	*gg*	*mm*
NZC	*aa*	*gg*	*m′m′*
NZW	*aa*	*GG*	*MM*
Normal	*aa*	*gg*	*MM*
NZB × NZC	*Aa*	*gg*	*mm′*
NZB × NZW	*Aa*	*Gg*	*mM*
NZB × normal	*Aa*	*gg*	*mM*
NZB.ch	*AA*	*gg*	*m′m′*

[a]From Warner (1978).

2 linked, and this gene is contributed by the NZW parent (Papoian and Talal, 1980; Knight *et al.,* 1977). While clearly a complex area, this model is not inconsistent with current observations. The ultimate objective, of course, will be to determine the cellular level at which "autoimmune" genes are expressed. Recently, several "primary" B-cell and T-cell defects in NZ mice have been proposed (discussed in a subsequent section). Analysis of these cellular defects in NZC and NZB.ch mice and genetic analysis in NZB hybrid mice would be critical because the cellular defects may represent the expression of the genetic loci described by Warner's model.

Numerous chromosomal aberrations have been observed in NZB mice. Hyperdiploidy was reported in the spleens of a majority of NZB but not in control strain DBA/2 spleens. Hyperdiploidy appeared to be inherited as a recessive trait and was not related to autoimmune immunological abnormalities (Raveche *et al.,* 1979). However, it has been suggested that aneuploid clones in the spleens of old NZB mice are at least potentially neoplastic (Friedman *et al.,* 1978). Additionally, a considerable increase in chromatid and chromosome breaks has been observed in embryonic fibroblast cultures from NZB mice. It is hypothesized that the excessive chromosome damage is related to the consistent production of xenotropic virus (Emerit *et al.,* 1979).

The influence of viral factors on the autoimmune disease of NZB has been studied by genetic analysis and will be discussed in the following section.

5. Viral Factors

In the mid-1960s, several laboratories reported detecting C-type particles budding from cell surfaces of NZB organs by electron microscopy (East *et al.,* 1967b; Yumoto and Dmochowski, 1967; Mellors and Huang, 1966). Murine C-type viruses contain an RNA core, an internal nucleoprotein that makes up the group-specific (gs) antigen(s), and an outer envelope coat. The gs antigen is specific for each animal species whereas the type-specific envelope antigen identifies a particular C-type virus within the species (Levy *et al.,* 1975). Three murine leukemia viruses (MLV) with different type-specific coats have been reported in NZB mice.

Mellors and Huang (1967) reported the successful transfer of specific characteristics of NZB disease to random-bred Swiss mice with cell-free extracts and filtrates of NZB spleens. However, these results are unconfirmed (Howie and Helyer, 1968; East *et al.,* 1967a,b; Holmes *et al.,* 1961). Although the virus was not isolated from their colony, they concluded that transmission was due to a Gross-type MLV based on the presence of Gross soluble antigen (GSA) and antibody in the circulation. It was suggested that these GSA–antibody complexes lead to nephritis. However, similar glomerular lesions are found in AKR mice that harbor a Gross MLV and in mice infected with Gross, Friend, and Moloney viruses (Levy, 1975). The severe autoimmune disease of NZB but its absence in the other infected strains has not been explained. In addition, it has been demonstrated that the majority of immune complexes in the kidneys of NZB mice are of the DNA–anti-DNA variety and not MLV (Dixon *et al.,* 1971; Lambert and Dixon, 1968). A Moloney-type MLV was recovered from established lines of lymphocytes from NZB and NZW mice (Lerner *et al.,* 1972). The virus cannot be demonstrated from fresh extracts and its isolation has not been reported by other laboratories (Levy, 1974). The third MLV reported in NZB mice was a xenotropic type C RNA virus that had neither Gross nor Moloney type-specific antigens and could not be propagated in mouse cells (Levy and Pincus, 1970). That this virus can be recovered from mice but cannot be propagated in mouse cells indicates that it is a true endogenous virus, as horizontal infection in the host cannot occur.

It has subsequently been demonstrated that NZB mice express xenotropic type C RNA virus throughout life and in high titer (Datta and Schwartz, 1977, 1976; Levy *et al.,* 1975). The correlation between high-titer xenotropic virus production and autoimmune disease has suggested a cause and effect relationship. This concept was strengthened when viral–antiviral immune complexes were recovered from the renal lesions in NZB and B/W mice (Yoshiki *et al.,* 1974; Mellors *et al.,* 1971, 1969) and by the suggestion that xenotropic virus caused disturbances of immunoregulation in NZB mice (Levy,1976). The lack of transmission of autoimmune disease by cell-free filtrates was explained by the fact that xenotropic viruses cannot productively infect mouse cells, only cells of heterologous species (Levy *et al.,* 1975). Recently, it has been demonstrated that high expression of xenotropic virus in NZB is a genetically determined trait controlled by two independently segregating, autosomal dominant loci (*Nzv-1* and *Nzv-2*) (Datta and Schwartz, 1977, 1976). This has facilitated genetic analysis of the relationship between xenotropic viruses and autoimmunity. The hybrid chosen to study this relationship was NZB × SWR because NZB are homozygous for the dominant alleles of viral expression and SWR are homozygous for the recessive alleles; also SWR do not spontaneously develop autoimmune disease (Datta and Schwartz, 1978). Analysis of the F_1, F_2, and backcross progeny of NZB × SWR mice demonstrated the following: (1) some progeny that expressed titers of virus as high as the NZB parent failed to develop signs of autoimmunity; (2) virus-negative mice found in crosses could still develop autoantibodies; and (3) there was no correlation between phenotypic expression of virus and the incidence of glomerular lesions. These studies indicated that although xenotropic virus and viral antigens may participate secondarily in the formation of immunopathogenic lesions in the NZB mouse and its crosses, they are not a primary requirement (Datta and Schwartz, 1978). The expression of xenotropic viral genes occurs in many cells undergoing differentiation and proliferation and the investigators suggested that the continuous expression of these genes in NZB mice may be a function of abnormal regulatory genes. Therefore, the immunologic and virologic abnormalities may be separate manifestations of a basic disturbance in gene regulation.

The murine RNA viral envelope glycoproteins (gp70) have also been implicated in the pathogenesis of lupus nephritis in NZ mice because of the high concentrations of gp70 in the sera of these mice (Lerner *et al.,* 1976; Strand and August, 1976; Yoshiki *et al.,* 1974) and because gp70 is deposited in diseased glomeruli along with host immunoglobulin, complement, and nuclear antigens and antibody (Croker *et al.,* 1975; Lambert and Dixon, 1968). Serum levels of antibodies against three antigenically different gp70 were measured in NZB, B/W, and other SLE-prone mice. It was concluded that anti-AKR or anti-Rauscher MLV gp70 antibodies are not involved in formation of gp70 immune complexes in these mice but antibodies to xenotropic gp70 are (Izui *et al.,* 1979). In another study, the presence of xenotropic gp70 and its relation to glomerulonephritis was analyzed. It was concluded that high serum levels of gp70 are not in themselves a cause of murine nephritis, and it is an abnormal immune response of the host to gp70 which is critical rather than the level or type of gp70 in the circulation (Andrews *et al.,* 1978). The gp70 of the NZB xenotropic virus exists as a normal serum protein in virtually all mice regardless of virologic status (Elder *et al.,* 1977). Thus the xenotropic virus of NZB mice may be considered an "autovirus" and gp70 a potential autoantigen. These antigens may play a secondary role in the autoimmune disease of NZB by supplying autoantigens in the same way as do erythrocyte autoantigens or DNA (Datta and Schwartz, 1978).

Viruses may have a secondary role in the autoimmune disease of NZB mice in another manner. Viral infections such as lymphocytic choriomeningitis and polyoma may cause an early induction and may enhance the titer of ANA, thus accelerating immune-complex glomerulonephritis. They may also cause glomerulonephritis without ANA production in nonautoimmune strains (Tonietti *et al.,* 1970). In contrast, infection with lactic dehydrogenase virus causes a reduction in ANA, diminished glomerulonephritis, and prolongation of survival in NZ mice (Dixon *et al.,* 1971, 1970). It should also be emphasized that these viral infections have no influence on antierythrocyte autoantibodies or on the severity of hemolytic anemia.

6. Immunologic Manipulations

6.1. Thymectomy

In both NZB and B/W mice there are major abnormalities of the thymus, including degeneration of the thymic epithelial cells during the neonatal period (Gershwin *et al.,* 1978; De Vries and Hijmans, 1967), and subsequent development of lymphocytic follicles within the thymic medullae (Burnet and Holmes, 1964; Burnet, 1962). These observations suggested that the thymus may be involved in the development of disease in these animals. Neonatal thymectomy of both NZB (East *et al.,* 1967a,b) and B/W mice (Steinberg *et al.,* 1970; Howie and Helyer, 1966) has been reported to accelerate the onset of disease. Alternatively, thymectomy of NZB mice 2–6 days after birth was observed to delay the onset of DAT positivity (Holmes and Burnet, 1966). However, thymectomy in NZB mice is associated with a very high incidence of early death from "wasting disease" and liver necrosis and therefore makes analysis of these animals difficult. Howie and Helyer (1966) did not observe DAT positivity in neonatally thymectomized CBA/T6 mice, although occasional DAT-positive mice have been detected in thymectomized A/J (Yunis *et al.,* 1967) and C57BL/6 mice (Teague *et al.,* 1970). In (NZB × NZC)F_1 hybrid mice neonatal thymectomy was reported not to accelerate onset of disease (Miyasato *et al.,* 1967). In general such observations suggest an acceleration of disease after neonatal thymectomy; however, possible differences related to sex were not considered. Recently, however, it was demonstrated that neonatal thymectomy of B/W mice caused a premature switch from IgM to IgG autoantibody to DNA and caused earlier death in male mice. In contrast, female mice lived significantly longer after thymectomy (Roubinian *et al.,* 1977b). These results imply that thymic regulatory influence in the male exerted a suppressive effect on disease expression, whereas the thymus appeared to exert a deleterious effect on disease expression in the female B/W mouse. The effect of thymectomy on tolerance induction has also been examined (Steinberg *et al.,* 1970). It was concluded that the thymus was not necessary for the induction of tolerance to bovine gamma globulin (BGG) or poly(I·C). However, thymus from young syngeneic mice had a positive effect on both tolerance to BGG and spontaneous anti-DNA antibody formation. Because neonatal thymectomy does not eliminate the influence of the thymus during gestation (Milich and Gershwin, 1977), another approach to analyzing thymic influence in NZB mice has been the development of congenitally athymic, nude NZB mice (Gershwin *et al.*, 1979b). NZB nude (*nu/nu*) mice survive less than 3 weeks and dic of a runting discasc much likc neonatally thymectomized NZB mice. However, they were shown to lack progenitor T-cell populations present in BALB/c nude mice, to have an earlier appearance of NTA than their nu/+ littermates, and to exhibit polyclonal B-cell activation independent of thymic status. These findings suggested both a T-cell deficiency due to an absence of T-precursor cells as well as an independent B-cell defect. Finally, these observations also suggest that the polyclonally activated B cells in NZB mice are not prone to either suppressor- or helper-induced T-cell regulation, whereas NTA production is subject to thymic influence.

6.2. Splenectomy

In adult NZB mice with established, florid disease, splenectomy results in a rapidly progressive anemia and early death probably because the spleen is a critical site of extramedullary hematopoiesis. In contrast, in younger animals with hemolytic disease, splenectomy has no significant effect (Helyer and Howie, 1963b). Splenectomy in animals under 4 months of age lessens the severity of the hemolytic disease but exaggerates the frequency of renal disease (Holmes and Burnet, 1963b). Splenectomy in animals under 4 months of age lessens the severity of the hemolytic disease but exaggerates the frequency of renal disease (Holmes and Burnet, 1963b). Splenectomy of B/W mice at 2 months of age had no influence on survival (Burnet and Holmes, 1965a). Neonatal splenectomy in B/W male mice resulted in earlier deaths but it did not affect the survival of B/W female mice (Roubinian *et al.*, 1977b). Neonatal splenectomy in NZB mice has not been reported. However, an NZB asplenic (*Dh*/+) mouse has been developed (Gershwin *et al.*, 1979a). It was observed that NZB *Dh*/+ mice develop autoimmune disease similar to that of their NZB +/+ littermates; however, the onset of NTA was delayed and the titer lower in asplenic NZB mice. Furthermore, NZB *Dh*/+ mice retained suppressor function, whereas NZB +/+ mice did not. Suppressor function was measured by the ability of thymocyte cell preparations to alter the reactivity of a graft-versus-host disease induced by 10-week-old NZB +/+ spleen cells in neonatal N:NIH(s) mice. Thus, autoimmune disease developed independently of NTA influence and without premature loss of suppressor cell function (Gershwin *et al.*, 1979a).

6.3. Transfer of Autoimmune Disease

Transfer of thymic grafts from NZB mice conferred a positive DAT on neonatally thymectomized control CBA/T6 mice (Helyer and Howie, 1963b), but not in (NZB/T6) or (NZB/C3H)F_1 hybrids unless they were irradiated (Holmes and Burnet, 1966). Immediate grafting of a normal thymus into NZ mice after neonatal thymectomy reestablished the usual pattern of disease in both NZB and B/W mice (Howie and and Helyer,

1966). Transfer of 2-week syngeneic thymus to neonatally thymectomized, B/W mice prevented the accelerated anti-DNA formation induced by thymectomy, but 10-week thymus was not effective (Steinberg *et al.*, 1970). Following the implantation of Millipore diffusion chambers containing NZB thymus, positive DAT reactions were reported in four different strains of nonautoimmune mice (Masters and Spurling, 1967). However, this finding was not confirmed by other investigators and may have been due to cell leakage (Tuffrey *et al.*, 1973).

Transfer of spleen cells from NZB mice with established disease was found to be capable of inducing DAT positivity in young syngeneic receipients (Kaye and Hook, 1964; Holmes *et al.*, 1961). Transfer of disease to H-2-incompatible recipients has not been successful (Braverman and Slesinski, 1968). Denman *et al.* (1969) transferred splenic and bone marrow cells from old NZB mice to normal BALB/c recipients (H-2 compatible) and successful transfer of DAT positivity occurred only in anti-T-cell-treated recipients. The decreased incidence of hemolytic anemia in normal BALB/c recipients with intact thymuses suggested that normal mice may be able to suppress the derepression and differentiation of the precursors of antierythrocyte autoantibody-forming cells, a capacity that is partially abrogated following antilymphocyte globulin; however, problems in interpretation in view of minor histocompatibility differences must be considered (DeHeer and Edgington, 1976a). Morton and Siegel (1974) transferred fetal liver and bone marrow from 20-day-old NZB mice to lethally irradiated H-2-compatible DBA/2 recipients and observed DAT positivity within 3 months of transfer in the fetal liver recipients and much later in the bone marrow recipients. These results suggested a hemopoietic stem-cell defect. Similar studies have been performed recently in NZB (DeHeer and Edgington, 1977) and B/W (Akizuki *et al.*, 1978) mice. In the studies of NZB mice, only the lethally irradiated, bone-marrow-reconstituted recipients of NZB marrow generated antierythrocyte PFC. Control NZB mice reconstituted with non-NZB marrow did not develop antierythrocyte PFC. The recipient mice received weekly sublethal doses of anti-Thy-1, which suppressed lymphoid regeneration and the capacity for an SRBC antibody response but had no effect on the development of antierythrocyte PFC. Thus it appears that B cells or stem cells, independent from T-helper populations, carry the potential for autoantibody production. Evidence that the lack of T-suppressor cells was not responsible for the response was demonstrated by the reciprocal transfer; non-NZB B cells transferred into irradiated NZB recipients did not transfer antierythrocyte PFC. The findings in B/W mice were also consistent with a B-cell defect; however, the reciprocal transfers were not performed. Reversal of the NZB syndrome by the transfer of nonautoimmune, histocompatible bone marrow into lethally irradiated NZB mice was demonstrated by suppression of ANA production, suppression of endogenous spleen colony formation, and normalization of the response to *in vivo* SRBC immunization (Morton and Siegel, 1979).

7. Cell-Mediated Immune Responses and T-Cell Subsets

Historically, NZ mice have been characterized by an age-associated decline in cell-mediated immune (CMI) function. However, evidence is accumulating that the impairment of CMI function in aging NZ mice may not be as prevalent as previously assumed, implying a selective dysfunction in subsets of T lymphocytes. The following is a review of a variety of studies of cellular immunity in NZ mice including recent, interesting observations relating to functional defects in T-cells subpopulations.

Young NZ mice (1–4 months) demonstrate normal numbers of recirculating (Zatz *et al.*, 1971) and theta-bearing lymphocytes (Stobo *et al.*, 1972b); normal rejection of allogeneic skin grafts (Stutman, 1972); induction of graft-versus-host (GVH) disease by spleen cells (Gelfand and Steinberg, 1973; Stutman *et al.*, 1968); and normal proliferative response of thymus and spleen to phytohemagglutinin (PHA) (Leventhal and Talal, 1970). In fact, increased CMI activity in young NZ mice has been reported, including hyperactivity in a mixed lymphocyte proliferative reaction (MLR) to allogeneic cells (Palmer *et al.*, 1976); enhanced ability to regress tumors induced by murine sarcoma virus at a younger age than control strains (Gazdar *et al.*, 1971); acquisition of the capacity of thymocytes to respond to concanavalin A (Con A) at an early age (Stobo *et al.*, 1972a); increased augmentation of an antibody response to SRBC after *in vivo* sensitization (Playfair, 1971, 1968); and heightened cross-reaction in antibody responses to erythrocytes from various species (Playfair and Marshall-Clarke 1973). Alternatively, the MLR to allogeneic cells in young NZB mice can range from normal to markedly diminished according to Smith and Pasternak (1978).

There is significant evidence suggesting that aged NZ mice have marked deficiencies in CMI. These include: decreased mitogenic responses to PHA (Falkoff *et al.*, 1978; Leventhal and Talal, 1970), Con A (Stobo *et al.*, 1972a), and allogeneic cells (Falkoff *et al.*, 1978; Rodey *et al.*, 1971; Leventhal and Talal, 1970); impaired ability to induce GVH disease (Cantor *et al.*, 1970; Stutman *et al.*, 1968) and to reject skin allografts (Gelfand and Steinberg, 1973) and tumors (Gazdar *et al.*, 1971; Mellors and Huang, 1966); a decreased cytotoxic response to spleen cells after alloimmunization *in vivo* or *in vitro* (Fernandes *et al.*, 1976; Gershwin *et al.*, 1974a); and decreased reactivity to vaccinia virus scarification (Owen *et al.*, 1975). The observed impairment in CMI may be related to reduced numbers of circulating T lymphocytes. Quantitation of these cells in NZ mice by surface marker analysis has shown: a depletion in long-lived recirculating lymphocytes in the peripheral blood, spleen, and lymph nodes (Zatz *et al.*, 1971; Denman and Denman, 1970); a decreased frequency of theta-positive lymphocytes, first apparent at approximately 6 months of age (Stobo *et al.*, 1972a; Stutman, 1972; Waksman *et al.*, 1972); a decreased incidence of surface-immunoglobulin-positive lymphocytes (Stobo *et al.*, 1972b) and mouse B-lymphocyte antigen positive cells (Waksman *et al.*, 1972). These cell depletions correlated with an increase of "null" cells, bearing neither theta nor immunoglobulin markers on their surface. Bhoopalam *et al.* (1971) have reported a decrease in the number of lymphocytes possessing surface IgG and IgM in aging NZ mice. Recently, an age-associated decrease in the incidence of T cells and of IgG Fc-receptor-bearing cells in B/W mice was reported (Theofilopoulous *et al.*, 1979a).

The cause of the impairment in CMI function in aging NZ mice has been analyzed by studying the restoration in spleen cells from old NZ mice, of the ability to mount a GVH reaction (Cantor *et al.*, 1970) and allograft rejection (Gelfand and Steinberg, 1973) with small numbers of spleen cells from young syngeneic mice. Restoration of these functions demonstrated that the old NZ mice were deficient in a spleen-cell subpopulation. The cells present in young animals and lacking in older mice were found to be recirculating, thymus-derived, corticosteroid-sensitive cells. Young spleen cells were found to synergize with splenic T cells present in old mice. This finding illustrates that effector functions reflect the synergistic activities of at least two subclasses of T cells. This point has been further demonstrated by the dissection of another cell-mediated response in aging NZB mice, specifically, loss of the ability to develop a cytotoxic response after alloimmunization *in vitro* as studied by the independent analysis of T-helper and T-killer cells (Falkoff *et al.*, 1978). T-helper and T-killer cells are distinguishable with respect to surface antigens, mitogen responsiveness, recirculation pattern, and alloantigenic differences (Bach *et al.*, 1976; Cantor and Weissman, 1976;). Indeed, spleen cells from old NZB mice were deficient in developing a cytotoxic response during *in vitro* alloimmunization and this deficiency coincided with a diminution of T-helper cells. However, normal levels of T-killer cells were present in these old NZB mice, suggesting that this cytotoxic subclass of T cells is intact in aging NZB mice. Similarly, it was concluded in a GVH assay that donor spleen cells from NZB mice up to 60 weeks of age retained precursors of cytotoxic effectors capable of exerting GVH reactions in irradiated adult recipients. Weak GVH responses were obtained in unirradiated neonatal recipients and it was concluded that T-helper cells may be unnecessary in adult hosts (Eastcott *et al.*, 1979). This T-cell–T-cell synergism is not always appreciated in experimental systems that measure terminal effector functions. For example, many of the systems that measure suppressor cells in NZB mice do so without distinguishing between an increase in helper activity and a decrease in suppressor activity. This is relevant as increased helper T activity has been reported in 4- to 6-month-old NZB mice (Falkoff *et al.*, 1978; Palmer *et al.*, 1976).

The decreased CMI activity has been related to several factors including the production of NTA in older NZ mice (Parker *et al.*, 1974; Gelfand and Steinberg, 1973), abnormal differentiation of T-cell subpopulations, abnormal thymic environment (Gershwin *et al.*, 1978; Dauphinee *et al.*, 1975; De Vries and Hijmans, 1967), and the absence of thymic hormones (Bach *et al.*, 1973). The quantitative increase in circulating NTA titers with age is associated with an absolute decrease in the number of NTA and complement-vulnerable T cells found in lymph nodes, spleen, and blood of NZB mice (Shirai *et al.*, 1972). NTA also directly alters T-cell migration patterns (Shirai *et al.*, 1973) and cellular immune responses (Gelfand *et al.*, 1974). Similarly, NTA was shown to decrease the response of normal spleen cells to Con A as compared to PHA (Ochiai *et al.*, 1976). The relation of these and other factors to suppressor function in NZ mice is discussed later.

Despite the numerous examples of abnormalities in CMI, several investigators have reported relatively normal CMI in NZ mice. In 2- to 14-month old NZB and B/W mice, Zinkernagel and Dixon (1977) demonstrated normal alloreactive cytotoxic T-lymphocyte (CTL) activity as compared to age-matched control strains. These latter observations included *in vivo* responses against EL-4 tumor cells comparable to control strains, vaccinia virus-specific CTL activity after systemic infection similar to control strains, T-dependent primary footpad swelling after local infection with LCM virus within a normal range, and adequate protection against *Listeria monocytogenes* infection. It was concluded that T-cell defects may be selective and either not detected in these systems or nonexistent. Similarly, 3- to 8-month-old NZB and B/W mice were found to demonstrate almost equivalent CTL activity against H-2-incompatible target cells as nonautoimmune strains of the same *H-2* haplotype (Theofilopoulos *et al.*, 1979b). Creighton *et al.* (1978) found no difference in either T-helper or T-suppressor functions when comparing young and old B/W mice. In another study, the response to allogeneic cells of old NZB mice was found to vary from normal to very low proliferation (Smith and Pasternak, 1978). Similarly, the decline in cytotoxic activity against alloantigens in aged NZB mice was found to be comparable to the decline in aged nonautoimmune mice and it was suggested that the defect was related to the physiologic aging process rather than to autoimmune disease (Hirano and Nordin, 1976). These discrepancies may reflect different experimental protocols, animals of different colonies, or the variation in clinical status of the mice at the time of experimentation.

Recently, some rather unique properties of the CMI response of NZ mice have been elucidated. It was demonstrated that spleen cells from 3- to 8-month-old NZB and B/W mice could generate CTL activity against H-2-identical allogeneic targets (i.e., BALB/c) of almost the same magnitude as that generated against H-2-incompatible target cells (Theofilopoulos *et al.,* 1979b; Botzenhardt *et al.,* 1978). Furthermore, NZ spleen cells sensitized with H-2-identical allogeneic cells could lyse syngeneic cells to a lesser extent. However, spleen cells sensitized with syngeneic cells were not active against syngeneic or allogeneic targets. The NZB (H-2^d) anti-H-2^d activity is directed against minor histocompatibility differences (Botzenhardt *et al.,* 1978); however, using congenic mice it was shown that the activity is to some extent recognized in the context of the *H-2* haplotype, as it was optimum when the NZ cells matched the target at the K or D end of H-2 (Theofilopoulos *et al.,* 1979b). The significance of this CTL reactivity with H-2-identical targets in autoimmune disease etiology is unknown. However, the investigators also studied MRL and BXSB mice and found no such reactivity. They suggested that because this abnormality in T-cell effector function was not common to all SLE-prone strains, a primary etiologic role was not likely. The finding that spleen cells from NZ mice could not be sensitized *in vitro* with syngeneic cells to develop CTL activity against syngeneic targets is at variance with other investigators who have observed syngeneic killing or proliferation in spleen cells from NZ mice. However, the target cells and methodologies were quite different. Stiller *et al.* (1973) showed that the spleen cells from old, diseased NZB mice were cytotoxic against autologous fibroblast monolayers, whereas cells from young NZB mice demonstrated cytotoxicity of intermediate degree. Allogeneic H-2-matched targets were not lysed. It has also been observed that young and old B/W spleen cells proliferate in the presence of syngeneic heart and kidney cells (Senitzer *et al.,* 1979); spleen cells from old B/W mice were activated by syngeneic heart and kidney at a time when polyclonal activation by Con A was lost.

NZB mice manifest a poor or reduced stimulation in a syngeneic mixed lymphocyte reaction (SMLR) assay, whereas all other strains tested exhibited strong proliferation (Smith and Pasternak, 1978). SMLR represents the response of T cells to non-T cells and this may be a manifestation of an immunoregulatory mechanism directed primarily at autoreactive B cells. This interesting trait was found to be recessive, as (NZB × BALB/c)F_1 hybrids exhibited strong SMLR. Defective autologous MLR in human subjects with SLE has also been observed (Sakane *et al.,* 1978).

Natural killer (NK) cell activity has been analyzed in NZB mice. Greenberg and Playfair (1973) reported increased activity after 3 months of age. In contrast, normal NK activity has been reported in NZB mice (Croker *et al.,* 1979). In B/W mice, it was suggested that excessive NK activity contributed to the maturation and acceleration of autoimmunity (Seamen *et al.,* 1980).

The loss of suppressor T cells in aging NZ mice has been suggested as the primary defect leading to the autoimmune disease in these animals (Talal and Steinberg, 1974). While it is clear that regulatory defects exist in NZ mice, the relative contribution of suppressor cells, the manner in which they are defined, and the methods used to demonstrate them have generated much confusion and controversy. It has been proposed that NZB and B/W mice lose suppressor cells for a variety of immunologic functions. For example, the PFC response to pneumococcal polysaccharide type III (SSSIII), a thymus-independent antigen, increases with age in NZB mice. This is in contrast to a progressive decrease in PFC response with age found in control BALB/c mice (Barthold *et al.*, 1974). This increase in SSSIII-reactive PFC was prevented by transfer of thymocytes from 1-month-old but not from 2-month-old syngeneic donors. Similar results have been reported in B/W mice using the thymus-independent antigen poly(I·C) (Chused *et al.*, 1973). These results have been interpreted as an accelerated loss of suppressor T-cell (Ts) activity with age, which permits excessive proliferation of specific B cells. Similarly, the GVH response of B/W mice was suppressed with spleen cells or thymocytes from young mice; old spleen cells were ineffective (Gerber *et al.*, 1974; Hardin *et al.*, 1973). The administration of 2-week-old syngeneic thymocytes to 4-week-old NZB mice suppressed the antierythrocyte response, whereas starting treatment at 10 weeks was ineffective (Gershwin and Steinberg, 1975). Thymus grafts from 2-week-old but not 10-week-old B/W mice transferred the ability to develop tolerance to BGG and prevented the accelerated anti-DNA formation induced by neonatal thymectomy (Steinberg *et al.*, 1970). These findings further suggest the loss of suppressor activity in aging NZ thymus. In addition, it was observed that as B/W mice aged they lost splenic T cells that could be activated by Con A to become suppressor cells and lost the ability to inhibit *in vitro* IgM synthesis of pokeweed mitogen (PWM)-activated spleen cells. However, aged B/W spleen cells retained the capacity to respond to suppressor signals from Con A-pulsed normal spleen cells and treatment with soluble immune response suppressor led to a reduction in autoimmune manifestations (Krakauer *et al.*, 1976). The cause of the proposed decline in suppressor functions is unknown; however, a genetic abnormality, thymic hormonal deficiency, and NTA have been implicated. Recently, NTA has been shown to have a selective cytotoxic effect on Ts cells (Shirai *et al.*, 1978; Klassen *et al.*, 1977a). Arguing against the role of NTA in depressing Ts cell activity is the fact that many immunologically normal inbred murine strains have a prevalence and titer of NTA as high as NZ mice. Similarly, antinuclear and other autoantibodies may occur prior to or in the absence of NTA (Goldblum *et al.*, 1974). However, NZB peripheral T cells subjected to analysis with the fluorescence-activated cell sorter indicate the presence of detectable intracellular but not surface IgM; this may represent rapidly endocytosed anti-T-cell autoantibody. Indeed, the immunoglobulin in NZB T cells was detected as early as 4 weeks of age. Therefore, the presence of NTA in young NZB mice may not be detectable by standard methods (Chused *et al.*, 1978).

The inability to generate Ts cells in NZ mice has not been universal. Primi *et al.* (1978) found normal Con A-activated Ts cells in old NZB mice. However, old NZB spleen cells were refractory to inhibition by Con A-activated young NZB or BALB/c cells. This indicated that successful suppression requires at least two functions: and active suppressor cell and the ability of the antibody-forming cell to recognize the suppressor message. Other studies have shown that while one mouse strain may be capable of producing a suppressor factor and not responding to it, a second strain may not produce a suppressor factor but may be capable of responding (Taniguchi *et al.*, 1976). An F_1 hybrid of these two strains would be capable of performing both functions. Manny *et al.* (1979) confirmed that old B/W Con A-activated spleen cells lacked suppressor function. By contrast, Con A-activated spleen cells from NZB mice, whether from old or young mice, showed normal suppressor capacity. This suggests that the "normal" NZW mouse contributes genes that modify suppressor-cell activity in the B/W F_1. Therefore, the defect in Con A-induced suppression in older NZ mice has been attributed to functional abnormalities of Ts cells in B/W mice (Krakauer *et al.,* 1976) or to the lack of acceptor sites for suppressor messages on the surface of B cells (Primi *et al.,* 1978). However, it has recently been demonstrated that young and old NZB and B/W mice retained the Con A-induced suppressor capacity and likewise retained the ability to respond to syngeneic and allogeneic suppressor messages (Theofilopoulos *et al.,* 1980). A bone marrow (BM) suppressor cell believed to be a pre-T lymphocyte has been described (Dauphinee and Talal, 1979). Analysis of NZB mice revealed normal BM pre-T-suppressor cells but spleen cells unresponsive to the BM-suppressor signal. In another model, suppressor-cell activation was produced by treatment with antitheta in B/W mice. Although spleen cells from both old and young B/W mice could be activated to exhibit suppressor activity, old BM spleen cells could not respond (Klassen *et al.,* 1977b).

Radioresistant, nonphagocytic, mononuclear cells, not sensitive to treatment with anti-Thy-1, have also been described to have suppressor activity; such suppressor activity was found to be elevated in old B/W mice (Michalski *et al.,* 1979; Roder *et al.,* 1977). It has similarly been suggested that increasing suppressor activity in aging mice may be associated with autoimmune disease and malignancy in these animals (Roder *et al.,* 1977, 1975). In addition, an excess production of low-molecular-weight suppressor substance by aging B/W spleen cells has been reported (Ranney and Steinberg, 1976).

Phenotyping studies have shown that T cells from NZB mice at age 2 weeks to 1 year had normal to supranormal numbers of Ly-23^+ cells, the phenotype expressed by suppressor cells (Cantor *et al.,* 1978). Furthermore, T cells from old mice had normal suppressive capacity in a functional assay of anti-SRBC PFC inhibition. More significantly, the principal T-cell subset in which NZB mice were found to be deficient expresses the Ly-123^+ cell-surface phenotype (Cantor *et al.,* 1978). Such cells have been found to be active in feedback suppression and probably receive activating signals from the Ly-1^+ helper cell. The reason synergistic T-cell interactions are most easily observed in studies of NZB mice may not be due to a deficit in one of the positively interacting T-cell populations, but rather to a lack of feedback inhibition and therefore a magnification of positive interaction effects (Gershon *et al.,* 1978) Indeed, it was proposed that the lack of activity of this Ly-123^+ cell is the basis for the major immunologic defect in the NZB mouse. There is a unique animal model to assess the influence of suppressor function. Hereditarily asplenic (*Dh*/+) NZB mice develop autoimmune disease to the same extent as their NZB +/+ littermates; however, they possess very low NTA titers and maintain easily detectable levels of suppressor function. Therefore, NZB *Dh*/+ mice develop autoimmune disease without the influence of NTA and without a premature loss of suppressor-cell function (Gershwin *et al.,* 1979a). Thus, loss of suppressor-cell activity cannot be implicated readily in autoimmune pathogenesis.

In summary, a variety of abnormalities in suppressor function in NZ mice have been observed. These wide-ranging and contrasting abnormalities have been defined by: an age-related decline in the ability to restore suppressor functions to a number of effector systems, a lack of artificially inducible suppressor cells, an inability to respond to normal levels of suppressor cells or substances, an increasing suppressor activity in aging animals, and a

TABLE IV
Abnormalities in Suppressor Function Observed in Old NZ Mice

Suppressor activity	NZB	NZW	Nonautoimmune
Con A-activated Ts[a]	+	−	+
Response to Con A Ts	−	+	+
Anti-Thy-1.2-activated Ts	ND[b]	+	+
Response to Thy-1.2 Ts	ND	−	+
BM suppressor	+	ND	+
Response to BMs[c]	−	ND	+
Null[d] suppressor cell	ND		0
Ly-23 Ts	+	ND	+
Ly-123	− or ↓	ND	+
SMLR[e]	−	ND	+
LMWS[f]	ND	↑	+
Suppression of effectors (generalized)	−	−	+

[a]Ts, T suppressor.
[b]ND, not determined.
[c]BMs, bone marrow suppressor.
[d]Null, radioresistant, nonphagocytic, mononuclear, non-T cell.
[e]SMLR, syngeneic mixed lymphocyte response.
[f]LMWS, low-molecular-weight suppressor substance.

deficiency in a subset of suppressor cells. A summary of these functionally and serologically defined abnormalities in suppressor activity in NZ mice is presented in Table IV. Although these assays of suppressor function are relatively undefined, their development may prove to be very useful in understanding the pathogenesis of the disease in NZ mice. The value may not necessarily be in the direct definition of cellular defects, but rather in providing assays in which *in vivo* correlates of these *in vitro* abnormalities can be tested. For example, do NZB.ch congenic mice, which demonstrate decreased autoimmune expression, exhibit increased or decreased suppressor function in these assays? Such experiments would allow the evaluation of the relative contribution of these observed suppressor defects to the expression of autoimmunity.

8. Humoral Immune Responses and B-Cell Subsets

Humoral responses to a variety of antigens have been extensively studied and NZ mice appear to exhibit complex abnormalities in responsiveness to heterologous antigens. These complexities mitigate against the characterization of the humoral response in NZ mice as hyper- or hyporesponsive to T-dependent or T-independent antigens. The responses to T-dependent antigens will be considered first.

Neonatal NZB mice were found to be hyperactive to SRBC immunization as compared to mice of other strains (Morton and Siegel, 1969; Evans *et al.*, 1968; Playfair, 1968). Adult levels of responsiveness were attained by NZB mice by 5 days of age. The precocious response to SRBC was thymus dependent. Similarly, increased numbers of background

PFC to SRBC in young, unimmunized NZB mice have been reported (Diener, 1966). In adult, DAT-positive mice, both a decreased number of anti-SRBC PFC and a delayed development of the primary response were observed after primary immunization. Hyperresponsiveness in young NZB mice to bovine serum albumin (BSA) (Weir *et al.,* 1968), BGG (Braverman, 1968b; Parker *et al.,* 1968), and subcellular liver antigens (Elson and Naysmith, 1969) has also been reported. However, an enhanced humoral response is not demonstrated against all antigens. Playfair (1968) observed normal responses to pig and chicken erythrocytes. NZB and B/W mice were reported to be hyperresponsive to SRBC and BSA but not to keyhole limpet hemocyanin or human gamma globulin (HGG) (Cerottini *et al.,* 1969). It appears that quantitative differences in the immune responses of NZB mice may be specific for the immunizing antigens. It was suggested that the genetically determined hyperactivity in NZ mice may involved antigens pertinent to their autoimmune disease, as indicated by the fact that young B/W mice demonstrate an enhanced production of ANA after immunization with DNA-mBSA (Cerottini *et al.,* 1969). NZ mice display qualitative as well as quantitative aberrations in humoral immunity. The temporal anti-SRBC response of normal mice is characterized by an increase in mean relative antibody affinity and a concomitant decrease in the heterogeneity of binding affinities. However, the anti-SRBC response of NZB mice was characterized by a temporal decrease in antibody affinity associated with clonal selection of cells possessing surface receptors of lower rather than higher affinity for antigen (DeHeer and Edgington, 1975). It has also been observed that the anti-X and the anti-HB autoantibody responses of NZB mice exhibit maturational defects (DeHeer and Edgington, 1976a). Elkerbout and Hijmans (1974) reported a prolonged primary and secondary antibody response to SRBC immunization in NZ mice associated with low-avidity antibody. Low-avidity antibody to human transferrin produced in NZB mice has also been reported (Petty and Steward, 1972). Similarly, the avidity of circulating anti-DNA antibody in B/W mice was observed to increase up to 5 months of age and thereafter to fall with increasing age (Steward *et al.,* 1975). It was suggested that low-avidity antibody is less efficient at immune elimination of antigen thatn high-avidity antibody and that this favors the persistence in the circulation of antigen-excess immune complexes which are deposited in the tissues.

Age-related changes in the humoral immune response of NZ mice to T-independent antigens have also been studied. The responses of 3- to 6-month-old NZB mice to SSSIII and bacterial levan were found to be depressed by Purves and Playfair (1973). However, an age-associated increase in antibody response to SSSIII, suggested to be due to a loss of suppressor function, has also been reported (Barthold *et al.,* 1974). Similarly, Chused *et al.* (1973) demonstrated an age-related increase in response to poly(I·C). The humoral responses to SSSIII and poly(I·C) are known to involve regulatory T cells (Baker *et al.,* 1970); however, antibody response to LPS is thought to be independent of thymic regulatory effect (Viet and Michael, 1972). Blankwater *et al.* (1975) observed a low PFC response to detoxified LPS in young NZB mice that declined with age relative to BALB/c controls. These latter observations led to the conclusion that the age-related decline in response to LPS was a consequence of a decrease in B-cell function unrelated to T-cell abnormalities. More recently, the immune response to both DNP-Ficoll (Moutsopoulos *et al.,* 1977) and TNP-Ficoll (Cohen and Ziff, 1977) was found to be within the normal range in young and old NZ mice. However, the PFC response to DNP-lysine-Ficoll in NZB mice was observed to exceed that of C3H controls by two- to fourfold. Further maturational abnormalities were demonstrated by NZB mice. For example, the mean antibody affinity progressively decreased, consistent with a clonal selection of cells secreting low-avidity anti-DNP in NZB mice (DeHeer and Edgington, 1980).

Another reported aspect of aberrant immune responsiveness in NZ mice has been their relative resistance to the induction and maintenance of immunologic tolerance to certain antigens. Staples and Talal (1969) observed that 6-week-old NZB and B/W mice were relatively resistant to tolerance induction by soluble, ultra-centrifuged, deaggregated BGG and HGG. Mice 3 weeks of age or younger developed temporary tolerance that was lost in 2–4 weeks (Staples *et al.*, 1970). In contrast, Cerottini *et al.* (1969) observed similar tolerance induction to HGG in NZB, A/J, BALB/c, and CBA mice, and suggested that resistance to induction of tolerance may not be a general phenomenon of NZB mice, but dependent on the dose of antigen or tolerogen. Normal B-lymphocyte high-dose tolerance to SSSIII and bacterial levan was reported in NZB mice (Purves and Playfair, 1973). Based on reconstitution experiments, induction of tolerance to SRBC could be achieved with NZB B lymphocytes but not with NZB T lymphocytes (Purves and Playfair, 1973; Playfair, 1971). Cyclophosphamide-induced tolerance is expressed by either thymus or bone marrow cells in nonautoimmune strains; however, in NZ mice it was found that only bone marrow cells expressed this form of induced unresponsiveness (Jacobs *et al.*, 1971, 1970). These experiments suggested that NZB T cells were unusually resistant to tolerance. Thus, tolerance in these animals appeared to be a short-term function of bone marrow and spleen and was restricted to the neonatal period.

The ability to localize aggregated protein antigens and immune complexes in germinal centers has been attributed to B cells and this function has been found deficient in NZB and B/W mice (Brown *et al.*, 1971). Another abnormality of NZ mice is an increased susceptibility to experimental induction of autoantibody formation (Milich and Gershwin, 1979a; Cooke and Playfair, 1977; Flad *et al.*, 1969). Young, disease-free NZB and B/W mice were found to be high responders in the production of a DAT-positive, hemolytic anemia induced by rat erythrocyte immunization, whereas NZW mice were low responders (Milich and Gershwin, 1979a).

A defect in NZB macrophages, involving antigen processing, has been suggested by Thomas and Weir (1972). NZB peritoneal macrophages were found to be more efficient in taking up and less effective in degrading BSA than macrophages from mice of control strains. In contrast, using a bacterial antigen, both phagocytic and bactericidal activities of NZB macrophages were found to be normal (Whaley and Singh, 1973). The blood clearance and liver, spleen, and kidney uptake of immune complexes were studied in B/W mice and no impairment was observed. The recovery from reticuloendothelial saturation was also within normal limits (Finbloom and Plotz, 1979).

Spleen cells of NZB mice have been observed to hyporespond to SRBC *in vitro* despite normal or supranormal *in vivo* responses (McCoombs *et al.*, 1974). This aberrant reponse was detected in NZB mice as early as 5 weeks of age and is believed to be a functional deficiency of adherent cells in the spleens of unprimed mice (McCoombs *et al.*, 1975). Alternatively, Roder *et al.* (1977) suggest that the deficient response to SRBC *in vitro* is related to the effects of a suppressor cell, the response being observed to increase with age. However, others report that macrophage function in NZ mice is normal, but that splenic T and B cells are hyporesponsive to SRBC *in vitro* (Kassan and Chused, 1977). While the mechanism of this paradoxical response is unclear, its relevance is illustrated by the fact that of numerous strains analyzed in this assay, three of the four lowest responding strains spontaneously develop autoimmune disease: NZB, A/J, and AHE/J (Kassan and Chused, 1977).

NZ mice also demonstrate stem-cell defects. NZB mice were found to show markedly elevated numbers of spleen colonies at all radiation doses in an endogenous colony formation assay (Warner and Moore, 1971). However, genetic analysis minimized the etiologic role of this abnormality in autoimmune disease. Indeed, there was no correlation of autoantibodies and heightened endogenous colony formation. Recently, an abnormality in the ability of stem cells of NZB mice to form myeloid colonies in response to a particular stimulus *in vitro* was described (Kincade *et al.*, 1979).

Myeloma tumors may be artificially induced by injecting mineral oil into the peritoneal cavity of inbred BALB/c mice. The observation that myelomas could be induced in NZB mice has allowed the comparison of two populations of myeloma immunoglobulins induced within a single species. Comparison of antigen-binding properties, immunoglobulin class distribution, and sequence data indicated that myeloma proteins from BALB/c and NZB mice constitute two populations of immunoglobulins with distinct functional and structural properties (Loh *et al.*, 1979; Morse *et al.*, 1978). Numerous NZB but not BALB/c myeloma proteins were found to bind DNA. Recently, a hybridoma developed from an unimmunized B/W mouse was shown to produce a monoclonal antibody that had a high affinity for ribosomal RNA (Eilat *et al.*, 1980).

Intrinsic and/or extrinsic B-cell defects may be operative in regard to the immunological abnormalities observed in NZB mice. Recently, numerous functional B-cell defects have been described and evidence is accumulating that these aberrations are intrinsic to B-cell subsets and not necessarily due to T-cell disorders observed in NZ mice. Spleen cells from very young NZB and B/W mice have been shown to spontaneously secrete excessive amounts of pentameric IgM in short-term culture (Moutsopoulos *et al.*, 1977). It was found that splenic B cells of NZ mice, in contrast to nonautoimmune strains, produce IgM in short-term cultures at birth and that by 6–8 weeks of age produce 40 times more IgM than normal-strain spleen cells. Furthermore, the IgM produced was not specific for autoantigens. No sex difference was observed between spleen cells of male and female 3- to 10-week-old NZB mice. BALB/c nude (*nu/nu*) spleen cells were shown to produce only slightly more IgM than normal strains and much less than NZB cells in this assay. These investigators also observed that serum IgM was detected in NZB and B/W mice on the first day of life in contrast to its appearance at 2 weeks in normal strains; from 2 to 10 weeks of age, NZ serum IgM levels were 19 times normal. These findings were confirmed by Manny *et al.* (1979) and expanded by the observations that: (1) there were two populations of IgM-secreting cells, an abnormal spontaneously activated hypersecretor population and a population that responded normally to LPS; (2) fetal liver contained the abnormal population; (3) the abnormality was attributed to two factors that segregated as independent phenotypes, an increased number of IgM-containing cells (dominant) and an increased secretion per cell (recessive); and (4) the hypersecretion of IgM by the abnormal B cells is T cell independent as treatment of fetal liver cells with fetal or adult thymocytes did not influence the results. It has also been reported that spleen cells from NZB mice had an elevated ratio of cell-surface IgM/IgD as early as 2 weeks after birth compared to BALB/c spleen cells (Cohen *et al.*, 1978). At 4 weeks of age, such ratios are equivalent to those seen for BALB/c controls between 4 weeks and 1 year. The IgM/IgD ratios of NZB splenocytes remained constant, whereas those of BALB/c decrease and reach adult levels at 6 weeks. The elevated ratio of surface IgM/IgD was found to reflect a substantial decrease in IgD and a modest increase in IgM. This pattern is characteristic of mitogenically or antigenically activated B cells from nonautoimmune strains (Chused *et al.*, 1978; Bourgois *et al.*, 1977). Another group has studied the spontaneous hypersecretion of IgM by lymphocytes in NZ mice by measuring background IgM PFC and sera antibody against hapten-coupled red blood cells (Izui *et al.*, 1978). It was found that NZ mice produced far more antihapten PFC in their spleens and greater concentrations of antihapten antibodies in their sera than immunologically normal strains; the increased nonspecific antibody production by the abnormal B cells correlated with the spontaneous development of anti-ssDNA, but not with serum levels of gp70. Significant elevations in antihapten PFC were observed by 2 weeks of age and continued through 4 to 5 months as compared to normal strains. Antihapten antibodies were determined to be of low avidity, of the IgM class, and nonspecific. This spontaneous antihapten response observed in NZ mice was comparable to the response of 2-month-old C57BL/6 mice to a high-dose injection of LPS. In addition, spleen cells from young NZ mice respond normally to polyclonal B-cell activators (PBA), such as LPS and PPD, with proliferation and production of nonspecific antibody. However, spleen cells from old (10 months) B/W mice responded to PBA with increased proliferation but no antibody generation, despite a normal PFC response to TNP-Ficoll immunization (Cohen and Ziff, 1977). This abnormal PBA response was not due to suppressor cells as anti-brain-associated theta-treated cells still failed to generate PFC. It was suggested that the failure to respond to PBA with PFC generation was a result of *in vivo* polyclonal B-cell activation in the course of autoantibody formation. In support of this, Gronowicz and Coutinno (1975) have shown that treatment of normal mice with LPS leads to an inability to generate PFC in response to LPS or PPD when spleen cells were subsequently challenged 4 days later; however, the capacity to proliferate was intact.

In a further attempt to characterize the B-cell status of NZB mice, LPS-induced B-cell colony formation has been examined. It was observed that NZB mice had statistically higher numbers of splenic B-cell colony-forming (CFU) cells compared to other strains from birth through 2 months of age (Ohsugi and Gershwin, 1979). An age-associated decline in splenic B-cell CFU was apparent and by 6 months of age this capacity was significantly reduced. Comparable values were obtained by cloning bone marrow and lymph node lymphocytes. Furthermore, the numbers of B-cell CFU from 17-day-old NZB liver were significantly higher than from control strains. Significantly, the numbers of B-cell CFU in asplenic NZB and in nude NZB mice were similar to those of their littermate controls, suggesting that this B-cell defect was unrelated to the presence of suppressor function and that it was thymus independent, respectively. Germfree NZB mice were identically studied and no differences were observed between conventional and germfree mice, indicating the lack of influence of gut flora. No sex difference was observed between B-cell CFU in male and female mice. Increased splenic B-cell CFU in NZB mice have also been reported by Kincade *et al.* (1979). Unusual anti-μ-resistant cells were also described. Whether the increased LPS-induced B-cell CFU observed in NZB mice represent the same B-cell subpopulation that appears to be polyclonally activated is unknown. However, CBA/N mice are known to lack a subpopulation of B cells capable of forming LPS-induced B-cell CFU in semisolid agar (Kincade, 1977). They also have other B-cell abnormalities related to an X-chromosome-linked defect (see Chapter 8, Volume 1). Recent studies of

(CBA/N × NZB)F_1 hybrids indicated that, with regard to abnormalities of IgM production, the NZB genetic defect acts almost exclusively upon that subset of B cells upon which the X-linked CBA/N defect acts (Nakajima *et al.*, 1979; Taurog *et al.*, 1979). In contrast to the abnormalities of IgM production in NZB mice, CBA/N mice bear and X-linked defect in the immune response to T-independent antigens such as SSSIII (Amsbaugh *et al.*, 1972), poly(I·C) (Scher *et al.*, 1973), and haptenated Ficoll derivatives (Cohen *et al.*, 1976). This defect is associated with low serum IgM levels (Amsbaugh *et al.*, 1974), reduced numbers of splenic B cells (Scher *et al.*, 1975), and a decreased incidence of B cells with a low density of surface membrane IgM (Scher *et al.*, 1976). Therefore, if the spontaneous hypersecretion of IgM as seen in the NZB strain involved the subset of B cells that the CBA/N strain lacks (CBA/N × NZB)F_1 females would express the NZB abnormality and the F_1 males of this cross would not. It was shown that (CBA/N × NZB)F_1 male B-cell function was identical to that of CBA/N males; (NZB × CBA/N) males and (CBA/N × NZB) females, which lack the defective X-linked gene, resembled the other (NZB)F_1 hybrids studied. It was shown that castration had no effect on total IgM secretion and most significantly that (CBA/N × NZB) males had retarded development of anti-erythrocyte antibodies relative to their female littermates. This suggested that the X-linked gene was capable of favorably modifying the expression of autoimmunity. Whether the same effects will be observed in NZB mice awaits the study of congenic mice bearing the CBA/N X chromosome on an NZB background (Taurog *et al.*, 1979). It has also been demonstrated in this hybrid cross that the expression of xenotropic virus is not linked to the hypersecretion of IgM (Nakajima *et al.*, 1979) and therefore hypersecreting B cells are not the primary site of production of xenotropic virus in NZB mice.

These studies have indicated the presence of an abnormality in the B lymphocytes of NZB mice. One of the main features of the disturbed B-cell function observed was the marked hypersecretion of IgM. This heightened state of B-cell activation may be: (1) spontaneous, a result of exogenous or endogenous polyclonal B-cell activation; (2) an intrinsic B-cell defect; (3) a consequence of inadequate T suppressor; or (4) increased T-lymphocyte helper function. The evidence that the spontaneous activation of B cells in NZ mice is T independent is substantial. B-cell activation due to a lack of suppressor cells is contraindicated by the following: suppressor T-cell function induced by Con A in 4-week-old B/W mice is as high as in normal strains (Krakauer *et al.*, 1976); the decline of suppressor T-cell functions described in GVH reactions (Hardin *et al.*, 1973) and in the response to SSSIII (Barthold *et al.*, 1974) is not demonstrable until long after the enhanced B-cell activation has been observed; NZB asplenic (*Dh*/+) mice, with normal suppressor activity, have the same elevated numbers of B-cell CFU as their NZB +/+ littermates (Gershwin *et al.*, 1979a). Finally, the addition of either fetal or adult thymocytes to cultured fetal liver cells failed to suppress the hypersecretion of IgM in NZB mice (Manny *et al.*, 1979). B-cell activation due to an enhanced T-helper effect is unlikely in light of numerous observations: (a) NZB nude (*nu*/*nu*) mice express the same heightened spontaneous polyclonal B-cell responses as their NZB +/+ littermate controls (Gershwin *et al.*, 1979b); (b) NZB *nu*/*nu* mice have four times as many plasma cells and secrete five times as much IgM as compared to outbred Swiss *nu*/*nu* mice (Chused *et al.*, 1978); (c) outbred, athymic nude mice have a very low incidence of autoimmune disease (Gershwin *et al.*, 1975); and (d) when adult NZB and NZW mice were thymectomized, lethally irradiated, and reconstituted with stem cells from syngeneic fetal liver, the spleens of NZB mice contained many

TABLE V
Characteristics of Abnormal B-Cell Subset(s) in NZ Mice

1.	Spontaneous IgM secretion in short-term *in vitro* cultures is elevated
2.	Appearance of spontaneous antihapten PFC and serum antibody is elevated
3.	Increased serum levels of IgM
4.	Increased incidence of LPS-induced B-cell colony-forming cells
5.	Low proportion of IgD relative to IgM on B-lymphocyte surface membrane
6.	Decreased antibody production upon PBA activation in old NZ mice
7.	Many abnormalities involve the same B-cell subset(s) lacking in CBA/N mice
8.	Early appearance of abnormalities, from birth to 2 weeks
9.	Thymic independence of abnormalities appears likely
10.	Abnormal B-cell activation appears to be sex independent

more plasma cells and secreted more IgM than NZW control spleens (Chused *et al.*, 1979). This latter finding demonstrates that the lack of suppressor T cells does not lead to B-cell activation on a nonautoimmune genetic background (NZW). However, a relationship between polyclonal activation of B cells and the expression of autoimmunity has been indicated. It was reported that PBA could induce the formation of anti-DNA anitbodies when injected into normal and athymic mice (Izui *et al.*, 1977; Fournie *et al.*, 1974). Moller *et al.* (1976) also observed that LPS administered to tolerant mice resulted in the induction of autoantibody directed against the antigen to which the mice had been rendered tolerant. Primi *et al.* (1977) have reported that PBA administered to normal mice stimulated the rapid formation of autoantibodies directed against spleen and peripheral blood cells. More specifically, the spontaneous, increased production of nonspecific antibody produced in NZ mice correlated with the spontaneous development of anti-ssDNA autoantibodies in these animals (Izui *et al.*, 1978). In addition, the (CBA/N × NZB) hybrid studies indicate that in the absence of spontaneously activated B-cell subset(s), autoimmune disease is repressed at least in the F_1 hybrids (Taurog *et al.*, 1979). The exogenous or endogenous nature of the polyclonally activated B cells in NZ mice is unknown. Numerous abnormalities of B-cell subsets in NZ mice have been demonstrated and and are summarized in Table V.

9. Pathogenesis

An age-associated decline in CMI function appears to occur in NZ mice, although the degree and range of this decline have recently been questioned (Theofilopoulos *et al.*, 1979a,b; Creighton *et al.*, 1978; Zingernagel and Dixon, 1977). Thymocyte defects have been described, particularly defects in NZ suppressor T-cell activity (Table IV). Although the assays developed to measure suppressor activity are often undefined in terms of the antigenic specificity and the T-cell subset(s) involved, they do indicate abnormalities of regulatory function in NZ mice. In addition, phenotyping of NZB T cells has revealed a deficiency in Ly-123^{+} cells (Cantor *et al.*, 1978). Furthermore, NZB *nu/nu* mice have a paucity of the T-cell precursors present in other inbred nude mice (Gershwin *et al.*, 1979b).

Complex humoral responses by NZ mice to heterologous antigens have been described, which may reflect genetically determined immune responses rather than a general tendency toward hyperresponsiveness in these animals. However, qualitative aberrations in the antibody maturational process to both heterologous and self antigens have been described (DeHeer and Edgington, 1980, 1975). Recently, a number of B-cell abnormalities have been observed. The main feature of the B-cell dysfunction is the polyclonal activation of B-cell subset(s), which appears to be genetically determined and independent of T-cell regulatory effects (Manny *et al.,* 1979; Ohsugi *et al.,* 1979; Izui *et al.,* 1978).

Uncertainty exists regarding the relative importance of B- and T-cell defects and their ontogenic relationship in the expression of autoimmunity in NZ mice. The T-cell abnormalities of NZ mice may be indicative of deregulation of spontaneously activated T cells as has been suggested for B cells, which would imply a common stem-cell defect. Alternatively, autoantibodies directed against T-cell receptors may alter the function of intrinsically normal T cells. The results of a fluorescence-activated cell sorter analysis of NZB peripheral T cells indicate the presence of intracellular NTA as early as 4 weeks of age and suggest that NZB T cells function in an excess of membrane-active autoantibody from a very early age.

Determination of the intrinsic state of NZB T cells would require the prevention of exposure to autoantibody (Chused *et al.,* 1978). Interestingly, NZB asplenic (*Dh*/+) mice display delayed onset and reduced titers of NTA as compared to their NZB +/+ littermate controls but nevertheless express autoimmune disease to the same extent (Gershwin *et al.,* 1979a). This seems to indicate the secondary importance of NTA and that NZB T cells are either intrinsically abnormal or not greatly implicated in disease expression.

Another approach to studying the relative importance of T and B cells in autoimmunity has been to examine the disease-producing potential of B cells in the absence of T-regulatory cells. This has been accomplished in two systems: the reconstitution of lethally irradiated, anti-Thy-1-treated NZB mice with syngeneic bone marrow cells and the development of NZB *nu*/*nu* mice. These systems have conclusively demonstrated that the lack of suppressor T cells alone is not a sufficient stimulus to induce autoimmune disease in nonautoimmune B cells (DeHeer and Edgington, 1977) and that NZB cells in the absence of suppressor or helper T-cell activity do possess the potential to induce autoantibody formation (Gershwin *et al.,* 1979b; DeHeer and Edgington, 1977). However, the potential for autoantibody production in the absence of adequate suppressor activity and in the presence of appropriate suppressor function are not analogous. Indeed, the reconstitution of BALB/c mice possessing intact thymuses with NZB marrow resulted in a very low expression of disease (Denman *et al.,* 1969). However, these findings are difficult to interpret due to minor histocompatibility differences. At present, the only acceptable model of NZB B cells in the presence of appropriate, syngeneic suppressor T function is the very young NZB mouse itself. In this "model," B cells are polyclonally activated; however, there are no autoantibodies produced or at least detected prior to 4 weeks of age. Furthermore, the IgM to IgG switch in autoantibody class must be explained by any model suggesting a primary B-cell defect. Similarly, the sex-hormone-dependent differences in expression of autoimmunity in (NZB × normal)F_1 hybrids and the fact that spontaneous polyclonal activity of B cells is hormone independent (Ohsugi and Gershwin, 1979; Taurog *et al.,* 1979) suggest that modifying factors may be necessary to extend polyclonal B-cell activation to autoantibody production.

The existing data are not inconsistent with the hypothesis that independent T- and B-cell defects are present in NZ mice; the simultaneous expression of both defects may be necessary for the development of autoimmune disease. Experimental evidence excludes increased helper T-cell function; therefore, the T-cell defect would necessarily be the absence of appropriate suppression. In this regard, polyclonal activation of B cells may be an autoimmune predisposing factor. Polyclonal activation of B cells has been observed in all SLE-prone murine strains tested (Izui *et al.,* 1978). In fact, this has been described as the only common denominator among these strains (Theofilopoulos *et al.,* 1979a,b). It is important to emphasize that the B-cell dysfunction is thymus and sex hormone independent. Thus, the putative suppressor defect would not function at the level of the polyclonally activated B cell. Alternatively, modifying factors unrelated to thymic regulatory circuits may be implicated in directing polyclonally activated B cells toward autoantibody formation. Examples would include the immunogenicity and availability of autoantigens, variation in immune-response genes, clonal repertoire, and the availability of *V*-region genes coding for high-avidity autoantibody (Chused *et al.,* 1979). As the spontaneous polyclonal activation of B cells is not affected by sex hormones, the observed sex influence on autoimmune disease must operate at the level of the modifying factors. In view of the fact that neonatal thymectomy affected male B/W mice differently than female mice (Roubinian *et al.,* 1977b), it is possible that sex hormones may exert their influence on autoimmune disease through the thymus. If this were the case, it would be tempting to speculate that androgens potentiate T-cell suppressor differentiation or function, whereas estrogens potentiate helper function. This would lend further support to the concept of involvement of T-regulatory function as a modifying factor in the development of autoimmune disease. NZB *nu/nu* mice demonstrate enhanced NTA activity as compared to their NZB +/+ littermate controls; nevertheless, their polyclonally activated B-cell activity is equivalent. This suggests a thymic influence with respect to NTA autoantibody production but the lack of such influence on polyclonal activation of B cells (Gershwin *et al.,* 1979b).

If this hypothesis of autoantibody formation were related to Warners genetic model (Table III), the polyclonal activation of B cells would be a likely candidate for the expression of the *A* gene, a dominant gene that predisposes to generalized autoimmunity. Preliminary genetic analysis has revealed that the hypersecretion of IgM is inherited in a polygenic manner; however, the number of IgM-secreting cells appears to be inherited as a dominant trait (Manny *et al.,* 1979). The *M* locus may be responsible for the expression of some T-regulatory function or other modifying factors. The NZB mouse would have an inoperative or nonfunctional allele at this locus. The wild-type allele would strongly suppress the expression of the *A* locus. The third locus (*G*) was described as an *Ir* gene predisposing to ANA production, which would be present in the NZW but not the NZB mouse. While the specific interaction of T and B cells to produce autoantibody and the genetic factors governing them are relatively unknown, this model of polyclonally activated B cells interacting with modifying factors may provide insight into the pathogenesis of autoimmune disease in these animals. Indeed, recent evidence for independent B-cell abnormalities present at birth, and for defective T-cell–T-cell interactions, suggests that the previously proposed theory of suppressor-cell deficiency may have been an oversimplification.

The influences of viral infection on pathogenesis of autoimmune disease in NZ mice have been discussed. Genetic analysis of xenotropic viral expression and the lack of correlation to disease expression has minimized the possible etiologic role that viral infection may play in autoimmune diseases of NZ mice.

Numerous abnormalities have been identified in NZ mice, including extremely high

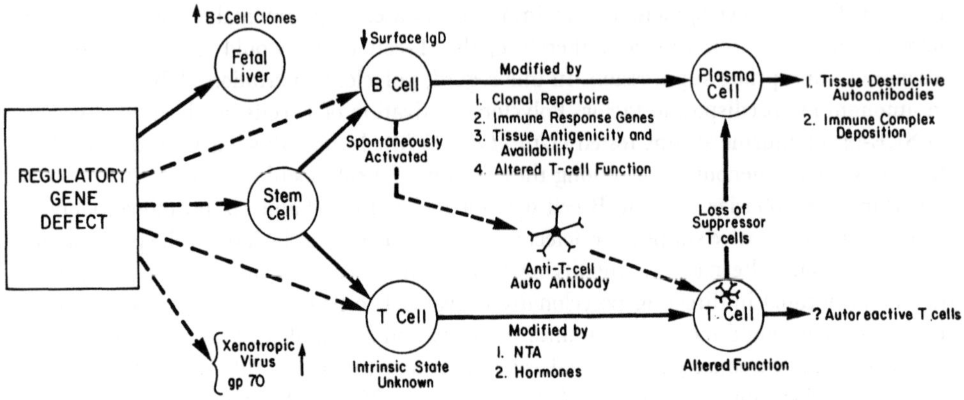

FIGURE 1. Proposed mechanisms for development of autoimmunity in New Zealand mice.

titers of xenotropic virus particles, excessive production of gp70, hypersecretion of IgM by B cells, possible impaired function of macrophages, stem-cell defects, increased incidence of B-cell colony-forming cells, premature thymic involution and thymic epithelial-cell dysfunction, a pronounced tendency toward autoimmunization, susceptibility to neoplasia, and chromosomal abnormalities. It has been suggested that NZB mice may have a defect(s) in fundamental genetic regulatory mechanisms and that impaired genetic regulation could be associated with activation of various cellular genes (i.e., "proviral genes" and "autoantibody genes") (Manny *et al.,* 1979). Figure 1 depicts possible mechanisms of autoimmune disease in NZ mice.

10. Modifiers of Disease

The influence of sex on the expression of autoimmune disease in NZ mice has been briefly discussed in previous sections of this review. A generalized enhanced immune responsiveness in females as compared to males has been observed in many inbred murine strains. This "superiority" of female responsiveness has been demonstrated by: an increased humoral response to a variety of antigens (Butcholar and Chapman, 1965; Stern and Davidshon, 1955); the relatively easier induction of tolerance in male mice (Dresser, 1962); and the enhanced rejection of skin allografts by females (Graff *et al.,* 1969; Kongshavn and Bliss, 1969). Additionally, a greater susceptibility to the experimental induction of autoimmune disease has been observed in female as compared to male mice in a variety of inbred strains (Milich and Gershwin, 1981). Female predominance in a series of human autoimmune diseases, including SLE, has long been apparent. In NZB mice, Raveche *et al.* (1976) observed that the incidence of NTA was greater in female than male mice and that castration of the male mice equilibrated this difference. In B/W mice, autoimmunity develops earlier and with greater severity in female mice than in males. It has been shown that sex hormones modulate the development of autoimmunity as determined by survival, concentration and class of autoantibody to nucleic acids, and severity of immune-complex nephritis

(Talal and Roubinian, 1978). Studies in nonautoimmune and B/W mice suggest that the immunoregulatory role of sex hormones is at least partially mediated through the thymus. The B-cell hypersecretion of IgM associated with NZ mice (Taurog *et al.*, 1979) and the increased incidence of LPS-induced B-cell colony-forming cells in NZB mice were demonstrated to be unaffected by sex or sex hormones (Ohsugi and Gershwin, 1979). Therefore, the observed influence of sex hormones on the autoimmune diseases of these animals is necessarily mediated through other mechanisms, possibly at the level of T-cell regulatory function. Androgen treatment as a mode of therapy in human autoimmune disease may be a possibility. Certainly in the murine B/W model, androgen treatment has been shown to be effective in amelioration of autoimmune disease (Roubinian *et al.*, 1977a).

The amount of thymic hormone in NZB and B/W mice has been reported to be abnormally low (Bach *et al.*, 1973). The administration of thymosin (a thymic hormone) to NZB mice was reported to restore some but not all T-helper-cell functions (Gershwin *et al.*, 1974b) and to restore normal thymocyte proliferation and suppressive function as measured by *in vitro* experiments (Dauphinee *et al.*, 1974). In contrast, Talal *et al.* (1975) reported the lack of influence of thymosin on the titer of antibodies to nucleic acids in aged B/W mice. Similarly, treatment with thymosin was reported not to be effective in diminishing the appearance of ANA or in prolonging survival in NZ mice (Gershwin *et al.*, 1976). Administration of thymosin (fraction V) to B/W mice was observed to accelerate the appearance of proteinuria and anti-DNA antibodies, to increase deposition of immunoglobulin in kidneys, and to shorten survival (Mehta *et al.*, 1978). More recently, the use of the pentapeptide thymopoietin 32-36 was shown to produce significant positive effects on a number of immunological parameters studied in NZB and B/W mice; however, survival time was not increased (Gershwin *et al.*, 1979c). The relative inefficiency of these thymic potentiating factors for prolonging survival may be indicative of the contribution of certain B-cell defects in the pathogenesis of autoimmune disease in NZ mice. Similarly, the positive influence observed on CMI functions and the lack of correlation with increased survival suggest the secondary nature of these defects.

An interesting approach to therapy has been the induction of tolerance to nucleic acids by treatment with synthetic double-stranded poly(I · C) plus cyclophosphamide, which led to reduced antibodies to nucleic acids (Steinberg and Talal, 1971). Similarly, treatment with tolerogen (isologous IgG–nucleoside conjugates) plus cortisone suppressed the appearance of anti-denatured-DNA antibodies and prolonged survival; however, antibody to native DNA was not decreased (Borel *et al.*, 1978).

The murine leukemia viruses are sensitive to interferon, but the potent interferon inducer Statolon had no effect on the disease of NZB (Russell, 1968) and B/W mice (Lambert and Dixon, 1970). Additionally, the administration to B/W mice of the antiviral agent 1-methylisatin from birth until death induced some delay in the time of development of proteinuria and ANA and an increase in the mean survival time; however, eventually all mice died of nephritis. It was concluded that an overlay of virus–antiviral immune complexes on the underlying immune-complex disease was suppressed (Gabriel, 1971). These results suggest the secondary pathogenic effect of viral infection in NZ mice.

Numerous immunosuppressive therapeutic agents have been tested in NZ mice. Both cortisone acetate (Giltinan *et al.*, 1965) and betamethazone phosphate (Casey and Howie, 1965) have been shown to induce remission in NZB mice with established disease and to prevent the development of autoantibody in young mice. The use of 6-mercaptopurine in young NZB mice did not delay or modify their hemolytic disease and its use is associated with an increased incidence of malignant thymic neoplasia; however, improvement in the renal disease of B/W mice was observed (Casey, 1968b). Long-term cyclophosphamide admisistration has been reported to markedly reduce the incidence of lethal kidney disease in B/W mice with a resultant greatly increased life span (Russell and Hicks, 1968). Corticosteroids and azathioprine in combination have also been found effective in the treatment of renal disease in B/W mice (Gelfand and Steinberg, 1972). The combination of three drugs, cyclophosphamide, azathioprine, and methylprednisolone, was superior to all single- or double-drug regimes with respect to proteinuria, anti-DNA antibodies, renal histology, and survival (Gelfand and Steinberg, 1972). Cyclophosphamide has been observed to be the most impressive single drug in suppressing lupus nephritis, although it has a significant oncogenic potential.

Numerous other treatments have been applied to NZ mice. NZB mice have been treated with antilymphocyte globulin with suppression of the hemolytic anemia (Denman *et al.,* 1971); however, no favorable effects on antinuclear antibodies, renal disease, or elevated IgM levels in B/W mice were noted (Denman *et al.,* 1970). Treatment of B/W mice with prostaglandin E_1 has been reported to suppress proliferative glomerulonephritis and to drastically improve survival, but no effect on the development of antinuclear antigens was observed (Krakauer *et al.,* 1978, Zurier *et al.,* 1977).

Levamisole is an antihelminthic drug that alters intracellular levels of cyclic nucleotides. Immune effects are predominantly on phagocytes and T cells with no direct evidence of direct effects on B-cell function (Symoens and Rosenthal, 1977). In an attempt to increase suppressor T-cell function, a number of studies have been carried out in NZ mice. Although the results have been varied, generally levamisole has been found to be ineffective and in some cases it may accelerate murine lupus (Hahn and Knotts, 1979). Protective effects have been reported and no correlation with increased suppressor T-cell functions has been observed (Vogler, 1977).

TABLE VI
General Effects of Therapeutic Modalities in NZ Mice

Treatment	Improve symptoms	Prolong survival
Thymic hormones	+	0
Nucleic acid tolerization	+	+
Interferon inducers, antivirals	+	±
Cortisone	+	+
Azathioprine	+(B/W)	0
6-Mercaptopurine	0	0
Cyclophosphamide	+	+
Antilymphocyte serum	+(NZB)	0
Prostaglandin E_1	+	+
Levamisole	+,−	+,−
Total lymphoid irradiation	+	+
Dietary restriction	+	+
Soluble immune-response suppressor	+	+
Androgen	+	+
Hypertransfusion/datinomycin	+	+

A recent approach to therapy has been the use of total lymphoid irradiation (TLI). It has been demonstrated that TLI can suppress renal disease and markedly increase survival in B/W mice when applied even after the onset of clinical symptoms (Kotzin and Strober, 1979; Slavin, 1979). Dietary restriction has been reported to prolong the life span of B/W mice and to alter their immune function, thereby decreasing autoimmune expression (Fernandes *et al.,* 1978). As discussed previously, Con A-activated spleen cells release a soluble immune-response suppressor into the supernatant. This material has been reported to favorably alter disease manifestations and to prolong survival in aged B/W mice (Krakauer *et al.,* 1976). Treatment of female B/W mice with androgens has also been reported to retard disease expression as discussed previously (Talal and Roubinian, 1978). Finally, in an effort to suppress erythropoiesis and perhaps to limit availability of DNA for immune-complex formation, Gabrielsen and Olsen (1977) hypertransfused B/W mice and reported a 9-week increase in survival time. An erythropoiesis-suppressive drug, Dactinomycin, also has been reported to favorably affect survival (Gabrielsen *et al.,* 1976). It was concluded that erythroblast DNA may be involved in the lupus nephritis of B/W mice. Table VI provides a summary of the various therapeutic regimes applied to NZ mice and their relative effectiveness.

11. Summary

In the past few years there has been an explosion of research information on NZ mice. This has reflected a general increase in knowledge and technology in cellular immunology and in lymphocyte biology. Continuing research in many specific areas promises to expand this increasing data base. The numerous abnormalities of CMI in NZ mice chronicled in the past are currently being analyzed in terms of cell synergism and cell–cell interactions rather than as simple effector mechanisms. Such analysis is indicating more selective rather than general impairment of CMI in aging NZ mice. Similarly, the allogeneic, H-2-identical cytotoxic activity expressed by NZ mice and its relevance to autoimmune disease are being explored. In the area of NZ suppressor functions, continued refinement and standardization of suppressor assays must be achieved before adequate interpretation of the present information can be made. Suppressor activity in aging NZ mice has been described as depressed, normal, and elevated. This is an area of acute interest. While the primary etiologic significance of T-cell regulatory defects is currently being minimized, their possible secondary role in the pathogenesis of autoimmune disease in NZ mice is actively being investigated.

The recent discovery of a significant B-cell abnormality present from birth in all SLE-prone murine strains has dramatically increased the probability that intrinsic B-cell hyperactivity is a primary etiologic factor in murine autoimmunity. The hypersecretion of IgM and the increased incidence of B-cell colony-forming units observed in NZ mice are being subjected to further genetic analysis in many laboratories. Similarly, the continued evaluation of the (CBA/N × NZB) hybrid promises to be very enlightening, particularly the development of congenic mice with the CB/N X chromosome on an NZB background. In addition, the development of H-2-congenic NZB strains and of immunoglobulin allotype-congenic NZB strains is proceeding. These strains will provide information on the possible roles of the MHC and V_H regions in autoimmune pathogenesis. The NZB.ch congenic strain has already proven to be of great value. The continued development and further characterization of immunologically mutant NZ mice have been and will continue to be of immense usefulness in elucidating the mechanisms of autoimmune development in these animals. While NZ mice express a variety of host interactions, recent genetic analysis suggests that viral infection is not a primary etiologic agent in their autoimmune disease. A well-defined clinical entity and a range of abnormalities have been delineated in NZ mice. The present goal is to define the cellular basis of autoimmunity through continued immunologic and genetic analysis of this important animal model of human autoimmune disease.

References

Abramson, N., Gelfand, E. W., Jandl, J. H., and Rosen, R. S., 1970, The interaction between human monocytes and red cells. Specificity for IgG subclasses and IgG fragments, *J. Exp. Med.* **132:**1207.

Akizuki, M., Reeves, J., and Steinberg, A., 1978, Expression of autoimmunity by NZB/NZW marrow, *Clin. Immunol. Immunopathol.* **10:**247.

Allison, A. C., Denman, A. M., and Barnes, R. D., 1971, Cooperating and controlling functions of thymus-derived lymphocytes in relation to autoimmunity, *Lancet* **2:**135.

Amsbaugh, D. F., Hansen, C. T., Prescott, B., Stashak, P. W., Asofsky, R., and Baker, P. J., 1972, Genetic control of the antibody response to type III pneumococcal polysaccharide in mice. I. Evidence that an X-linked gene plays a decisive role in determining responsiveness, *J. Exp. Med.* **136:**931.

Amsbaugh, D. F., Hansen, C. T., Prescott, B., Stashak, P. W., Asofsky, R., and Baker, P. J., 1974, Genetic control of the antibody response to type III pneumococcal polysaccharide in mice. II. Relationship between IgM immunoglobulin levels and the ability to give an IgM antibody response, *J. Exp. Med.* **139:**1499.

Andrews, B. S., Eisenberg, R. A., Theofilopoulos, A. N., Izui, S., Wilson, C. B., McConahey, P. J., Murphy, E. D., Roths, J. B., and Dixon, F. J., 1978, Spontaneous murine lupus-like syndrome. Clinical and immunopathological manifestations in several strains, *J. Exp. Med.* **148:**1198.

Bach, F. H., Bach, M. L., and Sondel, P. M., 1976, Differential function of major histocompatibility complex antigens in T-lymphocyte activation, *Nature (London)* **259:**273.

Bach, J. F., Dardenne, M., and Salomon, J. C., 1973, Studies on thymus products. IV. Absence of serum thymic activity in adult NZB and (NZB $\times$ NZW)F_1 mice, *Clin. Exp. Immunol.* **14:**247.

Baker, P. J., Barth, R. F., Stashak, P. W., and Amsbaugh, D., 1970, Enhancement of the antibody response to type III pneumococcal polysaccharide in mice treated with antilymphocyte serum, *J. Immunol.* **104:**1313.

Barnes, R. D., and Tuffrey, M., 1969, A transplacental factor in the disease of the autoimmune NZB/B1 mouse, *Lancet* **1:**240.

Barnes, R. D., Tuffrey, M., Kingman, J., and Risdon, R. A., 1972, The disease of the NZB mouse. Examinations of exchange ovum transplantation derived NZB and CFW mice, *Clin. Exp. Immunol.* **10:**493.

Barthold, D. R., Kysela, S., and Steinberg, A. D., 1974, Age-dependent loss of suppressor T cells in female NZB mice, *J. Immunol.* **112:**9.

Bhoopalam, N., Yakulis, V. J., Costea, N., and Heller, P., 1971, Membrane immunoglobulins of lymphocytes in NZB mice, *J. Immunol.* **107:**1501.

Bielschowsky, M., and Beilschowsky, F., 1962, Reaction of the reticular tissue of mice with autoimmune haemolytic anaemia to 2-aminofluorene, *Nature (London)* **194:**692.

Bielschowsky, M., and Bielschowsky, F., 1964, Observations on NZB/B_1 mice: Differential fertility in reciprocal crosses and transmission of autoimmune hemolytic anemia to NZB/B_1 hybrids, *Aust. J. Exp. Biol. Med. Sci.* **42:**561.

Bielschowsky, M., and D'Ath, E. F., 1971, The kidneys of NZB-Bl, NZC-Bl and NZY-Bl mice, *J. Pathol.* **103:**97.

Bielschowsky, M., Helyer, B. J., and Howie, J. B., 1959, Spontaneous haemolytic anaemia in mice of the NZB/Bl strain, *Proc. Univ. Otago Med. Sch.* **37:**9.

Blankwater, M., Levert, L., and Hijmans, W., 1974, Age-related decline in the antibody response to *E. coli* lipopolysaccharide in NZB mice, *Immunology* **28:**847.

Borel, Y., Lewis, R., Schwartz, J., and Stellar, B., 1978, Treatment of lupus nephritis in adult BW mice by cortisone-facilitated tolerance to nucleic acid antigens, *J. Clin. Invest* **61**:276.

Botzenhardt, V., Klein, J., and Ziff, M., 1978, Cytotoxic reactions of NZB spleen cells with lymphocytes of MHC identical strains, *J. Exp. Med.* **147**:1435.

Bourgois, A., Kitajima, K., Hunter, I. R., and Askonas, B. A., 1977, Surface imunoglobins of lipopolysaccharide-stimulated spleen cells. The behavior of IgM, IgD, and IgG, *Eur. J. Immunol.* **7**:151.

Braverman, I. M., 1968a, Study of autoimmune disease in New Zealand mice. I. Genetic features and natural history of NZB, NZY and NZW strains and NZB/NZW hybrids, *J. Invest. Dermatol.* **50**:483.

Braverman, I. M., 1968b, Development of tolerance in New Zealand mice, *Clin. Res.* **16**:318.

Braverman, I. M., and Slesinski, J., 1968, Study of autoimmune disease in New Zealand mice. III. Mechanisms underlying haemolytic anaemia of NZB mice, *J. Invest. Dermatol.* **51**:274.

Brown, J. C., De Jesus, D., and Holborow, E. J., 1971, The inability of NZB × B/W hybrid mice to localise altered IgG in splenic germinal centers, *Arthritis Rheum.* **14**:153.

Burnet, F. M., 1959, *The Clonal Selection Theory of Acquired Immunity,* Vanderbilt University Press, Nashville.

Burnet, F. M., 1962, The immunological significance of the thymus, *Australas. Ann. Med.* **11**:79.

Burnet, F. M., 1965, Mast cells in the thymus of NZB mice, *J. Pathol. Bacteriol.* **89**:271.

Burnet, F. M., and Holmes, M. C., 1964, Thymic changes in the mouse strain NZB in relation to the autoimmune state, *J. Pathol. Bacterior.* **88**:229.

Burnet, F. M., and Holmes, M. C., 1965a, The natural history of the NZB/NZWF$_1$ hybrid mouse. A laboratory model of system lupus erythematosus, *Australas. Ann. Med.* **14**:185.

Burnet, F. M., and Holmes, M. C., 1965b, Genetic investigations of autoimmune disease in mice, *Nature (London)* **207**:368.

Butcholar, J. R., and Chapman, B. A., 1966, The influence of sex upon the antibody response to an incompatible tumor, *Immunology* **9**:553.

Cantor, H., 1978, Lymphocytic communication in autoimmunity, in: *Genetic Control of Autoimmune Disease* (N. R. Rose, P. E. Bigazzi, and N. L. Warner, eds.), pp. 151–161, Elsevier/North-Holland, Amsterdam.

Cantor, H., and Boyse, E. A., 1975, Functional subclasses of T lymphocytes bearing different Ly antigens. I. The generation of functionally distinct T-cell subclasses in a differentiative process indepndent of antigen, *J. Exp. Med.* **141**:1376.

Cantor, H., and Weissman, I., 1976, Development and function of subpopulations of thymocytes and T lymphocytes, *Prog. Allergy* **20**:1.

Cantor, H., Asofsky, R., and Talal, N., 1970, Synergy among lymphoid cells mediating the graft-vs-host response. I. Synergy in graft-vs-host reactions produced by cells from NZB/Bl mice, *J. Exp. Med.* **131**:223.

Cantor, H., McVay-Boudreau, L., Hugenberger, J., Naidorf, K., Shen, F. W., and Gershon, R. K., 1978, Immunoregulatory circuits among T cell sets. II. Physiologic role of feedback inhibition *in vivo:* Absence in NZB mice, *J. Exp. Med.* **147**:116.

Carpenter, D. F. Steinberg, A. D., Schur, P. H., and Talal, N., 1970, The pathogenesis of autoimmunity in New Zealand mice. II. Acceleration of glomerulonephritis by polyinosinic-polycytidylic acid, *Lab. Invest.* **23**:628.

Casey, T. P., 1968a, The development of lymphomas in mice with autoimmune disorders treated with azothioprine, *Blood* **31**:396.

Casey, T. P., 1968b, 6-Mercaptopurine administration to NZB mice, *Aust. J. Exp. Biol. Med. Sci.* **46**:327.

Casey, T. P., 1968c, The use of immunosuppressant drugs in the autoimmune disorders of NZB and NZB × NZW hybrid mice, *Bibl. Haematol. (Basel)* **29**:485.

Casey, T. P., and Howie, J. B., 1965, Autoimmune haemolytic anaemia in NZB/Bl mice treated with the corticosteroid drug betamethazone, *Blood* **25**:423.

Cerottini, J. C., Lambert, P. H., and Dixon, F. J., 1969, Comparison of the immune responsiveness of NZB and NZB × NZW F$_1$ hybrid mice with that of other strains of mice, *J. Exp. Med.* **130**:1093.

Chou, S. T., and Herdson, P. B., 1969, Histological and fine structural abnormalities of the livers of NZB × NZW F$_1$ hybrid mice, *Br. J. Exp. Pathol.* **50**:250.

Chused, T. M., Steinberg, A. D., and Parker, L. M., 1973, Enhanced antibody response of mice to polyinosinic polycytidylic acid by anti-thymocyte serum and its age-dependent loss in NZB/W mice, *J. Immunol.* **111**:52.

Chused, T. M., Moutsopoulos, H. M., Sharrow, S. O., Hansen, C. T., and Morse, H. C., III, 1978, Mechanism of autoimmune disease in NZB mice, in: *Genetic Control of Autoimmune Disease* (N. R. Rose, P. E. Bigazzi, and N. L. Warner, eds.), p. 177, Elsevier/North-Holland, Amsterdam.

Chused, T. M., Moutsopoulos, H. M., Sharrow, S. O., and Hansen, C. T., 1979. Evidence of a primary B cell abnormality in NZB mice, in: *B Cells in Immune Responses* (M. Cooper, D. Mosier, I. Scher, and E. Vitteta, ed., Elsevier/North-Holland, Amsterdam.

Cohen, P. L., and Ziff, M., 1977, Abnormal polyclonal B cell activation in NZB/NZW F_1 mice, *J. Immunol.* **119:**1534.

Cohen, P. L., Scher, I., and Mosier, D. E., 1976, *In vitro* studies of the genetically determined unresponsiveness to thymus-independent antigens in CBA/N mice, *J. Immunol.* **116:**301.

Cohen, P. L., Ligler, F. S., Ziff, M., and Vitetta, E. S., 1978, Increased ratio of surface IgM/surface IgD on spleen cells from NZB mice, *Arthritis Rheum.* **21:**551.

Cooke, A., and Playfair, J., 1977, Hyper-responsiveness in NZB mice to experimental induction of anti-red cell autoantibody, *Clin. Exp. Immunol.* **27:**538.

Costea, N., Yakulis, V., and Heller, P., 1970, Cold reacting antibody in NZB-Bl mice, *Blood* **35:**583.

Creighton, D., Crowell, J. S., Jr., and Zinkernagel, R. M., 1978, Cell-mediated immunity in autoimmune mice, *Fed. Proc.* **37:**1380.

Croker, B. P., Del Villano, B. C., Jensen, F. C., Lerner, R. A., and Dixon, F. J., 1975, Immunopathogenicity and oncogenicity of murine leukemia viruses. I. Induction of immunologic disease and lymphoma in (BALB/c × NZB)F_1 mice by Scripps leukemia virus, *J. Exp. Med.* **140:**1028.

Croker, B. P., Zinkernagel, R., and Dixon, F., 1979, Natural cytotoxic cells in NZB mice: Spontaneous *in vivo* and *in vitro* primed activity and structural characterization, *Clin. Immunol. Immunopathol.* **12:**410.

Datta, S. K., and Schwartz, R. S., 1976, Genetics of expression of xenotropic virus and autoimmunity in NZB mice, *Natural (London)* **263:**412.

Datta, S. K., and Schwartz, R. S., 1977, Mendelian segregation of loci controlling xenotropic virus production in NZB crosses, *Virology* **83:**449.

Datta, S. K., and Schwartz, R. S., 1978, Genetic, viral and immunologic aspects of autoimmune disease of NZB mice, in: *Genetic Control of Autoimmune Disease* (N. R. Rose, P. E. Bigazzi, and N. L. Warner, eds.), p. 193, Elsevier/North-Holland, Amsterdam.

Dauphinee, M., and Talal, N., 1979, Failure of NZB spleen to respond to prethymic bone marrow suppressor cells, *J. Immunol.* **122:**936.

Dauphinee, M. J., Talal, N., Goldstein, A. L., and White, A., 1974, Thymosin corrects the abnormal DNA synthetic response of NZB mice, *Proc. Natl. Acad. Sci. USA* **71:**2637.

Dauphinee, M. J., Palmer, D. W., and Talal, N., 1975, Evidence for an abnormal micro-environment in the thymus of New Zealand Black mice, *J. Immunol.* **115:**1054.

DeHeer, D. H., and Edgington, T. S., 1974, Clonal heterogeneity of the anti-erythrocyte autoantibody responses of NZB mice, *J. Immunol.* **113:**1184.

DeHeer, D. H., and Edgington, T. S., 1975, Aberrant maturational characteristics of the immune responses of NZB mice to autologous and heterologous erythrocyte antigens, *Cell. Immunol.* **19:**183.

DeHeer, D. H., and Edgington, T. S., 1976a, Cellular events associated with the immunogenesis of anti-erythrocyte autoantibody responses of NZB mice, *Transplant. Rev.* **31:**116.

DeHeer, D. H., and Edgington, T. S., 1976b, Specific antigen-binding and antibody-secreting lymphocytes associated with the erythrocyte autoantibody responses of NZB and genetically unrelated mice, *J. Immunol.* **116:**1051.

DeHeer, D. H., and Edgington, T., 1977, Evidence for a B lymphocyte defect underlying the anti-x erythrocyte autoantibody response of NZB mice, *J. Immunol.* **118:**1858.

DeHeer, D. H., and Edgington, T., 1980, Aberrant maturation of the immune response to the thymus-independent antigen DNP-Lys-Ficoll by NZB mice, *Clin. Immunol. Immunopathol.* **15:**472.

Denman, A. M., and Denman, E. J. 1970, Proliferative activity in the lymphatic tissue of germ-free NZ black mice, *Int. J. Cancer* **6:**108.

Denman, A. M., Russell, A. S. an Deman, E. J., 1969, Adoptive transfer of the disease of New Zealand black mice to normal mouse strains, *Clin. Exp. Immunol.* **5:**567.

Denman, A. M., Russell, A. S., and Denman, E. J., 1970, Renal disease in (NZB × NZW)F_1 hybrid mice treated with antilymphocytic antibody, *Clin. Exp. Immunol.* **6:**325.

Denman, A. M., Russell, A. S., Loewi, G., and Denman, E. J., 1971, Immunopathology of New Zealand black mice treated with antilymphocytic globulin, *Immunology* **20:**973.

De Vries, M. J., and Hijmans, W., 1966, A deficient development of the epithelium and autoimmune disease in NZB mice, *J. Pathol. Bacteriol.* **81**:487.

De Vries, M. J., and Hijmans, W., 1967, Pathological changes of thymic epithelial cells and autoimmune disease in NZB, NZW and (NZB × NZW)F_1 mice, *Immunology* **12**:179.

Diener, E., 1966, The immune response in NZB and (NZB × C3H)F_1 hybrid mice measured by the haemolysin plaque assay technique, *Int. Arch. Allergy* **30**:120.

Dixon, F. J., Tonietti, G., Oldstone, M. B. A., Aoki, T., and McConahey, P. J., 1970, The effect of chronic viral infections on NZ mice, *6th Immunopathol. Int. Symp.*

Dixon, F. J., Oldstone, M. B., and Tonietti, G., 1971, Pathogenesis of immune complex glomerulonephritis of New Zealand mice, *J. Exp. Med.* **134**:65.

Dresser, D. W., 1962, Specific inhibition of antibody production. I. Protein overloading paralysis, *Immunology* **5**:161.

Dubois, E. L., and Arterberry, J. D., 1966, Etiology, in: *Lupus Erythematosus* (E. L. Dubois, ed.), p. 78, McGraw–Hill, New York.

Dubois, E. L., Horowitz, R. E., Demopoulos, H. B., and Teplitz, R., 1966, NZB/NZW mice as a model of systemic lupus erythematosus, *J. Am. Med. Assoc.* **195**:285.

East, J., 1970, Immunopathology and neoplasms in New Zealand mice and SJL-J mice, *Prog. Exp. Tumour Res.* **13**:84.

East, J., de Sousa, M. A. B., and Parrott, D. M., 1965, Immunopathology of New Zealand black (NZB) mice, *Transplantation* **3**:711.

East, J., de Sousa, M. A. B., Parrot, D. M., and Jaquet, H., 1967a, Consequence of neonatal thymectomy in New Zealand black mice, *Clin. Exp. Immunol.* **2**:203.

East, J., Prosser, P. R., and Holborow, E. J., 1967b, Autoimmune reactions and virus-like particles in germ-free NZB mice, *Lancet* **1**:755.

Eastcott, J., Broitman, S., and Bennet, M., 1979, Vigorous graft vs. host reactions in irradiated adult mice induced by NZB lymphocytes, *Fed. Proc.* **14**:5195.

Eilat, D., Natale, P., Steinberg, A., and Schechter, A., 1977, Properties of t-RNA-specific antibodies from NZB/NZW mice, *J. Immunol.* **118**:1016.

Eilat, D., Asofsky, R., and Laskov, R., 1980, A hybridoma from an autoimmune NZB/NZW mouse producing monoclonal antibody to ribosomal-RNA, *J. Immunol.* **124**:766.

Elder, J. H., Jensen, F. C., and Bryant, M. L., 1977, Polymorphism of the major envelope glycoprotein (gp70) of murine C-type viruses: Virion associated and differentiation antigens encoded by a multi-gene family, *Nature (London)* **267**:23.

Elkerbout, F. A. S., and Hijmans, W., 1974, Relative avidity of antibodies towards sheep red blood cells in New Zealand black mice, *Immunology* **26**:901.

Elson, C. J., and Naysmith, J. D., 1969, The incidence of spontaneously occurring anti-liver antibodies in New Zealand black mice, *Immunology* **16**:537.

Emerit, I., Bernard, C., and Housset, E., 1979, Chromosome breakage and xenotropic C-type virus in embryonic cultures from NZB mice, *Cytogenet. Cell Genet.* **23**:264.

Evans, M. M., Williamson, W. G., and Irvine, W. J., 1968, The appearance of immunological competence at an early age in New Zealand black mice, *Clin. Exp. Immunol.* **3**:375.

Falkoff, R., Scavulli, J., and Dutton, R., 1978, Independent analysis of T helper and T killer cell function in young and old NZB mice, *J. Immunol.* **121**:897.

Fernandes, G., Yunis, E. J., and Good, R. A., 1976, Age and genetic influence on immunity in NZB and autoimmune-resistant mice, *Clin. Immunol. Immunopathol.* **6**:318.

Fernandes, G., Friend, P., Yunis, E., and Good, R. A., 1978, Influence of dietary restriction on immunologic function and renal disease in (NZB × NZW)F_1 mice. *Proc. Natl. Acad. Sci. USA* **75**:1500.

Finbloom, D., and Plotz, P., 1979, Studies of RE function in the mouse with model immune complexes, *J. Immunol.* **123**:1600.

Flad, H. D., Playfair, J. H., Ghaffar, A., and Miescher, P. A., 1969, Strain differences in the liver autoantibody response of inbred mice, *Proc. Soc. Exp. Biol. Med.* **131**:121.

Fournie, G. J., Lambert, P. H., and Miescher, P. A., 1974, Release of DNA in circulating blood and induction of anti-DNA antibodies after injection of bacterial lipopolysaccharides, *J. Exp. Med.* **140**:1189.

Friedman, J., Fialkow, P., Bryant, J., Reddy, A., and Salo, A., 1978, Neoplastic behavior of chromosomally abnormal clones in NZB mice, *Int. J. Cancer* **22**:458.

Gabriel, R., 1971, 1-Methylisatin 3-thiosemicarbazone treatment of NZB × NZW hybrid mice, *Br. J. Exp. Pathol.* **52**:271.

Gabrielsen, A., and Olsen, C., 1977, Prolonged survival of hypertransfused NZB/NZW mice, *Immunology* **33**:449.

Gabrielsen, A. E., Lubert, A. S., and Olsen, C. T., 1976, Supression of murine lupus erythematosus by Dactinomycin, *Nature (London)* **264**:439.

Gazdar, A. F., Beitzel, W., and Talal, N., 1971, The age related responses of New Zealand mice to a murine sarcoma virus, *Clin. Exp. Immunol.* **8**:501.

Gelfand, M. D., and Steinberg, A. D., 1972, Therapeutic studies in NZB/NZW F_1 mice. II. Relative efficacy of azathioprine, cyclophosphamide and methyl prednisolone, *Arthritis Rheum.* **15**:247.

Gelfand, M. D., and Steinberg, A. D., 1973, Mechanism of allograft rejection in New Zealand mice. I. Cell synergy and its age-dependent loss. *J. Immunol.* **110**:1652.

Gelfand, M. D., Parker, L. M., and Steinberg, A. D., 1974, Mechanism of allograft rejection in New Zealand mice. II. Role of serum factor, *J. Immunol.* **113**:1.

Gerber, N. C., Hardin, J. A., Chused, T. M., and Steinberg, A. D., 1974, Loss with age in NZB/W mice of thymic suppressor cells in the graft-vs-host reaction, *J. Immunol.* **113**:1618.

Gershon, R. K., 1974, T cell control of antibody production, in: *Contemporary Topics in Immunobiology* (M. D. Cooper and N. L. Warner, eds.), p. 1, Plenum Press, New York.

Gershon, R. Eardley, D., Naidorf, K., and Cantor, H., 1978, Association of defective feedback suppressor T cell activity with autoimmunity in NZB mice, *Arthritis Rheum.* **21**(Suppl. 5):180.

Gershwin, M. E., and Steinberg, A. D., 1975, Suppression of autoimmune hemolytic anemia in New Zealand (NZB) mice by syngeneic young thymocytes, *Clin. Immunol. Immunopathol.* **4**:38.

Gershwin, M. E., Klingenstein, R. J., Chused, T. M., and Steinberg, A. D., 1974a, *In vitro* and *in vivo* cytotoxicity to EL-4 sarcoma in New Zealand mice, *J. Immunol.* **113**:1968.

Gershwin, M. E., Ahmed, A., Steinberg, A. D., Thurman, G. B., and Goldstein, A. L., 1974b, Correction of T cell function by thymosin in New Zealand mice, *J. Immunol.* **113**:1068.

Gershwin, M. E., Merchant, B., Gelfand, M. D., Vickers, J., Steinberg, A. D., and Hansen, C. T., 1975, The natural history and immunopathology of outbred athymic (nude) mice, *Clin. Immunol. Immunopathol.* **4**:324.

Gershwin, M. E., Steinberg, A. D., Ahmed, A., and Derkay, C., 1976, Study of thymic factors. II. Failure of thymosin to alter the natural history of NZB and NZB/NZW mice, *Arthritis Rheum.* **19**:862.

Gershwin, M. E., Ikeda, R. M., Kruse, W. L., Wilson, F., Shifrine, M., and Spangler, W., 1978, Age-dependent loss in New Zealand mice of morphological and functional characteristics of thymic epithelial cells, *J. Immunol.* **120**:971.

Gershwin, M. E., Castles, J. J., Ikeda, R. M., Erickson, K., and Montero, J., 1979a, Studies of congenitally immunologic mutant New Zealand mice. II. Absence of T cell progenitor populations and B cell defects of congenitally athymic (nude) New Zealand black (NZB) mice, *J. Immunol.* **122**:2020.

Gershwin, M. E., Castles, J. J., Ikeda, R. M., Erickson, K., and Montero, J., 1979b, Studies of congenitally immunologic mutant New Zealand mice. I. Autoimmune features of hereditarily asplenic (Dh/+) NZB mice; reduction of naturally occurring thymocytotoxic antibody and normal suppressor function, *J. Immunol.* **122**:710.

Ghaffar, A., and Playfair, H. J. L., 1971, The genetic basis of autoimmunity in NZB mice studied by progeny testing, *Clin. Exp. Immunol.* **3**:479.

Gilliam, J., 1975, The significance of cutaneous immunoglobulin deposits in lupus erythematosus and NZB/NZW F_1 hybrid mice, *J. Invest. Dermatol.* **65**:154.

Giltinan, P. J., Holmes, M. C., and Burnet, F. M., 1965, Cortisone acetate treatment of haemolytic anaemia in NZB mice, *Aust. J. Exp. Biol. Med. Sci.* **43**:523.

Goldblum, R., Pillarisetty, R., and Talal, N., 1974, Independent appearance of NTA and anti-RNA antibodies in BW mice, *Immunology* **28**:121.

Graff, R. J., Lappi, M. A., and Snell, G. D., 1969, The influence of the gonads and adrenal glands on the immune response to skin grafts, *Transplantation* **7**(2):105.

Greenberg, A., and Playfair, J., 1973, Natural killer cell activity in NZB mice, *Clin. Exp. Immunol.* **15**:445.

Gronowicz, E., and Coutinno, A., 1975, Functional analysis of B cell heterogeneity, *Transplant, Rev.* **24**:3.

Hahn, B., and Knotts, L., 1979, Alteration of lymphocyte function in NZB/NZW mice. IV. Response to levamisole, *Arthritis Rheum.* **22**:236.

Harbeck, R., Hoffman, A., Hoffman, S., Shucord, D., and Carr, R., 1978, A naturally occurring antibody in NZ mice cytotoxic to dissociated cerebellar cells, *Clin. Exp. Immunol.* **31**:313.

Hardin, J. A., Chused, T. M., and Steinberg, A. D., 1973, Suppressor T cells in the graft-versus-host reaction, *J. Immunol.* **111**:650.

Helyer, B. J., and Howie, J. B., 1961, Presence of lupus erythematosus test in crossbred strain of mice NZB/BL–NZY/BL, *Proc. Univ. Otago Med. Sch.* **39**:3.

Helyer, B. J., and Howie, J. B., 1963a, Renal disease associated with positive lupus erythematosus tests in a cross bred strain of mice, *Nature (London)* **197**:197.

Helyer, B. J., and Howie, J. B., 1963b, Spontaneous autoimmune disease in NZB/Bl mice, *Br. J. Haematol.* **9**:119.

Hicks, J. D., 1966, Vascular changes in the kidneys of NZB mice and F_1 NZB × NZW hybrids, *J. Pathol. Bacteriol.* **91**:479.

Hicks, J. D., and Burnet, F. M., 1966, Renal lesions in the "autoimmune" mouse strains NZB and F_1 NZB × NZW, *J. Pathol. Bacteriol.* **91**:467.

Hirano, T., and Nordin, A., 1976, Age associated decline in the *in vitro* development of cytotoxic lymphocyte in NZB mice, *J. Immunol.* **117**:1093.

Hoffman, S., Hoffman, A., Shucord, D., and Harbeck, R., 1978, Antibodies to dissociated cerebellar cells in NZ mice as demonstrated by immunofluorescence, *Brain Res.* **142**:477.

Holborow, E. J., Barnes, R. D. S., and Tuffrey, M., 1965, A new red-cell autoantibody in NZB mice, *Nature (London)* **207**:601.

Holmes, M. C., and Burnet, F. M., 1963a, The natural history of autoimmune disease in NZB mice, *Ann. Intern. Med.* **59**:265.

Holmes, M. C., and Burnet, F. M., 1963b, The influence of splenectomy in NZB mice, *Aust. J. Exp. Biol. Med. Sci.* **41**:449.

Holmes, M. C., and Burnet, F. M., 1964, The inheritance of autoimmune disease in mice. A study of hybrids of the strains NZB and C3H, *Heredity* **19**:419.

Holmes, M. C., and Burnet, F. M., 1966, Thymic changes in NZB mice and hybrids, in: *Thymus: Experimental and Clinical Studies* (G. E. W. Wolstenholme and R. Porter, eds.), Churchill, London.

Holmes, M. C., Gorrie, J., and Burnet, F. M., 1961, Transmission by splenic cells of an autoimmune disease occurring spontaneously in mice, *Lancet* **2**:638.

Howie, J. B., and Helyer, B. J., 1965, Autoimmune disease in mice, *Ann. N.Y. Acad. Sci.* **124**:167.

Howie, J. B., and Helyer, B. J., 1966, The influence of neonatal thymectomy and thymus grafting on spontaneous autoimmune disease in mice, in: *Thymus: Experimental and Clinical Studies* (G. E. W. Wolstenholme and R. Porter, eds.), Churchill, London.

Howie, J. B., and Helyer, B. J., 1968, The immunology and pathology of NZB mice, *Adv. Immunol.* **9**:215.

Howie, J. B., and Simpson, L. D., 1976, The immunopathology of the NZB mice and their hybrids, in: *Textbook of Immunopathology,* 2nd ed. (P. A. Miescher and H. J. Mueller-Eberhard, eds.), pp. 247–280, Grune & Stratton, New York.

Izui, S., Lambert, P. H., Fournie, G. J., Turler, H., and Miescher, P. A., 1977, Features of systemic lupus erythematosus in mice injected with bacterial lipopolysaccharides. Identification of circulating DNA and renal localization of DNA–anti-DNA complex, *J. Exp. Med.* **145**:1115.

Izui, S., McConahey, P., and Dixon, F. J., 1978, Increased spontaneous polyclonal activation of B lymphocytes in mice with spontaneous autoimmune disease, *J. Immunol.* **121**:2213.

Izui, S., McConahey, P., Theofilopoulos, A., and Dixon, F., 1979, Association of circulating retroviral gp 70–anti-gp 70 immune complexes with murine systemic lupus erythematosus, *J. Exp. Med.* **149**:1099.

Jacobs, M. E., Gordon, J. K., and Talal, N., 1970, The role of the NZB/NZW thymus and bone marrow in cyclophosphamide induced tolerance, *Arthritis Rheum.* **13**:326.

Jacobs, M. E., Gordon, J. K., and Talal, N., 1971, Inability of the NZB/NZW F_1 thymus to transfer cyclophosphamide-induced tolerance to sheep erythrocytes, *J. Immunol.* **107**:359.

Jerne, N. K., 1971, The somatic generation of immune recognition, *Eur. J. Immunol.* **1**:1.

Kamm, D. P., Miyasato, F., and Pollak, V. E., 1966, Serum proteins in NZB/Bl and related strains of mice, *Proc. Soc. Exp. Biol. Med.* **122**:903.

Kassan, S., and Chused, T., 1977, Impairment of primary *in vitro* response of NZ mouse spleen cells to a thymic-dependent antigen, *Cell. Immunol.* **20**:135.

Katz, D. H., Paul, W. E., and Benacerraf, B., 1973, Carrier function in antihapten antibody responses. VI.

Establishment of experimental conditions for either inhibitory or enhancing influences of carrier-specific cells on antibody production, *J. Immunol.* **110:**107.

Kaye, D., and Hook, E. W., 1964, Influence of autoimmune haemolytic anaemia and susceptibility to *Salmonella* infection, *Proc. Soc. Exp. Biol. Med.* **117:**135.

Kessler, H. S., 1968, A laboratory model for Sjögren's syndrome, *Am. J. Pathol.* **52:**671.

Kincade, P. W., 1977, Defective colony formation by B lymphocytes from CBA/N and C3H/HeJ mice, *J. Exp. Med.* **145:**249.

Kincade, P., Lee, G., Fernandes, G., Moore, M., Williams, N., and Good, R., 1979, Abnormalities in clonable B lymphocytes and myeloid progenitors in autoimmune NZB mice, *Proc. Natl. Acad. Sci. USA* **76:**3464.

Klassen, L. W., Krakauer, R. S., and Steinberg, A. D., 1977a, Selective loss of suppressor cell function in New Zealand mice induced by NTA, *J. Immunol.* **119:**830.

Klassen, L. W., Zarate, A., Gelfand, M. C., and Steinberg, A. D., 1977b, Suppressor cell activation by anti-Thy 1.2 in NZB/WF$_1$ mice, *J. Immunol.* **119:**2067.

Knight, J. G., Adams, D. D, and Purves, H. D., 1977, The genetic contribution of the NZB mouse to the renal disease of the NZB × NZW hybrid, *Clin. Exp. Immunol.* **28:**352.

Kongshavn, P. A. L., and Bliss, J. Q., 1969, Sex differences in survival of H-2 incompatible skin grafts in mice treated with antilymphocyte serum, *Nature (London)* **226:**451.

Kotzin, B., and Strober, S., 1979, Reversal of NZB/NZW disease with lymphoid irradiation, *J. Exp. Med.* **150:**371.

Krakauer, K., Torrey, S., and Zurier, R., 1978, Prostaglandin E$_1$ treatment of NZB/W mice, *Clin. Immunol. Immunopathol.* **11:**256.

Krakauer, R. S., Waldmann, T. A., and Strober, W., 1976, Loss of suppressor T cells in adult NZB/NZW mice, *J. Exp. Med.* **144:**662.

Lambert, P. H., and Dixon, F. J., 1968, Pathogenesis of the glomerulonephritis of NZB/W mice, *J. Exp. Med.* **127:**507.

Lambert, P. H., and Dixon, F. J., 1970, Genesis of antinuclear antibody in NZB/W mice: Role of genetic factors and of viral infections, *Clin. Exp. Immunol.* **6:**829.

Lambert, P., and Oldstone, M., 1973, Host IgG and complement deposits in the choroid plexus during spontaneous immune complex disease, *Science* **180:**408.

Lerner, R. A., Jensen, F., and Kennel, S. J., 1972, Karyotypic virologic and immunologic analyses of two continuous lymphocyte lines established from New Zealand black mice: Possible relationship of chromosomal mosaicism to autoimmunity, *Proc. Natl. Acad. Sci. USA* **69:**2965.

Lerner, R. A., Wilson, C. B., Del Villano, B. C., McConahey, P. J., and Dixon, F. J., 1976, Endogenous oncornaviral gene expression in adult and fetal mice: Quantitative, histologic and physiologic studies of the major viral glycoprotein gp 70, *J. Exp. Med.* **143:**151.

Leventhal, B. G., and Talal, N., 1970, Response of NZB and NZB/NZW spleen cells to mitogenic agents, *J. Immunol.* **104:**918.

Levy, J. A., 1974, Autoimmunity, neoplasia and C-type viruses, *Am. J. Clin. Pathol.* **62:**261.

Levy, J. A., 1976, Endogenous C-type viruses: Double agents in natural life processes, *Biomedicine* **24**(2):84.

Levy, J. A., and Pincus, T., 1970, Demonstration of biological activity of a murine leukemia virus of New Zealand black mice, *Science* **170:**326.

Levy, J. A., Kazan, P., and Varnier, O., 1975, Murine xenotropic type-C virus. I. Distribution and further characterization of the virus in NZB mice, *J. Virol.* **16:**844.

Liburd, E., Russel, A., and Dossetor, J., 1974, An antibody to isogenic fibroblasts in the serum of diseased NZB mice, *Clin. Exp. Immunol.* **16:**637.

Linder, E., 1977, Autoantibodies against the inner aspect of erythrocyte membranes in NZB mice, *Clin. Exp. Immunol.* **27:**531.

Linder, E., and Edgington, T. S., 1972a, Antigenic specificity of anti-erythrocyte autoantibody responses by NZB mice: Identification and partial characterization of two erythrocyte surface autoantigents, *J. Immunol.* **180:**1615.

Linder, E., and Edgington, T. S., 1972b, Cell surface autoantigenic expressions and their circulating analogues in the pathogenesis of immune injury in NZB mice, *Fed. Proc.* **31:**766.

Linder, E., and Edgington, T. S., 1973, Immunobiology of the autoantigody response. I. Circulating analogues of erythrocyte autoantigens and heterogeneity of the autoimmune responses of NZB mice, *Clin. Exp. Immunol.* **13:**279.

Lindsey, E. S., Donaldson, G. W., and Woodruff, M. F. A., 1966, Erythrocyte survival in normal mice and in mice with autoimmune haemolytic anaemia, *Clin. Exp. Immunol.* **8:**210.

Loh, E., Black, B., Riblet, R., Weigert, M., Hood, J., and Hood, L., 1979, Myeloma proteins from NZB and BALB/c mice, *Proc. Natl. Acad. Sci. USA* **76:**1395.

McCoombs, C., Hom, J., Talal, N., and Mishell, R. I., 1974, Decreased response of cultured New Zealand mouse spleen cells to sheep erythrocytes, *J. Immunol.* **112:**326.

McCoombs, C., Hom, J., Talal, N., and Mishell, R. I., 1975, Functional deficiency of splenic adherent cells in New Zealand black mice, *J. Immunol.* **115:**1695.

McGiven, A. R., and Ghose, T., 1968, Antinuclear factor in NZB–NZW mice: Incidence and *in vitro* effects, *Clin. Exp. Immunol.* **3:**657.

McGiven, A. R., and Ironside, P. N., 1968, Elution of antinuclear factor from renal lesions of NZB/NZW mice, *Clin. Exp. Immunol.* **3:**655.

Manny, N., Datta, S., and Schwartz, R., 1979, Synthesis of IgM by cells of NZB and SWR mice and their crosses, *J. Immunol.* **122:**1220.

Masters, J. M., and Spurling, C. L., 1967, Implantation of diffusion chambers containing NZB thymus, *Blood* **30:**569.

Mehta, J., Knotts, L., Groig, C., Burkholder, S., Miller, C., and Hahn, B., 1978, Effect of altered lymphocyte function on immunologic disorders in NZB/NZW mice, *Arthritis Rheum.* **21:**196.

Mellors, R. C., 1965, Autoimmune disease in NZB/Bl mice. I. Pathology and pathogenesis of model system of spontaneous glomerulonephritis, *J. Exp. Med.* **122:**25.

Mellors, R. C., 1966a, Autoimmune and immunoproliferative disease of NZB/Bl mice and hybrids, *Int. Rev. Exp. Pathol.* **5:**217.

Mellors, R. C., 1966b, Autoimmune disease in NZB/Bl mice, *J. Exp. Med.* **123:**1024.

Mellors, R. C., and Huang, C. Y., 1966, Immunopathology of NZB/Bl mice. V. Virus-like (filtrable) agent separable from lymphoma cells and identifiable by electron microscopy, *J. Exp. Med.* **124:**1031.

Mellors, R. C., and Huang, C. Y., 1967, Immunopathology of NZB/Bl mice. VI. Virus separable from spleen and pathogenic for Swiss mice, *J. Exp. Med.* **126:**53.

Mellors, R. C., Aoki, T., and Heubner, R. J., 1969, Further implication of murine leukemia-like virus in the disorders of NZB mice, *J. Exp. Med.* **129:**1045.

Mellors, R. C., Shirai, T., and Aoki, T., 1971, Wild-type Gross leukemia virus and the pathogenesis of the glomerulonephritis of New Zealand mice, *J. Exp. Med.* **133:**113.

Michalski, J., McCombs, C., and Talal, N.,1979, Suppressor cells and immunodeficiency in (NZW)F_1 hybrid mice, *Eur. J. Immunol.* **9:**440.

Milich, D. R., and Gershwin, M. E., 1977, T cell differentiation and the nude mouse, *Dev. Comp. Immunol.* **1:**289.

Milich, D. R.,and Gershwin, M. E., 1979a, Murine autoimmune hemolytic anemia induced via xenogeneic erythrocyte immunization. I. Qualitative characteristics and strain variation, susceptibility to induction, *Clin. Immunol. Immunopathol.* **14:**176.

Milich, D. R., and Gershwin, M. E., 1979b, Murine autoimmune hemolytic anemia induced via xenogeneic erythrocyte immunization. II. Antigenic specificity and clonal heterogeneity and cellular correlates of the response, *Clin. Immunol. Immunopathol.* **14:**184.

Milich, D. R., and Gershwin, M. E., 1981, Murine autoimmune hemolytic anemia induced via xenogeneic erythrocyte immunization. III. Influence of sex, *Clin. Immunol. Immunopathol.,* **18:**1.

Miyasato, F., Manaligod, M., Pietra, G. G., and Pollak, V. E., 1967, Effects of neonatal thymectomy on spontaneous murine autoimmune disease, *Proc. Soc. Exp. Biol. Med.* **125:**1180.

Moller, G., Gronowicz, E., Persson, U., Coutinho, A., Moller, E., Hammarstrom, L., and Smith, E., 1976, Spleen cells from animals tolerant to a thymus-dependent antigen can be activated by lipopolysaccharide to synthesize antibodies against the tolerogen, *J. Exp. Med.* **143:**1429.

Morse, H., Riblet, R., Asofsky, R., and Weigert, M., 1978, Plasmacytoma of the NZB mouse, *J. Immunol.* **121:**1969.

Morton, J. I., and Siegel, B. V., 1969, Response of NZB mice to foreign antigen and development of autoimmune disease, *J. Reticuloendothelial Soc.* **6:**78.

Morton, J. I., and Siegel, B. V., 1974, Early autoantibody formation in lethally irradiated or drug-treated H-2-compatible recipients of pre-autoimmune NZB bone marrow or fetal liver cells, *Transplantation* **17:**624.

Morton, J., and Siegel, B., 1979, Transplantation of autoimmune potential. IV. Reversal of the NZB autoimmune syndrome by bone marrow transplantation, *Transplantation* **27**:133.

Moses, E., and Fuchs, S., 1974, Linkage between immune response potential to DNA and X chromosome, *Nature (London)* **249**:167.

Moutsopoulos, H. M., Boehm-Truitt, M., Kassan, S. S., and Chused, T. M., 1977, Demonstration of activation of B lymphocytes in New Zealand mice at birth by an immunoradiometric assay for murine IgM, *J. Immunol.* **119**:1639.

Nakajima, P., Datta, S., Schwartz, R., and Huber, B., 1979, Localization of spontaneously hyperactive B cells in NZB mice to a specific B cell subset, *Proc. Natl. Acad. Sci. USA* **76**:461.

Norins, L. C., and Holmes, M. C., 1964, Antinuclear factor in mice, *J. Immunol.* **93**:148.

Nossal, G. J. V., 1965, Self recognition, *Ann. N.Y. Acad. Sci.* **124**:37.

Ochiai, T., Ahmed, A., Scher, I., Sell, K. W., and Steinberg, A. D., 1976, Functional characterization of a naturally occurring antibody cytotoxic for a subpopulation of splenic T cells, *Transplantation* **22**:1.

Ohsugi, Y., and Gershwin, M. E., 1979, Studies of congenitally immunologic mutant New Zealand mice. II. Growth of B lymphocyte clones in congenitally athymic (nude) and hereditarily asplenic (Dh/+) NZB mice: A primary B cell defect, *J. Immunol.* **123**:1260.

Owen, D., Hill, A., and Argetn, S., 1975, Reaction of mouse strains to skin test for ectromelia using an allied virus as inoculum, *Nature (London)* **254**:598.

Palmer, D., Dauphinee, M., Murphy, E., and Talal, N., 1976, Hyperactive T cell function in young NZB mice: Increased proliferative responses to allogeneic cells, *Clin. Exp. Immunol.* **23**:578.

Pansky, B., and Freimer, M., 1974, Spontaneous carditis in the NZB mouse and its hybrids, *Arthritis Rheum* **17**:403.

Papoian, R., and Talal, N., 1980, Ability of NZW but not NZB antigen-presenting cells to support T cell proliferative responses to DNA methylated albumin, *J. Immunol.* **124**:515.

Papoian, R., Pillarisetty, R., and Talal, N., 1977, Immunological regulation of spontaneous antibodies to DNA and RNA. II. Sequential switch from IgM to IgG in NZB/NZW F_1 mice, *Immunology* **32**:75.

Parker, J., Staples, P. H., and Talal, N., 1968, Inability to induce tolerance in NZB and B/W mice, *Fed. Proc.* **27**:695.

Parker, L. M., Chused, T. M., and Steinberg, A. D., 1974, Immunofluorescent studies on thymocytoxic antibody from New Zealand black mice, *J. Immunol.* **112**:285.

Pertschuk, L., Katariva, S., and Rainford, E., 1976, Relationship of subepidermal immune complex to renal deposits in the NZB/NZW F_1 mouse, *Dermatologica* **153**:7.

Petty, R. E., and Steward, M. W., 1972, Relative affinity of antiprotein antibodies in New Zealand mice, *Clin. Exp. Immunol.* **12**:343.

Playfair, J. H. L., 1968, Strain differences in the immune response of mice. I. The neonatal response to sheep red blood cells, *Immunology* **15**:35.

Playfair, J. H. L., 1971, Strain differences in the immune response of mice. II. A raised threshold in NZB thymus cells, *Immunology* **21**:1037.

Playfair, J. H. L., and Marshall-Clark, S., 1973, Induction of red cell auto-antibodies in normal mice, *Nature New Biol.* **243**:213.

Plescia, O. J., and Braun, W., (eds.), 1968, *Nucleic Acids in Immunology,* Springer, New York.

Primi, D., Hammarstrom, L., Smith, C. I. E., and Moller, G., 1977, Characterization of self-reactive B cells by polyclonal B-cell activators, *J. Exp. Med.* **145**:21.

Primi, D., Hammarstrom, L., and Smith, E., 1978, Genetic control of lymphocyte suppression. I. Lack of suppression in aged NZB mice is due to a B cell defect, *J. Immunol.* **121**:2241.

Purves, E. C., and Playfair, J. H. L., 1973, Normal tolerance characteristics of the antibody-forming cell precursors of the NZB mouse, *Clin. Exp. Immunol.* **15**:113.

Raff, M. C., and Cantor, H., 1971, Subpopulations of thymus cells and thymus-derived lymphocytes, in: *Progress in Immunology* (B. Amos, ed.), p. 83, Academic Press, New York.

Ranney, D. F., and Steinberg, A. D., 1976, Differences in the age-dependent release of a low molecular weight suppressor (LMWS) and stimulators by normal and NZB/W lymphoid organs, *J. Immunol.* **117**:1219.

Raveche, E. S., Klassen, L. W., and Steinberg, A. D., 1976, Sex differences in formation of anti-T cell antibodies, *Nature (London)* **263**:415.

Raveche, E. S., Steinberg, A. D., Klassen, L. W., and Tjio, J. H., 1978, Genetic studies in NZB mice. I. Spontaneous autoantibody production, *J. Exp. Med.* **147**:1487.

Raveche, E., Tjio, J., and Steinberg, A., 1979, Genetic studies in NZB mice. II. Hyperdiploidy in the spleen of NZB mice and their hybrids, *Cytogenet. Cell Genet.* **23:**182.

Roder, J. C., Bell, D. A., and Singhal, S. K., 1975, T cell activation and cellular cooperation in autoimmune NZB/NZW F_1 hybrid mice, *J. Immunol.* **115:**466.

Roder, J., Bell, D., and Singhal, S., 1977, Regulation of the immune response in autoimmune BW F_1 mice, *Cell. Immunol.* **29:**272.

Rodey, G. E., Good, R. A., and Yunis, E. J., 1971, Progressive loss *in vitro* of cellular immunity with ageing in strains of mice susceptible to autoimmune disease, *Clin. Exp. Immunol.* **9:**305.

Roubinian, J. R., Papoian, R., and Talal, N., 1977a, Androgenic hormones moderate autoantibody responses and improve survival in murine lupus, *J. Clin. Inves.* **59:**1066.

Roubinian, J., Papoian, R., Pillarisetty, R., Sawada, S., and Talal, N., 1977b, Immunological regulation of spontaneous antibodies to DNA and RNA. III. Early effects of neonatal thymectomy and splenectomy, *Immunology* **33:**399.

Russell, A. S., 1968, Statolon in murine autoimmune disease, *Lancet* **2:**407.

Russell, P. J., and Hicks, J. D., 1968, Cyclophosphamide treatment of renal disease in (NZB × NZW)F_1 mice, *Lancet* **1:**440.

Sakane, T., Steinberg, A. D., and Green I., 1978, Failure of autologous mixed lymphocyte reactions between T and non-T cells in patients with systemic lupus erythematosus, *Proc. Natl. Acad. Sci. USA* **75:**3463.

Scher, I., Frantz, I., and Steinberg, A. D., 1973, The genetics of the immune response to synthetic double stranded RNA in a mutant CBA mouse strain, *J. Immunol.* **110:**1396.

Scher, I., Ahmed, A., Strong, D. M., Steinberg, A. D., and Paul, W. E., 1975, X-linked B-lymphocyte immune defect in CBA/HN mice. I. Studies of the function and composition of spleen cells, *J. Exp. Med.* **141:**788.

Scher, I., Sharrow, S. O., and Paul, W. E., 1976, A-linked B-lymphocyte defect in CBA/N mice. III. Abnormal development of B-lymphocyte populations defined by their density of surface immunoglobulin, *J. Exp. Med.* **144:**507.

Schur, P. H., Stollar, D. B., Steinberg, A. D., and Talal, N., 1971, Incidence of antibodies to double-stranded RNA in systemic lupus erythematosus and related diseases, *Arthritis Rheum.* **14:**342.

Seaman, W., Blackman, M., Greenspan, J., and Talal, N., 1980, Effect of ^{89}Sr on immunity and autoimmunity in NZB/NZW $_1$ mice, *J. Immunol.* **124:**812.

Senitzer, D., Cafruny, W., Poeder, R., and Freimer, E., 1979, Autoimmune cell-mediated responses to syngeneic cells in NZB/NZW F_1 mice, *J. Immunol.* **123:**660.

Shirai, T.,and Mellors, R. C., 1971, Natural thymocytotoxic autoantibody and reactive antigen in New Zealand black and other mice, *Proc. Natl. Acad. Sci. (Wash.)* **68:**1412.

Shirai, T., and Mellors, R. C., 1972, Natural cytotoxic autoantibody against thymocytes in NZB mice, *Clin. Exp. Immunol.* **12:**133.

Shirai, T., Yoshiki, T., and Mellors, R. C., 1972, Age-decrease of cells sensitive to an autoantibody-specific for thymocytes and thymus-dependent lymphocytes in NZB mice, *Clin. Exp. Immunol.* **12:**455.

Shirai, T., Yoshiki, T., and Mellors, R. C., 1973, Effect of natural thymocytotoxic autoantibody of NZB mice and of specifically prepared antilymphocyte serum on the tissue distriution of Cr-labeled lymphocytes, *J. Immunol.* **110:**517.

Shirai, T., Hayakawa, K., Okumura, K. O., and Tada, T., 1978, Differential cytotoxic effect of natural thymocytotoxic autoantibody of NZB mice on functional subsets of T cells, *J. Immunol.* **120:**1924.

Slavin, S., 1979, Successful treatment of autoimmune disease in (NZB/NZW)F_1 female mice by using fractionated total lymphoid irradiation, *Proc. Natl. Acad. Sci. USA* **76:**5274.

Smith, J., and Pasternak, R., 1978, Syngeneic mixed lymphocyte reaction in mice: Strain distribution, kinetics, participating cells, and absence in NZB mice, *J. Immunol.* **121:**1889.

Sontheimer, R., and Gilliam, J., 1979, Regional variation in the deposition of subepidermal immunoglobulin in NZB/W F_1 mice: Association with epidermal DNA synthesis, *J. Invest. Dermatol.* **72:**25.

Staples, P. J., and Talal, N., 1969, Relative inability to induce tolerance in adult NZB and NZB/NZW F_1 mice, *J. Exp. Med.* **129:**123.

Staples, P. J., Steinerg, A. D., and Talal, N., 1970, Induction of immunological tolerance in older New Zealand mice repopulated with young spleen, bone marrow, or thymus, *J. Exp. Med.* **131:**123.

Steinberg, A. D., and Talal, N., 1971, Suppression of antibodies to nucleic acids with polyinosinic polycytidylic acid and cyclophosphamide in murine lupus. *Clin Exp. Immunol.* **7:**687.

Steinberg, A. D., Baron, S. H., and Talal, N., 1969, The pathogenesis of autoimmunity in New Zealand mice.

I. Induction of anti-nucleic acid antibodies in polyinosinic-polycytidylic acid, *Proc. Natl. Acad. Sci. (Wash.)* **63**:1102.

Steinberg, A. D., Law, L. W., and Talal, N., 1970, The role of the NZB/NZW F_1 thymus in experimental tolerance and autoimmunity, *Arthritis Rheum.* **13**:369.

Steinberg, A. D., Pincus, T., and Talal, N., 1971, The pathogenesis of autoimmunity in New Zealand mice. 3. Factors influencing the formation of anti-nucleic acid antibodies, *Immunology* **20**:523.

Stern, K., and Davidsohn, I., 1955, Effect of estrogen and cortisone on immune hemoantibodies in mice of inbred strains, *J. Immunol.* **74**:479.

Steward, M. W., and Hay, F. C., 1976, Changes in immunoglobulin class and subclass of anti-DNA antibodies with increasing age in NZB/NZW F_1 hybrid mice, *Clin. Exp. Immunol.* **26**:363.

Steward, M., Katz, F., and West, N., 1975, The role of low affinity antibody in immune complex disease, *Clin. Exp. Immunol.* **21**:121.

Stiller, C. R., Russell, A. S., McConnachie, P., Dossetor, J. B., and Diener, F., 1973, Cell mediated immunity in NZB mice, *Clin. Exp. Immunol.* **15**:445.

Stobo, J. D., Talal, N., and Paul, W. E., 1972a, Lymphocyte classes in New Zealand mice. I. Ontogeny and mitogen responsiveness of thymocytes and thymus-derived lymphocytes, *J. Immunol.* **109**:692.

Stobo, J. D., Talal, N., and Paul, W. E., 1972b, Lymphocyte classes in New Zealand mice. II. Decreased frequency of immunoglobulin-bearing lymphocytes and increased frequency of lymphocytes lacking detectable (θ) or immunoglobulin determinants, *J. Immunol.* **109**:701.

Stollar, B. D., 1970, Double-helical polynucleotides in immunochemical recognition of differing conformations, *Science* **169**:609.

Strand, M., and August, J. T., 1976, Oncornavirus envelope glycoprotein in serum of mice, *Virology* **75**:130.

Stutman, O., 1972, Lymphocyte subpopulations in NZB mice: Deficit of thymus-dependent lymphocytes, *J. Immunol.* **109**:602.

Stutman, O., Yunis, E. J., and Good, R. A., 1968, Deficient immunologic functions of NZB mice, *Proc. Soc. Exp. Biol. Med.* **127**:1204.

Symoens, J., and Rosenthal, M., 1977, Levamisole in the modulation of the immune response: The current experimental and clinical state, *J. Reticuloendothelial Soc.* **175**:221.

Talal, N., 1970, Immunologic and viral factors in the pathogenesis of systemic lupus erythematosus, *Arthritis Rheum.* **13**:887.

Talal, N., and Roubinian, J., 1978, Sex hormones and autoimmunity, in: *Genetic Control of Autoimmune Disease* (N. R. Rose, P. E. Bigazzi, and N. L. Warner, eds.), Vol. 1, pp. 163–176, Elsevier/North-Holland, Amsterdam.

Talal, N., and Steinberg, A., 1974, The pathogenesis of autoimmunity in NZB mice, in: *Current Topics in Microbiology and Immunology* (W. Arber *et al.*, eds.), Vol. 64, pp. 79–103, Springer-Verlag,Berlin.

Talal, N., Dauphinee, M., and Pillarisetty, R., 1975, Effect of thymosin on thymocytic proliferation and autoimmunity in NZB mice, *Ann. N.Y. Acad. Sci.* **249**:438.

Taniguchi, M., Tada, T., and Tokuhisa, T., 1976, Properties of the antigen-specific suppressive T-cell factor in the regulation of antibody response of the mouse. III. Dual gene control of the T-cell-mediated suppression of the antibody response, *J.Exp. Med.* **114**:20.

Taurog, J., Moutsopoulos, H., Rosenberg, Y., Chused, T., and Steinberg, A., 1979, CBA/N X-linked B cell defect prevents NZB B-cell hyperactivity in F_1 mice, *J. Exp. Med.* **150**:31.

Teague, P. O., Yunis, E. J., Rodey, G., Fish, A. J., Stutman, O., and Good, R. A., 1970, Autoimmune phenomena and renal disease in mice. Role of thymectomy, aging and involution of immunologic capacity, *Lab. Invest.* **22**:121.

Theofilopoulos, A. N., Eisenberg, R. A., Bourdon, M., Crowell, J. S., Jr., and Dixon, F. J., 1979a, Distribution of lymphocytes identified by surface markers in murine strains with SLE-like syndromes, *J. Exp. Med.* **149**:516.

Theofilopoulos, A., Showler, D., Katz, D.,and Dixon, F., 1979b, Patterns of immune reactivity in autoimmune murine strains, *J. Immunol.* **122**:2319.

Theofilopoulos, A., McConahey, P., Izui, S., Eisenberg, R., Pereira, B., and Creighton, D., 1980, A comparative immunologic analysis of several murine strains with autoimmune manifestations, *Clin. Immunol. Immunopathol.* **15**:258.

Thomas, H. I. J., and Weir, D. M., 1972, The *in vitro* activity of peritoneal exudate cells from New Zealand black mice (NZB/Bl) in the presence of I^{125}-BSA (bovine serum albumin), *Clin. Exp. Immunol.* **12**:263.

Tonietti, G., Oldstone, M. B. A., and Dixon, F. J., 1970, The effect of induced chronic viral infections on the immunologic diseases of New Zealand mice, *J. Exp. Med.* **132:**89.

Tuffrey, M., Kingman, J., and Barnes, R., 1973, The failure to transfer the NZB disease following implantation of NZB thymus-containing chambers, *Eur. J. Immunol.* **3:**519.

Veit, B. C., and Michael, J. G., 1972, The lack of thymic influence in regulating the immune response to *Escherichia coli* 0127 endotoxin, *J. Immunol.* **109:**547.

Viklicky, V., and Polackova, M., 1969, Mast cell incidence and cell proliferation in the lymphoid organs of NZB mice, *Folia Biol.* **15:**432.

Vogler, C., 1977, Effect of levamisole on the NZB/NZW mouse model for lupus eythematosus, *Clin. Res.* **25:**637 (abstract).

Waksman, B. H., Raff, M. C., and East, J., 1972, T and B lymphocytes in New Zealand black mice. An analysis of the theta TL and NBLA markers, *Clin. Exp. Immunol.* **11:**1.

Warner, N. L., 1973, Genetic control of spontaneous and induced anti-erythrocyte autoantibody production in mice, *Clin. Immunol. Immunopathol.* **1:**353.

Warner, N. L., 1974, Immunoglobulins in NZB/Bl mice. II. Phenotypic restriction of anti-red cell autoantibody production in NZB hybrid mice, *Clin. Immunol. Immunopathol.* **2:**556.

Warner, N. L., 1978, Genetic aspects of immunologic abnormalities in New Zealand mouse strains, *Arthritis Rheum.* **21** (Suppl. 5):106.

Warner, N., and Moore, M., 1971, Defects in hematopoietic differentiation in NZB and NZC mice, *J. Exp. Med.* **134:**313.

Warner, N. L., and Wistar, R. J., 1968, Immunoglobulins in NZB-Bl mice. I. Serum immunoglobulin class of erythrocyte autoantibody, *J. Exp. Med.* **127:**169.

Weir, D. M., McBride, W., and Naysmith, J. D., 1968, Immune response to a soluble protein antigen in NAB mice, *Nature (London)* **219:**1276.

Whaley, K., and Singh, H., 1973, *In vitro* studies on the phagocytosis of *Staphylococcus aureus* by peritoneal macrophages of New Zealand mice, *Immunology* **25:**24.

Yachnin, S., 1962, Non antigenicity of synthetic polyribonucleotides and apurinic acid, *Nature (London)* **196:**1319.

Yosiki, T., Mellors, R. C., and Strand, M., 1974, The viral envelope glycoprotein of murine leukemia virus and the pathogenesis of immune complex glomerulonephritis of New Zealand mice, *J. Exp. Med.* **140:**1011.

Yumoto, T., and Dmochowski, L., 1967, Studies on the relationship of murine leukemia virus to autoimmune disease of mice prone to leukemia, *Medical Recordings Annual* **60:**133.

Yunis, E. J., Hong, R., Grewe, M. A., Martinex, C., Cornelius, E., and Good, R. A., 1967, Post-thymectomy wasting associated with autoimmune phenomena. Antiglobulin-positive anemia in A and C57BL/6 Ks mice, *J. Exp. Med.* **125:**947.

Zatz, M. M., Mellors, R. C., and Lance, E. M., 1971, Changes in lymphoid populations of aging CBA and NZB mice, *Clin. Exp. Immunol.* **8:**491.

Zinkernagel, R. M., and Dixon, F. J., 1977, Comparison of T cell-mediated immune responsiveness of NZB (NZB × NZW)F_1 hybrid and other murine strains, *Clin. Exp. Immunol.* **29:**110.

Zurier, R., Damjonov, I., Sayadoff, D., and Rothfield, N., 1977, Prostaglandin E_1 treatment of NZB/NZW F_1 hybrid mice. II. Prevention of glomerulonephritis, *Arthritis Rheum.* **20:**1449.

5
Hemopoietic Abnormalities in New Zealand Black and Motheaten Mice

PAUL W. KINCADE

1. Introduction

The two strains of mice to be compared in this chapter develop elevated numbers of immunoglobulin (Ig)-secreting cells and the products of these cells include autoantibodies. The importance of this observation is underscored by the fact that it is common to several other autoimmune strains and their hybrids and by experiments suggesting that it may occur independently of T lymphocytes. In the case of motheaten mice, a single recessive mutation is responsible for death at an early age. In contrast, several genes may necessarily interact under the influence of hormones to produce the comparatively mild disease of NZB mice. The objective of this chapter will be to review certain observations pertaining to hemopoietic and B-cell abnormalities in NZB and motheaten mice. Speculation will run along the line of possible relationships between B-lymphocyte activation and defects in non-lymphoid cells. Cell transfer experiments and *in vitro* studies that bear on this question will receive particular emphasis. No attempts will be made to review the immunopathologic and hormonal studies done with NZB mice, and for the sake of argument, it will be assumed that T lymphocytes are not primarily responsible for these disorders. Those subjects are thoroughly discussed in Chapter 4 of this volume. Consideration of immune-defective animals such as these can provide insights into many aspects of the development and regulation of lymphoid and myeloid cells, and this should be true irrespective of whether particular abnormalities are directly involved in autoimmune phenomena.

PAUL W. KINCADE • Sloan–Kettering Institute for Cancer Research, Donald S. Walker Laboratory, Rye, New York 10580.

2. New Zealand Black Mice

2.1. Stem-Cell Abnormalities

A number of investigators have found it possible to transfer autoimmune and other phenomena peculiar to New Zealand mice to normal mice by irradiation and grafting. The simplest interpretation of these results is that hempoietic cells, and in particular, multipotential stem cells, of New Zealand mice carry the potential for disease. This could be specious, for the transferred suspensions were complex mixtures of stem cells, committed progenitor cells of various lineages, and even some elements of microenvironments which affect hemopoietic cell differentiation. It remains problematic as to which of these lineages and stages of development are affected by genes necessary for autoimmunity. A brief consideration of these experiments may, however, suggest new ways to investigate this question.

Murine stem cells are commonly measured by the *in vivo* spleen colony-forming unit (CFU-s) assay devised by Till and McCulloch (1961). Limiting numbers of hemopoietic cells are injected into lethally irradiated, syngeneic recipients and macroscopic foci of proliferating cells are enumerated in the spleen 8–10 days later. These foci are erythroid, granulocytic, megakaryocytic, or of mixed morphology, and there is a varying degree of renewal of CFU-s within these colonies that can proliferate on secondary transfer. It is believed that each colony derives from a single CFU-s and that at least some of these cells are pluripotential (Metcalf and Moore,1971). On the other hand, one study suggests that some CFU-s might be restricted, for example, to myeloid or myeloid and T-lymphocyte differentiation rather than being capable of forming all types of lymphoid and hemopoietic cells (Abramson *et al.,* 1977). Analyses of CFU-s suggest that the highest incidence of stem cells in adult animals is in bone marrow, that different populations of stem cells may differ in proliferative potential, and that a majority of stem cells are normally noncycling (Metcalf and Moore, 1971).

The incidences and total numbers of CFU-s in NZB marrow and spleen overlap values obtained with normal strains of mice, but striking elevations were found in numbers of endogenous spleen colonies (Warner and Moore, 1971; Siegel and Morton, 1976); that is, large numbers of foci developed in the spleens of irradiated NZB mice without any injection of donor cells. It was suggested that this apparent resistance of NZB CFU-s to irradiation may be related to the ability of these mice to survive exposures that would be lethal for other mice (Morton and Siegel, 1971). Endogenous CFU-s from normal and NZB mice can be suppressed with cycle-active drugs, and this would indicate that stem cells in NZB spleens are atypically activated (Boggs *et al.,* 1973; Siegel and Morton, 1976). A number of substances, including some released in stimulated lymphocyte cultures, are capable of quickly inducing CFU-s to cycle, so this phenomenon could be secondary to autoimmune processes taking place in NZB tissues (Wright *et al.,* 1979; Cerny, 1974). However, young NZB mice had higher than normal numbers of endogenous CFU-s, and these were also high in the related New Zealand Chocolate (NZC) strain of mice, which do not develop autoantibodies, and in irradiated normal mice reconstituted with NZB bone marrow (Siegel and Morton, 1976; Warner and Moore, 1971; Morton and Siegel, 1978a). NZC spleens were very poor environments for focal proliferation of CFUs from either NZB or NZC marrow. (Certain other similarities to NZB abnormalities will be discussed below.) These findings could relate to intrinsic defects in uncommitted stem cells of NZB and NZC mice.

However, the differentiative and proliferative potential of endogenous CFU-s transferred to normal mice has not yet been determined.

Elevated numbers of endogenous stem cells, development of antibodies to nuclear and erythrocyte self-antigens, and B-cell hyperactivity have all been transferable from NZB mice to MHC-matched, irradiated, normal mice (Morton and Siegel, 1978a; Morton and Siegel, 1974a; DeHeer and Edgington, 1978; Chused *et al.,* 1978). Unusual aspects of myeloid progenitor cells to be described below are also found in such recipients of NZB cells (Table I). Bone marrow and spleen cell grafts include mature B as well as T lymphocytes. Experiments involving thymectomized or anti-Thy-1-treated recipients and fetal liver or anti-Thy-1-plus-complement-treated bone marrow cells for grafting indicate that mature T cells are not essential for adoptive expression of these NZB characteristics (DeHeer and Edgington, 1978; Chused *et al.,* 1978). This is a conclusion that accords with studies of athymic nude mice having NZB genetic backgrounds (Gershwin *et al.,* 1979).

It is far more difficult to evaluate the importance of other cell types in the transferred inocula. Most investigators assume that immunocompetent cells regenerate in irradiated, grafted mice within a few weeks as direct descendants of donor multipotential stem cells. However, our recent studies of reconstituted CBA/N strain mice suggest that over this interval a majority of newly formed B lymphocytes are derived from one or more types of surface-immunoglobulin-negative (sIg^-) precursors that are not closely related to CFU-s (Paige *et al.,* 1979). These committed precursors are presumably the same as "pre-B"-cell populations defined by other functional assays and by their synthesis of cytoplasmic, but not surface display of, IgM heavy chains (Miller *et al.,* 1975; Melchers, 1977; Raff *et al.,* 1976). In one study grafts of NZB fetal liver, but not bone marrow, to irradiated DBA/2 mice led to the development of Coomb's-positive erythrocytes in the recipients (Morton and Siegel, 1974b). This is interesting in light of the fact that fetal liver pre-B cells were found to be more efficient than those of adult bone marrow in colonizing unirradiated CBA/N mice. We have also found that these populations differ in mean sedimentation rate at unit gravity (Paige *et al.,* 1979; Paige *et al.,* 1981). It is probable, therefore, that in all experiments performed to date, immediate precursors of B cells, which had already been conditioned by NZB microenvironments, merely continued their differentiation after transfer to

TABLE I
Abnormalities in Clonable Lymphocytes and Myeloid Progenitor Cells of NZB Mice and Transfer to Irradiated Normal Recipients

	Bone marrow granulocyte–macrophage colonies			Lymph node B-Lymphocyte colonies	
	WEHI-3[a]	LCCM	LCCM + serum Fx	LPS alone	LPS + Anti-μ
NZB/*Umc*	2 ± 1	148 ± 9	144 ± 5	3458 ± 97	164 ± 33
CBA/H	182 ± 5	140 ± 10	37 ± 1	2147 ± 65	0
BDF_1 grafted with NZB fetal liver	14 ± 2	186 ± 11	97 ± 8	4427 ± 218	44 ± 25

[a]Nonlymphoid colonies in cultures of bone marrow cells were stimulated with semipurified WEHI-3 or L-cell conditioned medium (LCCM) with or without a macroglobulin inhibitor isolated from normal serum (serum Fx). Lymph node cells were cultured for B-lymphocyte colonies with or without 25 μg/ml goat anti-mouse μ-chain antibodies. All results are given as colonies per 10^5 cultured cells ± S.E.

the normal recipients. Multipotential stem cells are present in low frequency (10^{-3}) in hemopoietic organs, and their only practical source that is not known to also contain pre-B cells is embryonic yolk sac. Stem cells in that tissue can differentiate to B and T lymphocytes on transfer to adult tissues, but they do so inefficiently, and markers are required to distinguish donor from host cells (Paige *et al.,* 1979). Similarly, it can be argued that committed progenitor cells of other hemopoietic lineages are responsible for the NZB defects that develop in the grafted normal recipients. However, in some cases it has been possible to do the reciprocal of these experiments, i.e., to diminish NZB peculiarities by irradiating and grafting them with normal cells (Morton and Siegel, 1978b; DeHeer and Edgington, 1978). This at least demonstrates that factors associated with irradiation-resistant tissues of NZB mice are not dominant over normal grafts, and we can assume for the moment that NZB defects are attributable to one or more transferable cells in fetal liver and in the Thy-1^- component of adult marrow.

Stem-cell survival is short in conventional liquid cultures, but in a system devised by Dexter and co-workers (1977), hematopoiesis can often be maintained *in vitro* for several months. Bone marrow is aspirated into flasks containing Fischer's medium supplemented with selected horse serum. Normally, within 2–3 weeks the numbers of nonadherent stem cells and progenitor cells gradually decline in these cultures and a monolayer, including distinctive fat cells, is formed. Stem-cell maintenance and differentiation can be maintained much longer if these cultures are then recharged with fresh marrow cells. There are strain variations in the survival times of cells in these cultures, and in this respect NZB was the poorest of 21 strains thus far studied (M. A. S. Moore, A. P. C. Sheridan, and P. W. Kincade, unpublished observations). NZB marrow cells were cocultured with cells from mice with genetically determined anemias in order to localize this abnormality. Homozygous Sl/Sl^d mice have known defects in hematopoietic microenvironments (Mecalf and Moore, 1971), and their cells survive poorly either when cultured alone or when mixed with NZB marrow. A quite different result was obtained when W/W^v marrow cells were employed. These mice are thought to have defects restricted to stem-cell replication (Metcalf and Moore, 1971), and their cells also do not survive well in culture. However, when W/W^v and NZB cells were mixed and cultured, an apparent complementation between defects was obtained. The mixed cultures survived much longer than either of the two types of cells cultured alone. This suggests that some element necessary for a functioning *in vitro* microenvironment is present in marrow aspirates and that it is atypical in NZB mice. If such a deficient cell type is radiosensitive, it could quite possibly have contributed to the results from the grafting experiments discussed above. In other words, some cells, other than stem cells and progenitor cells, may have been transferred and these cells could have influenced the development of immunocompetent cells.

2.2. Myeloid Progenitor-Cell Abnormalities

In vitro cloning techniques are now available for enumerating and characterizing committed precursors of erythroid, megakaryocytic, eosinophilic, granulocytic, and monocyte–macrophage cells (Metcalf, 1977). Each of these assays depends on one or more factors, such as erythropoietin or other colony-stimulating activities (CSA), and these stimuli are produced by a variety of normal and neoplastic cells. The cells that clone in the presence of these substances are heterogeneous in terms of size and buoyant density, and if subsets exist, neither the number of CSA to which each is responsive nor their differentiative potential is fully understood (Williams and Jackson, 1977; Williams and Eger, 1978). Studies of granulocyte–macrophage progenitors by the culture colony-forming unit (CFU-c) technique with several different CSA have revealed some interesting anomalies of NZB hemopoietic cells.

One of the most widely used sources of CSA for studying murine CFU-c is serum from mice injected 3 hr previously with lipopolysaccharide (LPS) preparations of bacterial endotoxin. With such stimuli, normal-to-elevated numbers of colonies were obtained in cultures of NZB bone marrow and spleen, and these were normally cycling as assessed by their sensitivity to thymidine suiciding (Metcalf, 1972; Metcalf and Stevens, 1972). Our laboratory has extensively used culture medium conditioned by WEHI-3 myelomonocytic leukemia cells as as source of CSA. With a crude preparation, NZB CFU-c numbers were consistently low over a broad age span; when this material was partially purified by passage over DEAE, even lower responses were measured in these mice (Kincade *et al.*, 1979) (Table I). In contrast, when either medium conditioned by L cells or endotoxin serum was used as the source of CSA, NZB marrow cells not ony responded well but the plateaus of colony numbers seen with normal mice were not obtained. With the most potent preparations of these CSA, NZB colonies were still increasing after constant numbers of normal colonies were attained. This could mean either that NZB mice have very large numbers of cells responsive to these two factors and/or that they are unaffected by any inhibitors present in these preparations.

Responses of normal and NZB CFU-c to L-cell CSA are completely and specificially blocked by inclusion in the culture of an antiserum prepared against the purified factor (Stanley and Guilbert, 1980). The DEAE-nonadherent fraction of WEHI-3 CSA, to which NZB marrow cells are poorly responsive, is antigenically distinct from L-cell CSA and the antiserum does not inhibit colony formation by normal bone marrow cells (E. R. Stanley and N. Williams, unpublished observations; P. W. Kincade and E. R. Stanley, unpublished observations). This has made it possible to determine that there are at least three types of granulocyte–macrophage CSA. The first two are typified by L-cell and semipurified WEHI-3 and are distinguishable from each other by their opposite effects on NZB marrow as well as by their serological differences. That a third type exists follows from our recent analysis of CSA produced by WEHI-274 cells (P. W. Kincade and N. L. Warner, unpublished observations). This stimulus works well with NZB cells but is only partially blocked by the anti-L-cell serum. It is possible that many other types of CSA will be found and NZB mice will be of value in distinguishing them from the WEHI-3-type CSA.

Semipurified WEHI-3 CSA elicits large numbers of colonies in cultures of CBA/H, CBA/*cum*, CBA/N, C57BL/10, C57BL/6, C57BL/6.*Ig*^{e}, DBA/1, DBA/2, SJL, BALB/c, SWR, C3H, MRL/*Lpr*, BXSB, and NZW bone marrow. We have recently found that there are two strains besides NZB that are very poor responders to this stimulus. One of these is the NZC strain, which, as previously discussed, has high numbers of endogenous stem-cell colonies but does not develop autoimmune disease (Warner and Moore, 1971). The other is the C58/J strain, which is known to have a very high incidence of spontaneous leukemia and to produce xenotropic virus (Storer, 1966; Stephenson *et al.*, 1975). Hybrid F_1 offspring of the cross NZB × CBA or NZB × NZW tend to respond with values intermediate between the parental strains (Kincaide *et al.*, 1979). These observations suggest that this characteristic may be governed by one or more codominant or semidominant genes and that it is not sufficient for immunologic dysfunction. It is now necessary to study backcross generations to determine if this response characteristic is essential for expression of autoimmunity of the NZB type. At present it is only possible to note that spontaneous B-cell activation and certain of the autoimmune responses are also intermediate in heterozygous individuals (Moutsopoulos *et al.*, 1977; Izue *et al.*, 1978; Raveche *et al.*, 1978; Manny *et al.*, 1979).

The fraction of WEHI-3 CSA that does not adhere to DEAE is known to preferentially stimulate neutrophil granulocyte colonies (Williams and Eger, 1978; Williams *et al.*, 1978), and L-cell CSA elicits almost exclusively monocyte–macrophage colonies (Stanley and Guilbert, 1980). Besides the disparate responses of NZB CFU-c to these two stimuli, colonies developing in crude WEHI-3 were often predominantly of macrophage type. This suggested initially that the abnormality might be restricted to a particular subset of CFU-c that has limited differentiative potential. Although the physical properties of progenitor cells from these mice have not yet been studied, an apparently normal blood picture and the presence of adequate numbers of granulocyte colonies in endotoxin-serum-stimulated cultures do not support this notion. We have also found that NZB and CBA mice have comparable levels of serum CSA after LPS injection. Macrophages from adult NZB mice have been found to be defective in primary *in vitro* immune responses and the same has been suspected for NZC macrophages (McCombs *et al.*, 1975; Herrod and Warner, 1972; Warner, 1978). Deficiencies in macrophage spreading have only been found in older New Zealand mice, and the reticuloendothelial function of (NZB × NZW)F_1 hybrids is reportedly normal (Johnson *et al.*, 1978; Finbloom and Plotz, 1979). It would be interesting to obtain macrophages from very young NZB mice and to determine their functional capabilities.

There are other substances besides CSA that influence proliferation and differentiation of CFU-c, and these may be important to hematopoietic regulation *in vivo*. Preparations of CSA that elicity predominantly macrophage colony formation stimulate mature macrophages to release prostaglandins of the E series (PGE) and a feedback loop is completed by the fact that PGE preferentially inhibits monocyte–macrophage production *in vitro* (Kurland and Bockman, 1978; Williams, 1979; Pelus *et al.*, 1979). NZB CFU-c were more resistant than normal to PGE inhibition when L-cell CSA, a stimulator of macrophage colony formation, was used, and the reverse was true when the stimulus was crude WEHI-3 conditioned medium (Kincade *et al.*, 1979). Normal mouse serum contains several potent inhibitors of *in vitro* hemopoietic cells and immune responses (Metcalf and Russell, 1976). We have recently found that NZB CFU-c stimulated by L-cell CSA are relatively resistant to the effects of a high-molecular-weight fraction of serum that is excluded by Sepharose 6B (Table I). As yet we have no evidence that these findings are of physiologic significance, but we do know that the peculiarities of NZB CFU-c are apparent with cultures of fetal liver long before immunologic abnormalities are demonstrable.

2.3. B-Lymphocyte Abnormalities

Immunologic function of NZB mice is reviewed in detail elsewhere in this volume (Chapter 4). Consequently, a brief consideration of unusual aspects of their humoral immune system will suffice for this discussion. Perhaps most relevant to autoimmune disease are observations suggesting that B cells from these mice spontaneously differentiate to secrete large amounts of IgM. That this occurs independently of T cells follows from observations of athymic NZB mice, transfer experiments using thymectomized mice, and studies showing B-cell differentiation in cultures of embryonic and neonatal NZB cells (Gershwin *et al.,* 1979; DeHeer and Edington, 1978; Chused *et al.,* 1978; Manny *et al.,* 1979). It has not, however, been excluded that nonlymphoid cells may participate in spontaneous B-cell activation and this question will ultimately have to be addressed by some type of limiting-dilution analyses. Studies have, however, been initiated with cloning techniques that provide insight into the frequency and functions of certain B-cell sets in these mice.

We have found a semisolid agar cloning assay originated by Metcalf *et al.* (1975) to be a simple and reliable method for enumerating and characterizing murine B cells. In this system cells are dispersed in a gel matrix containing appropriate mitogens and 2-mercaptoethanol (Kincade *et al.,* 1976). One of three different potentiators must also be used to achieve reproducible cloning efficiencies and a linear relationship between numbers of cells plated and numbers of resulting colonies (Metcalf, 1976; Kurland *et al.,* 1977). These potentiators are intact sheep erythrocytes (SRBC), lipopolysaccharide (LPS) of endotoxin, or feeder layers of adherent macrophages. The numbers of colonies obtained with combinations of these potentiators can equal the sum obtained with each one used alone and in some cases it is clear that the B-cell populations responding under the different culture conditions are not the same. For example, surface-Ig-positive (sIg^+) cells in newborn spleen, adult bone marrow, spleen, lymph nodes, and peripheral blood all clone with about equal efficiency in the presence of LPS (Kincade *et al.,* 1978). However, bone marrow B cells and, in particular, the newly formed sIg^+ cells form colonies poorly in the presence of SRBC (Lala *et al.,* 1979; P. W. Kincade, unpublished observations). Abnormalities in NZB B cells are apparent in LPS-containing but not in SRBC-containing semisolid cultures.

A considerable amount is known about the characteristics of B cells from normal mice that form colonies *in vitro.* Essentially all of them, regardless of their distribution and stage of development, are $sIgM^+$ and Lyb-2^+ (a murine B-cell specific alloantigen). In adult tissues most such B cells are Ia^+ and Qa-2^+, and three subsets can be distinguished on the basis of their sIgD expression and function. A first category is $sIgD^-$ and this probably includes immature as well as post-antigen-stimulated memory B cells. A second is $sIgD^+$ and it can be prevented from mitogen-induced clonal proliferaton by nanogram quantities of anti-δ antibodies added to the cultures. All of the B cells in neonatal spleen and adult bone marrow that bear IgD are of this second type. Finally, a third subset is $sIgD^+$ but is unaffected by antibody directed to these receptors. The sIg^- precursors of B cells in fetal tissues and in adult bone marrow have little, if any, colony-forming ability, immature $sIgD^-$ B cells do not acquire sIgD *in vitro,* and terminal differentiation to secretion occurs only to a small degree within the cultures (Kincade *et al.,* 1978, 1980a,b). These observations suggest that the low-cell-density conditions that pertain in the semisolid agar support proliferative but not differentiative processes. In contrast, if sIg^- pre-B cells from adult marrow are held in relatively crowded liquid cultures, they promptly acquire the ability to form colonies (Kincade *et al.,* 1980a).

Clonable B cells are as heterogeneous as all other B cells in terms of buoyant density, size, and expression of Fc and C receptors (Metcalf *et al.,* 1976). The most notable category

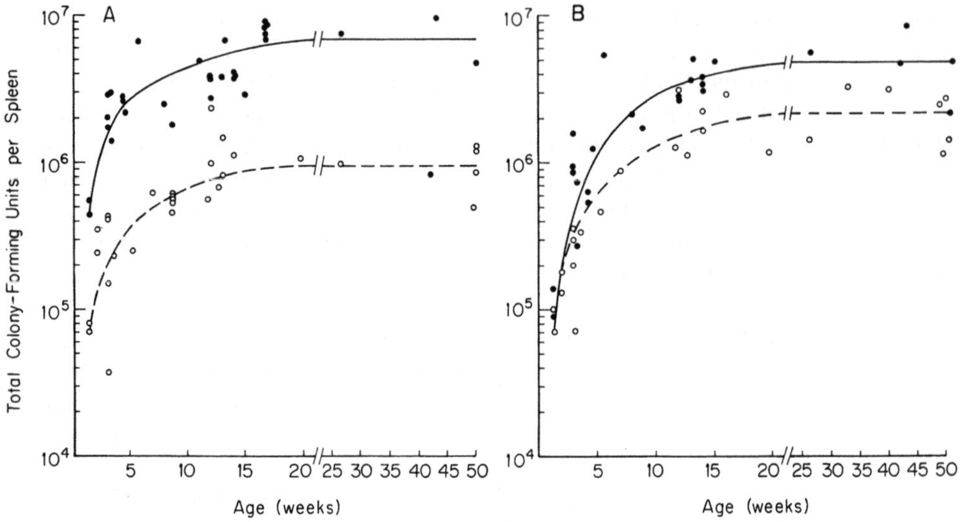

FIGURE 1. Comparison of clonable B lymphocytes in NZB (●) and CBA/H (○) speens in cultures potentiated with LPS (A) or SRBC (B). Colony-forming units per spleen were calculated by multiplying numbers of total nucleated cells per spleen times the average incidence of colonies in four replicate cultures of 2 × 10^4 cells.

of B cells not detected in this assay is that of the partially B-cell-deficient CBA/N strain of mice (Kincade, 1977). This finding is germane to the present discussion because male NZB mice bearing the defective X^{id} gene of CBA/N mice do not have abnormally high numbers of spontaneously activated B cells (Nakajima *et al.*, 1979). If intrinsic abnormalities in particular categories of B cells are responsible for autoimmune responses, these could be the same as those detected by the semisolid agar cloning assay. The incidence and total numbers of clonable B cells are high in young NZB tissues and remain so throughout adult life (Kincade *et al.*, 1979; Ohsugi and Gershwin, 1979). However, this is only striking if LPS is used to potentiate the cultures, and cloning of NZB and CBA B cells is relatively similar in the presence of SRBC (Fig. 1). B-cell proliferation under either of these conditions is independent of helper or suppressor T cells or macrophages (Kurland *et al.*, 1977; Kincade *et al.*, 1978), and this makes it possible to look for intrinsic abnormalities in individual B cells. The relative amount of IgD on NZB B-cell populations is known to be low (Cohen *et al.*, 1978; Chused *et al.*, 1978), and we tested their sensitivity in culture to monoclonal anti-δ antibodies (Kincade *et al.*, 1979). Neither the proportion of anti-δ-sensitive B cells nor the amounts of antibody required for their inhibition were remarkable, and this suggests that IgD receptors on at least one B-cell subset function normally in regulation.

B cells of these mice were also not atypical in their sensitivities to prostaglandin E or α fetoprotein. We had previously found that essentially all clonable B cells in all tissues and stages of development from a number of normal strains of mice, including NZC, were sensitive to anti-IgM antibodies. NZB B cells were remarkable in that a substantial number were not inhibited by antibodies directed to IgM heavy chains in LPS-potentiated cultures. Numbers of these unusual anti-μ-resistant B cells were maximal at around 17 weeks of age, and they were also most prevalent in lymph nodes. It is unlikely that these B cells

TABLE II
Adherence of B Cells to Antibody-Coated Dishes[a]

	Unfractionated control		Cells recovered from anti-μ-coated dishes		Cells recovered from anti-δ-coated dishes	
	Total Ig^+ cells	Clonable B cells	Total Ig^+ cells	Clonable B cells	Total Ig^+ cells	Clonable B cells
CBA bone marrow	1.00	11,800	0.02 [98]	170[99]	0.17 [83]	9,500 [19]
NZB bone marrow	0.45	26,600	0.05 [89]	2325[91]	0.22 [51]	8,966 [66]
CBA spleen	3.78	107,000	0.02 [99]	112[99]	0.64 [83]	6,129 [94]
NZB spleen	4.06	334,000	0.05 [99]	2893[99]	0.63 [84]	24,046 [93]
CBA lymph nodes	1.01	43,700	0.01 [99]	<62[100]	0.06 [94]	493 [99]
NZB lymph nodes	1.45	189,300	0.05 [97]	145[99]	0.08 [94]	3,836 [98]

[a]Cells were separated by incubation in antibody-coated plastic petri dishes. Details of the method and specificity controls showing minimal nonspecific cell loss are presented elsewhere (Kincade *et al.*, 1980b). Cells that did not adhere to the dishes were examined by immunofluorescence and cultured in the presence of LPS, and the results are given as total cells staining with anti-κ antibodies (Ig^+) $\times 10^{-6}$ and total colony-forming B cells recovered [percent depletion]. CBA/H and NZB/*Umc* mice were 5-month-old females.

had lost their sIgM as they were normally and specifically adherent to anti-μ- or anti-δ-coated plastic dishes (Table II). Polyvalent anti-Ig antibodies inhibited all of the colonies and a high proportion of the anti-μ-resistant clones were inhibited when anti-δ was also added (data not shown). All of the B cells that cloned in SRBC-containing cultures were anti-μ sensitive (Table III) so the atypical cells must be included in the populations that were potentiated by LPS. These findings suggest that a particular B-cell subset is somewhat expanded in NZB mice and that some of these acquire a subtle defect in their sensitivity to regulation. The distribution of these cells serves to indicate that they are not immature, and they are most prevalent when polyclonal B-cell activation is thought to be maximal

TABLE III
Importance of Culture Conditions for Cloning Anti-μ-Resistant B Cells

	Potentiated with LPS		Potentiated with SRBC	
	Control	Anti-μ	Control	Anti-μ
NZB/*Umc*				
Bone marrow	226 ± 16[a]	5 ± 3	74 ± 10	0
Spleen	3340 ± 219	1040 ± 205	2010 ± 98	0
Lymph nodes	1893 ± 48	847 ± 57	1100 ± 45	0
CBA/H				
Bone marrow	118 ± 26	0	71 ± 10	0
Spleen	1070 ± 73	0	1565 ± 111	0
Lymph nodes	437 ± 19	0	450 ± 31	0

[a]Data are from a single experiment comparing 20-week-old mice and are expressed as B-lymphocyte colonies/10^5 cultured cells ± S.E. Purified goat anti-mouse μ-chain antibodies were used as a final concentration of 25 μg/ml.

(Izui *et al.*, 1978). It has also been reported that B cells from old NZB mice are abnormally resistant to Con A-elicited suppressor factors (Primi *et al.*, 1978).

3. Motheaten Mice

3.1. Genetics and Pathology

The known complexity of regulatory networks in the immune system suggests that a variety of fail-safe mechanisms must operate to maintain homeostasis. Motheaten mice provide an example of a single mutation that results in severe immunologic dysfunction and early death. Identification of the product of this mutant gene and its normal function may therefore be of considerable basic importance. The disorder has been attributed to a recessive gene located on chromosome 6 and two spontaneous mutations were described by Schultz and Green of The Jackson Laboratory (1975, 1976). Homozygous mutant mice develop skin lesions soon after birth, and they die before 8 weeks of age. Irregular deposits of immunoglobulin are present in glomeruli; with some motheaten mice, these can be demonstrated in the thymus, skin, and lungs (Green and Shultz, 1975; Shultz and Green, 1976; Shultz *et al.*, 1978; P. W. Kincade, unpublished observations). Histologic analysis reveals a deficiency of lymphocytes in spleen, lymph nodes, and Peyer's patches and in particular the T-independent areas of these tissues. Plasma cells are abundant and myelopoiesis is expanded in the spleen. Mortality in these animals has not been associated with any one infectious agent and is not improved by specific-pathogen-free husbandry (Shultz *et al.*, 1978; Lutzner and Hansen, 1976). Heterozygous individuals can be identified by mating or prepared by ovarian transplant. These animals appear to be normal until late in life when IgM levels become elevated and they develop reticulum-cell tumors in unusually high frequency (Sidman *et al.*, 1978b; L. Shultz, personal communication).

3.2. B-Lymphocyte Abnormalities

The most atypical features of motheaten mice pertain to the humoral immune system. In the spleen the incidence of B cells is subnormal at all ages, and at 5 weeks of age motheaten mice have approximately one-fourth the total number found in littermate controls. B cells as a whole tend to have less surface IgM in young motheaten mice, subsequent development of populations bearing sIgD and complement receptors is slight, and redistribution (capping) of sIg is subnormal (Sidman *et al.*, 1978a; Davidson *et al.*, 1979). Functional B cells, as assessed by cloning in semisolid agar, are usually much more depressed than numbers of total sIg^+ cells, and this is particularly true for B cells from the lymph nodes and spleen. The few residual clonable cells in these mice did not, however, seem unusual in terms of their sensitivity to anti-δ antibodies added to the cultures, pretreatment with anti-Ig antibodies, or treatment with anti-Ia serum and complement (Kincade, 1978). At all ages stimulation of motheaten cells by B-cell mitogens in liquid cultures is poor, and their *in vitro* responses to the T-cell-independent anitgens FITC-Ficoll and FITC-LPS are essentially nil (Kincade, 1978; Sidman *et al.*, 1978b; Davison *et al.*, 1979). Primary antibody responses *in vivo* to the T-dependent antigen SRBC averaged 7% of normal (Shultz and Green, 1976).

A number of findings indicate that large numbers and multiple clones of motheaten B cells differentiate to actively secreting cells as an early age (Shultz and Green, 1976; Shultz and Zurier, 1978; Sidman *et al.*, 1978b; Davidson *et al.*, 1979). By 3 weeks of age, serum immunoglobulin levels (especially IgM) are elevated. These sera contain progressively increasing amounts of viral antigen, antibody to double-stranded DNA, and antibody to thymocytes. The latter is of the IgM class and it binds to the cortisone-sensitive fraction of thymocytes from most strains of mice, but not to their peripheral T cells. A rare, λ2 type of light chain was also demonstrated in serum from these mice, and their tissues contain high numbers of large sIg^+ cells, plasma cells, and plaque-forming cells that lyse TNP- or FITC-conjugated SRBC.

3.3. T-Lymphocyte Abnormalities

Thymus-dependent areas of motheaten tissues are relatively spared by their disorder and their numbers of Thy-1-bearing cells are essentially normal (Green and Shultz, 1975; Sidman *et al.*, 1978a). However, T cells from affected mice give suboptimal graft-versus-host responses on transfer to newborn F_1 mice, they stimulate poorly with Con A *in vitro*, and they do not immunize well with irradiated allogenic cells *in vitro* to become cytolytic effectors (Shultz and Green, 1976; Kincade, 1978; Sidman *et al.*, 1978b; Davidson *et al.*, 1979). The apparent disparity between T-cell numbers and their functional capability in motheaten mice suggests that subset typing by means of Lyt alloantisera and additional functional analyses could be instructive.

3.4. Nonlymphoid Cell Abnormalities

The incidence of stem cells as measured by the CFU-s assay is normal or elevated in motheaten mice (Shultz and Zurier, 1978). However, lethally irradiated, normal recipients of bone marrow from these mice survived no longer than ungrafted, irradiated mice. Both had a mean survival time of approximately 2 weeks. This result is open to several interpretations. Autoaggressive cells or their precursors in the transferred marrow cell suspension could have rapidly killed the immunologically crippled, irradiated recipients. Assuming that no infectious agents were transferred, an alternative possibility is that stem-cell defects preclude sufficient hemopoietic reconstitution for survival. Observations that tend to minimize this latter possibility were made by grafting motheaten bone marrow cells into mutant W/W^v mice without irradiation (Shultz and Zurier, 1978; P. W. Kincade, unpublished observations). W/W^v mice have macrocytic anemia associated with stem cells that are defective in their replication. Unlike normal mice, these mice rapidly become stem cell chimeras after grafting with normal hemopoietic cells, and erythroid as well as myeloid cells of donor type are formed (Harrison and Astle, 1976). Motheaten bone marrow cells were equivalent to control cells in curing the anemia of W/W^v recipients, but they then led to autoantibody production, increased serum immunoglobulin levels, and death by 20 weeks after grafting. It therefore appears that some type of motheaten marrow cell can expand and differentiate sufficiently after adoptive transfer and can disrupt immune function and kill the recipient. Homozygous mutant mice cannot be readily identified until several days after birth. This has made it impractical to study the fate of fetal motheaten cells after transfer to normal recipients. Thorough studies have as yet not been made of macrophages and their precursors in motheaten mice. However, in co-culture experiments, spleen cells from these mice suppressed mitogen-stimulated proliferation and differentiation responses of normal B cells, and this suppressor activity was partially and variably removed by allowing the motheaten cells to adhere to plastic (Sidman *et al.*, 1978b). This was taken as suggestive evidence for aberrant macrophage function which might be related to the poor specific immune responses of these mice.

4. Synthesis and Conclusions

The information discussed is insufficient to ascribe the diseases of NZB and motheaten mice to particular abnormalities of any kind, including those of hemopoietic cells. However, available methodology should be adequate to resolve possible interrelationships between the various peculiarities of these mice. For example, it is not yet clear if polyclonal B-cell activation is due to intrinsic defects of NZB B cells and if interactions with nonlymphoid cells are necessary. It is known that the immediate precursors of B cells in normal bone marrow and fetal liver will complete their maturation *in vitro* (Melchers, 1977; Lau *et al.*, 1979; Kincade *et al.*, 1980a). Therefore, it should be possible to culture limiting numbers of NZB pre-B cells in the presence of macrophages, T cells, or thymus filler cells from NZB or normal syngeneic mice and determine if (and under what circumstances) spontaneous B-cell activation and IgM secretion take place.

The transition from pre-B to B cells is normally a time of extreme suseptibility to tolerization, and using recently developed techniques this characteristic could be reexamined with NZB mice (Metcalf and Klinman, 1977). It has been shown that these and other autoimmune strains of mice have large numbers of cells secreting antibodies to various commonly used haptens (Izui *et al.*, 1978; Chused *et al.*, 1978). However, studies should be done using hapten inhibition, isoelectric focusing, or other well-established techniques to determine how heterogeneous this antibody is. In other words, whether the secreted antibody is truly polyclonal. Relative restriction of this antibody might suggest that the B cells are partially activated by available antigens before being driven to terminal differentiation. This would be an extension of studies suggesting polyclonal responses to certain erythrocyte antigens (DeHeer and Edgington, 1976), and it is noteworthy that thymocytotoxic antibodies made by NZB and motheaten mice have specificity for certain types of T lymphocytes (Shirai *et al.*, 1978; Shultz and Zurier, 1978).

Cell transfer experiments are among the most common of approaches used by immunologists and experimental hematologists and the studies described above demonstrate that most of the NZB and motheaten characteristics are reproduced in normal recipients of hemopoietic cells from these mice. Cell separation methodology is becoming increasingly sophisticated and hopefully it will soon be possible to resolve which of the many cell types present in the grafts were responsible for this adoptive expression. New monoclonal and conventional antibodies that may reveal distinctive antigens on multipotential stem cells and committed progenitor cells of various kinds may be particularly useful in this regard. Grafting may also be a practical solution to the limitations on experiments with motheaten mice, those being that affected individuals are very small in size and they die at an early age due to their susceptibility to infection and the severity of their disease. As the condition is recessive, only one-fourth of the offspring of known heterozygous breeders are motheaten. It is difficult, for these reasons, to maintain a sufficient and constant supply of experimental animals. However, bone marrow from one affected mouse can be used to graft a number of irradiated adult, syngeneic normal mice (Shultz and Zurier, 1978). Reciprocal grafting of normal cells to motheaten mice has not been described, and although it may be technically difficult, one or more types of normal donor cells might be found to have a dominant influence on the diseased recipient.

It is not difficult to propose mechanisms through which defects in nonlymphoid cells could contribute to autoimmune disease. Aside from the well-known role of macrophages in antigen processing and presentation, soluble products released from these cells can have potent effects on immune responses *in vitro* (Unanue, 1978; Aarden *et al.,* 1979). It has been suggested that macrophages can enhance pre-B to B-cell maturation and that they can activate B cells directly (Hoffmann *et al.,* 1979; Opitz *et al.,* 1976), but synergy between macrophages and their factors with antigens or mitogens is better known (Kurland *et al.,* 1977; Hoffman *et al.,* 1979). That is, once B cells are caused to divide, macrophage products can augment their differentiation to Ig-secreting cells. Macrophages can also release immunosuppressive substances such as prostaglandins and nominally immunosuppressive substances such as thymidine (Kurland and Bockman, 1978; Pelus *et al.,* 1979; Unanue, 1978). It would be instructive to quantify release of these substances from the macrophages of NZB and motheaten mice, particularly if these substances could be obtained from very young animals. An interesting mechanism that might exacerbate B-cell activation is suggested by recent findings that macrophages can process denatured immunoglobulin to produce a polyclonal stimulator of B cells (Morgan and Weigle, 1980). Granulocytes contain numerous proteases that are mitogenic for B cells themselves and some of these proteases can activate the alternative pathway of complement (Kaplan and Bona, 1974; Schorlemmer and Allison, 1976). These components, in turn, can be mitogenic for B cells and can also cause more lysosomal enzyme release (Hartmann and Bokish, 1975; Goldstein *et al.,* 1973). Inappropriate release of these or other lymphokines could directly or indirectly activate B cells polyclonally and drive them to antibody secretion.

Abnormalities have been found in or attributed to T cells, thymic epithelium, B cells, macrophages, and stem cells in NZB mice. This complexity has led to the suggestion that some central genetic aberration might be involved and, indeed, chromosomal abnormalities and a slow DNA repair rate have been reported for these mice (Manny *et al.,* 1979; Paffenholz, 1978; Fialkow *et al.,* 1978; Reddy *et al.,* 1978). However, a recent analysis of F_1 and backcross matings between NZB and DBA/2 mice did not reveal an association between karyotypic abnormalities and immunologic disturbances (Raveche *et al.,* 1979). It is therefore likely that all of the peculiarities of these mice will eventually be ascribed to one of several, possibly interacting, genes. Sex hormones can influence the expression of some of the NZB characteristics and one group attempted to overcome this problem by using gonadectomized animals (Raveche *et al.,* 1978). Another difficulty is that a characteristic that appears to segregate as a single gene in one type of mating may behave quite differently in another (Warner, 1978). This strongly suggests that two or more genes must interact to produce some of the NZB anomalies. There are at least two mechanisms that could be evoked to explain this. Single defects might not be seen because of normal compensatory responses. A weakness in this second apparatus might therefore be necessary for phenotypic expression of the first defective gene. Another explanation is suggested by findings that the ontogeny, density, and/or distribution of cell-surface alloantigens can be influenced by other than the structural genes (Boyse *et al.,* 1967; Boubelik *et al.,* 1975; Kincade *et al.,* 1980b). This means that although an individual has the potential to display a given cell-surface receptor, the amount of it that is expressed, as well as its cellular distribution, may be modulated by other genes.

Perhaps the most valuable aspect of genetic studies is that they allow one to discount the importance of certain abnormalities and to recognize associations between others. We have recently adapted a method for aspirating bone marrow from anesthetized mice. This not only makes it possible to study several parameters with individual backcross mice but to select those bearing hemopoietic defects for further matings. Recombinant inbred strains of mice derived from NZB and normal strains crosses will no doubt also be extremely useful in dissecting this problem. The motheaten model has appeal if only because it is governed by a single recessive gene and with it the challenge will be to determine how such severe defects in immunoregulation result from this single mutation.

ACKNOWLEDGMENT

The author's research was supported by Grants CA-17404 and CA-08748 and Research Career Development Award AI-00265 from the United States Public Health Service.

References

Aarden, L. A., Brunner, T. K., Cerottini, J. C., Dayer, J. M., deWeck, A. L., Dinarello, C. A., DiSabato, G., Farrar, J. J., Gery I., Gillis, S., Handschumacher, R. E., Henney, C.S., Hoffman, M. K., Koopman, W. J., Krane, S. M., Lachman, L. B., Lefkovits, I., Mishell, R. I., Mizel, S. B., Oppenheim, J. J., Paetkau, V., Plate, J., Rollinghoff, M., Rosenstreich, D., Rosenthal, A. S., Rosenwasser, L. J., Schimpl, A., Shin, H. S., Simon, P. L., Smith, K. A., Wagner, H., Watson, J. D., Wecker, E., and Wood, D. D., 1979, Revised nomenclature for antigen-nonspecific T cell proliferation and helper factors, *J. Immunol.* **123**:2928.

Abramson, S., Miller, R. G., and Phillips, R. A., 1977, The identification in adult bone marrow of pluripotent and restricted stem cells of the myeloid and lymphoid systems, *J. Exp. Med.* **145**:1567.

Boggs, S. S., Boggs, D. R., Neil, G. L., and Sartiano, G., 1973, Cycling characteristics of endogenous spleen colony-forming cells as measured with cytosine arabinoside and methotrexate, *J. Lab. Clin. Med.* **82**:727.

Boubelik, M., Lengerova, A., Bailey, D. W., and Matousek, V., 1975, A model for genetic analysis of programmed gene expression as reflected in the development of membrane antigens, *Dev. Biol.* **47**:206.

Boyse, E. A., Stockert, E., and Old, L. J., 1967, Modification of the antigenic structure of the cell membrane by thymus-leukemia (TL) antibody, *Proc.Natl. Acad. Sci. USA* **58**:954.

Cerny, J., 1974, Stimulation of bone marrow hemopoietic stem cells by a factor from activated T cells, *Nature (London)* **249**:63.

Chused, T. M., Moutsopoulos, H. M., Sharrow, S. O., Hansen, C. T., and Morse, H. C., III, 1978, Mechanism of autoimmune disease in New Zealand black mice, in: *Genetic Control of Autoimmune Disease* (N. R. Rose, P. E. Bigazzi, and N. L. Warner, eds.), pp. 177–190, Elsevier/North-Holland, Amsterdam.

Cohen, P., Ziff, M., and Vitetta, E. S., 1978, Characterization of a B cell defect in the NZB mouse manifested by an increased ratio of surface IgM to IgD, *J. Immunol.* **121**:973.

Davidson, W. F., Morse, H. C., III, Sharrow, S. O., and Chused, T. M., 1979, Phenotypic and functional effects of the motheaten gene on murine B and T lymphocytes, *J. Immunol.* **122**:864.

DeHeer, D. H., and Edgington, T. S., 1976, Cellular events associated with the immunogenesis of anti-erythrocyte autoantibody responses of NZB mice, *Transplant. Rev.* **31**:117.

DeHeer, D. H., and Edgington, T. S., 1978, Evidence for a B lymphocyte defect underlying the anti-X anti-erythrocyte autoantibody response of NZB mice, *J. Immunol.* **118**:1858.

Dexter, T. M., Allen, T. D., and Lajtha, L. G., 1977, Conditions controlling the proliferation of haemopoietic stem cells *in vitro, J. Cell. Physiol.* **91**:335.

Fialkow, P. J., Bryant, J. I., and Friedman, J. M., 1978, Specificity of acquired clonal chromosome abnormalities in New Zealand black mice, *Int. J. Cancer* **21**:505.

Finbloom, D. S., and Plotz, P. H., 1979, Studies of reticuloendothelial function in the mouse with model immune complexes. II. Serum clearance, tissue uptake and reticuloendothelial saturation in NZB/W mice, *J. Immunol.* **123**:1600.

Gershwin, M. E., Castles, J. J., Erickson, K., and Ahmend, A., 1979, Studies of congenitally immunologic mutant New Zealand mice. II. Absence of T cell progenitor populations and B cell defects of congenitally athymic (nude) New Zealand black (NZB) mice, *J. Immunol.* **122**:2020.

Goldstein, I., Hoffstein, S., Gallin, J., and Weissmann, G., 1973, Mechanisms of lysosomal enzyme release from human leukocytes: Microtubule assembly and membrane fusion induced by a complement component, *Proc, Natl. Acad. Sci. USA* **70**:2916.

Green, M. C., and Shultz, L. D., 1975, Motheaten, an immunodeficient mutant of the mouse. I. Genetics and pathology, *J. Hered.* **66**:250.

Harrison, D. E., and Astle, C. M., 1976, Population of lymphoid tissues in cured W-anemic mice by donor cells, *Transplantation* **22**:42.

Hartmann, K. U., and Bokish, V. A., 1975, Stimulation of murine B lymphocytes by isolated C3b, *J. Exp. Med.* **142**:600.

Herrod, H. G., and Warner, N. L., 1972, Defects in the immune response of NZC mice, *Proc. Soc. Exp. Biol. Med.* **140**:1254.

Hoffmann, M. K., Koenig, S., Mittler, R. S., Oettgen, H. F., Ralph, P., Galanos, C., and Hämmerling, U., 1979, Macrophage factor controlling differentiaion of B cells, *J. Immunol.* **122**:497.

Izui, S., McConahey, P. J., and Dixon, F. J., 1978, Increased spontaneous polyclonal activation of B lymphocytes in mice with spontaneous autoimmune disease, *J. Immunol.* **121**:2213.

Johnson, D.R, Fernandes, G., and Douglas, S. D., 1978, Age related decline in cytoplastmic spreading of mouse peritoneal macrophages, *Dev. Comp. Immun.* **2**:347.

Kaplan, J. G., and Bona, C., 1974, Proteases as mitogens: The effect of trypsin and pronase on mouse and human lymphocytes, *Exp. Cell Res.* **88**:388.

Kincade, P. W., 1977, Defective colony formation by B-lymphocytes from CBA/N and C3H/HeJ mice, *J. Exp. Med.* **145**:249.

Kincade, P. W., 1978, Incidence and characteristics of functional B lymphocytes in motheaten mice, in: *Genetic Control of Autoimmune Disease* (N. R. Rose, P. E. Bigazzi, and N. L. Warner, eds.), pp. 241–248, Elsevier/North-Holland, Amsterdam.

Kincade, P. W., Ralph, P., and Moore, M. A. S., 1976, Growth of B-lymphocyte clones in semisolid culture is mitogen dependent, *J. Exp. Med.* **143**:1265.

Kincade, P. W., Paige, C. J., Parkhouse, R. M. E., and Lee, G., 1978, Characterization of murine colony-forming B cells. I. Distribution, resistance to anti-immunoglobulin antibodies, and expression of Ia antigens, *J. Immunol.* **120**:1289.

Kincade, P. W., Lee, G., Fernandes, G., Moore, M. A. S., Williams, N., and Good, R. A., 1979, Abnormalities in clonable B lymphocytes and myeloid progenitors in autoimmune NZB mice, *Proc. Natl. Acad. Sci. USA* **76**:3464.

Kincade, P. W., Lee, G., Scheid, M. P., and Blum, M. D., 1980a, Characterization of murine colony-forming B cells. II. Limits to *in vitro* maturation, Lyb-2 expression, resolution of IgD^+ subsets, and further population analysis, *J. Immunol.* **124**:947.

Kincade, P. W., Flaherty, L., Lee, G., Watanabe, T., and Michaelson, J., 1980b, Qa antigen expression on functional lymphoid, myeloid, and stem cells in adult mice, *J. Immunol.* **124**:2879.

Kurland, J. I., and Bockman, R., 1978, Prostaglandin E production by human blood monocytes and mouse peritoneal macrophages, *J. Exp. Med.* **147**:952.

Kurland, J. I., Kincade, P. W., and Moore, M. A. S., 1977, Regulation of B lymphocyte clonal proliferation by stimulatory and inhibitory macrophage-derived factors, *J. Exp. Med.* **146**:1420.

Lala, P. K., Johnson, G. R., Battye, F. L., and Nossal, G. J. V., 1979, Maturation of B lymphocytes. I. Concurrent appearance of increasing Ig, Ia, and mitogen responsiveness, *J. Immunol.* **122**:334.

Lau, C. Y., Melchers, F., Miller, R. G., and Phillips, R. A., 1979, *In vitro* differentiation of B lymphocytes from pre-B cells, *J. Immunol.* **122**:1273.

Lutzner, M. A., and Hansen, C. T., 1976, Motheaten: An immunodeficient mouse with markedly less ability to survive than the nude mouse in a germfree environment, *J. Immunol.* **116**:1496.

McCombs, C., Hom, J., Talal, N., and Mischell, R. I., 1975, Functional deficiency of splenic adherent cells in New Zealand black mice, *J. Immunol.* **115**:1695.

Manny, N., Datta, S. K., and Schwartz, R. S., 1979, Synthesis of IgM by cells of NZB and SWR mice and their crosses, *J. Immunol.* **122**:1220.

Melchers, F., 1977, B lymphocyte development in fetal liver. I. Development of reactivities to B cell mitogens *"in vivo"* and *"in vitro," Eur. J. Immunol.* **7**:476.

Metcalf, D., 1972, Effect of thymidine suiciding on colony formation *in vitro* by mouse hematopoietic cells, *Proc. Soc. Exp. Biol. Med.* **139**:511.

Metcalf, D., 1976, Role of mercaptoethanol and endotoxin in stimulating B lymphocyte colony formation *in vitro, J. Immunol.* **116**:635.

Metcalf, D., 1977, *Hemopoietic Colonies,* Springer-Verlag, New York.

Metcalf, D., and Moore, M. A. S., 1971, *Haemopoietic Cells,* American Elsevier, New York.

Metcalf, D., and Russell, S., 1976, Inhibition by mouse serum of hemopoietic colony formation, *Exp. Hematol.* **4**:353.

Metcalf, D., and Stevens, S., 1972, Influence of age and antigenic stimulation on granulocyte and macrophage progenitor cells in the mouse spleen, *Cell Tissue Kinet.* **5**:433.

Metcalf, D., Nossal, G. J. V., Warner, N. L., Miller, J. F. A. P., Mandel, T. E., Layton, J. E., and Gutman, G. A., 1975, Growth of B-lymphocyte colonies *in vitro, J. Exp. Med.* **142**:1534.

Metcalf, D., Wilson, J. W., Shortman, K., Miller, J. F. A. P., and Stocker, J., 1976, The nature of the cells generating B-lymphocyte colonies *in vitro, J. Cell. Physiol.* **88**:107.

Metcalf, E. S., and Klinman, N. R., 1977, *In vitro* tolerance induction of bone marrow cells: A marker for B cell maturation, *J. Immunol.* **118**:2111.

Miller, R. G., Gorczynski, R. M. Lafleur, L., MacDonald, H. R., and Phillips, R. A., 1975, Cell separation analysis of B and T lymphocyte differentiation, *Transplant. Rev.* **25**:59.

Morgan, E. L, and Weigle, W. O., 1980, Regulation of Fc fragment- induced murine spleen cell proliferation, *J. Exp. Med.* **151**:1.

Morton, J. I., and Siegel, B. V., 1971, Radiation sensitivity of New Zealand black mice and the development of autoimmune disease and neoplasia, *Proc. Natl. Acad. Sci. USA* **68**:124.

Morton, J. I., and Siegel, B. V., 1974a, Transplantation of autoimmune potential. I. Development of antinuclear antibodies in H-2 histocompatible recipients of bone marrow from New Zealand black mice, *Proc. Natl. Acad. Sci. USA* **71**:2162.

Morton, J. I., and Siegel, B. V., 1974b, Early autoantibody formation in lethally irradiated or drug treated H-2 compatible recipients of preautoimmune NZB bone marrow or fetal liver cells, *Transplantation* **17**:621.

Morton, J. I., and Siegel, B. V., 1978a, Transplantation of autoimmune potential. III. Immunological hyperresponsiveness and elevated endogenous spleen colony formation in lethally irradiated recipients of NZB bone marrow cells, *Immunology* **34**:863.

Morton, J. I., and Siegel, B. V., 1978b, Transplantation of autoimmune potential. IV. Reversal of the NZB autoimmune syndrome by bone marrow transplantation, *Transplantation* **27**:133.

Moutsopoulos, H. M., Boehm-Truitt, M., Kassan, S. S., and Chused, T. M., 1977, Demonstration of activation of B lymphocytes in New Zealand black mice at birth by an immunoradiometric assay for murine IgM, *J. Immunol.* **119**:1639.

Nakajima, P. B., Datta, S. K., Schwartz, R. S., and Huber, B. T., 1979, Localization of spontaneously hyperactive B cells of NZB mice to a specific B cell subset, *Proc. Natl. Acad. Sci. USA* **76**: 4613.

Ohsugi, Y., and Gershwin, M. E., 1979, Studies of congenitally immunologic mutant New Zealand mice. III. Growth of B lymphocyte clones in congenitally athymic (nude) and hereditarily asplenic (Dh/+) NZB mice: A primary B cell defect, *J. Immunol.* **123**:1260.

Opitz, H. G., Opitz, U., Lemke, H., Huget, R., and Flad, H. D., 1976, Polyclonal stimulation of lymphocytes by macrophages, *Eur. J. Immunol.* **6**:457.

Paffenholz, V., 1978, Correlation between DNA repair of embryonic fibroblasts and different life span of 3 inbred mouse strains, *Mech. Ageing Dev.* **7**:131.

Paige, C. J., Kincade, P. W., Moore, M. A. S., and Lee, G., 1979, The fate of fetal and adult B-cell progenitors grafted into immunodeficient CBA/N mice, *J. Exp. Med.* **150**:548.

Paige, C. J., Kincade, P. W., Shinefeld, L. A., and Sato, V. L., 1981, Precursors of murine B lymphocytes. Physical and functional characterization and distinctions from myeloid stem cells, *J. Exp. Med.*, **153**:154.

Pelus, L. M., Broxmeyer, H. E., Kurland, J. I., and Moore, M. A. S., 1979, Regulation of macrophage and granulocyte proliferation; Specificities of prostaglandin E and lactoferrin, *J. Exp. Med.* **150**:154.

Primi, D., Hammarstrom, L., and Smith, C. I. E., 1978, Genetic control of lymphocyte suppression. I. Lack of suppression in aged NZB mice is due to a B cell defect, *J. Immunol.* **121**:2241.

Raff, M. C., Megson, M., Owen, J. J., and Cooper, M. D., 1976, Early production of intracellular IgM by B-lymphocyte precursors in mouse, *Nature (London)* **249**:224.

Raveche, E. S., Steinberg, A. D., Klassen, L. W., and Tjio, J. H., 1978, Genetic studies in NZB mice. I. Spontaneous autoantibody production, *J. Exp. Med.* **147**:1487.

Raveche, E. S., Tjio, J. H., and Steinberg, A. D., 1979, Genetic studies in NZB mice. II. Hyperdiploidy in the spleen of NZB mice and their hybrids, *Cytogenet. Cell Genet.* **23**:182.

Reddy, A. L., Fialkow, P. J., and Salo, A., 1978, Ultraviolet radiation-induced chromosomal abnormalities in fetal fibroblasts from New Zealand black mice, *Science* **201**:920.

Schorlemmer, H. U., and Allison, A. C., 1976, Effects of activated complement components on enzyme secretion by macrophages, *Immunology* **31**:781.

Shirai, T., Hayakawa, K., Okumura, K., and Tada, T., 1978, Differential cytotoxic effect of natural thymocytotoxic autoantibody of NZB mice on functional subsets of T cells, *J. Immunol.* **120**:1924.

Shultz, L. D., and Green, M. C., 1976, Motheaten, an immunodeficient mutant of the mouse. II. Depressed immune competence and elevated serum immunoglobulins, *J. Immunol.* **116**:936.

Shultz, L. D., and Zurier, R. B., 1978, "Motheaten": A single gene model for stem cell dysfunction and early onset autoimmunity, in: *Genetic Control of Autoimmune Disease* (N. R. Rose, P. E. Bigazzi, and N. L. Warner, eds.), pp. 229–240, Elsevier/North-Holland, Amsterdam.

Shultz, L. D., Sidman, C. L., and Unanue, E., 1978, Immunologic dysfunction in motheaten mice: Immunodeficiency, autoimmunity and hyperimmunoglobulinemia in a short-lived mutant, in: *Comparative and Developmental Aspects of Immunity and Disease* (M. E. Gershwin and E. L. Cooper, eds.), pp. 260–269, Pergamon Press, New York.

Sidman, C. L., Shultz, L. D., and Unanue, E., 1978a, The mouse mutant "motheaten." I. Development of lymphocyte populations, *J. Immunol.* **121**:2392.

Sidman, C. L., Shultz, L. D., and Unanue, E., 1978b, the mouse mutant "motheaten." II. Functional studies of the immune system, *J. Immunol.* **121**:2399.

Siegel, B. V., and Morton, J. I., 1976, Tumor virus effects on immunocyte precursor cells: Hemopoietic stem cell behavior and leukemogenic susceptibility, *Ann. N.Y. Acad. Sci.* **276**:442.

Stanley, E. R., and Guilbert, L. J., 1980, Regulation of macrophage production by a colony stimulating factor, in: *Mononuclear Phagocytes—Functional Aspects* (R. van Furth, ed.), p. 417, Nijhoff, The Hague.

Stephenson, J. R., Reynolds, R. K., Tronick, S. R., and Aaronson, S. A., 1975, Distribution of three classes of endogenous type-C RNA viruses among inbred strains of mice, *Virology* **67**:404.

Storer, J. B., 1966, Longevity and gross pathology at death in 22 inbred mouse strains, *J. Gerontol.* **21**:404.

Till, J. E., and McCulloch, E. A., 1961, A direct measurement of the radiation sensitivity of normal mouse bone marrow cells, *Radiat. Res.* **14**:213.

Unanue, E. R., 1978, The regulation of lymphocyte functions by the macrophage, *Immunol Rev.* **40**:227.

Warner, N. L., 1978, Genetic aspects of immunologic abnormalities in New Zealand mouse strains, *Arthritis Rheum.* **21**:S106.

Warner, N. L., and Moore, M. A. S., 1971, Defects in hematopoietic differentiation in NZB and NZC mice, *J. Exp. Med.* **134**:313.

Williams, N., 1979, Preferential inhibition of murine macrophage colony formation by prostaglandin E, *Blood* **53**:1089.

Williams, N., and Eger, R. R., 1978, Purification and characterization of clonable murine granulocyte–macrophage precursor cell populations, in: *Hematopoietic Cell Differentiation* (D. W. Golde, M. Cline, D. Metcalf, and C. F. Fox, eds.), pp. 385–398, Academic Press, New York.

Williams, N., and Jackson, H., 1977, Analysis of populations of macrophage–granulocye progenitor cells stimulated by activities in mouse lung-conditioned medium, *Exp. Hematol.* **5**:523.

Williams, N., Eger, R. R., Moore, M. A. S., and Mendelsohn, N., 1978, Diferentiation of mouse bone marrow progenitor cells to neutrophil granulocytes by an activity separated from WEHI-3 cell conditioned medium, *Differentiation* **11**:59.

Wright, E. G., Lord, B. I., Dexter, T. M., and Lajtha, L. G., 1979, Mechanisms of haemopoietic stem cell proliferation control, *Blood Cells* **5**:247.

6

Lymphoproliferation (*lpr*) and Other Single-Locus Models for Murine Lupus

EDWIN D. MURPHY

1. Introduction

Systemic lupus erythematosus (SLE) and rheumatoid arthritis are the classical types of generalized autoimmune disease in man (Burnet, 1972). In 1959, Bielschowsky and coworkers reported that strain NZB mice die prematurely of hepatosplenomegaly and autoimmune hemolytic anemia, the first recognition of a spontaneous autoimmune disease in an experimental animal. Helyer and Howie (1963) reported that the (NZB × NZW)F_1 hybrid developed a spontaneous autoimmune disease more closely resembling human SLE, characterized by antinuclear antibodies, glomerulonephritis, and death secondary to renal failure. These murine models became the standard laboratory models for the experimental study of spontaneous autoimmune disease. Murphy and Roths (1977, 1978a,b) reported the first single-gene model for autoimmunity with lymphoproliferation controlled by the mutant gene *lpr*, lymphoproliferation. It has become the model of choice for the study of both SLE (Andrews *et al.*, 1978; Eisenberg *et al.*, 1978; Dixon, 1979) and rheumatoid arthritis (Andrews *et al.*, 1978; Abe *et al.*, 1980).

Huston and Steinberg (1979) have reviewed the literature on animal models of human SLE and have stressed the evidence for the multifactorial basis for SLE, including genetic, immunological, and viral factors. Dixon (1979) has analyzed the contributions of the murine models to our understanding of the pathogenesis of SLE and has concluded that a variety of quite different and independent genetic abnormalities may operate via different routes toward an end result of more or less similar distortion of the immunological system. He sees the task ahead as "the identification of the genetically determined molecular influ-

EDWIN D. MURPHY • The Jackson Laboratory, Bar Harbor, Maine 04609.

ences that so pervert immune function and cause disease." Theofilopoulos *et al.* (1980b) have summarized the comprehensive comparative analysis of the immunologic abnormalities in murine strains with spontaneous autoimmune disease that has been conducted in the Department of Immunopathology at the Scripps Clinic and Research Foundation.

This task of identification and isolation of genes in order to determine the mechanism of gene action at the molecular level requires more than the study of inbred strain differences. The most powerful tool is a mutant gene, on an inbred strain background, that can induce autoimmune disease in other inbred strain backgrounds. Such truly isolated genes can be studied in direct contrast with normal alleles at the same locus in congenic strains. They can be studied for interrelationships with known genes controlling the immune network, and they can be used as a genetic probe to begin the dissection of the genetic mechanisms in complex polygenic models such as NZB and (NZB × NZW)F_1 (Warner, 1977), MRL/Mp-+/+ (Murphy and Roths, 1978c), and BXSB/Mp (Murphy and Roths, 1979a). Coisogenic and congenic inbred strains differing at a single locus provide the powerful tool of tissue exchange in the analysis of the mechanism of gene action.

This chapter reviews the available information on the mutant gene *lpr* and strain MRL. Recently at the Jackson Laboratory another mutation for lymphoproliferation and autoimmunity has occurred at a different locus, *gld* (generalized lymphoproliferative disease) in strain C3H/HeJ (Murphy *et al.,* 1979). The Y chromosome of strain BXSB, which accelerates autoimmune disease in BXSB and its F_1 hybrids (Murphy and Roths, 1979a), can be used in similar fashion to a mutant gene in genetic studies.

Most of the studies on strain MRL and the mutant gene *lpr** published before 1980 have been based on the closely related substrains MRL/l* (genotype, *lpr/lpr*) and MRL/n (genotype, +/+). Beginning in mid-1978, these substrains have been replaced by the congenic MRL/Mp-*lpr/lpr* and MRL/Mp-+/+. Strain MRL/Mp was developed by transferring *lpr* from MRL/l to MRL/n (now designated MRL/Mp-+/+) (see Section 3.2.2). Thus the genotypes *lpr/lpr, lpr/*+, and +/+ are available for study on the same genetic background.

2. Development of MRL Substrains

The MRL substrains developed as a by-product of a series of crosses involving inbred strains AKR/J, C57BL/6J, C3H/Di, and LG/J (Murphy and Roths, 1978c) (Fig. 1). The series of crosses was begun in 1960 by Dr. M. M. Dickie in order to transfer a mutation for achondroplasia *(cn)* from the high leukemic background of strain AKR to a background without early incidence of leukemia. A new problem of dental malocclusion developed and was not eliminated until the final backcrosses to strain LG/J. From this point

*Inbred strain designations and gene symbols, which may combine letters and numerals, are unusually vulnerable to misreading and typographical errors. The designations *lpr* and MRL/l contain the lower-case symbol "l" (el), which in many typescripts cannot be distinguished from "one" or sans serif capital "I." Both of these errors have occurred in the literature.

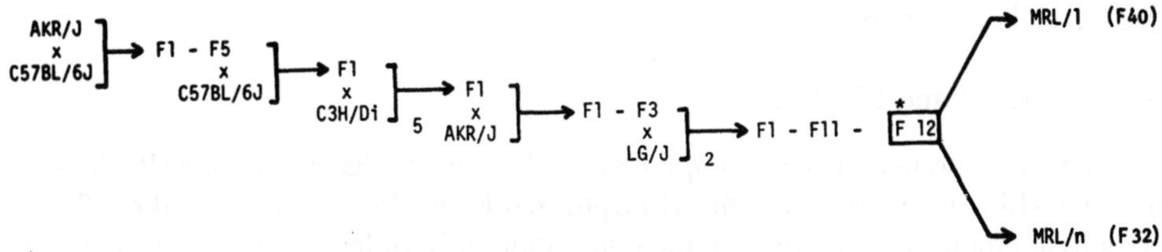

FIGURE 1. Development of substrains MRL/l and MRL/n. F indicates generations of brother × sister matings. Number at bottom of bracket indicates number of backcross generations. (*) The first case of early-age lymphadenopathy occurred in the female sibling of this breeding pair.

rigid inbreeding (brother × sister) was carried out by P. W. Lane. It is estimated from the breeding history that the new composite genome is derived 75.0% from LG, 12.6% from AKR, 12.1% from C3H, and 0.3% from C57BL/6. In the 12th generation of inbreeding, Lane observed that 5 of 12 offspring developed massive generalized lymph node enlargement early in life. In the stem-line mating F_{12} the female developed the lymphadenopathy, the male did not. At this stage, Dr. M. Cherry received breeding stock for H-2 and Ly typing. Starting with F_{13} it became evident that a breeding line with a 100% incidence of lymphadenopathy had developed. The line was transferred at F_{15} to Dr. E. D. Murphy for study of the lymphadenopathy, and became substrain MRL/l. The development of an unaffected line derived from the F_{12} stem-line mating was started by Lane. The line was transferred to Murphy at F_{16} and with continued selection became substrain MRL/n. The two substrains, MRL/l and MRL/n, separated at the 13th generation of inbreeding share at least 89% of their genomes. The original mutant gene *cn* was eliminated from the MRL substrains. It is maintained by Lane in a related inbred substrain that was derived from the MRL stem line at F_{11}.

3. Genetics

3.1. Autosomal Recessive Gene, *lpr*

Breeding tests involving F_1, F_2, and reciprocal backcrosses between MRL/l and MRL/n confirmed the hypothesis of a single autosomal recessive gene controlling the massive lymphoproliferation (Murphy and Roths, 1978a,b). The gene was named lymphoproliferation, *lpr* (Murphy and Roths, 1977). The odds are overwhelming that the gene occurred as a mutation during the recent period of inbreeding, by F_{10}. Genetic linkge of *lpr* has not yet been established (June 1981). Approximately 47% of the autosomal genome had been tested as of March 1980. Twenty-seven linkage markers have been employed (gene symbols indicated in parentheses following the chromosome number): 1(*Idh-1, Pep-3*); 2(*Sd, a, Ra*); 3(*Va*); 4(*Pt, Gpd-1*); 5(*Hm, Pgm-1, W*); 6(*mi*); 7(*Gpi-1, c, Hbb*); 8(*Es-1, e*); 9(*Mod-1*); 10(*Sl*); 11(*wa-2, sh-2, Re, Es-3*); 13(*Xt*); 15(*Ca*); 17(*T*); 18(*Tw*).

3.2. MRL Substrains

3.2.1. MRL/l and MRL/n

The identification of the gene *lpr* permitted the establishment of the MRL/l (*lpr/lpr*) and MRL/n (+/+) substrains (Murphy and Roths, 1977). Separated after 12 generations of brother × sister mating, these substrains differed from one another in an estimated maximum of 11% of their genomes (Green, 1966). Continued inbreeding of the separate substrains reached F_{25} for MRL/l and F_{24} for MRL/n by January 1975. The backgrounds of the substrains were sufficiently close for precisely controlled studies of differences attributable to *lpr*. While skin grafts were not rejected, the 11% difference in genome could include factors that interfere with *in vivo* cell transfers and *in vitro* cell interactions.

3.2.2. Congenic Inbred Lines, MRL/Mp-*lpr/lpr* and +/+

The isolation of *lpr* in substrain MRL/l permitted the ready development of a pair of congenic inbred lines by transfer of the mutant gene from MRL/l to mice derived from the MRL/n stem line. The original substrain MRL/n was redesignated MRL/MP-+/+ at F_{28} and has reached F_{42} (June 1981). By early 1978 the gene *lpr* had been transferred by 5 cycles of cross-intercross matings to MRL/n (MRL/Mp-+/+) with an estimated residual difference between the lines MRL/Mp-*lpr/lpr* and MRL/Mp-+/+ of less than 1% of the genome (Murphy and Roths, 1978b). By early 1980, 10 cycles had been completed, reducing the residual difference to less than 0.1%. In mid-1978 the congenic MRL/Mp-*lpr/lpr* replaced the MRL/l mice supplied to other investigators from the breeding colony of E. D. Murphy at the Jackson Laboratory.

3.2.3. Identified Genes

3.2.3a. Coat Color Genes. These include nonagouti *(a/a)* and albino *(c/c)*.

3.2.3b. Polymorphic Isozyme and Biochemical Loci. Those typed for MRL/l are: *Amy-1*^{a}, *Car-2*^{a}, *Dip-1*^{b}, *Es-1*^{b}, *Es-2*^{b}, *Es-3*^{c}, *Es-10*^{a}, *Got-1*^{a}, *Got-2*^{b}, *Gpi-1*^{a}, *Gpd-1*^{b}, *Gr-1*^{a}, *Hbb*^{a}, *Id-1*^{a}, *Ldr-1*^{a}, *Mod-1*^{a}, *Mod-2*^{b}, *Np-1*^{a}, *Pgm-1*^{a}, and *Pgm-2*^{a} (Murphy and Roths, 1978c).

3.2.3c. Cell-Membrane Alloantigenic Determinants. These determinants of MRL/l and MRL/n are H-2^k, Thy-1.2, TL$^-$, Ly-1.2, Ly-2.1, and Ly-3.1 (Murphy and Roths, 1978b). Morgan and McKenzie (1980) have confirmed these phenotypes for MRL/Mp-*lpr/lpr* and +/+ and identified Ly-4.2, Ly-5.2, Ly-6.2, and Ly-7.2. R. Rich (personal communication) has typed MRL/Mp as Qa-1^b. Y. Obata (personal communication) has typed MRL/l and MRL/n as Gix$^+$.

3.3. Other Congenic Inbred Strains with *lpr*

The isolation of *lpr* permitted the transfer of the gene to standard inbred strains used in cancer and immunogenetic research (Murphy and Roths, 1979b). By early 1980 the transfers had reached the backcross generations indicated in parentheses: AKR/J (8),

BALB/cJ (6), C3H/HeJ (7), C57BL/6J (9), C57BL/10SnJ (5), NZB/BlNJ (8), and SJL/J (11). It is now possible to rank the strain background effects in terms of diminishing degree of lymphoproliferation: SJL, NZB, MRL, C3H, AKR, C57BL/6; and increasing longevity: SJL, NZB, MRL, C57BL/6, C3H, and AKR. These new congenic inbred strains with *lpr* make it possible to begin to separate primary gene effects from background modifications and to study interactions with other identified genes controlling the immune system.

4. Clinical Features

MRL mice of the *lpr/lpr* genotype die early in adult life with massive generalized enlargement of lymph nodes. Erythematous skin lesions, necrosis of the tips of the ears, and swollen feet may also be present.

4.1. Longevity

The mean life span of 28 MRL/l females was 120 ± 4 days and of 20 MRL/l males 154 ± 3 days (Fig. 2); the mean life span of 14 MRL/n females was 513 ± 32 days and of 14 MRL/n males 646 ± 33 days (Murphy and Roths, 1978b). In each substrain the females died at a significantly earlier age. Similar life spans have been observed in other

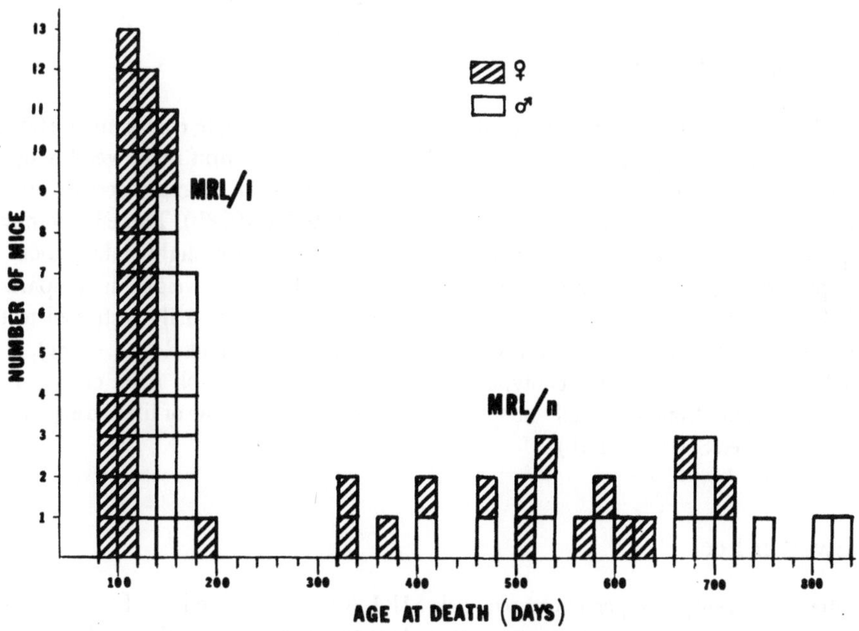

FIGURE 2. Longevity of MRL/l *(lpr/lpr)* and MRL/n (+/+) mice.

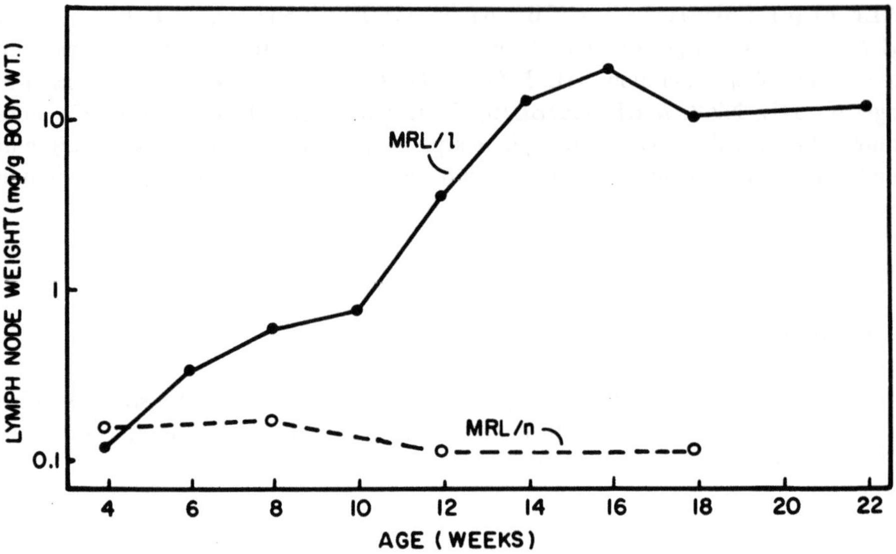

FIGURE 3. Mean weights of axillary, lumbar, and renal lymph nodes corrected for body weight in female MRL/l and MRL/n mice ($n = 5 - 7$ mice per age group).

laboratories (Andrews *et al.*, 1978). The life span of MRL/Mp-*lpr/lpr* judged from breeders does not appear to differ from that of MRL/l (E. D. Murphy and J. B. Roths, unpublished observations).

4.2. Lymphoproliferation

All MRL mice of *lpr/lpr* genotype develop massive lymph node enlargement starting by 8 weeks of age and progressing to over 100 times control lymph node weights by 16 weeks of age (Fig. 3). During the last week or so of life there is an average decrease of one-third from the maximum lymph node size (Murphy and Roths, 1978b). There is ordinarily an approximately sevenfold enlargement of the spleen and only slight enlargement of Peyer's patches. By 22 weeks of age there is a doubling of thymic weight in comparison with MRL/n controls. The associated cellular proliferation is primarily restricted to the medulla as the cortex is atrophic (Andrews *et al.*, 1978; Murphy and Roths, 1978c).

MRL mice of the +/+ genotype do not have multiple lymph node enlargement unless they develop "reticulum cell" neoplasms during the second year of life (Murphy and Roths, 1978a) (see Section 5.1.1b).

4.3. Swollen Feet

Fifteen to twenty-five percent of old, sick MRL/l mice had swelling of the joints and surrounding tissues of the hind feet and lower legs (Andrews *et al.*, 1978; Murphy and Roths, 1978c).

5. Pathology

5.1. Lymph Nodes

The generalized lymph node enlargement of MRL/l mice is as massive as in cases of lymphocytic leukemia or "reticulum cell" sarcomas in the mouse (Fig. 3) (see Section 4.2) (Murphy, 1966).

5.1.1. Histology

The enlarged lymph nodes of MRL/l mice are characterized by a predominantly lymphocytic proliferation with an admixture of histiocytes, plasma cells, and immunoblasts (Murphy and Roths, 1977, 1978a,c) (Figs. 4–6). The cellular infiltrate blurs the nodal architecture. During the last week or so of life, along with the terminal decrease in lymph node size there is extensive necrosis and hemorrhage (Andrews *et al.,* 1978; Murphy and Roths, 1978c).

5.1.1a. Question of Malignancy. The massive lymphoproliferation, blurring of lymph node architecture, and extensive perivascular infiltration in many organs raise the diagnostic question of a malignant lymphoma, particularly "reticulum cell" sarcoma, type B (Murphy, 1966). Dissemination beyond the perivascular area into the parenchyma of major organs has not been observed. There is no leukemic blood picture. Five attempts at transplantation of the lymphoid masses from MRL/l to MRL/n and F_1 hybrids have been negative (Murphy and Roths, 1978c), as have five attempts at transplantation from MRL/Mp-*lpr/lpr* to the congenic MRL/Mp-+/+ (E. D. Murphy, unpublished observations). Whether the lymphoproliferative process represents a prelymphomatous stage will be determined by the ultimate fate of the transplants in the congenic mice.

5.1.1b. Question of the Role of Autoimmunity in Subsequent Lymphoma Development. It is commonly assumed that the "reticulum cell" tumors that occur in the second year of life in NZB mice are caused by autoimmunity (Huston and Steinberg, 1979). However, the actual incidence of this tumor in NZB mice is no higher than in other inbred strains that lack evidence of early autoimmune disease (Murphy, 1979). The MRL/l and MRL/Mp-*lpr/lpr* mice usually do not live long enough for lymphoma or other tumor development. Approximately half of MRL/n (MRL/Mp-+/+) mice die with malignant tumors, of which one-third are "reticulum cell" neoplasms (Murphy and Roths, 1978a). The role of autoimmunity is no clearer than in the case of strain NZB. Unequivocal answers should be possible in long-surviving *lpr/lpr* and *lpr*/+ mice in comparison with congenic +/+ on standard inbred strain backgrounds such as C57BL/6, C3H, BALB/c, etc. It will be necessary to delay death from immune-complex glomerulonephritis in the *lpr/lpr* mice.

5.2. Thymus

In MRL/l mice there is a progessive decline in thymic weight between 6 and 12 weeks of age, followed by enlargement to twice the size of MRL/n thymuses by 22 weeks of age. There is an early progressive atrophy of the cortex (Andrews *et al.,* 1978; Murphy and Roths, 1978c). The later enlargement is restricted to the medulla.

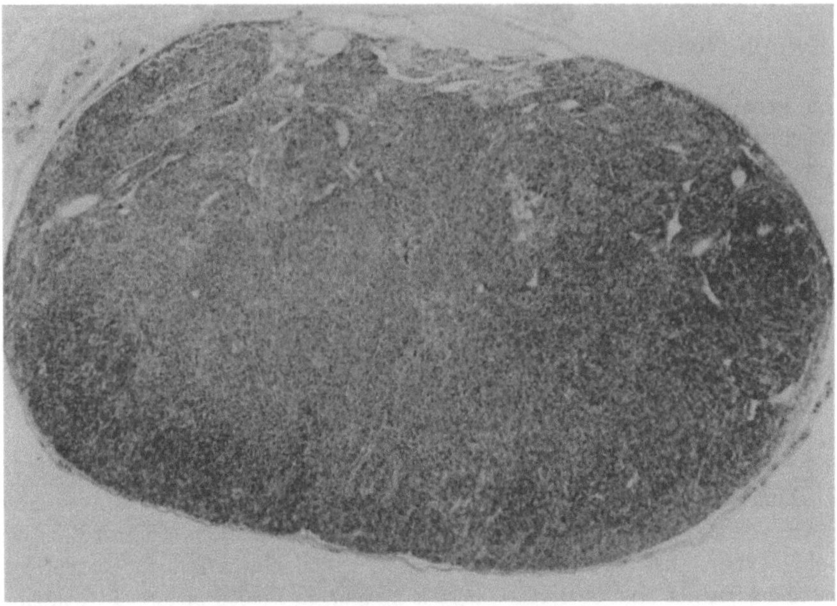

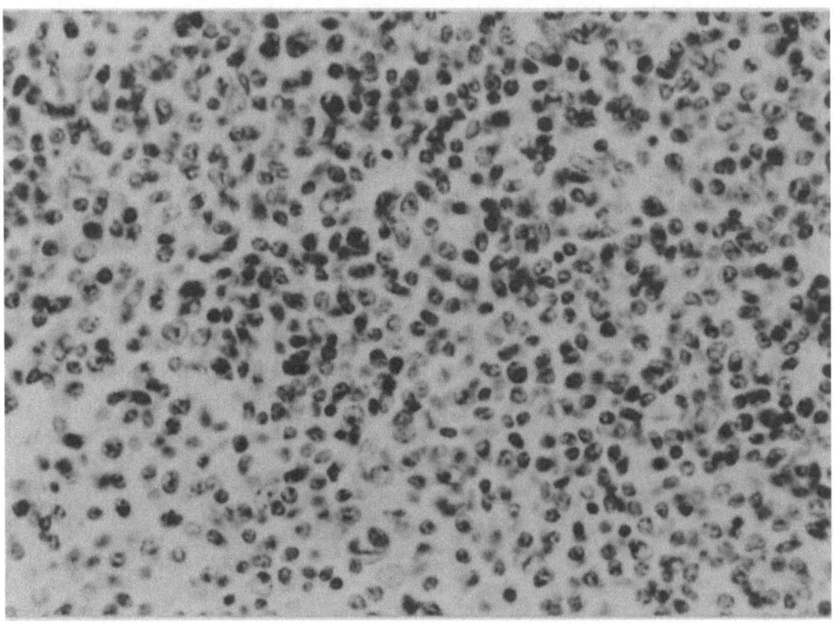

FIGURE 4. Low-power view of enlarged lymph node of 4-month-old MRL/l female mouse. Note blurring of architecture. Hematoxylin and eosin.

FIGURE 5. Proliferation of lymphocytes with admixed histiocytes, plasma cells, and immunoblasts. Hematoxylin and eosin.

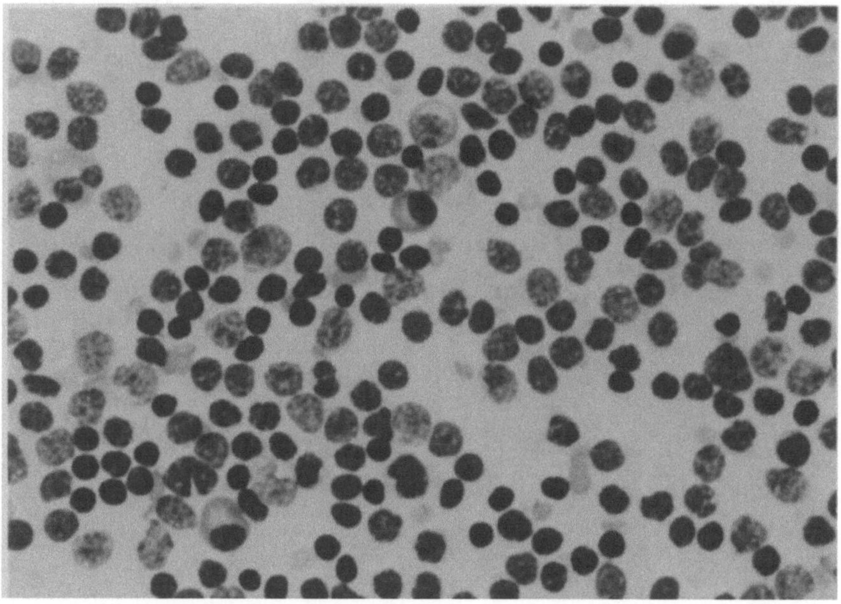

FIGURE 6. Imprint of enlarged lymph node showing mixed cellular population of lymphocytes, histiocytes, plasma cells, and immunoblasts. May–Grünwald–Giemsa.

5.3. Kidney

5.3.1. Clinical Pathology

The usual cause of death in MRL/l mice is glomerulonephritis (Andrews *et al.,* 1978; Murphy and Roths, 1978c). Proteinuria averaging ninefold that of MRL/n mice occurred in 6 of 12 MRL/l females (Murphy and Roths, 1978a). Andrews *et al.,* (1978) reported urinary protein values of 2.6 to 3.8 mg/day in 3- to 6-month-old MRL/l males and females. They also reported terminal anasarca in approximately 20% of MRL/l males and females and in BXSB males. In the reference colonies at the Jackson Laboratory anascara is uncommon in strain MRL/l but very common in BXSB males (more than 60%) (E. D. Murphy, unpublished). Clinically affected 4-month-old MRL-*lpr/lpr* females have double the blood urea nitrogen level, 40.9 mg/dl, compared with age- and sex-matched MRL-+/+ controls, 19.9 mg/dl, ($p < 0.025$; unpublished data).

5.3.2. Histopathology

Strain MRL/l mice develop a largely subacute proliferative glomerulonephritis (Andrews *et al.,* 1978; Murphy and Roths, 1978a,c). The glomerular lesions involve deposition of periodic acid–Schiff (PAS)-positive material and proliferation of both endothelial and mesangial cells with occasional crescent formation and basement membrane thickening (Andrews *et al.,* 1978) (Fig. 7). Severe renal lesions show capsular cell proliferation, tubular damage, and casts.

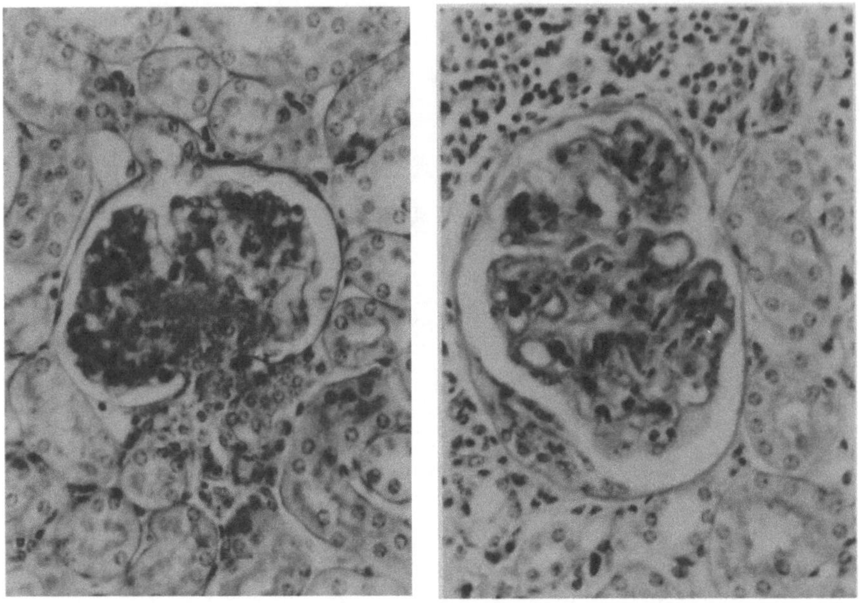

FIGURE 7. Glomerulus of 5-month-old MRL/Mp-*lpr/lpr* female mouse showing deposition of PAS-positive material and increased cellularity. Periodic acid–Schiff reaction.

FIGURE 8. Glomerulus of 17-month-old MRL/Mp-+/+ female mouse showing crescent formation and thickening of basement membranes with "wire-loop" effect around capillaries. Periodic acid–Schiff reaction.

Strain MRL/n (MRL/Mp-+/+) mice develop a chronic glomerulonephritis (Murphy and Roths, 1978a,c) (Fig. 8).

5.3.3. Immunopathology

5.3.3a. Immunofluorescence Studies. The renal disease in strain MRL/l is an immune-complex glomerulonephritis (Andrews *et al.,* 1978; Murphy and Roths, 1978a, c). Andrews *et al.* (1978) have compared the renal lesions in (NZB × NZW)F_1 females and in MRL/l and BXSB males. Granular deposits of mouse IgG and C3 of variable intensity were present in glomerular capillary walls, mesangia, and tubulointerstitial areas in the three autoimmune models. Granular IgG and C3 deposits increased in MRL/l from 2 to 5 months of age. The deposits often tended to predominate in the capillary wall where the finely granular deposits were closely approximated. Murine retroviral envelope glycoprotein (gp70), detected with a goat and anti-Rauscher gp70 antiserum, appeared in graunular glomerular and tubular basement membrane deposits. However, the deposits were often much less striking and distributed differently from the IgG and C3 deposits, especially in MRL/l. In all of the murine lupus models the glomeruli contain heavy deposits of immune complexes in which gp70 and DNA, presumed antigens, can be detected by immunofluorescence (Dixon, 1979). Equivocal staining with antiserum to murine retroviral core protein (p30) was seen in rare glomeruli of occasional mice in each strain.

5.3.3b. Renal Eluates. Dixon (1979) has summarized the results of studies on the immune-complex glomerulonephritis of the major models of murine lupus. Immunoglobulin eluates from the diseased glomeruli contain significant amounts of antibody to nuclear antigens, including DNA. While these eluates do not contain free antibodies to gp70, presumably because of interference by the large amount of naturally occurring gp70 in the kidney tubules, they do contain gp70–anti-gp70 complexes. It appears that two of the major pathogenic immune responses in murine SLE are those to DNA, plus perhaps other nuclear antigens, and those to gp70 (see Sections 6.4.5a–c).

5.4. Vascular System

5.4.1. Generalized Arterial Disease

An acute polyarteritis with a necrotizing, cellular inflammatory response (Fig. 9) is common in strain MRL/l (Andrews *et al.*, 1978; Murphy and Roths, 1978a,c). Accinni and Dixon (1979) have drawn attention to a related degenerative arterial and arteriolar lesion without significant cellular inflammation (Fig. 10) that is more frequently observed in murine lupus. The vessels show deposits of eosinophilic, PAS-positive material in their walls, accompanied by intimal reaction and medial degeneration. These lesions are asso-

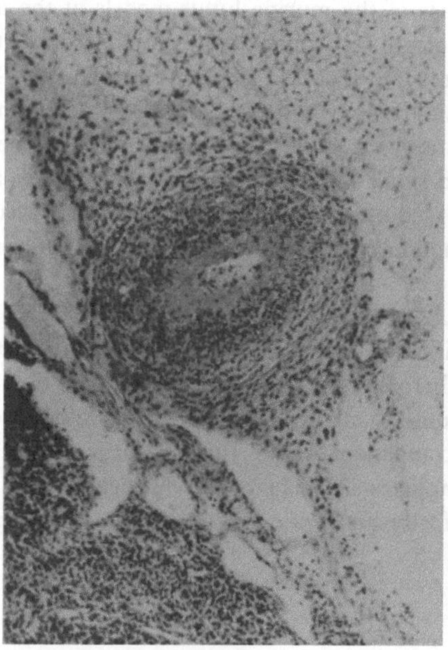

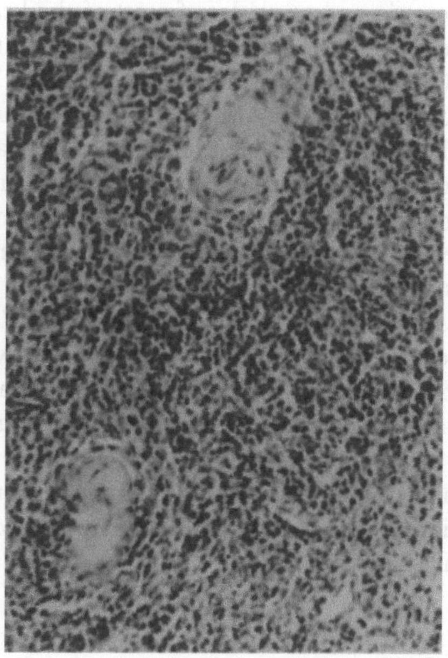

FIGURE 9. Acute polyarteritis with cellular inflammatory response in 6-month-old MRL/l male mouse. Hematoxylin and eosin.

FIGURE 10. Degenerative arteriolar lesion without cellular inflammatory response in spleen of 6-month-old MRL/l female mouse. Hematoxylin and eosin.

ciated with deposition of antigen–antibody complexes and lead to thrombotic and obliterative vascular changes, with infarcts, particularly in the heart. Infarcted areas in lymph nodes and spleen are also associated with arterial disease in MRL/l, MRL/n, BXSB males, and BXSB females (E. D. Murphy, unpublished observations). The degenerative arterial lesion is particulary common in the more slowly progressive murine lupus in MRL/n and BXSB females. However, polyarteritis may also occur.

5.4.2. Degenerative Coronary Artery Disease and Myocardial Infarction

Andrews *et al.* (1978) observed that 15–30% of (NZB × NZW)F_1, MRL/l, and BXSB male mice had myocardial infarcts involving either ventricle and judged extensive enough to be a contributing cause of death. Accinni and Dixon (1979) have studied the coronary arteries in these strains by light, immunofluorescence, and electron microscopy.

5.4.2a. Light Microscopy. Vascular lesions were most frequently observed in the medium and small coronary arteries and arterioles in all three strains. They were local and segmental in distribution and characterized by deposits of amorphous, eosinophilic, PAS-positive material in the vessel walls without associated inflammatory reaction. The amorphous deposits partially and sometimes completely obliterated the lumens of small arteries and arterioles. Intimal hyperplasia was occasionally seen. Infarcts could be multiple and involved the ventricles, rarely the atria.

5.4.2b. Immunofluorescence Microscopy. Deposits of mouse IgG and C3 were present in the walls of medium and small coronary arteries, arterioles, venules, and capillaries in the atria and ventricles of most of the hearts of the murine lupus models at the ages studied. All of the mice had advanced immunologic renal disease. A combined deposition of IgG and C3 was generally observed, with less frequent deposition of retroviral gp70.

5.4.2c. Electron Microscopy. Focal, granular, electron-dense deposits similar to those found in the glomerular capillaries of the same mice were seen beneath the endothelial basement membrane and between medium and small arteries, arterioles, and venules, and along capillary walls. Swelling and alteration of the endothelium were sometimes accompanied by aggregation of platelets, fibrin formation, and vascular occlusion in small arterioles and capillaries.

5.4.2d. Coronary Arteries in Human SLE. Accinni and Dixon (1979) cite the frequency of myocardial infarction in patients with long-standing SLE. The coronary artery lesions are generally described as classical atherosclerosis. The authors suggest that the predisposing inherent insusceptibility of the mouse to atherogenesis limits the vascular disease in murine lupus to the immunologic component. They further suggest that the repeated vascular deposition of immune complexes associated with incidental immune reactions might be a significant component of atherosclerotic disease in immunologically normal individuals.

5.5. Lungs

The lungs of MRL/l mice show extensive perivascular and peribronchial lymphoproliferation composed of lymphocytes, plasma cells, and histiocytes (Andrews *et al.*, 1978; Murphy and Roths, 1978c). Patches of atelectasis and areas with exudate, containing macrophages, are common and more severe bronchopneumonia may occur. Pneumonia also occurs in BXSB males and (NZB $\times$ NZW)F_1 mice (Andrews *et al.*, 1978).

5.6. Joints

Strain MRL/l mice develop a 15–25% incidence of swelling of the hind feet and lower legs (Andrews *et al.*, 1978; Murphy and Roths, 1978c; Abe *et al.*, 1980). Histologically there is an acute to chronic inflammatory process, with destruction of articular cartilage, proliferation of synovia, pannus formation, and joint effusions. The periarticular tissues show acute and chronic inflammation with occasional foci of necrosis reminiscent of rheumatoid nodules (Andrews *et al.*, 1978). Among the mouse models for lupus this arthritis is found only in MRL/l. This strain is also unique in that it is the only strain that develops significant levels of rheumatoid factor (Andrews *et al.*, 1978; Dixon, 1979; Eisenberg *et al.*, 1978b) (see Section 6.4.3). With both serologic and arthritic changes resembling those of rheumatoid arthritis, the MRL/Mp-*lpr/lpr* mouse should provide an important spontaneous model for the study of this human disease.

6. Serum Factors

6.1. Serum Proteins

Total serum proteins and electrophoretic quantification of serum protein classes were determined for 22 MRL/l mice, 4 months of age, and for 15 age-matched MRL/n controls (Murphy and Roths, 1978a,c). Total serum proteins were increased from 6.0 $\pm$ 0.15 g/dl in controls to 7.0 $\pm$ 0.2 g/dl in MRL/l. Albumin was reduced from 3.9 to 3.2 g/dl. The alpha globulins were unchanged.There was a twofold increase in beta globulins and at least a fivefold increase in gamma globulins.

6.1.1. Immunoglobulins

The major immunoglobulin classes were measured by radial immunodiffusion. In MRL/l sera there were twofold increases in IgA, IgM, and IgG2b concentrations (Murphy and Roths, 1978a,c). In 10 of 11 sera the concentration of both IgG1 and IgG2a exceeded the working limits of the immunodiffusion plates. It can be estimated that clinically affected MRL/l mice have at least six times as much IgG1 and IgG2a as controls (Table I). Excesses in IgG2a levels were invariably associated with elevated IgG1 levels. This is consistent with our finding of broad-based gamma-region electrophoretic patterns suggesting a polyclonal gammopathy. Andrews *et al.*, (1978) reported IgG concentrations in MRL/l mice that averaged nearly five times those of BALB/c controls at 2–3 months of age and eight times higher than controls at 4–5 months. In a survey of five models for murine lupus, the highest mean IgG concentration occurred in MRL/l. Andrews *et al.* (1978) reported a 43% incidence of monoclonal gamma globulins in MRL/l mice, which is much higher than in the other murine models.

TABLE I
Serum Immunoglobulin Levels in MRL/Mp-lpr/lpr, MRL/Mp-+/+, and Control Mice

Strain-genotype	Age[b]	*N*	Serum immunoglobulins[a] IgA	IgG1	IgG2a	IgG2b	IgM
MRL/Mp-*lpr/lpr*	14–20	11	65 ± 15	>1332 ± 128[c]	>1527 ± 73[c]	205 ± 51	55 ± 7
MRL/Mp-+/+	16–20	4–6	27 ± 13	199 ± 27	261 ± 69	93 ± 39	35 ± 5
C57BL/6J-+/+	16–18	4–5	13 ± 2	58 ± 16	167 ± 39	134 ± 42	25 ± 8
C3H/HeJ-+/+	21–24	8	21 ± 1	67 ± 5	125 ± 32	19 ± 4	26 ± 2

[a]Serum immunoglobulins measured by single radial immunodiffusion. Values expressed as mg Ig/100 ml serum (mean ± S.E.M.).
[b]Age in weeks.
[c]MRL/Mp-*lpr/lpr* serum has IgG1 and IgG2a levels that exceed the working limits of these plates (Meloy Laboratories, Springfield, Virginia).

6.2. Autoantibodies

6.2.1. Antinuclear Antibodies

The sera of MRL/l mice were negative for antinuclear antibodies (ANA) at 4 weeks of age (Murphy and Roths, 1978a). At a dilution of 1:10, the sera were positive in 3 of 9 MRL/l mice at 8 weeks and in 9 of 9 by 20 weeks of age. By contrast, only 2 of 10 MRL/n were positive by 20 weeks of age and 5–6 at 40 weeks. In a comparison of murine lupus models, Andrews *et al.* (1978) found the highest ANA geometric mean titers in MRL/l mice 4–5 months of age with the next highest titers in (NZB × NZW)F_1 females 9 months of age, followed by BXSB males at 5–6 months of age. In all the strains tested, a peripheral or rim pattern of nuclear fluorescence was always present at the highest positive serum dilution. The homogeneous pattern was sometimes seen at lower dilutions.

6.2.2. Anti-DNA Antibodies

6.2.2a. Anti-Double-Stranded (ds) DNA. Levels of anti-dsDNA antibodies were highest in the serum of MRL/l mice of both sexes (4–5 months of age) in comparison with (NZB × NZW)F_1 females (9 months), NZB (9 months), and BXSB (4–5 months), in that order (Andrews *et al.*, 1978).

High levels of antibodies to dsDNA were detected in supernatants of cultured spleen cells from 4-month-old MRL/l and BXSB male mice and 8-month-old (NZB × NZW)F_1 female mice in the absence of an exogenous source of antigen (Sawada and Talal, 1979).

6.2.2b. Anti-Single-Stranded (ss) DNA. Levels of anti-ssDNA antibodies can be found in low concentrations in the serum of immunologically normal mice. However, higher levels were found at 2 months of age in MRL/l mice, (NZB × NZW)F_1 females, and NZB females (Andrews *et al.*, 1978).The highest levels were found in MRL/l (4–5 months of age), then (NZB × NZW)F_1 females, NZB females, and BXSB males.

6.2.3. Anti-Sm Antibody

The anti-Sm "marker" antibody is highly restricted to human SLE. It is rarely found in other human autoimmune diseases. A survey of five murine models for lupus, MRL/l,

MRL/n, (NZB × NZW)F_1, NZB, and BXSB, and seven control strains was carried out by Eisenberg *et al.* (1978). With the exception of one pool from (NZB × NZW)F_1 female mice, anti-Sm was found only in the MRL/l and MRL/n strains. In the MRL/l mice the marker was absent at 1–3 months of age, and was present in 37% of males and 10% of females at 4–5 months. In strain MRL/n some mice of both sexes were positive at 1 to 3 months, and 35% of males and 45% of females at 4–6 months of age. It is noteworthy that in strain MRL/n the marker is present long before the murine lupus is fully developed. The MRL substrains, and especially the congenics, should provide the first experimental models for exploration of this unique autoantibody in SLE.

6.2.4. Antithymocyte Antibody

Among five mouse strains or hybrids that develop murine lupus, NZB and (NZB × NZW)F_1 had high levels of natural thymocytotoxic autoantibodies (NTA) starting early in life (Eisenberg *et al.,* 1979a). The levels in BXSB males overlapped with (NZB × NZW)F_1 males at 1–7 months of age. The levels in BXSB females were somewhat lower and both MRL/l and MRL/n males and females had very low levels. NTA were found in all of 14 other mouse strains tested, with particularly high levels in SM, RF, I, NC, and 129-GIX$^+$. Females were usually more reactive than males. The NTA of mice appear to be mostly IgM molecules. In the autoimmune mice they are partly cryoprecipitable, but there is no specific concentration of NTA activity in the cryoglobulins.

6.2.5. Antierythrocyte Antibody

In their survey of five autoimmune strains or hybrids, Andrews *et al.* (1978) reported the highest levels of direct antierythrocyte antibody in NZB and (NZB × NZW)F_1 females. BXSB males ranged next with 18% positive at 5–6 months. The incidence in MRL/l males and females reached 4 and 11% respectively.

6.3. Complement

Andrews *et al.* (1978) reported that the concentration of hemolytic complement fell with age in all types of mice with a lupuslike syndrome. The progressive fall began with the onset of disease and progressed as the immunologic abnormalities increased. The data on the more rapidly progressing syndromes in MRL/l, BXSB males, and (NZB × NZW)F_1 females are charted.

6.4. Immune Complexes

6.4.1. Circulating Immune Complexes

Andrews *et al.* (1978) found the highest levels of circulating immune complexes (IC) in MRL/l males and females, which at 2–3 months of age had mean IC levels of 90 μg equivalent aggregated mouse IgG/ml serum and at 4–5 months, nearly 900 μg. The mean levels in BXSB males reached 178 μg at 5 months, (NZB × NZB)F_1 females, 144 μg at

9 months, and NZB females, 115 μg at 9 months. The measurement of IC was unrelated to the hyperimmunoglobulin levels in these mice. Removal of serum cryoglobulins did not significantly reduce the IC levels.

6.4.2. Cryoglobulins

Andrews *et al.* (1978) found the most striking concentrations of cryoglobulins in MRL/l mice of both sexes, the values rising from 173 μg/ml at 2 months to 2180 μg/ml at 5 months. Control levels for BALB/c mice were 134 μg/ml at 3 months and 206 μg/ml at 9 months. (NZB × NZW)F_1 females had slightly elevated levels of approximately 200 μg/ml at 3 months and slightly greater than 300 μg/ml at 6 and 9 months. NZB females had large amounts of cryoglobulin with concentrations of approximately 350 μg/ml at 3 and 6 months, and 570 μg/ml at 9 months. BXSB males had elevated cryoglobulins of 472 μg/ml at 2 months followed by decreases to 260 μg/ml at 3 months and 275 μg/ml at 5 months.

IgM and IgG1 were most frequently found in cryoprecipitates of all mice tested, followed by IgG2A. C3 was found in 50% of the cryoglobulins of MRL/l, 20% of NZ mice, and only 11% in those of BXSB.

6.4.3. Rheumatoid Factor

Andrews *et al.* (1978) found significant elevations of IgM rheumatoid factor (RF) only in MRL/l mice (approximately one-half) and in a single (NZB × NZW)F_1 female. Eisenberg *et al.* (1979b) found that sera from a majority of old, sick MRL/l mice interact with other MRL/l sera to form visible immunoprecipitates and to fix complement. The principle of these serum–serum interactions appears to be an IgG-RF present as an immune factor. The authors have not found IgG-RF in sera from other autoimmune strains.In MRL/l, IgG-RF could also be found in serum cryoprecipitates. The authors believe that at least part of the cryoglobulin reaction is due to RF specificity. The levels of IgG-RF in individual MRL/l sera did not correlate with the presence or strength of serum–serum reactivity.

It is of particular interest that of all the autoimmune mice, only the MRL/l develop chronically swollen feet . Histologically these demonstrate acute and chronic inflammation in both the joints and the periarticular tissue resembling the changes of rheumatoid arthritis. Thus, MRL/l mice provide a good model for human rheumatoid arthritis, in which serum or synovial fluid IgG–IgG complexes are frequently present and in which extraarticular disease is correlated with the presence of such complexes or with IgG-RF in general (Eisenberg *et al.*, 1979b; Izui and Eisenberg, 1980).

6.4.4. Retroviral gp70 and Complexes

6.4.4a. Free gp70. NZB, (NZB × NZW)F_1, MRL/l, and BXSB male mice have relatively high serum concentrations of the retroviral envelope protein, gp70 (Andrews *et al.*, 1978; Izui *et al.*, 1979). However, several immunologically normal strains (NZW, DBA/2, and 129) have equally high levels. C57BL/6, C3H, and BALB/c have low levels.

There is no correlation between the levels of gp70 in the autoimmune models and the progression of glomerulonephritis or of autoimmune disease (Izui *et al.*, 1979).

6.4.4b. Heavy gp70. Izui *et al.* (1979) have found a heavy form of gp70 that varies in sedimentation rate from 9 to 19 S in sucrose density gradient analysis of the sera of MRL/l, BXSB males (NZB × NZW)F_1, and NZB. The gp70 was complexed to Ig and these complexes appeared with the onset of renal disease and persisted throughout its course. The MRL/l mice and BXSB males develop the heavy gp70 in their sera earlier than the other models. The MRL/n mice, which develop glomerulonephritis much later in life, show no heavy gp70 until 1 year of age when significant quantities have been observed in females. Unlike the 5 S gp70, the heavy gp70 increases not only in incidence but also in amount with the progression of disease in the four major SLE models. Normal strains including those with high free gp70 serum levels do not develop the sedimenting gp70, at least during the first 8 to 10 months of life.

The fact that the heavy gp70 is selectively absorbed with anti-IgG antibodies or with *Staphylococcus aureus* protein A (staph A) suggests that it is complexed with antibodies. The rise in incidence and quantity of these gp70 immune complexes with the progression of disease, suggest that Ig-complexed heavy gp70 may be involved as one of many complexes in the pathogenesis of glomerulonephritis of mice with SLE (Izui *et al.*, 1979). The Ig-complexed, heavy form of gp70 is a new useful marker in murine SLE.

6.4.5. Immune Complexes in Renal Eluates

6.4.5a. Antibody Activity in Renal Eluates. The concentrations of ANA activity in IgG eluted from kidneys in relation to the concentrations of ANA activity in serum IgG were 2- to 10-fold higher in individual eluates of (NZB × NZW)F_1 kidneys and 2- to 6-fold higher in eluates from both MRL/l and BXSB kidneys (Andrews *et al.*, 1978). Antibodies against dsDNA were considerably more concentrated in (NZB × NZW)F_1 renal eluates (25 to 31-fold greater than in serum) while MRL/l eluates showed only a 1- to 6-fold increase and BXSB eluates ranged from 1- to 12-fold more concentrated than in serum. Antibodies to Rauscher gp70 and to AKR gp70 were not found concentrated in Ig eluates from any of these SLE models.

6.4.5b. Retroviral gp70 Complexes. Heavy gp70 was found in renal eluates from 5-month-old MRL/l, 6-month-old BXSB, 8-month-old (NZB × NZW)F_1, and 10-month-old NZB mice (Izui *et al.*, 1979). This heavy gp70 was selectively absorbed by staph A or anti-mouse IgG antibodies.

Izui *et al.* (1979) consider the circulating gp70 immune complexes a potential source of tissue injury in mice with SLE. However, they do not imply that gp70 immune complexes are the primary cause of glomerulonephritis in these mice. A number of antibodies reactive with and forming complexes with native or altered autologous antigens seem to be involved in the pathogenesis of murine SLE. The pathological importance of the significant concentrations of anti-DNA antibodies in renal eluates is underlined by the decreased anti-DNA producion, decreased nephritis, and prolonged survival of (NZB × NZW)F_1 hybrids following induction of tolerance to DNA antigen (Borel *et al.*, 1973; Parker *et al.*, 1974).

6.4.5c. Antibodies to dsDNA. Antibodies to dsDNA in the serum and glomerular deposits of DNA–antiDNA immune complexes have been studied by isoelectric focusing in (NZB × NZW)F_1, MRL/l, and BXSB males (Ebling and Hahn, 1980). Significant quantities of DNA antibodies appeared by 3 months of age in MRL/l mice of both sexes and by 4 months in BXSB males and (NZB × NZW)F_1 females. MRL/l mice had the largest number of DNA-binding bands, up to 30. The large array of anti-DNA antibodies from all three models had isoelectric points ranging from pH 5.5 to 9.0. In contrast, eluates from three different pools of kidneys MRL/l males, MRL/l females, and (NZB × NZW)F_1 females with proteinuria and high serum binding of DNA) showed restricted heterogeneity of the DNA binding bands. After acid elution there were only three bands of IgG that bound DNA and focused between pH 8.0 and 9.0. The authors suggest that anti-DNA with isoelectric points from pH 8.0 to 9.0 may be more pathogenic than other subpopulations. Experiments in which these particular subpopulations are suppressed or are introduced into healthy mouse strains or premorbid lupus strains are in progress.

7. Immunologic Mechanisms

7.1. B-Cell Frequency and Markers

7.1.1. Enumeration of B (Ig-Bearing) and T Cells

Murphy and Roths (1978a,c) reported the low frequency of B cells and high frequency of T cells in the massively enlarged lymph nodes of MRL/l. The proportions of B and T lymphocytes in the mesenteric nodes of MRL/l and MRL/n mice were determined from ages 6 through 18 weeks by immunofluorescence, using surface immunoglobulin and the Thy-1.2 alloantigen as markers. At 6 weeks of age the mesenteric nodes of both substrains weighed approximately 35 mg and contained similar percentages of B cells (MRL/l, 32%, MRL/n, 28%) and T cells (MRL/l, 62%, MRL/n, 66%). At 18 weeks of age the mesenteric node of MRL/n had increased to a mean weight of 53 mg with no change in the proportion of lymphocyte populations. In contrast, the mesenteric node of MRL/l mice increased in weight to 2200 mg and the proportions of B and T lymphocytes had changed to 7 and 88%, respectively. Thus, there was a decreased frequency of B cells, but in view of the 60-fold increase in lymph node size and lymphocyte population, there was an absolute increase in number of B cells.

Theofilopoulos *et al.* (1979a) determined the changes in B- and T-cell composition of NZB, (NZB × NZW)F_1, MRL/l, MRL/n, and BXSB mice at 1 month of age and with characteristic autoimmune disease (4.5–8 months). In control C57BL/6 mice the frequencies and absolute numbers of Ig-bearing cells in mesenteric node and spleen remained stable with age. NZB and (NZB × NZW)F_1 mice had reduced frequencies and numbers of Ig-bearing cells in spleens and lymph nodes with increased age and disease. In MRL/l mice there was an even greater reduction in frequencies, but not absolute numbers, of these B cells, especially in lymph nodes, by 4.5 months of age. BXSB males, but not females, had a substantial increase in the frequency and absolute numbers of Ig-bearing cells associated with advanced disease. The frequency of Ig-bearing cells in the thymus increased with age in NZB, (NZB × NZW)F_1, MRL/l, MRL/n, and BXSB males.

7.1.2. Enumeration of C3d Receptor-Bearing Cells

The frequencies of C3d receptor-bearing cells in lymphoid organs, but not bone marrow, were lower than those of the Ig-bearing cells in NZB, (NZB × NZW)F_1, MRL/l, MRL/n, BXSB, and C57BL/6 (Theofilopoulos *et al.*, 1979a). However, they reflected the same trends with the exception of BXSB males, which showed a decline with age of C3d receptor-bearing cells while the Ig-bearing cells increased in number. MRL/l but not MRL/n mice showed a consistent decline in frequency of cells with either marker in spleen, mesenteric node, and peripheral blood.

7.1.3. Enumeration of IgG Fc Receptor-Bearing Cells

The proportion of IgG Fc receptor-bearing cells in lymphoid organs as similar in C57BL/6 and NZB, (NZB × NZW)F_1, MRL/l, MRL/n, and BXSB males when young (Theofilopoulos *et al.*, 1979a). The frequencies of these cells decreased with age and disease in spleen, mesenteric node, and peripheral blood of MRL/l, BXSB males, (NZB × NZW)F_1, and NZB.

7.2. T-Cell Frequency and Markers

7.2.1. Enumeration of T Cells

Murphy and Roths (1978c) demonstrated that the massive lymphoproliferation in strain MRL/l was composed of approximately 90% T cells. The frequency curve of MRL/l mesenteric node T cells rose from the normal level of 62% at 6 weeks of age to 88% at 18 weeks, when the nodes were about 60-fold enlarged. Theofilopoulos *et al.* (1979a) also found a clear increase in frequencies and a massive increase in absolute numbers of T cells in the mesenteric nodes and in the spleens of MRL/l mice by 4.5 months of age. In contrast, there was a decline in frequency and number of T cells in the lymphoid organs of NZB and (NZB × NZW)F_1 females and BXSB males with advanced age and disease. MRL/n showed a modest increase in frequency and number by 8 months of age. As the control C57BL/6 mice aged, their T-cell values remained stable.

7.2.2. Enumeration of Ly Alloantigen-Bearing Cells

Theofilopoulos *et al.* (1979a) have determined the frequency of the major Ly subsets in lymph node cells of MRL/l females at 6 and 20 weeks of age and compared them to similar data from 12-week-old C57BL/6 females. The distributions of the two large subsets, Ly-1$^+$ and Ly-123$^+$, and the much smaller Ly-23$^+$ subset were similar in the C57BL/6 and the young MRL/l females. By 20 weeks of age there was an extreme reduction in percentage of Ly-123$^+$ cells and Ly-23$^+$ cells in the MRL/l mice and a concomitant increase in Thy-1$^+$, Ly-1$^-$, Ly-23$^-$ (null) cells. The accumulation of Ly null cells in MRL/l nodes followed a straight line with age. The authors stress that the inability to lyse cells in a direct cytotoxicity assay may reflect low antigen expression rather than absence of Ly antigens in the case of the Ly null cells.

7.2.3. I-J-Positive Cells

Theofilopoulos *et al.* (1979a) have reported for MRL/l spleens an increaed frequency and number of cells bearing Ia antigens controlled by the *I-J* subregion of the *H-2* complex, using anti-I-J^k antiserum and fluorescence microscopy on populations of isolated T cells. BXSB male mice showed no change in the number of cells bearing *I-J^b*-controlled determinants. Allen and Niederhuber (1980, personal communication) have found an apparent increase in the number of I-J^+ cells in the spleens of MRL/Mp-*lpr/lpr* by cytotoxicity assay. They were unsuccessful in isolating an *I-J* gene product from these cells. The presence of large amounts of murine leukemia viral antigens made the interpretation of the results difficult.

7.3. Immunocompetency of B Cells

7.3.1. Polyclonal B-Cell Activation

Izui *et al.* (1978) have found spontaneous polyclonal B-cell activation as early as 2 weeks of age in NZB and (NZB × NZW)F_1 mice and by 1 month of age in MRL/l and BXSB (male) mice. These mice spontaneously developed far more (> 3 SD) antihapten antibody-forming cells in spleens and greater concentrations of antihapten antibodies in sera, than the control strains AKR, BALB/c, C3H, C57BL/6, DBA/1J, DBA/2J, LG/J, NZW, and young female BXSB. This increased nonspecific antibody production correlates well with the spontaneous development of anti-ssDNA antibodies. It is suggested that the generalized B-cell activation that has been observed in patients with SLE and in mice with a similar disorder may be a critical factor in the spontaneous development of autoantibodies.

7.3.2. Antibody Responses *in Vivo*

7.3.2a. Intact Young and Old Mice. Creighton *et al.* (1979b) analyzed the immunocompetency of young (2 months) and "old" (4–6 months) (NZB × NZW)F_1, MRL/l, and BXSB mice by qualitative and quantitative comparisons of antibody responses to defined hapten–carrier conjugates. Reference strains included BALB/c, C3H, C57BL/6, NZW, and MRL/n. While 2-month-old MRL/l mice responded within normal limits, as did 2- and 6-month-old (NZB × NZW)F_1 and 2- and 6-month-old BXSB mice, 4-month-old MRL/l mice fail to produce detectable antibodies of either IgG or IgE classes after primary or secondary immunization with 2,4-dinitrophenyl conjugates with hen ovalbumin (DNP-OVA), or keyhole limpet hemocyanin (DNP-KLH). MRL/n mice at 2 months of age had a high primary IgG anti-DNP antibody response. The authors suggest that because isolated B and T lymphocytes from 4-month-old MRL/l mice showed functions within the normal range when analyzed in the fresh environment of young recipients, the poor capacity for antibody production in the clinically diseased MRL/l mice might be a reflection of disturbed internal environment.

7.3.2b. Adoptively Primed B Cells. Creighton *et al.* (1979b) assessed immunocompetency of splenic B lymphocytes adoptively primed to DNP-KLH after transfer into young irradiated, KLH-primed syngeneic recipients. Isolated B lymphocytes from 2- and 4-month-old MRL/l donor mice developed IgG antibody responses that were significantly lower than those of C3H donor B cells only at the lowest dose.

7.4. Immunocompetency of T Cells

7.4.1. Helper-Cell Function *In Vivo*

Creighton *et al.* (1979b) determined the cooperative ability of KLH-primed splenic T cells in facilitating the development of secondary adoptive responses to DNP-KLH, when transferred together with young syngeneic DNP-OVA-primed B cells into irradiated syngeneic recipients. Adequate helper activity was shown for both IgG and IgE antibody responses by T cells derived from 2- and 4-month-old MRL/l mice. There were no pertinent differences in comparison with C3H controls.

7.4.2. Suppressor Function *in Vivo*

Creighton *et al.* (1979b) assessed the capacity of young and old (NZB × NZW)F_1, MRL/l, BXSB, and control mice to develop antigen-specific suppressive mechanims both in the intact animal and after adoptive transfer of their T cells into syngeneic recipients. All of the mice were able to develop OVA-specific suppressive antibodies effective in diminishing IgG and IgE antibody production. Old MRL/l females differed from the other autoimmune models only in having slightly delayed kinetics in their development of suppressive antibody. All of the autoimmune mice generated effective suppressor T lymphocytes capable of adoptively suppressing responses of both the IgG and the IgE classes.

7.4.3. Suppressor–Helper T-Cell Interaction *in Vitro*

Gershon *et al.* (1978) have demonstrated that MRL/l helper T cells are insensitive to suppression in an *in vitro* system, in contrast to MRL/n helper T cells. Thus, when primed MRL/n T cells (including helper T cells) are cultured with unprimed MRL/l spleen cells (including suppressor T cells), the *in vitro* antibody response to sheep red blood cells is suppressed. In contrast, no suppression is observed when primed splenic MRL/l T cells are cultured with unprimed MRL/l spleen cells. The authors suggest that this breakdown in immunoregulation may lead to unchecked helper T-cell activity inducing autoantibody production by B lymphocytes (Cantor and Gershon, 1979).

7.4.4. T-Dependent Antibody Responses *in Vitro*

The responses of the MRL substrains to sheep erythrocytes (SRBC) was measured *in vitro* using the complement-dependent direct hemolytic plaque assay. Twelve MRL/l and MRL/n pairs were tested at 15 weeks of age. MRL/l mice generated an average of only 53 plaques/10^6 spleen cells compared with 338 for controls ($p < 0.005$, unpublished data). The IgM plaque-forming response of 8-week-old MRL/l mice to horse erythrocytes was only 17% of control levels and by 18 weeks only 5% (Wilson and Braley-Mullen, 1980, personal communication).

7.4.5. Response to T-Cell Mitogens

The maximum response of spleen cells from MRL/l mice to phytohemagglutinin (PHA) was 10% of that of control MRL/n mice as early at 10 weeks of age. The maximum response of lymph node cells was 8% of controls (unpublished results). Reduced responsiveness to PHA and concanavalin A has been observed by Ruppert *et al.* (1980) and by Dumont (1980, personal communication).

7.5. Natural Killer Cells

Clark *et al.* (1980) analyzed MRL/Mp-*lpr/lpr* and +/+ control mice for their level of natural killer (NK) cell activity against YAC-1 tumor target cells. Natural cell-mediated cytotoxicity levels were low in young mice of both genotypes. NK activity increases significantly in *lpr/lpr* mice beginning at 8 weeks of age compared to congenic normals. Cudkowicz (1980, personal communication) also has evidence of increased NK activity of splenic effector cells in MRL-*lpr/lpr* compared to +/+ mice as well as in BXSB/Mp males compared to females at 10 weeks of age. Clark *et al.* speculate that the cells responsible for increased NK activity in *lpr/lpr* mutant mice may be identical to the population of Ly-1^-, Ly-23^-, Thy-1^+ cells that increase at similar ages (Theofilopoulis *et al.*, 1979a).

7.6. Adherent Cells

Spleen and lymph node cells from clinically affected MRL/l mice are relatively unresponsive in the mixed lymphocyte culture reaction compared to MRL/n controls. Removal of adherent cells is effective in restoring the proliferative response (Ruppert *et al.*, 1980).

8. Viral Studies

Endogenous retroviruses and their gene products have been implicated in the pathogenesis of murine lupus. A variety of chronic viral infections, such as lymphocytic choriomeningitis, polyoma, and murine leukemia virus, can accelerate the development of a lupuslike disease in animals with appropriate genetic backgrounds (Dixon *et al.*, 1971; Dixon, 1979). The deposition of the retroviral glycoprotein coat molecule, gp70, in the diseased glomeruli along with host immunoglobulins, complement, and nuclear antigens has been a critical observation (Lambert and Dixon, 1968; Yoshiki *et al.*, 1974). Yet the phenotypic expression of endogenous xenotropic viruses was unrelated to glomerulonephritis in crosses between strains NZB and SWR (Datta *et al.,* 1978). The role of retroviral gp70 in the pathogenesis of glomerulonephritis in murine lupus apparently involves the gp70–anti-gp70 complexes that appear in the circulation and in the glomeruli of NZB, (NZB × NZW)F_1, MRL/l and male BXSB mice (Izui *et al.,* 1979).

8.1. Endogenous Retroviruses

8.1.1. Ecotropes and Xenotropes

Dixon *et al.* (1978) have summarized their qualitative and quantitative analysis of retroviruses in (NZB × NZW)F_1, MRL/l, and BXSB mice. Studies employing nucleic acid hybridization using viral probes for isolation of infectious virus in tissue culture have shown significant differences in the retroviral genetic material carried by the three lupus models. MRL/l contains an AKR-type ecotropic viral genome at about half the concentration found in AKR. A xenotropic viral genome can be demonstrated with difficulty in tissue culture studies. BXSB, on the basis of hybridization studies, contains no AKR ecotropic viral genomes and has a xenotropic genome that is expressed rarely and with difficulty in tissue culture experiments. (NZB × NZW)F_1 mice contain an AKR-type and a xenotropic genome that is relatively easily expressed as infectious virus in tissue culture studies. Thus, the retroviral information incorporated in the genomes of the three models is quite different.

8.1.2. C-Type Virions in Epithelia

Croker *et al.* (1977) compared spontaneous retrovirus expression by enumeration with electron microscopy of C-type virus particles in tissues of autoimmune models and of a number of control strains. All strains tested, with the exception of AKR, contained low spontaneous levels of C-type virions in their lymphoid tissues. A survey of a wide variety of epithelial tissues in NZB mice revealed large numbers of C-type particles, especially in the epithelial cells of the gallbladder and the vas deferens. No particles were detected in these epithelia in BALB/c or C57BL/10 mice. Large numbers of particles were found in all the autoimmune models, NZB, (NZB × NZW)F_1, MRL/l, BXSB, and MRL/n, but also in immunologically normal NZW, LG/J, and DBA/2J controls. The authors conclude that the viral expression possibly provides antigens that contribute to autoimmune disease, but are clearly not sufficient to cause the disease.

8.1.3. Viral Antigens

Dixon (1979) has summarized the conclusions of a series of studies on the possible role in murine lupus of the retroviral glycoprotein coat molecule, gp70, that is found free in the circulation of all mice. Mice that develop lupus apparently have the unique ability to respond to this endogenous molecule with a formation of immune complexes, which, together with the complexes involving DNA, appear to constitute the major pathogenic immune responses in murine SLE (see Section 6.4).

9. Cell Transfer Studies

Transplantation of hematopoietic stem cells from bone marrow and spleen provides a biological approach to the question of the genetic and cellular basis for autoimmunity. Congenic inbred strains permit cell transfer studies without interference by reactions to background antigenic differences between graft and host.

9.1. Congenic MRL/Mp-*lpr/lpr* and +/+

Young adult MRL/Mp-*lpr/lpr* and +/+ mice were lethally irradiated and reconstituted with 5–10 million nucleated bone marrow cells. The four chimeric groups, + → +, + → *lpr*, *lpr* → +, and *lpr* → *lpr*, were followed to determine longevity and expression of major phenotypic characters of *lpr* at 14 weeks after marrow transfer.

9.1.1. *lpr/lpr* Recipients

The + → *lpr* chimera showed significantly prolonged survival (> 40 weeks) in comparison with *lpr* → *lpr* (17 weeks) (Table II). In the + → *lpr* chimera there was suppression of the lymphadenopathy and splenomegaly and serum IgG levels were significantly lower than in *lpr* → *lpr* (Tables II and III).

9.1.2. +/+ Recipients

The longevity of irradiated +/+ mice was significantly reduced when repopulated with *lpr/lpr* stem cells (12 weeks) in comparison with syngeneic +/+ cells (> 39 weeks)

TABLE II
Effect of Stem-Cell Transfer on Longevity of MRL Mice[a]

Chimeric group	*N*	Incidence of lymphadenopathy	Survival time[b] (days) (posttreatment)	p^d
+/+ → +/+	12	0/12 (0%)	>275 ± 20[c]	<0.001
lpr → +/+	13	0/13 (0%)	87 ± 11	
+/+ → *lpr*	12	1/12 (8%)[e]	>281 ± 21[c]	<0.001
lpr → *lpr*	9	6/9 (67%)	122 ± 12	
mixed[f] → *lpr*	7	5/7 (71%)	203 ± 27	

[a]Four- to five-week-old MRL/Mp-*lpr/lpr* and congenic +/+ mice received 900–1000 R of ionizing radiation and were reconstituted by intravenous injection of 5–10 million nucleated marrow cells from 6- to 8-week-old donor mice.
[b]Mean ± S.E.M.
[c]Six of twelve mice of chimeric group +/+ → +/+ and one of twelve of group +/+ → *lpr* still surviving at time of analysis. Current longevity used for calculation of survival times.
[d]*p* determined by Student's *t* test for independent means.
[e]Incidence of lymphadenopathy significantly lower, $p < 0.01$ (chi-square analysis), than in *lpr* → *lpr* group.
[f]1:1 ratio of +/+ and *lpr/lpr* marrow cells donated.

TABLE III
Effect of Stem-Cell Transfer on the lpr Phenotype[a]

Chimeric group	N	Lymph node and spleen weights[b]			Serum values[b]	
		Axillary	Mesenteric	Spleen	IgG1	ANA
1. +/+ → +/+	5	4.6 ± 0.3	29 ± 3	80 ± 6	317 ± 60[c]	1.8 ± 0.7[d]
2. *lpr* → +/+	5	3.3 ± 0.7	57 ± 14	44 ± 4	400 ± 51	4.0 ± 0.0
3. +/+ → *lpr*	6	7.2 ± 0.4	54 ± 5	119 ± 7	245 ± 63	2.2 ± 0.7
4. lpr → *lpr*	7	48 ± 14	214 ± 61	264 ± 21	734 ± 56	3.7 ± 0.3
t tests[e]	1 vs. 2	NS	NS	< 0.005	NS	< 0.025
	3 vs. 4	< 0.025	< 0.05	< 0.001	< 0.001	NS

[a]Four-week-old MRL/Mp-*lpr/lpr* and congenic +/+ mice received 1000-R whole-body ^{137}Cs gamma irradiation followed by reconstitution with 5 million nucleated bone marrow cells from 6-week-old donors.
[b]Absolute organ weights and serological values expressed as the mean ± S.E.M.
[c]IgG1 levels expressed as mg/100 ml serum.
[d]ANA scored as intensity (scale 0–4) using 1/10 diluted serum in an indirect fluorescence antibody technique using rat liver substrate.
[e]Probability values determined by Student's *t* test for independent means. NS, nonsignificance at the $p < 0.05$ level.

(Table II). The lymphoproliferative phenotype was not observed in the *lpr* → + chimera. The clinical course and pathology resembled that of advanced graft-versus-host reaction with cell depletion in lymph nodes and spleen. F. J. Dixon has made similar observations (personal communication). This system may well provide clues to the basic mechanism of action of the *lpr* mutant gene.

9.2. Strain BXSB

Similarly designed stem cell transfer experiments were carried out between sexes in strain BXSB and between (NZB × BXSB)F_1 reciprocal hybrid males. The early expression of lymphoproliferation and reduced longevity of BXSB males was reproduced in BXSB females that had been lethally irradiated and repopulated with male stem cells. Conversely, the longevity of BXSB males was greatly increased by female stem cells and lymphadenopathy and splenomegaly were prevented. Similar transfer of phenotype expression in (NZB × BXSB)F_1 hybrid males indicated that the Y chromosome of BXSB was the determining factor for accelerated autoimmune disease. Eisenberg *et al.* (1979c) have reported similar results of bone marrow transfer between male and female BXSB mice.

10. Therapy

10.1. X Irradiation

Single doses of whole-body X irradiation of MRL/l mice increased the latent period for palpable lymphadenopathy and survival (unpublished data). Fourteen unirradiated *lpr/lpr* mice developed enlarged lymph nodes by 14 weeks of age and survived to a mean age of 24 weeks. Mice exposed to 200, 400, and 600 R at 6 weeks of age survived to 31, 40, and 40 weeks, respectively. A dose of 400 R was more effective in increasing survival when *lpr/lpr* mice were irradiated at 10 weeks of age, compared to 6 or 3 weeks. Only two-thirds of this group developed palpable lymph nodes with a doubling of latent period. Survival nearly doubled to 44 weeks of age. A. N. Theofilopoulos *et al.* (personal communication) have found that 82% of MRL/l mice that received 300 rads of whole-body irradiation at 3 months of age remained alive at 9 months of age. All MRL/l mice that received total lymphoid irradiation (3400 rads in 17 divided doses over a 3- to 4-week period), with shielding of skull, long bones, and tail, survived to 9 months of age. None of the irradiated animals had enlarged lymph nodes or splenomegaly and only 11–16% of them had developed glomerulonephritis.

10.2. Cell Replacement Therapy

MRL/Mp-*lpr/lpr* mice that have been irradiated and reconstituted with congenic +/+ bone marrow survive to > 40 weeks of age compared to 17 weeks for recipients of *lpr/lpr* bone marrow (see Section 9.1.1).

10.3. Thymectomy

Thymectomy of MRL/l mice at 8 to 9 and 20 to 21 days of age decreased the incidence of palpably enlarged lymph nodes from 100% to 17 and 50%, respectively. None of the sham-operated control mice but 48 and 64% of the 8- to 9- and 20- to 21-day thymectomized mice survived to 7 months of age (E. D. Murphy, unpublished).

Four months following neonatal thymectomy, MRL/Mp-*lpr/lpr* mice showed no clear-cut lymphadenopathy, had significantly reduced proteinuria, and had one-third the level of antibodies to native DNA. Ninety percent of thymectomized females and all males survived to 7 months of age compared to 0 and 20% for sham-operated controls (Steinberg *et al.*, 1980).

10.4. Thymic Hormones

MRL/l and MRL/n mice immunized with SRBC are low responders in the rosette-forming cell assay. Both extracted thymosin and synthetic thymopoietin significantly increased the ability of both MRL/l and MRL/n spleen and thymus cells to form rosettes (Abe *et al.*, 1980).

10.5. Prostaglandins

The administration of pharmacologic doses of prostaglandin E_1 to MRL/Mp-*lpr/lpr* mice was effective in prolonging survival, retarding development of immune-complex glomerulonephritis, and preventing peripheral T-cell hyperplasia (Kelly and Winkelstein, 1980). Twice-daily 200-μg doses of PGE_1 resulted in reduced amount of circulating gp70 immune complexes in MRL/l mice (S. Izui *et al.*, personal communication).

10.6. Androgen Treatment

Steinberg *et al.* (1980) has determined that androgen treatment (testosterone or 5 α-dihydrotestosterone) beginning at 2 weeks of age decreases by approximately one-half the serum levels of anti-DNA autoantibody and proteinuria measured at 4 months of age. Survival to 7 months of age was 60% for females and 70% for males. Lymphadenopathy was not significantly reduced.

11. Other Single-Locus Models

11.1. Strain BXSB and Its Y Chromosome

Recombinant inbred strain BXSB/Mp develops a spontaneous lupuslike syndrome which in contrast to that of other models is strikingly accelerated in males (Murphy *et al.*, 1977; Murphy and Roths, 1978c, 1979a). The autoimmune disease syndrome is characterized by moderate lymph node enlargement, splenomegaly, hemolytic anemia, hypergammaglobulinemia, immune-complex glomerulonephritis, and antinuclear, antierythrocytic, and thymocytotoxic autoantibodies. The accelerated disease in males occurs also in F_1 hybrids, but only when the male parent is BXSB (Murphy and Roths, 1978c, 1979a). Thus, there is evidence for a Y-chromosome-associated accelerating factor, possibly a gene on the Y chromosome of BXSB. The strain was derived from a cross between a C57BL/6J female and an SB/Le male. The Y chrmosome of BXSB is derived from the SB/Le male (Murphy and Roths, 1978c).

In contrast with the *lpr* models, the BXSB/Mp males develop a Coomb's-positive hemolytic anemia, intensified in (NZB × BXSB)F_1 males in which erythrocyte counts are 4×10^6 per mm^3 at 20 weeks of age, and circulating erythrocyte autoantibody is present in 16 of 17 mice between 16 and 20 weeks of age (Murphy and Roths, 1979a). High-titered ANA (1:200) and, in contrast with *lpr*, thymocytotoxic autoantibody are found in 15 of 16 BXSB males tested at 16–20 weeks of age. The hemolytic anemia is characterized by protoporphyrin levels of 543 mg/100 ml blood compared with 39 mg/100 ml in reciprocal male controls, at 5 months of age.

The gamma globulin fraction comprised 17% of total serum protein in clinically affected BXSB males, and 12% in clinically affected females (second year of life), compared with 2% in 5-month-old C57BL/6J males as controls. In the males there was approximately a threefold increase in IgM and IgG2b and a ninefold increase in IgA and IgG1, compared with 4-month-old C57BL/6 as controls. IgG2a was undetectable in a population of 11 BXSB males. The values for three clinically affected females showed similar proportionate increases for IgM, IgG2b, IgA, and IgG1, with half the control values for IgG2a (Murphy and Roths, 1978c).

Spleen cells of (NZB × BXSB)F_1 hybrid males were significantly less responsive to the T-cell mitogen PHA and more responsive to the B-cell mitogen LPS at preclinical, early clinical, and advanced clinical stages in comparison with age-matched reciprocal controls (Murphy and Roths, 1979a). The response of lymph node cells to PHA and LPS was significantly different from control values only during the advanced clinical stage. The Con A response of spleen cells but not lymph node cells of (NZB × BXSB)F_1 hybrid males was significantly reduced as early as 10 weeks of age. Spleen cells of (NZB × BXSB)F_1 hybrid males underwent spontaneous *in vitro* proliferation reaching seven times the control levels by 17 weeks of age.

In strain BXSB males and later in females there is moderate lymph node enlargement by a proliferation of lymphocytes with admixed plasma cells and histiocytes (Murphy and Roths, 1978c). The architecture is blurred in an advanced stage of more diffuse infiltration. The lymph nodes of (NZB × BXSB)F_1 males can be as large as 40 times the size of those of reciprocal hybrid control males. The massive lymphoproliferation involves the same mixed lymphocyte, plasma cell, and histiocyte population, but is even more diffuse with greater blurring of architecture (Murphy and Roths, 1979a). Immunoblasts are numerous. Microvascular proliferation may occur. In the BXSB mice there are occasionally limited eosinophilic (PAS-positive) interstitial deposits. The predominantly B-cell lymphoproliferation of BXSB is proposed as a good model for human immunoblastic lymphadenopathy.

Strain BXSB was included among the models of murine lupus compared in studies in the Department of Immunopathology at Scripps. Reported similarities to and differences from the MRL models have been cited throughout this chapter. For further information see Theofilopoulous *et al.* (1980a,b) and Creighton *et al.* (1979a).

11.2. Mutant Gene *gld*

The new mutation *gld* (generalized lymphoproliferative disease) that occurred in strain C3H/HeJ produces an early-onset autoimmune disease with massive lymphoproliferation (Murphy *et al.*, 1979). It is not allelic with *lpr* and appears to be linked to *Dip-1* on chromosome 1.

There are similarities to and major differences from the *lpr* models. The lymphoproliferation is massive with lymph node weights 8-fold heavier by 12 weeks of age and 50-fold by 20 weeks. There is a predominant mixture of T and null cells. The clinical course is greatly extended with median survival at 13 months rather than 4 to 5 months for *lpr* on several backgrounds. There is a decline in lymph node weights to 20 times normal by 30 weeks of age. The renal changes are less severe and do not appear to progress as rapidly as in *lpr*. The histology of the lymph nodes resembles that of *lpr*. Peripheral blood leukocyte counts are increased 5-fold in 24-week-old *gld/gld* mice. There is hypergammaglobulinemia with greater than 10-fold increases in serum IgA and IgG2b, and 3- to 6-fold elevations in IgG1 IgG2a, and IgM. Antinuclear antibody has been detected in all of six sera (1/10 dilution) from 24-week-old *gld/gld* mice by immunofluorescence, using rat liver substrate. In a pool of 10 sera from 24-week-old *gld/gld* mice there was high-intensity fluorescence for ANA at a dilution of 1/200. A control pool from eight coisogenic C3H-+/+ mice was negative at a 1/1 dilution. The gene *Lps* on Chromosome 4, which produces the defective response of strain C3H/HeJ to bacterial lipopolysaccaride (LPS), may be involved.

ACKNOWLEDGMENTS

This work was supported in part by NIH Research Grants CA-22948 from the National Cancer Institute and AG-00250 from the National Institute on Aging, and Contract CB-74174 from the National Cancer Institute.

References

Abe, C., Mitsunaga, K., and Shiokawa, Y., 1980, Spontaneous polyarthritis in MRL/l mice, *Ryumachi,* **20:**233.

Accinni, L., and Dixon, F. J., 1979, Degenerative vascular disease and myocardial infarction in mice with lupus-like syndrome, *Am. J. Pathol.* **96:**477.

Andrews, B. S., Eisenberg, R. A., Theofilopoulos, A. N., Izui, S., Wilson, C. B., McConahey, P. J., Murphy, E. D., Roths, J. B., and Dixon, F. J., 1978, Spontaneous murine lupus-like syndromes, clinical and immunopathological manifestations in several strains, *J. Exp. Med.* **148:**1198.

Bielschowsky, M., Helyer, B. J., and Howie, J. B., 1959, Spontaneous haemolytic anemia in mice of the NZB/Bl strain, *Proc. Univ. Otago Med. Sch.* **37:**9.

Borel, Y., Lewis, R. M., and Stollar, B. D., 1973, Prevention of murine lupus nephritis by carrier dependent induction of immunologic tolerance to denatured DNA, *Science* **182:**76.

Burnet, F. M., 1972, *Auto-Immunity and Auto-Immune Disease,* Davis, Philadelphia.

Cantor, H., and Gershon, R. K., 1979, Immunological circuits: Cellular composition, *Fed. Proc.* **38:**2058.

Clark, E. A., Windsor, N. T., Sturge, J. C., and Stanton, T. H., 1980, Mutations that influence natural cell-mediated cytotoxicity in rodents, in: *Natural Cell-Mediated Immunity against Tumors* (R. B. Herberman, ed.), pp. 417–429, Academic Press, New York.

Creighton, W. D., Zinkernagel, R. M., and Dixon, F. J., 1979a, T cell-mediated immune responses of lupus-prone BXSB mice and other murine strains, *Clin. Exp. Immunol.* **37:**181.

Creighton, W. D., Katz, D. H., and Dixon, F. J., 1979b, Antigen-specific immunocompetency, B cell function, and regulatory helper and suppressor T cell activities in spontaneously autoimmune mice, *J. Immunol.* **123:**2627.

Croker, B. P., McConahey, P. J., Murphy, E. D., and Dixon, F. J., 1977, Immunopathogenicity and oncogenicity of murine leukemia virus. III. Quantitation of spontaneous virus expression, *J. Natl. Cancer Inst.* **50:**199.

Datta, S. K., Manny, N., Andrzejewski, C., André-Schwartz, J., and Schwartz, R. S., 1978, Genetic studies of autoimmunity and retrovirus expression in crosses of New Zealand Black mice. I. Xenotropic virus, *J. Exp. Med.* **147:**854.

Dixon, F. J., 1979, The pathogenesis of murine systemic lupus erythematosus, Rous-Whipple award lecture, *Am. J. Pathol.* **97:**10.

Dixon, F. J., Oldstone, M. B. A., and Tonietti, G., 1971, Pathogenesis of immune complex glomerulonephritis of NZ mice, *J. Exp. Med.* **134:**655.

Dixon, F. J., Andrews, B. S., Eisenberg, R. A., McConahey, P. J., Theofilopoulos, A. N., and Wilson, C. B., 1978, Etiology and pathogenesis of a spontaneous lupus-like syndrome in mice, *Arthritis Rheum.* **21**(Suppl. 5):564.

Ebling, F., and Hahn, B. H., 1980, Restricted subpopulation of DNA antibodies in kidneys of mice with systemic lupus: Comparison of antibodies in serum and renal eluates. *Arthritis Rheum.* **23:**392.

Eisenberg, R. A., Tan, E. M., and Dixon, F. J., 1978, Presence of anti-Sm reactivity in autoimmune mouse strains, *J. Exp. Med.* **147:**582.

Eisenberg, R. A., Theofilopoulos, A. N., Andrews, B. S., Peters, C. J., Thor, L., and Dixon, F. J., 1979a, Natural thymocytotoxic autoantibodies in autoimmune and normal mice, *J. Immunol.* **122:**2272.

Eisenberg, R. A., Thor, L. T., and Dixon, F. J., 1979b, Serum–serum interactions in autoimmune mice, *Arthritis Rheum.* **22:**1074.

Eisenberg, R. A., Izui, S., McConahey, P. J., Peters, C. J., Theofilopoulos, A. N., and Dixon, F. J., 1979c, Cell transfer studies in autoimmune BXSB mice, *Fed. Proc.* **38:**1210 (abstract).

Gershon, R. K., Horowitz, M., Kemp, J. D., Murphy, D. B., and Murphy, E. D., 1978, The cellular site of immunoregulatory breakdown in the *lpr* mutant mouse, in: *Genetic Control of Autoimmune Disease* (N. R. Rose, P. E. Bigazzi, and N. L. Warner, eds.), pp. 223–227, Elsevier/North Holland, Amsterdam.

Green, E. L., (ed.), 1966, Breeding systems, in: *Biology of the Laboratory Mouse,* 2nd ed., pp. 11–22, McGraw-Hill, New York.

Helyer, B. J., and Howie, J. B., 1963, Renal disease associated with positive lupus erythematosus tests in a cross-bred strain of mice, *Nature (London)* **197:**197.

Huston, D. P., and Steinberg, A. D., 1979, Animal models of human systemic lupus erythematosus, *Yale J. Biol. Med.* **52:**289.

Izui, S., and Eisenberg, R. A., 1980, Circulating anti-DNA–rheumatoid factor complexes in MRL/l mice, *Clin. Immunol. Immunopathol.* **15:**536.

Izui, S., McConahey, P. J., and Dixon, F. J., 1978, Increased spontaneous polyclonal activation of B lymphocytes in mice with spontaneous autoimmune disease, *J. Immunol.* **121:**2213.

Izui, S., McConahey, P. J., Theofilopoulos, A. N., and Dixon, F. J., 1979, Association of circulating retroviral gp70–anti-gp70 immune complexes with murine systemic lupus erythematosus, *J. Exp. Med.* **149:**1099.

Kelly, V., and Winkelstein, A., 1980, PGE_1 prevents lymphoproliferation and renal disease in MRL/l mice, *Fed. Proc.* **39:**471 (atstract).

Lambert, P. H., and Dixon, F. J., 1968, Pathogenesis of glomerulonephritis of NZB/W mice, *J. Exp. Med.* **127:**507.

Morgan, G. M., and McKenzie, I. F. C., 1980, Cell membrane alloantigenic determinants of several unusual mouse strains, *J. Immunogenet.* **7:**393.

Murphy, E. D., 1966, Characteristic tumors, in: *Biology of the Laboratory Mouse* (E. L. Green, ed.) 2nd ed., pp. 521–562, McGraw-Hill, New York.

Murphy, E. D., 1979, Reticulum cell sarcomas, in: *Inbred and Genetically Defined Strains of Laboratory Animals, Part 1, Mouse and Rat* (P. L. Altman and D. D. Katz, eds.), pp. 169–173, Feder. Am. Soc. Exptl. Biol. Bethesda, Md.

Murphy, E. D., and Roths, J. B., 1977, A single gene model for massive lymphoproliferation with autoimmunity in new mouse strain MRL, *Fed. Proc.* **36:**1246 (abstract).

Murphy, E. D., and Roths, J. B., 1978a, A single gene model for massive lymphoproliferation with immune complex disease in the new mouse strain MRL, in: *Proceedings of the 16th International Congress on Hematology, 1976,* pp. 69–72, Excerpta Medica, Amsterdam.

Murphy, E. D., and Roths, J. B., 1978b, New inbred strains, *Mouse News Letter* **58:**51.

Murphy, E. D., and Roths, J. B., 1978c, Autoimmunity and lymphoproliferation: Induction by mutant gene *lpr,* and acceleration by a male-associated factor in strain BXSB mice, in: *Genetic Control of Autoimmune Disease* (N. R. Rose, P. E., Bigazzi, and N. L. Warner, eds.), pp. 207–221, Elsevier/North-Holland, Amsterdam.

Murphy, E. D., and Roths, J. B., 1979a, A Y-chromosome associated factor producing accelerated autoimmunity and lymphoproliferation in strain BXSB, *Arthritis Rheum.* **22:**1188.

Murphy, E. D., and Roths, J. B., 1979b, Mutant, inbred, and hybrid models for autoimmunity and lymphoproliferation, *Mouse News Letter* **60:**53.

Murphy, E. D, Roths, J. B., and Lane, P. W., 1977, Massive lymphoproliferation and autoimmunity controlled by single genes, *Proc. Am. Assoc. Cancer Res.* **18:**157 (abstract).

Murphy, E. D., Roths, J. B., and Eicher, E. M., 1979, Another mutation producing generalized lymphoproliferation, in: *50th Annual Report,* pp. 23–24, The Jackson Laboratory, Bar Harbor, Maine.

Paker, L. P., Hahn, B. H., and Osterland, C. K., 1974, Modification of NZB/NZW F_1 autoimmune disease by development of tolerance to DNA, *J. Immunol.* **113:**292.

Ruppert, B., Maisel, A., and Ford, R. J., 1980, In vitro lymphoproliferation of cells derived from MRL mice, *Fed. Proc.* **39:**778 (abstract).

Sawada, S., and Talal, N., 1979, Characteristics of in vitro production of antibodies to DNA in normal and autoimmune mice, *J. Immunol.* **122**:2309.

Steinberg, A. D., Roths, J. B., Murphy E. D., Steinberg, R. T., and Raveche, E. S., 1980, Effects of thymectomy or androgen administration upon the autoimmune disease of MRL/Mp-*lpr/lpr* mice, *J. Immunol.* **125**:871.

Theofilopoulos, A. N., Eisenberg, R. A., Bourdon, M., Crowell, J. S., Jr., and Dixon, F. J., 1979a, Distribution of lymphocytes identified by surface markers in murine strains with systemic lupus erythematosus-like syndromes, *J. Exp. Med.* **149**:516.

Theofilopoulos, A. N., Shawler, D. L., Katz, D. H., and Dixon F. J., 1979b, Patterns of immune reactivity in autoimmune murine strains. I. Cell-mediated immune response induced by *H-2* identical and *H-2* incompatible stimulator cells, *J. Immunol.* **122**:2319.

Theofilopoulos, A. N., Shawler, D. L., Eisenberg, R. A., and Dixon, F. J., 1980a, Splenic immunoglobulin-secreting cells and their regulation in autoimmune mice, *J. Exp. Med.* **151**:446.

Theofilopoulos, A. N., McConahey, P. J., Izui, S., Eisenberg, R. A., Pereira, A. B., and Creighton, W. D., 1980b, A comparative immunologic analysis of several murine strains with autoimmune manifestations, *Clin. Immunol. Immunopathol.* **15**:258.

Warner, N. L., 1977, Genetic aspects of autoimmune disease in animals, in: *Autoimmunity: Genetic, Immunologic, Virologic, and Clinical Aspects* (N. Talal, ed.), pp. 33–62, Academic Press, New York.

Yoshiki, T., Mellors, R. C., Strand, M., and August, J. T., 1974, The viral envelope glycoprotein of murine leukemia virus and the pathogenesis of immune complex glomerulonephritis of New Zealand mice, *J. Exp. Med.* **140**:1011.

7

Canine Systemic Lupus Erythematosus

FRED WILLIAM QUIMBY

1. Introduction

Dogs afflicted with systemic lupus erythematosus (SLE) were first described by Lewis *et al.* 1963. In his original description the principal pathologic alterations were hemolytic anemia and thrombocytopenia; these remain important attributes of the disease in both dogs and humans. Since that time selective breeding experiments and pedigree analysis of canine families prone to develop SLE have added much to our understanding of this disease.

This chapter will review the salient features of SLE as it occurs in dogs, and discuss how each of these features supports a genetic hypothesis for SLE.

2. Incidence of SLE in Dogs

In two independent studies involving a large number of clinical admissions of dogs in the metropolitan Boston area, the incidence of canine SLE was established as 1 per 5000 admissions (Lewis, 1972; Quimby *et al.,* 1978a). The only breed that is preferentially affected is the Shetland sheepdog, which appears to have an incidence in certain families approaching 10% (F. M. Quimby, submitted for publication). In one study, the disease appeared to occur in an equal number of male and female dogs; however, this population contained only half the number of intact female admissions as intact males due to the common occurrence of ovariohysterectomy in young female dogs (approximately 50% in this study group). Retrospective analysis of the records of the 19 cases of canine SLE selected from this study showed that only one case involved a castrated female.This finding suggests that the absence of an intact reproductive tract may impart a protective effect in female dogs (Quimby *et al.,* 1980). A similar finding has been documented in the New Zealand

FRED WILLIAM QUIMBY • Division of Laboratory Animal Medicine and Department of Pathology, Cornell University Medical College, New York, New York 10021.

mouse (Talal, 1976). In this model for SLE the characteristic pattern of disease seen in intact female mice is modified by early ovariectomy to a pattern of disease characteristic of male mice.

Lewis *et al.* (1965) suggested that most exacerbations of hemolytic anemia occur during the winter months (in Massachusetts); this same observation has been made by the author regarding the onset of both hemolytic anemia and glomerulonephritis in canine SLE. This may be related to the finding that many dogs with lupus develop cryoglobulins with DNA-binding activity (Schwartz *et al.,* 1978). In Shetland sheepdogs the incidence of dermal manifestations is increased in the summer months. This is in keeping with the longer duration of exposure, and intensity of ultraviolet radiation during that season.

3. Clinical Findings in Canine SLE

SLE in dogs is well recognized as a multisystemic disease, which may involve any organ system. Early investigators recognized hemolytic anemia and thrombocytopenia as lesions responsible for clinical signs in the majority of canine patients afflicted with SLE (Lewis *et al.,* 1963). Shortly after this discovery, Lewis *et al.* (1965) published findings indicating that nearly 50% of dogs with SLE have nephritis. Since that time, numerous investigators have documented the clinical and pathologic manifestations of SLE in dogs. The present list of organ involvement includes: polyarthritis, exfoliative dermatitis, butterfly rash, behavioral changes often accompanied by seizures, cardiac conduction abnormalities (in several instances accompanied by myocarditis), and primary peribiliary cirrhosis. In some animals exacerbations of dermatitis are causally related to exposure to strong sunlight. Many clinical signs are referable to lesions present in a particular organ; however, many others appear nonspecific, and could involve a wide variety of clinical and pathologic abnormalities, e.g., fever, lethargy, and loss of appetite. Due to the variety of clinical manifestations, the clinician must include SLE in the differential diagnosis of a large percentage of common canine afflictions.

The course of the disease varies greatly between dogs. The duration of illness has been documented in one case to be 14 years; however, most cases have less than a 3-year survival. In one study 20% of the dogs with hemolytic anemia died within the first week of treatment (Lewis *et al.,* 1963). Many dogs have repeated exacerbations and remissions. Each new exacerbation may be heralded by different clinical manifestations.

The sex, breed, and age at onset are not prognostic factors. Once a diagnosis is established, the disease is usually severe, and requires prompt treatment. In general, the long-term prognosis for dogs with SLE is grave.

4. Serologic Findings in Canine SLE

Autoantibodies against a variety of nuclear constituents are a primary feature of SLE in BW mice (Lewis, 1973), humans (DeHoratius *et al.,* 1975), and dogs. Lewis (1972, 1973) has described the presence of antibodies directed to DNA, RNA, and DNA–histone complexes in dogs with SLE. Cryoglobulins containing antibodies to DNA may occur

(Schwartz *et al.,* 1978). A variety of tests, including the Farr assay, the double-antibody assay, and the glass-fiber filter assay, have been used to detect anti-DNA-binding antibodies in canine serum. The use of an appropriate assay for measuring anti-DNA antibodies has been emphasized by the finding that many normal dogs contain an alpha-2 globulin with DNA-binding capacity (Thoburn *et al.,* 1972). In general, dogs with SLE are found to have consistently higher levels of antibodies to denatured DNA than to native DNA. When dogs have active SLE their sera usually contain high titers (> 1:1000) of antinuclear antibody, which produces a "rim" staining pattern. This is in contrast to the lower titer (< 1:100) and primarily diffuse staining pattern seen for antinuclear antibody in canine rheumatoid arthritis, immunologic thrombocytopenic purpura, and autoimmune hemolytic anemia (Quimby *et al.,* 1980). Although antibodies with specificity to DNA have been eluted from kidney specimens originating from cases of canine lupus nephritis, the precise relationship between levels of anti-DNA autoantibodies and activity of the disease is presently unknown.

Natural lymphocytotoxic antibodies (NLA) have been found in 75% of dogs with SLE. Studies have demonstrated that these autoantibodies always bind homologous lymphocytes best at 39°C, that lymphocyte lysis requires the presence of complement (both *in vivo* and *in vitro*), and that NLA are directed against a subpopulation of canine thymus-derived lymphocytes (F. M. Quimby, submitted for publication). Many dogs with SLE have decreased numbers of circulating lymphocytes; we have shown that a transient depletion of T-rosette-forming lymphocytes occurs in normal dogs infused with homologous NLA. While T-cell depletion may be affected by the presence of NLA, the precise relationship between NLA production and disease activity is unknown. It is interesting that in a closed colony of dogs bred to develop SLE, a high percentage of dogs have NLA early in life. Many of these dogs appear to have an inordinate number of infections caused by opportunistic invaders. A generalized NLA-mediated suppression of immunity may profoundly influence the ability of these animals to resist infection.

Antibodies against red cells have been associated with the hemolytic anemia of canine SLE. The autoantibody nature of these antierythrocyte antibodies has been confirmed by demonstrating the binding specificity of red cell eluates for homologous canine erythrocytes (Lewis *et al.,* 1965).

Both warm (IgG) and cold (IgM) agglutinins have been described in dogs (Slappendel *et al.,* 1975; Quimby and Kaplan, 1979), the class of autoantibody having a profound influence on the clinical course of the disease (Slappendel, 1978). Antibodies against homologous immunoglobulin G (rheumatoid factor) are frequently seen associated with symmetrical polyarthritis in canine SLE. These autoantibodies do not bind heterologous IgG and differ from human rheumatoid factor in their poor reactivity in the Rose–Waaler test (F. M. Quimby, unpublished observations). Antithyroid antibodies are frequently found in dogs with SLE (Lewis *et al.,* 1965), but impaired thyroid function is rare. Antibodies against both thyroglobulin and the microsomal antigens of the thyroid have been described (Quimby *et al.,* 1978a); however, pathologic lesions in the thyroid gland and altered thyroid function appear more commonly in dogs with antimicrosomal antibodies. Table I lists the incidence of autoantibodies found in 16 related Shetland sheepdogs with autoimmune disease and a high incidence of SLE.

Circulating immune complexes are found in most dogs with active SLE (Terman *et al.,* 1979). Many genetically susceptible dogs have low, but significant concentrations of

TABLE I
Incidence of Autoantibodies in 16 Related Shetland Sheepdogs with Autoimmune Disease

Autoantibody	Percentage
Antinuclear antibody	75
Lupus erythematosus factor	31
Rheumatoid factor	18
Antierythrocyte	18
Antilymphocyte	68

circulating immune complexes long before the onset of clinical signs (Poskitt and Poskitt, 1978). In animals with active nephritis or polyarthritis, the serum concentration of immune complexes is much higher. Despite the presumptive role of circulating immune complexes in canine SLE, little is known regarding complex size, the association between circulating complexes and complement, or the relationship between the presence of circulating complexes and the activity of the disease.

5. Pathologic Findings in Canine SLE

Dogs afflicted with SLE have a wide variety of pathologic lesions including: glomerulonephritis, chronic synovitis, hemolytic anemia, thrombocytopenia, and dermatitis. The incidence of organ involvement in 16 related Shetland sheepdogs with a high incidence of SLE may be seen in Table II.

The glomerulonephritis of canine SLE may be either the membranous or the membrano-proliferative type. In kidneys demonstrating a membrano-proliferative lesion, antibody–antigen–complement complexes occur along the endothelial side of the glomerular basement membrane (Slauson and Lewis, 1979). These aggregates have a typical granular distribution when visualized by immunofluorescence (Fig. 1). Eluates from nephritic kid-

TABLE II
Incidence of System Involvement in 16 Related Shetland Sheepdogs with Autoimmune Disease

System	Incidence (%)
Skin	43
Red cells	18
Kidneys	12
Platelets	12
Heart	12
Central nervous	12
Joints	6
Lymph nodes	6

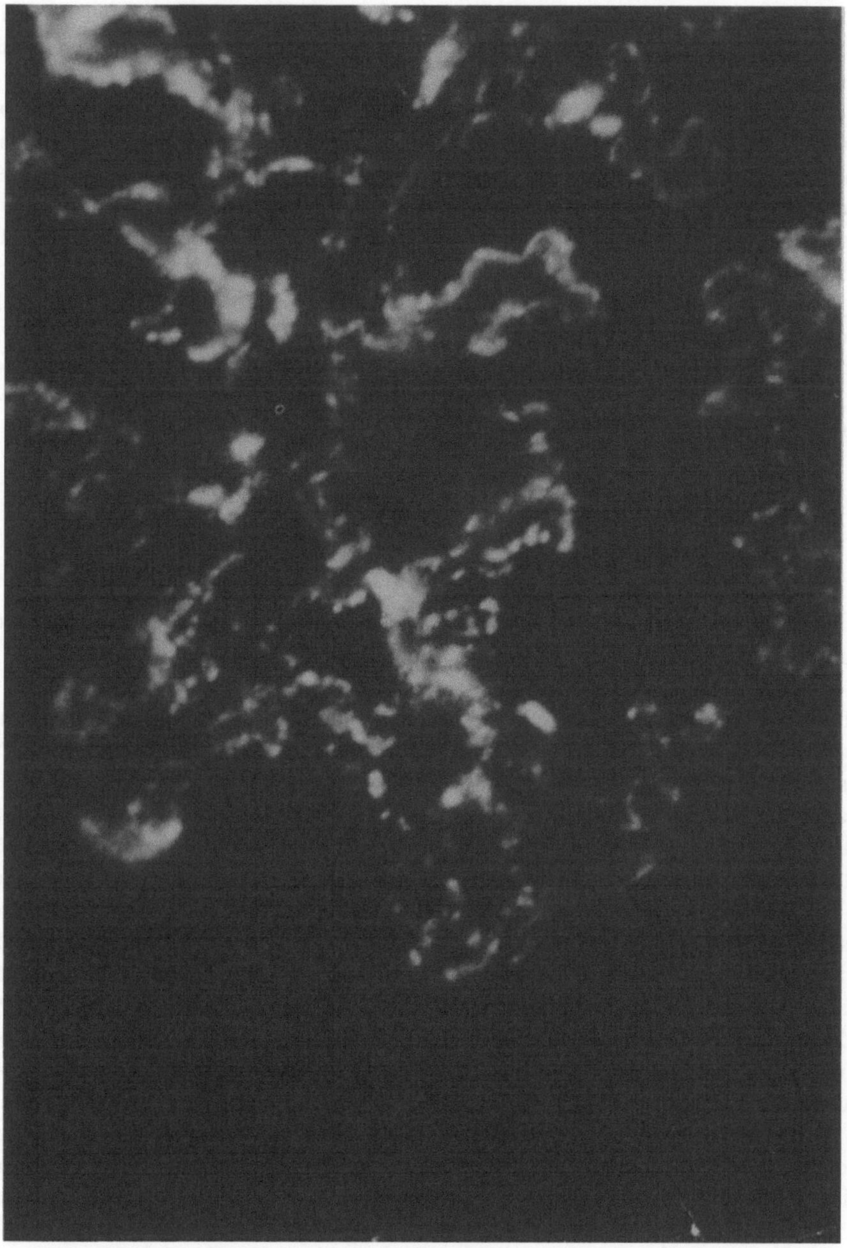

FIGURE 1. A typical glomerulus from a dog with lupus nephritis. The kidney section was treated with fluorescein-conjugated rabbit anti-dog IgG to demonstrate the granular pattern of immune complexes in this organ.

neys contain anti-DNA antibodies. While most dogs with SLE do not have evidence of pannus formation in diseased joints, they do have many of the synovial manifestations of rheumatoid arthritis including hyperplasia of synovial cells, chronic synovitis, and perivascular lymphoid cuffing. The joint fluid often contains an abundance of both phagocytes and lupus erythematosus cells (Quimby *et al.,* 1978a).

The histopathologic lesions seen in skin afflicted with a malar eruption include parakeratosis, telangiectasia of vessels in the papillary layer of the dermis, and the accumulation of mononuclear inflammatory cells in the dermis (Lewis *et al.,* 1965). The discoid skin lesions consist of hyperkeratotic plugs, subepidermal edema, basal cell swelling, and a sparse inflammatory infiltrate. These lesions are associated with the deposition of IgG at the dermal–epidermal junction of affected skin only.

A variety of other lesions have been seen associated with canine SLE. These include the formation of germinal centers in the thymus gland (Lewis and Schwartz, 1971), hyperplastic germinal centers in lymph nodes, fibrinoid degeneration of splenic arteries (Lewis *et al.,* 1965), ulcerations of the gastrointestinal tract (Quimby *et al.,* 1978a), myocarditis and thyroiditis (Lewis *et al.,* 1965).

6. Virologic Findings in Canine SLE

For years investigators have actively engaged in the pursuit of evidence that would implicate a horizontally transmitted agent in the disease SLE (Nardo and Norton, 1971). In a search for such an agent, cell-free filtrates of canine splenic tissue were inoculated into newborn dogs and mice. Canine and murine recipients of spleen from either SLE-afflicted dogs or close contacts of SLE-afflicted dogs routinely developed antinuclear antibody. Normal splenic tissue did not "induce" autoantibodies. The implication from these experiments is that a filterable horizontally transmitted agent has some role in canine SLE (Lewis *et al.*, 1973a). Certain of the mice receiving filtrates from dogs with SLE subsequently developed lymphoid malignancies. One such tumor produced both a retrovirus and a monoclonal anti-DNA antibody (Quimby *et al.,* 1975). The retrovirus (SP104) was characterized as an endogenous B-tropic virus of murine origin with the capacity to induce antinuclear antibody when passaged into other mice (Quimby *et al.,* 1978b). Activation of endogenous mouse retroviruses has been shown to occur by several other viruses (Kotler *et al.,* 1978; Reed and Rapp, 1978). Perhaps of most interest is that antisera made in rabbits to the disrupted SP104 virus bound selectively to a small, but significant number of peripheral blood lymphocytes from human patients with SLE (Lewis *et al.,* 1974) and canine patients with SLE (Lewis, unpublished observations) but not to lymphocytes from normal humans and dogs. These studies support the contention of Mellors and Mellors (1976) that antigens similar to those found in retrovirus appear in higher concentrations in the tissues of humans afflicted with SLE.

A further test of the hypothesis that transmissible agents are involved in canine SLE was made by rearing a litter of genetically susceptible dogs in a gnotobiotic environment. Despite the finding that three previous litters from the same mating developed autoantibodies by 6 months of age, none of the gnotobiotic animals seroconverted during their 4-

year residency in an isolator. After removal and transfer of the dogs back to the lupus colony, only one dog survived the first month in a conventional microbial environment. This dog developed antinuclear antibodies, lupus erythematosus (LE) cells, and autoimmune thyrogastric syndrome after 6 months (Quimby *et al.*, 1978a). This evidence suggests the existence of a transmissible agent acting in concert with certain genetic factors in canine SLE (Quimby and Schwartz, 1981).

Prompted by these observations, several investigators examined the possibility that the presumptive agent may cross species barriers and infect both dogs and man. Beaucher *et al.* (1977) reported findings on two families each having several members afflicted with SLE. Each family had a pet dog. Serum from these dogs was compared to other normal dogs in a double-blind study to evaluate anti-DNA antibodies. Significant levels were found only in dogs from the lupus families. Subsequent tests designed to evaluate cross-species seroconversion have been applied to eight families, in which either the dog or a human patient had SLE. In no instance did these tests reveal findings similar to Beaucher *et al.* (F. M. Quimby, unpublished results).

DeHoratius and Messner (1975) have shown an increased incidence of natural lymphocytotoxic antibody (NLA) in nonconsanguineous household contacts of SLE patients, again implying that certain autoantibodies may be acquired via horizontal transmission of an infectious agent. This antibody has not been seen in the canine colony contacts of canine patients with SLE, nor has it been seen in normal humans with close contact to dogs afflicted with SLE (F. M. Quimby, unpublised observations).

Despite all of the foregoing, the evidence implicating an infectious agent in the etiopathogenesis of canine or human SLE remains uncertain (Schwartz, 1975; Christian, 1976). Although similar studies in mice with SLE have demonstrated the presence of retroviral antigens in the immune complexes of murine SLE (Yoshiki *et al.*, 1974), other investigators have shown that neither the viral antigens nor the complete virions are essential for the development of SLE (Datta *et al.*, 1978). Regardless of the outcome of future studies implicating viruses in SLE, it appears highly likely that certain genetic factors will be required in order for such an agent to participate in the pathogenesis of this disease.

7. Genetic Findings in Canine SLE

In 1965 a breeding colony of dogs was established to evaluate the mechanism of inheritance of SLE in dogs. Dogs were selected as breeding pairs only if they had two consecutive positive LE cell tests. In each generation the offspring were serologically evaluated. The presence of positive LE cells occurred in the progeny with a greater incidence than was expected for a single autosomal dominant mode of inheritance; therefore, an interaction between genes and environment was postulated (Lewis and Schwartz, 1971). Although not used for breeder selection, the incidence of antinuclear antibody was high. Antinuclear antibody occurred in approximately the same percentage of dogs in each of five generations. The overall incidence of clinical SLE was 1 in 320 dogs, as compared to a frequency of 1 in 5000 for randomly selected dogs, but the incidence of other autoimmune diseases occurring in this colony was striking, approaching 34% in one family. Dogs with Sjögren's syndrome, juvenile diabetes mellitus, celiac disease, Hashimoto's thyroiditis (Quimby *et al.*, 1979), thyrogastric disease, and primary biliary cirrhosis (Quimby and Zink, 1978) were readily detected in these colonies. The incidence, distribution, and variety of these diseases suggested a very complex mode of inheritance.

In 1978 several Shetland sheepdogs were identified with clinical serologic and pathologic evidence of SLE. This finding led to a large-scale investigation of the distribution of SLE in shelties. After careful evaluation of breeding records from 120 dogs, and the subsequent clinical and/or serologic evaluation of 80 dogs, a pattern of inheritance similar to that seen in the inbred colony was obvious. Certain families were afflicted by no less that five different autoimmune disease, with the incidence of discoid lupus erythematosus being approximately 10% (Quimby, submitted for publication).

8. A Genetic Hypothesis

The diversity of autoimmune diseases that arose in these two instances, i.e., the inbred SLE colony and the noninbred shelty families, seems best explained by the existence of two classes of genes: (1) those that code for a general disposition to autoimmunity (class 1 genes) and (2) those that determine the phenotype of the disease (class 2 genes). Class 1 genes would relate to a defect in immunoregulation, whereas class 2 genes would influence the particular manifestations of disease, e.g., nephritis, rash, arthritis, and myositis. Alleles of class 2 genes could specify, for instance, the size of immune complexes, the function of complement, or the susceptibility to a horizontally transmitted agent. Both classes of these postulated genes may be complex, and consist of multiple, unlinked dominant and recessive loci (Quimby and Schwartz, 1978). This hypothesis explains why dogs in an inbred colony have a high incidence of autoantibodies (class 1 genes) but a much lower incidence of overt autoimmune disease. It also explains how two clinically normal indviduals may produce offspring with overt autoimmune disease. This hypothesis would explain why a dog with SLE bred to a normal sire, with different class 2 genes, may produce offspring with rheumatoid arthritis and thyroiditis. Furthermore, it explains the variety of autoimmune diseases seen in both inbred dogs, and noninbred dogs with a high incidence of class 1 genes. The hypothesis gains support from the finding by Miller and Schwartz (1979) that the clinically normal first-degree relatives of SLE patients have a 26% incidence of impaired suppressor T-cell function. This suggests that at least one important immunoregulatory factor is involved in SLE and is inherited by their offspring (class 1 genes?). These results in humans combined with numerous studies performed in mice and dogs support the view that the disease, SLE, has a genetic basis in all three species.

ACKNOWLEDGMENT

Work performed in this paper was supported in part by Grant AM-09351 from the United States Public Health Service.

References

Beaucher, W. N., Graman, R. H., and Condemi, J. J., 1977, Familial lupus erythematosus. Antibodies to DNA in household dogs, *N. Engl. J. Med.* **296:**982.

Christian, C. L., 1976, Systemic lupus erythematosus and type C RNA viruses, *N. Engl. J. Med.* **295:**501.

Datta, S. K., McConahey, P. J., Manny, N., Theophilopoulos, N., Dixon, F. J., and Schwartz, R. S., 1978, Genetic studies of autoimmunity and retrovirus expression in crosses of New Zealand black mice. II. The viral envelope glycoprotein gp 70, *J. Exp. Med.* **147:**872.

DeHoratius, R. J., and Messner, R. P., 1975, Lymphocytotoxic antibodies in family members of patients with systemic lupus erythematosus, *J. Clin. Invest.* **55:**1254.

DeHoratius, R. J., Pillarisetty, R., Messner, R., and Talal, N., 1975, Anti-nucleic antibodies in systemic lupus erythematosus patients and their families, *J. Clin. Invest.* **56:**1149.

Kotler, M., Lerer, I., Moyal, Z. B., and Spira, C., 1978, Simultaneous production of mouse endogenous virus and Rous sarcoma virus by Schmidt–Ruppin virus infected mouse cells, *J. Gen. Virol.* **40:**141.

Lewis, R. M., 1972, Animal model of human disease: Canine systemic lupus erythematosus, *Am. J.Pathol.* **69:**537.

Lewis, R. M., 1973, Autoimmune disease in New Zealand mice, *Comp. Pathol. Bull.* **5:**2.

Lewis, R. M., and Schwartz, R. S., 1971, Canine systemic lupus erythematosus: Genetic analysis of an established breeding colony, *J. Exp. Med.* **134:**417.

Lewis, R. M., Henry, W. B., Thornton, G. W., and Gilmore, C. E., 1963, A syndrome of autoimmune hemolytic anemia and thrombocytopenia in dogs, *Sci. Proc. J. Am. Vet. Med. Assoc.* **1:**140.

Lewis, R. M., Schwartz, R. S., and Henry, W. B., 1965, Canine systemic lupus erythematosus, *Blood* **25:**143.

Lewis, R. M., Andre-Schwartz, J., Harris, G. S., Hirsh, M. S., Black, P. H., and Schwartz, R. S., 1973a, Canine systemic lupus erythematosus: Transmission of serologic abnormalities by cell-free filtrates, *J. Clin. Invest.* **52:**1893.

Lewis, R. M., Stollar, B. D., and Goldberg, F. B., 1973b, A rapid sensitive test for the detection of antibodies to DNA, *J. Immunol. Methods* **3:**365.

Lewis, R. M., Tannenberg, W., Smith, C., and Schwartz, R. S., 1974, C-type viruses in systemic lupus erythematosus, *Nature (London)* **252:**78.

Mellors, R. C., and Mellors, J. W., 1976, Antigen related to mammalian type C RNA viral p30 protein is located in renal glomeruli in human systemic lupus erythematosus, *Proc. Natl. Acad. Sci. USA* **73:**233.

Miller, K. B., and Schwartz, R. S., 1979 Familial abnormalities of suppressor-cell function in systemic lupus erythematosus, *N. Engl. J. Med.* **301:**803.

Nardo, J. M., and Norton, W. L., 1971, Evidence for infectious etiologies in connective tissue disorders, *Semin. Arthritis Rheum.* **1:**116.

Poskitt, T., and Poskitt, P., 1978, The L1210 radioimmune assay for detecting circulating immune complexes, *Immunol. Commun.* **7:**543.

Quimby, F. W., and Kaplan, A. V., 1979, A radiolabelled staphylococcus protein A assay for warm agglutinin autoimmune hemolytic anemia, *Clin. Res.* **27:**587.

Quimby, F. W., and Schwartz, R. S., 1978, The etiopathogenesis of systemic lupus erythematosus, *Pathobiol. Annu.* **8:**35.

Quimby, F. W., and Schwartz, R. S., 1981, Systemic lupus erythematosus and associated diseases in laboratory animals, in: *Clinical Aspects of Immunology* (P. Lachmann, ed.), Blackwell, Oxford, in press.

Quimby, F. W., and Zink, N., 1978, Autoimmunity, multiple primary neoplasia and liver disease in a family of dogs, *Proc. 29th Annu. Meet. Am. Assoc. Lab. Anim. Sci.,* Abstract No. 116.

Quimby, F. W., Edmonds, H., and Lewis, R. M., 1975, The influence of isolation and serial passage of tumor cells and their products in murine myelomatosis, *Proc. 26th Annu. Meet. Am. Assoc. Lab. Anim. Sci.,* Abstract No. 132.

Quimby, F. W., Jensen, C., Nawrocki, D., and Schollin, P., 1978a, Selected autoimmune diseases in the dog, *Vet. Clinics N. Am.* **8:**665.

Quimby, F. W., Gebert, R., Datta, S., Andre-Schwartz, J., Tannenberg, W. J., Lewis, R. M., Weinstein, I. B., and Schwartz, R. S., 1978b, Characterization of a retrovirus that cross-reacts serologically with canine and human systemic lupus erythematosus, *Clin. Immunol. Immunopathol.* **9:**194.

Quimby, F. W., Schwartz, R. S., Poskitt, T., and Lewis, R. M., 1979, A disorder of dogs resembling Sjögren's syndrome, *Clin. Immunol. Immunopathol.* **12**:471.

Quimby, F. W., Brushwein, M., Smith, C., and Lewis, R. M., 1980, The efficacy of immunoserodiagnostic procedures in the recognition of canine immunologic diseases, *Am. J. Vet. Res.* **41**:1662–1666.

Reed, C. L., and Rapp, F., 1978, Induction of murine $_{p}30$ by superinfecting Herpes viruses, *J. Virol.* **19**:1028.

Schwartz, R. S., 1975, Viruses and systemic lupus erythematosus, *N. Engl. J. Med.* **293**:132.

Schwartz, R. S., Quimby, F. W., and Andre-Schwartz, J., 1978, Canine systemic lupus erythematosus: Phenotypic expression of autoimmunity in a closed colony, in: *Genetic Control of Autoimmune Disease* (N. R. Rose, P. E. Bigazzi, and N. L. Warner, eds.), pp. 287–294, Elsevier/North-Holland, Amsterdam.

Slappendel, R. J., 1978, Hemolytic anemia in the dog, in: *A Thesis,* pp. 96–108, Drukkerijelinkwijkbv, Utrecht.

Slappendel, R. J., Van Erp, C. L. G. M., Goudswaard, J., and Bethlehem, M., 1975, Cold hemagglutinin disease in a toy pinscher dog, *Tijdschr. Diergeneeskd.* **100**:445.

Slauson, D. O., and Lewis, R. M., 1979, Comparative pathology of glomerulonephritis in animals, *Vet. Pathol.* **16**:135.

Talal, N., 1976, Disordered immunologic regulation and autoimmunity, *Transplant Rev.* **31**:240–263.

Terman, D. S., Moore, D., Collins, J., Johnson, B., Person, D., Templeton, J., Poser, R., and Quimby, F. W., 1979, Detection of immune complexes in sera of dogs with rheumatic and neoplastic diseases by ^{125}I-Clq binding test, *J. Comp. Pathol.* **89**:221.

Thoburn, R., Hurvitz, A. I., and Kunkel, H. G., 1972, A DNA-binding protein in the serum of certain mammalian species, *Proc. Natl. Acad. Sci. USA* **69**:3327.

Yoshiki, T., Mellors, R. C., Strand, M., and August, J. T., 1974, The viral envelope glycoprotein of murine leukemia virus and the pathogenesis of immune complex glomerulonephritis of New Zealand mice, *J. Exp. Med.* **140**:1011.

8
Combined Immunodeficiency of Arabian Foals

TRAVIS C. MCGUIRE AND LANCE E. PERRYMAN

1. Introduction

Investigations into the cause of death of two sibling Arabian foals during the early 1970s led to the discovery of combined immunodeficiency (CID) in Arabian and part-Arabian foals (McGuire and Poppie, 1973; McGuire *et al.,* 1974; Poppie and McGuire, 1977). The defect undoubtedly existed prior to 1973, but the immunologic basis went unrecognized as emphasis was placed on the infectious agents isolated from affected foals (Johnson and Hutchins, 1967; McChesney *et al.,* 1970). Present evidence shows that affected foals lack significant immunologic function and die of infections before 5 months of age. The defect is transmitted as an autosomal recessive trait (Thompson *et al.,* 1975; Poppie and McGuire, 1977; Perryman and Torbeck, 1980) and closely resembles severe CID of children. Foals with CID occur throughout the United States (McChesney *et al.,* 1974; McGuire *et al.,* 1974; Ardans *et al.,* 1977) and have been reported from Australia (Thompson *et al.,* 1975; Studdert, 1978), Canada (Clayton, 1976; Henry and Gagnon, 1976), and England (Whitwell, 1978). This review will focus on the immunologic and biochemical findings in CID foals as well as results with immunotherapy.

2. Clinical Findings

2.1. Hematology

The most significant hematologic finding in CID foals is lymphopenia. In 159 CID cases studied, the peripheral blood lymphocyte count was <1000 per mm^3 (Perryman and

---

TRAVIS C. MCGUIRE AND LANCE E. PERRYMAN • Department of Microbiology and Pathology, College of Veterinary Medicine, Washington State University, Pullman, Washington 99164.

McGuire, 1980) while normal foals had a mean of 4119 per mm^3 and a range of 1590–8550 per mm^3 (McGuire *et al.,* 1974). Even though blood lymphocytes are few in number, some are always present and a few of these are enlarged with a vacuolated cytoplasm. The numbers and appearance of circulating erythrocytes, neutrophils, monocytes, and eosinophils are normal in CID foals prior to acquiring infections (McGuire *et al.,* 1974, 1975a; Banks and McGuire, 1975).

2.2. Onset of Infections and Time to Death

At birth CID foals resemble normal foals in size and all other external aspects. In a study of 66 confirmed CID cases, the beginning of clinical signs caused by infections ranged from 2 days to 2 months of age (Perryman *et al.,* 1978; Perryman, 1979). This variability in the onset of infections is caused by a number of factors, the most important being the quantity and quality of passively transferred maternal antibodies. Foals obtain all their maternal antibodies from the colostrum ingested after birth and none cross the placenta (reviewed by McGuire and Crawford, 1973). In non-CID foals, more than 10% fail to obtain or absorb adequate amounts of maternal immunoglobulins and these otherwise normal foals are temporarily more susceptible to infections than foals having adequate antibody transfer (McGuire *et al.,* 1975b, 1977). Those CID foals with failures in antibody transfer are immediately susceptible to a variety of infections, whereas those with adequate transfer are more resistant to pathogens in the local environment for several weeks (Poppie and McGuire, 1976). As maternal antibody clears from the CID foal's circulation, susceptibility to infections increases.

Survival of CID foals also varies and in 66 foals ranged from 15 days to 4.5 months of age (Perryman *et al.,* 1978). In the absence of immunotherapy, no reported cases of CID have lived past 5 months of age.

Although the causes of infections in CID foals are numerous, 55 of 66 had clinical evidence of respiratory infections and 59 had lesions in the lungs (Perryman *et al.,* 1978). The most common respiratory infectious agent was adenovirus (43 of 66), but almost all foals had bacterial-induced suppurative bronchopneumonia as well. In 22 of the 66 CID foals *Pneumocystis carinii* was found, usually in conjunction with other infections. Other lesions included adenovirus-induced pancreatitis, bacterial-caused hepatic necrosis and inflammation, omphalophlebitis, and enteritis. CID foals with an intestinal coccidian of the genus *Cryptosporidium* and concomitant adenoviral enteritis were reported (Snyder *et al.,* 1978). *Listeria monocytogenes* was isolated from the brain of one CID foal with acute necrotizing and granulomatous inflammation of the brain, heart, and adrenal glands (Clark *et al.,* 1978).

3. Analysis of B-Lymphocyte Function

3.1. Serum Immunoglobulin Class Measurement

Interpretation of serum immunoglobulin values in young foals, as in young children, is hindered by the presence of maternal immunoglobulins. Normal foals have detectable amounts of IgM and occasionally very small amounts of IgG at birth (Helms and Allen, 1971; McGuire and Crawford, 1973; Rejnek *et al.,* 1973). After nursing, most foals absorb enough IgGab (referred to as IgG), IgG subclasses IgG(T) and aggregating immunoglobulin, IgA, and IgM so that their serum amounts resemble those of adults (McGuire and Crawford, 1973). These maternal immunoglobulins are gradually catabolized and replaced by those synthesized by the foal.

In contrast to normal foal sera, CID foals lack IgM or any other immunoglobulin class at birth (McGuire *et al.,* 1974). After nursing and receiving an adequate passive transfer of immunoglobulin, CID foal sera cannot be distinguished from normal foal sera. Once absorption of immunoglobulins from colostrum is complete, no increase in amounts of any serum immunoglobulin class has been noted. Usually by 1 month of age CID foals have metabolized all their maternal IgM and no IgM can be detected in their sera. Sometime later, IgA and IgG(T) may be gone but no CID foals have lived long enough to metabolize all their maternal IgG (McGuire *et al.,* 1974; Perryman and McGuire, 1980). Even though no evidence for IgA production and secretion is available, free secretory component is present in external secretions (Buening *et al.,* 1978).

3.2. Evaluation of B Lymphocytes

Examination of the few lymphocytes isolated from CID spleen and lymph nodes with fluorescein-labeled goat anti-equine gamma globulin revealed a marked decrease in the number of immunoglobulin-bearing lymphocytes (McGuire *et al.,* 1975a). In contrast to studies with normal foals, no cells with surface immunoglobulin were found in the peripheral blood of CID foals even though sufficient numbers of cells that resembled lymphocytes were examined. The small number of immunoglobulin-coated cells noted in the tissue samples may have been macrophages, as intact immunoglobulin was used to prepare the fluorescent reagent. Lymphocytes from spleen and lymph nodes of other CID foals should be examined using a fluorescein-labeled $F(ab')_2$ anti-equine immunoglobulin reagent (Winchester *et al.,* 1975), which would avoid the problem of binding to cells lacking surface immunoglobulin, but having Fc receptors.

3.3. Results of Immunization

The absence of lymphocytes with surface immunoglobulin in CID foals, the absence of serum IgM at birth and after 1 month of age, the absence of serum IgA and IgG(T) if the foal lives long enough to metabolize the amounts obtained from colostrum, and the continual decline of serum IgG all suggest that no immunoglobulin is being made. This assumption is supported by the absence of antibody formation by 28 CID foals after immunization with sheep erythrocytes and/or keyhole limpet hemocyanin in incomplete Freund's adjuvant (McGuire *et al.,* 1975a; Perryman *et al.,* 1979a,b). Sera from CID foals lacked "natural" antibody to rabbit erythrocytes while it was easily detectable in normal foal sera. Therefore, when all available data on CID foals are considered including recent work from Australia (Lew *et al.,* 1980), there is no evidence for specific antibody formation.

4. Analysis of T-Lymphocyte Function

4.1. Lymphocyte Proliferation Assays

Mononuclear cell preparations from untreated CID foals containing cells resembling lymphocytes do not proliferate *in vitro* when stimulated with phytohemagglutinin, pokeweed mitogen, concanavalin A, or lipopolysaccharide, with the exception of two cases (McGuire *et al.,* 1975a; Perryman *et al.,* 1979a; Lew *et al.,* 1980). The latter two CID foals spontaneously developed phytolectin-responsive lymphocytes at approximately 3 weeks of age. This response was 10–40% of the stimulation noted with lymphocytes from age-matched normal foals (Perryman *et al.,* 1979a). Considering all the CID foal cases reported, the proliferation by mononuclear cells in these two foals is the only positive reaction to any functional immunologic test.

Peripheral blood mononuclear cells from seven CID foals stimulated allogenic lymphocytes from normal horses in mixed lymphocyte culture reactions (Perryman and McGuire, 1978). However, none of the CID foals had cells that were able to respond to allogeneic stimulation. CID foal cells acting as stimulators in the reaction could either be macrophages or lymphocytes. One-way mixed lymphocyte culture reactions using lymphocyte- and macrophage-enriched populations from normal horses as stimulator cells showed that lymphocytes were superior (Perryman and McGuire, 1978). These data indicate that CID foals produce lymphocytes that possess histocompatibility antigens, but are unable to proliferate.

4.2. Delayed Hypersensitivity Skin Tests

Skin testing of six CID foals was done after sensitization with five daily applications of 2 mg of dinitrochlorobenzene (DNCB) in dimethylsulfoxide. Challenge with 0.4 mg DNCB in olive oil 9–14 days after sensitization failed to elicit a response, whereas 85 normal horses and foals had an easily detectable increase in skin thickness (Hodgin *et al.,* 1978).

Injection of 50 μg of phytohemagglutinin into the skin of normal horses and foals causes skin thickening characterized by the slow (24 hr) development of a predominantly mononuclear cell infiltrate that closely resembles a delayed hypersensitivity response to antigen (Hodgin *et al.,* 1978). At least 35 CID foals failed to react to phytohemagglutinin injection (McGuire *et al.,* 1975a; Hodgin *et al.,* 1978; Perryman *et al.,* 1979a). Further evidence for T-lymphocyte involvement in the response of horses to intradermal injection of phytohemagglutinin was provided by a positive reaction in a B-lymphocyte-deficient horse with normal T-lymphocyte function (McGuire *et al.,* 1976b).

5. Analysis of Other Host Defense Systems

5.1. Complement

Serum complement function was evaluated in a small number of CID foals by three functional tests. (1) The ability of serum to cause binding of IgM-coated erythrocytes to equine monocytes was determined. This test is dependent on the activation and binding of C3. (2) Enhanced neutralization of IgG-sensitized equine arteritis virus was measured. Neutralization in this system is enhanced by the initial components of the complement system (C1,4,2,3) and does not occur in the absence of C1 (Radwan and Crawford, 1974). (3) The amount of serum required to lyse 50% of antibody-coated erythrocytes was assayed. No differences were found between normal and CID foal sera in mediating these complement-dependent reactions (McGuire *et al.,* 1975a).

5.2. Neutrophil and Monocyte Function

Sheep erythrocytes coated with equine IgG or IgM and complement bound to both monocytes and neutrophils from CID and normal foals (Banks and McGuire, 1975). CID monocytes and neutrophils also bound glutaraldehyde-treated erythrocytes and phytolectins in amounts similar to cells from normal foals. In addition, no abnormalities were noted in the phagocytic capacity of cells from CID foals.

6. Histophathologic Changes in Lymphoid Tissues

6.1. Absence of Lymphoid Structures in Spleen and Lymph Nodes

Changes in the spleens of 80 CID foals and 1 CID fetus were characterized by three morphologic features: (1) absence of germinal centers and primary follicles, (2) absence of periarteriolar lymphocytic sheaths, and (3) lack of plasma cells and a paucity of lymphocytes (McGuire *et al.,* 1976a; L. E. Perryman, unpublished cases). Normal foal spleens have easily identifiable germinal centers and follicles (Fig. 1). One useful feature to distinguish between CID spleens and other cases having lymphocyte depletion was total absence in CID spleens of B-lymphocyte-dependent germinal centers and follicles, including absence of the stromal scaffold for these structures (Fig. 2). The scaffold could always be found in non-CID spleens regardless of the amount of lymphocyte depletion, as in the case of lethally irradiated foals (Fig. 3). The only exception to this rule was seen in a Thoroughbred horse with agammaglobulinemia (McGuire *et al.,* 1976b). The CID spleens were also smaller than normal, causing the trabeculae to be narrowly spaced.

The appearance of lymph nodes from 80 CID foals and 1 CID fetus varied. However, there were generally four basic alterations (McGuire *et al.,* 1976a; L. E. Perryman, unpublished cases): (1) B-lymphocyte-dependent areas including germinal centers and follicles present in lymph nodes from normal foals (Fig. 4) were absent in lymph nodes from CID foals (Fig. 5); (2) T-lymphocyte-dependent paracortical areas of the nodes were sparsely populated with lymphocytes; (3) there was an increase in the macrophage populations especially in the medullary sinuses; (4) plasma cells were absent. The absence of lymphoid follicles and the small number of lymphocytes in CID foal lymph nodes were generally distinctive enough to allow differentiation from non-CID cases (Fig. 5). However, lymph nodes from some non-CID foals with generalized bacterial infections lacked well-defined lymphoid follicles and germinal centers, making it difficult to differentiate from CID foal lymph nodes. Peyer's patches in the CID cases examined had changes similar to those described in lymph nodes.

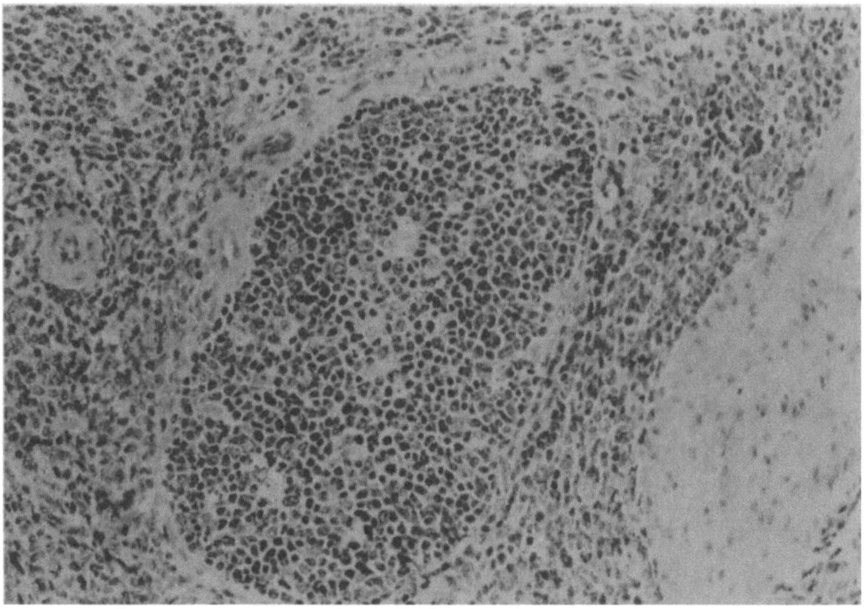

FIGURE 1. Lymphoid follicle with an adjacent arteriole in the spleen of a 1-month-old normal foal (hematoxylin and eosin, ×217).

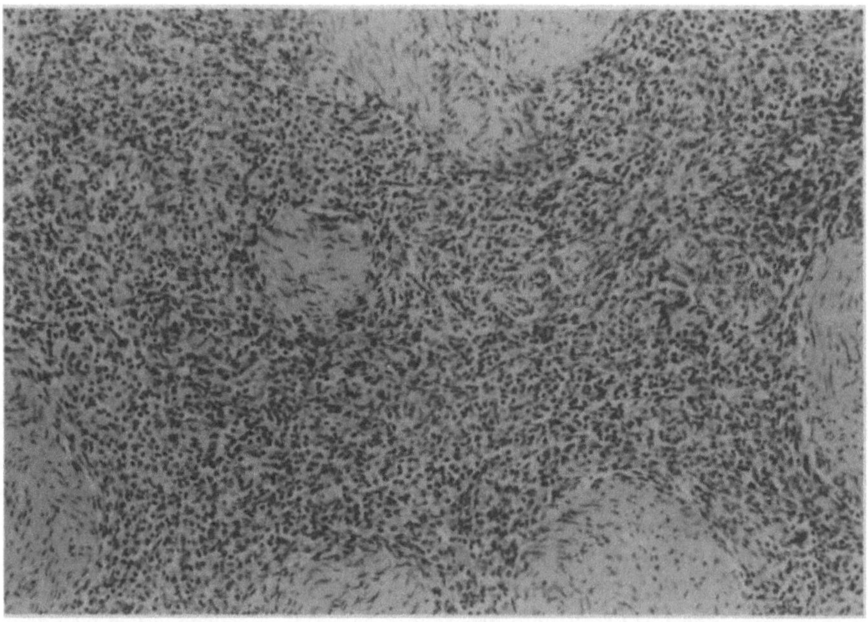

FIGURE 2. Spleen from a 1.5-month-old foal with CID. Note the absence of any organized lymphoid structure (hematoxylin and eosin, ×136).

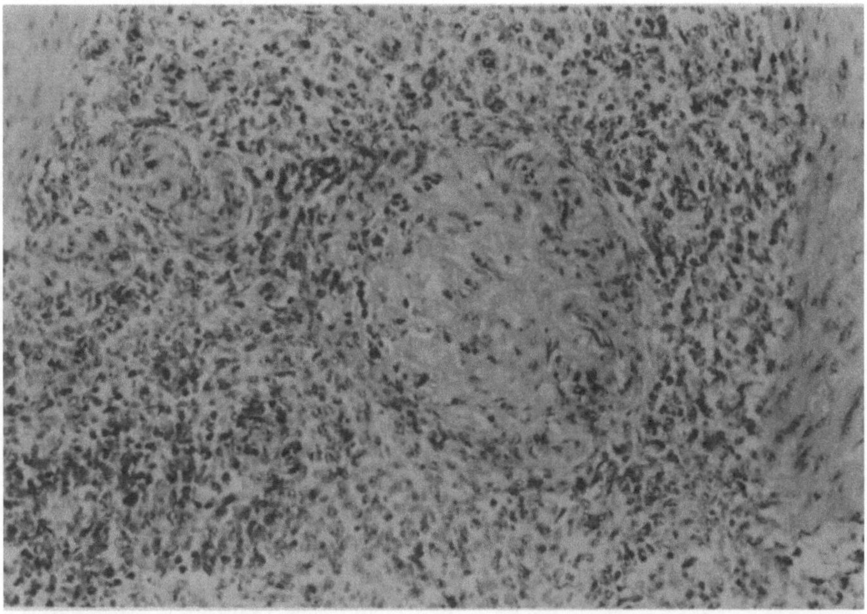

FIGURE 3. Spleen from a 1-month-old foal 10 days after lethal X irradition of 840 rads. In the absence of lymphocytes, the scaffold of the lympoid follicle remains. These structures have never been observed in spleens from foals with CID (hematoxylin and eosin, ×217).

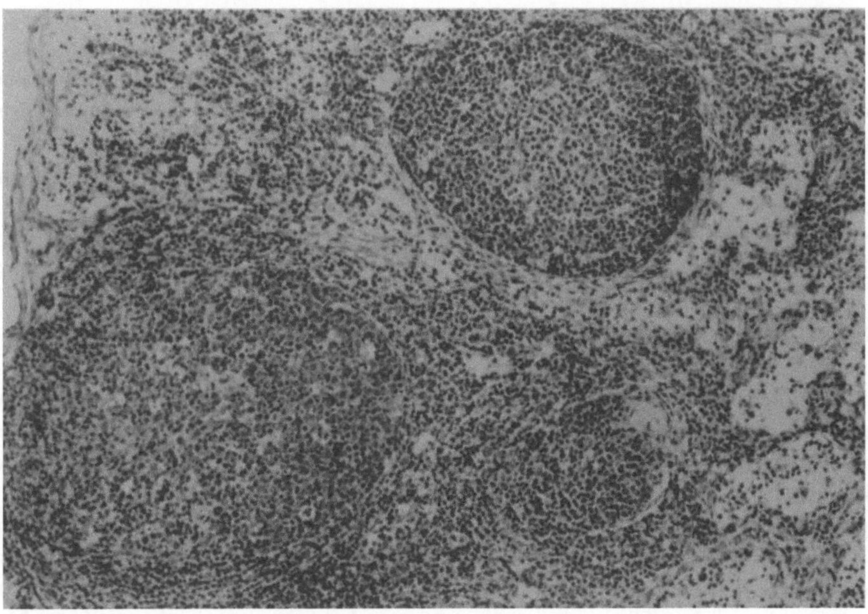

FIGURE 4. Peripheral lymph node from a 1-month-old normal foal. The outer capsule can be seen on the left side of the figure (hematoxylin and eosin, ×54).

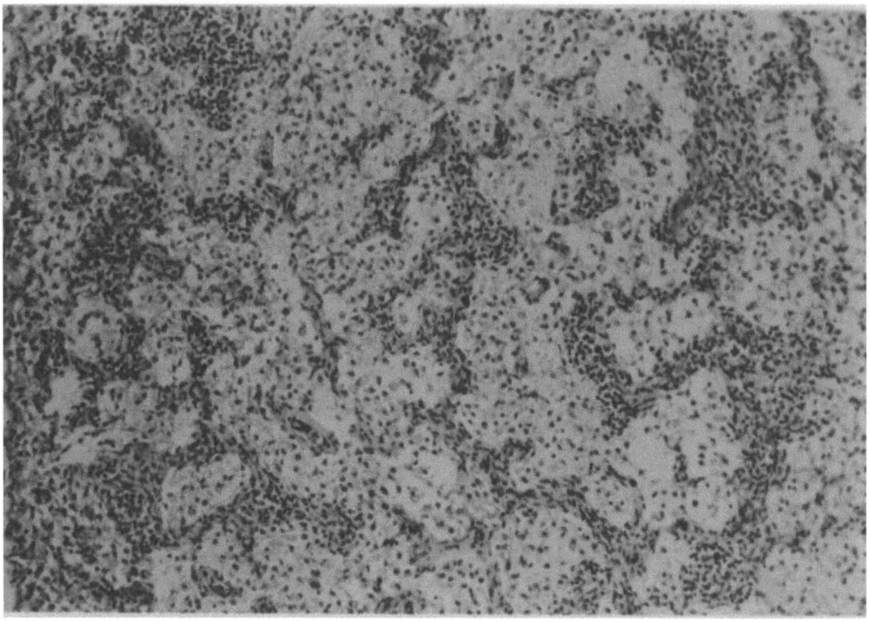

FIGURE 5. Peripheral lymph node from a 1-month-old CID foal. The outer capsule can be seen on the left side of the figure. In contrast to Fig. 4, there is a paucity of lymphocytes and no organized lymphoid structures (hematoxylin and eosin, ×109).

6.2. Thymic Hypoplasia

The thymus has been located in all cases of CID in which a thorough search has been made. Microscopic examination of 40 CID foal thymuses and 1 CID fetal thymus revealed extreme hypoplasia with very few lymphoid-appearing cells (McGuire *et al.*, 1976a; L. E. Perryman, unpublished cases). Morphologic features varied, although a few general characteristics were discernible. All thymuses had an outer capsule and a few had interlobular fibrous septa dividing the organ into lobules. In contrast to normal foal thymuses (Fig. 6), a distinct cortex and medulla were lacking. The usual case had lymphoid cell islands and branches surrounded by adipose tissue that extended to the outer capsule (Fig. 7). An epithelial component with Hassall's corpuscles was present without exception in the lymphoid cell islands or branches (Fig. 8).

The appearance of individual thymuses seemed to represent various stages in development. For instance, some had epithelial-cell-lined arborized spaces that may have been persistent central thymic canals. In many instances, these spaces were filled with necrotic and mineralized debris (Fig. 9). Outside the branches were lymphoid cells surrounded by adipose tissue. Other CID thymuses had a similar appearance except that the open areas within the branches lacked a defined epithelial lining. These thymuses possibly represent a more mature stage of development with epithelial cells losing the orientation to the central thymic canals and becoming more disperse. Some thymuses contained branches without central spaces probably representing complete obliteration of the canals. A few thymuses did not fit into the described pattern. Two had an embryonal appearance in that they con-

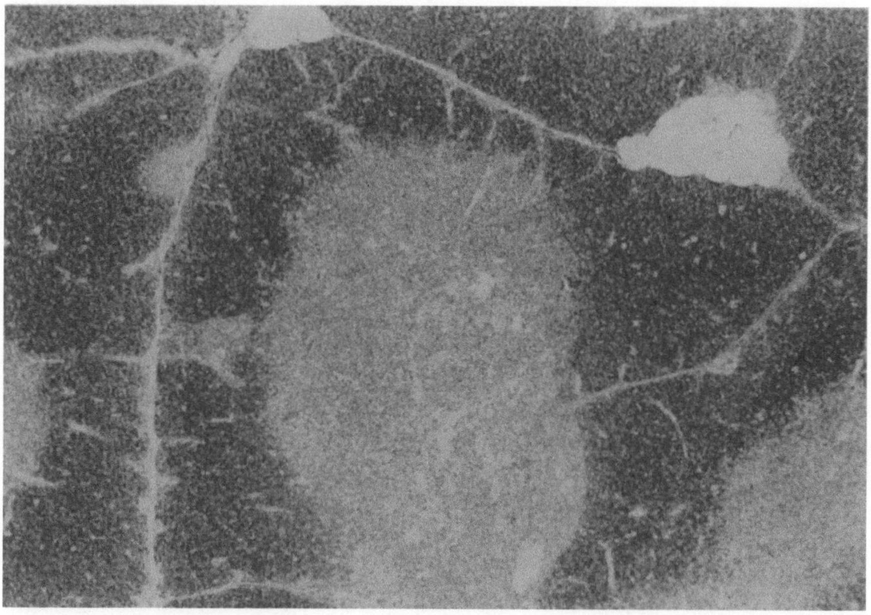

FIGURE 6. Thymus from a 1-month-old normal foal (hematoxylin and eosin, ×43).

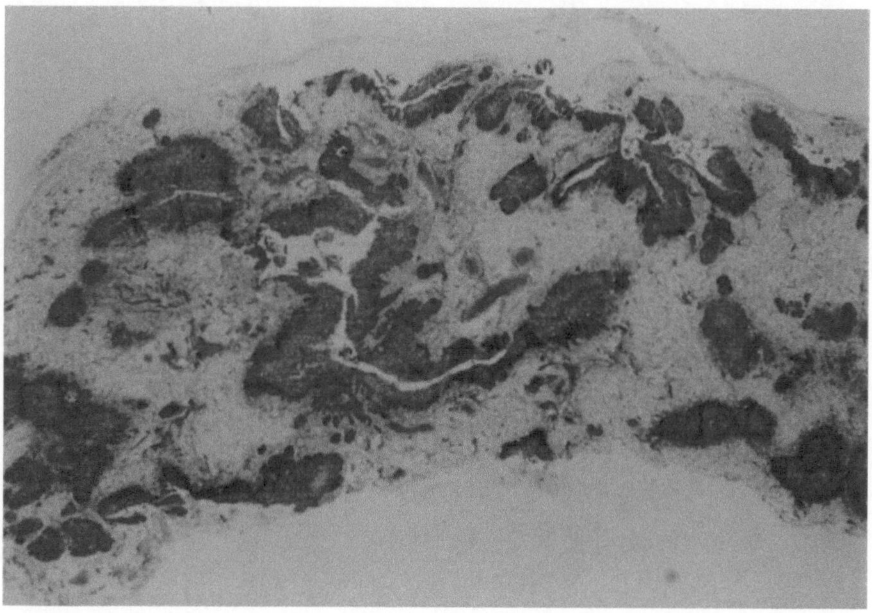

FIGURE 7. Hypoplastic thymus from a 1-month-old CID foal (hematoxylin and eosin, ×17).

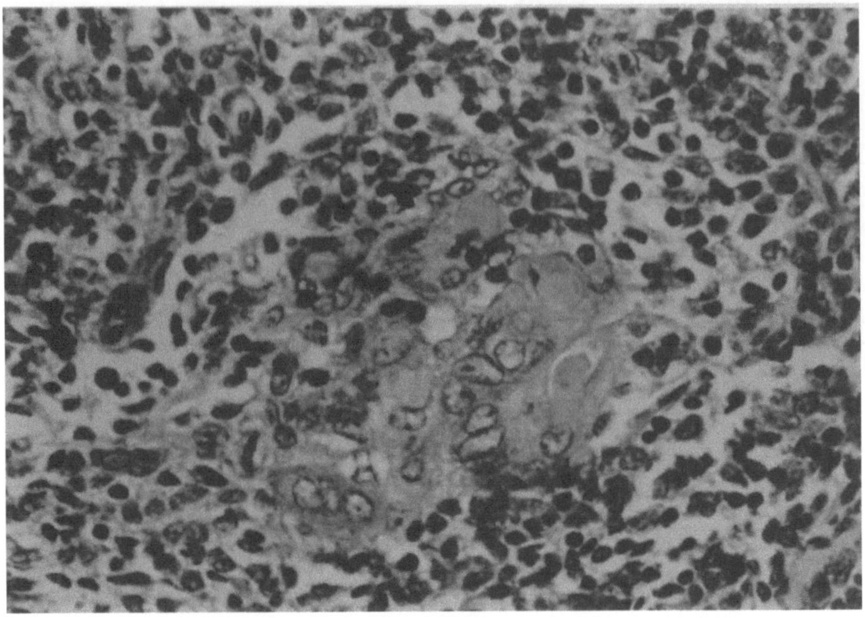

FIGURE 8. Hassall's corpuscle in a CID foal thymus (hematoxylin and eosin, × 435).

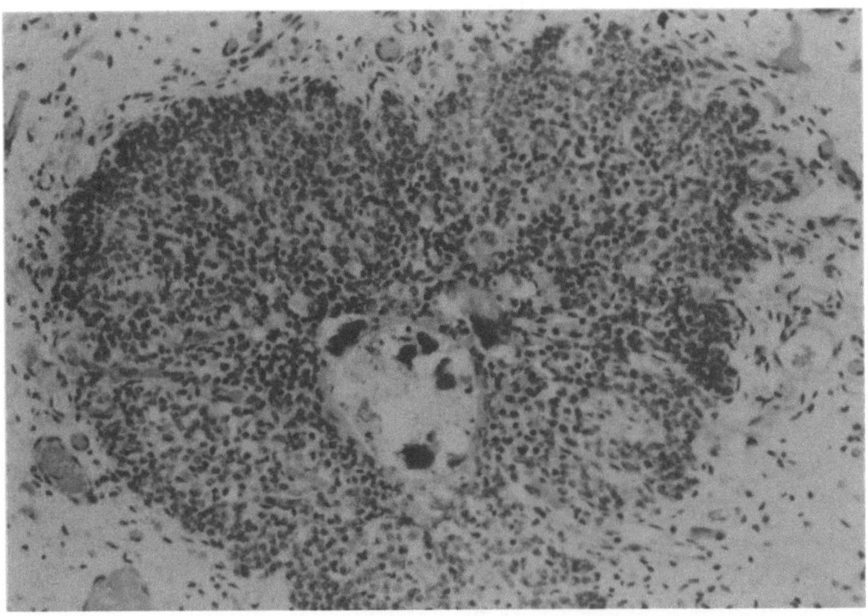

FIGURE 9. Thymus from a CID foal containing an epithelial-cell-lined space filled with necrotic and mineralized debris (hematoxylin and eosin, ×190).

tained connective tissue composed of spindle-shaped cells. However, Hassall's corpuscles were included in these areas and they were bordered by some cells resembling lymphocytes. Two other thymuses had only small amounts of identifiable lymphoid and epithelial components. Within these foci were hemorrhage, epithelial giant cells, individual epithelial cells, and both individual and clumped lymphocytes. These latter morphologic variations were unexplainable.

The extreme morphologic heterogeneity in thymuses from CID foals with a common genetic defect indicates that caution should be used in interpreting the morphology of thymuses from CID children where the syndrome is known to have different inheritance patterns (Borzy *et al.,* 1979).

Electron microscopic examination of CID foal thymuses showed that residual lymphocytes were separated from other structures by a basement membrane as in the case of normal thymuses (McGuire *et al.,* 1976a). Epithelial cells were present in both normal and CID thymuses and were identified by the presence of cytoplasmic tonofibrils and desmosomes. Macrophages were present in thymuses from CID foals and these cells did not contain excessive amounts of phagocytized debris or lymphocytes.

The presence of a morphologic thymic structure in CID foals with only a few lymphocytes, but lacking evidence of atrophic involution, suggests the hypoplasia is caused by an absence or inability of committed lymphocytes from the bone marrow to populate the organ. This suggestion is supported by the hypoplasia in the thymus from a CID fetus aborted 3 months before term (McGuire *et al.,* 1976a).

In attempts to determine the functional capacity of the thymic remnant in CID foals, thymic extracts and sera were evaluated for thymic hormone activity by several different procedures (Splitter *et al.,* 1979a,b): (1) thymic extracts given to lethally irradiated, bone-marrow-reconstituted mice were able to increase the plaque-forming-cell response to sheep erythrocytes; (2) ultrafiltrates of serum from CID foals that contained material of less than 50,000 daltons were able to cause a population of human bone marrow cells to express receptors for sheep erythrocytes, a characteristic of T lymphocytes; (3) the same ultrafiltrate of serum stimulated cyclic AMP synthesis in mouse thymocytes; (4) additionally, CID serum ultrafiltrates were assayed by a rosette-inhibition procedure in the presence of azothiaprine for circulating serum factor (Splitter *et al.,* 1979a). The circulating serum factor was present in CID foal sera, but amounts were lower than those from normal foal sera. These data suggest that the epithelial component of hypoplastic CID thymuses secretes some of the thymic hormones necessary to T-lymphocyte development and further indicate that lymphoid stem cells should reconstitute T- and B-lymphocyte functions in CID foals. So far, unequivocal evidence for immunologic reconstitution without the presence of fetal thymus is unavailable (Ardans *et al.,* 1977; Perryman *et al.,* 1979b; Perryman, 1981).

7. Criteria for Diagnosis of CID in Foals

Four other immunodeficiency disorders have been described in foals that require differential diagnosis from CID.

1. Ten to twenty percent of foals of all breeds fail, for various reasons, to obtain or absorb sufficient maternal immunoglobulins from colostrum. These foals are hypo-

gammaglobulinemic for a few weeks after birth (McGuire *et al.,* 1977; Perryman and McGuire, 1980). During this time these foals have to make a primary immune response to each infectious organism and are more susceptible to infections. Failure of passive immunoglobulin transfer can be differentiated from CID by lymphocyte counts and serum IgM measurements. CID foals are lymphopenic and lack IgM, whereas foals with failure of passive transfer have normal lymphocyte numbers, produce IgM *in utero,* and continue to make it throughout life. It should be kept in mind that CID foals lack IgM at birth, obtain IgM from colostrum, and then lack IgM after maternal IgM has been cleared from the circulation.

2. Transient hypogammaglobulinemia has been described in two Arabian foals having increased susceptibility to infection (McGuire *et al.,* 1975b; Perryman and McGuire, 1980). These foals had detectable serum IgM, but very low amounts of IgG at an age when they should have made their own IgG. As IgM amounts and lymphocyte counts were in the normal range, the cases were easily differentiated from CID.
3. Selective IgM deficiency has been found in 19 foals including 13 Arabian foals and 6 foals of other breeds (Perryman *et al.,* 1977; Perryman and McGuire, 1980). These foals lack serum IgM or have very low amounts and are susceptible to infections especially of the respiratory tract. In contrast to CID foals, their lymphocyte counts are within the normal range, as are serum IgG, IgA, and IgG(T) amounts. Foals with selective IgM deficiency often survive past 5 months, a time at which all unreconstituted CID foals are dead.
4. Agammaglobulinemia with T-lymphocyte function has been described in a Thoroughbred horse (Banks *et al.,* 1976; McGuire *et al.,* 1976b) and two Standardbred horses (Deem *et al.,* 1979; Perryman and McGuire, 1980). Two of these horses lived several months beyond the usual 5-month CID age limit, and had normal lymphocyte counts and measurable delayed hypersensitivity responses.

Considering the foal disorders that must be differentiated from CID, the following minimal criteria must be met for diagnosis: (1) absence of serum IgM; (2) <1000 lymphocytes/mm^3 of blood; (3) histopathologic changes in spleen, lymph nodes, and thymus compatible with CID; and (4) absence of a delayed hypersensitivity skin response or lack of a proliferative response to phytolectin stimulation by isolated mononuclear cells.

8. Documentation of Autosomal Recessive Inheritance

Initial arguments for an autosomal recessive mode of inheritance for CID in foals were based on the observations that both males and females were affected, the disorder was sublethal, and parents of affected foals had normal immune responses and resistance to infections (McGuire and Poppie, 1973). Based on a study of 17 foals dying of infections in Australia, Thompson *et al.* (1975) concluded that CID was transmitted as an autosomal recessive trait. Criteria for CID diagnosis in most of the 17 cases were based on fatal adenoviral infection, which is rarely fatal in normal foals. In a prospective breeding study, 17 mares, all of which previously had CID foals, were bred to 1 of 10 stallions that had sired CID foals for a total of 22 matings (Poppie and McGuire, 1977). Seven of the twenty-two offspring had CID, a result compatible with autosomal recessive inheritance. Another 2-year prospective breeding trial with 26 mares and 1 stallion, all of which were parents of CID foals, produced a total of 53 offspring. Fifteen of the fifty-three foals had CID, seven being males and eight females (Perryman and Torbeck, 1980). Finally, the sex has been recorded in 155 cases of CID diagnosed by our laboratory and 77 were males and 78 females (Perryman and McGuire, 1980). Therefore, all available data are compatible with CID in foals being inherited as an autosomal recessive trait.

9. Studies on the Biochemical Basis of the Defect

9.1. Presence of Adenosine Deaminase

Giblett *et al.* (1972) described two children with severe CID that had no detectable adenosine deaminase (ADA) in their erythrocytes. Subsequent studies showed that ADA was missing from all tissues of children with severe CID that lacked erythocyte ADA (Pickering *et al.,* 1974). The enzyme deficiency was transmitted as an autosomal recessive trait and did not occur in children with severe CID having a sex-linked mode of inheritance (Pickering *et al.,* 1974). It now appears that approximately 50% of those children with autosomal recessive inheritance of severe CID have ADA deficiency (Hirschhorn and Martin, 1978). Even though the precise relationship between ADA deficiency and severe CID remains to be defined, available evidence suggests that the enzyme absence causes the immunodeficiency and several mechanisms have been proposed (reviewed by Hirschhorn and Martin, 1978).

Amounts of ADA in the tissues of CID and normal foals were measured and no differences between them were noted (McGuire *et al.,* 1976c). Eleven different organs were examined including lymph nodes, spleen, and thymus. In order to confirm that ADA was present in the progeny of CID bone marrow stem cells, mononuclear cells containing lymphocytes and monocytes were assayed. Results with mononuclear cells from CID and normal foals gave similar results as did assays with platelets. The approximate molecular weight of ADA from tissues of normal and CID foals was 37,000. In initial studies no ADA was detected in erythrocytes from either CID or normal foals, while the assay system would easily detect ADA in human erythrocytes (McGuire *et al.,* 1976c). Subsequent studies by other workers demonstrated ADA in both CID and normal erythrocytes in amounts about 50% of those in mice (Castles *et al.,* 1977). These differences in results are unexplained and probably result from sensitivity differences in the assay systems (Castles *et al.,* 1977).

Because available data demonstrates normal amounts of ADA in tissues and cells from CID foals, two other enzymes of the same purine salvage pathway were examined. Tissues (McGuire *et al.,* 1976c) and erythrocytes (Castles *et al.,* 1977) from CID foals contained normal amounts of nucleoside phosphorylase. Xanthine oxidase was present in the same quantities in liver and spleen from one CID foal and one normal foal (McGuire *et al.,* 1976c).

9.2. Evaluation of Purine Metabolism

Approximately 50% of CID children with an autosomal recessive mode of inheritance have normal amounts of ADA (Hirschhorn and Martin, 1978). The cause of CID in foals may be the same as the cause of CID in ADA normal children. Because at least three enzymes in purine metabolism, including ADA, nucleoside phosphorylase, and ecto-5′-nucleotidase, have been associated with immunodeficiencies (reviewed by Hirschhorn and Martin, 1978), various aspects of purine metabolism in CID foals were studied (Magnuson and Perryman, 1979). Lymphocytes from 2 of 30 CID foals tested had detectable proliferative responses when stimulated with mitogens and the responses of these lymphocytes were compared with those from normal foals in the presence of adenosine (Magnuson and Perryman, 1979). Adenosine inhibited the uptake of [^{3}H]thymidine incorporation by phytohemagglutinin-stimulated lymphocytes from both normal and CID foals in a dose-dependent manner. Significantly less adenosine was required to inhibit proliferation in CID foal lymphocytes when compared to normal foal lymphocytes. Uridine reversed the inhibitory effect that adenosine had on phytohemagglutinin-stimulated [^{3}H]thymidine uptake by normal horse and human lymphocytes. In contrast, uridine had no effect on the inhibitory effect of adenosine on CID foal lymphocytes (Magnuson and Perryman, 1979).

Evaluation of nucleotide levels by high-pressure liquid chromatography in normal, carrier (heterozygotes), and CID erythrocytes revealed no differences (Magnuson and Perryman, 1979). Incubation of erythrocytes with adenosine caused the total nucleotide pool to increase by 1.5- to 2-fold in all cases. In normal and carrier erythrocytes, the nucleotide pool increase was accompanied by a significant decrease in the ATP:(AMP + ADP + ATP) ratio, while this ratio changed only slightly in similarly treated erythrocytes from CID foals.

These alterations of purine metabolism in CID foals do not pinpoint the cause of the immunodeficiency, although they do provide the basis for subsequent analyses of purine metabolism and interconnecting pathways.

10. Immunotherapy

10.1. General Considerations

Histocompatible bone marrow transplants reconstitute immunologic function in children with severe CID and represent the treatment of choice. The functional lymphocytes populating these children after transplantation are donor derived. In such cases it seems that lymphocyte precursor cells are being provided by injected bone marrow cells and that the recipient environment for differentiation and proliferation is adequate. Because the thymus plays a vital role in the maturation of T lymphocytes, reconstitution by bone marrow cells suggest that the thymus is intact in such patients. Heterogeneity in CID of children clearly exists as evidenced by inheritance patterns (at least sex-linked and autosomal recessive) and biochemical defects (autosomal recessive–ADA deficient, and autosomal recessive–ADA normal). Additionally, there is discordant clinical expression, not only in CID children having different inheritance patterns, but also differences in the clinical expression of CID between siblings (Giblett *et al.,* 1972; Hirschhorn and Martin, 1978). The introduction of heterogeneity at all levels in the CID syndrome introduces variables at the therapy level influencing the choice and success of treatment procedures (Pahwa *et al.,* 1978). For example, some immunologic reconstitution of CID children occurred with fetal thymus injection (Ammann *et al.,* 1973) and in other instances with implantation of cultured thymic epithelium (Hong *et al.,* 1976). The latter treatment clearly suggests that appropriate precursor cells are present in the treated children but lack the appropriate thymic influences for development. These studies have been extended by evaluating the potential of CID bone marrow cells to respond to thymic hormones before selecting the method of treatment. These considerations and the ability to reconstitute immunologic function in selected cases with enzyme replacement have complicated the selection of a therapeutic procedure.

10.2. Reconstitution with Fetal Thymus and Liver Cells

In the absence of histocompatible bone marrow, fetal liver and/or fetal thymus transplants have been used to reconstitute CID children (Buckley, 1977). The reliability of fetal tissues for reconstitution is poor when compared to the use of bone marrow. Unsatisfactory results may result from inadequate knowledge including the optimum fetal age of the donor, the number of cells, and the route of administration.

Fetal tissues have been evaluated for immunologic reconstitution of CID foals (Perryman *et al.,* 1979b; Perryman, 1981). In an initial study, six CID foals were given fetal liver cells intravenously or intraperitoneally and fetal thymuses subcutaneously (Perryman *et al.,* 1979b). One foal receiving tissues from a 68-day fetus had no detectable effects from the treatment and died 74 days later. Four CID foals were given liver cells and thymus from fetuses between 90 and 110 days of gestational age. Three of these had graft-versus-host reactions, the fourth having no demonstrable effect. One of the foals with a graft-versus-host reaction had a slight increase in lymphocytes and lymphocyte response to mitogens. The sixth foal was given tissues from an 80-day fetus and had evidence of engraftment. This transplanted CID foal lived to 11.5 months of age before death due to physical exertion following entrapment under a pen partition. The 11.5 months is in contrast to untreated CID foals, which die before 5 months of age. Evidence for immunologic function in the foal surviving the 11.5 months included phytolectin-responsive lymphocytes, synthesis of the four major equine immunoglobulin classes, and specific IgA antibody secretion following immunization. The fetal donor and CID recipient were both males and karyotypic evaluation of lymphocytes could not be used to determine the origin of the functional lymphocytes. Erythrocyte typing and mixed lymphocyte cultures were also inconclusive in determining the origin of the proliferating lymphocytes. The presence of immunologic function and the survival time of the CID foal argue for reconstituion by the fetal tissues.

In a subsequent study, one of three CID foals given liver cells and thymus from 80-day fetuses developed functional T lymphocytes, a graft-versus-host reaction, and survived to 8 months of age (Perryman, 1981; Perryman and Liu, 1980).

In vitro studies to determine the appearance of mixed-lymphocyte-culture-responsive and phytolectin-responsive lymphocytes in equine fetal tissues were made (Perryman *et al.,* 1979b, 1980). Mixed-lymphocyte-culture-responsive cells were present in the thymus by

100 days' gestation while phytolectin-responsive cells were present by 80 days. Results with fetal liver were erratic. The *in vivo* and *in vitro* data indicate the future transplants with equine fetal liver and thymus should use fetuses of around 80 days' gestational age to avoid graft-versus-host reactions and obtain immunologic function.

10.3. Other Treatment Methods

Washed, irradiated erythrocytes given to some ADA-deficient CID children have resulted in clinical inprovement and detectable immunologic function (Polmar *et al.,* 1976). Even though CID foals have functional ADA, it seemed possible that other enzyme deficiencies might be replaced or alleviated by erythrocyte administration. Therefore, two CID foals were given a liter of washed, irradiated normal horse erythrocytes weekly for 2 to 3 weeks (Perryman *et al.,* 1979a). Evaluation of clinical and immunologic parameters revealed no improvement from this treatment.

One attempt to reconstitute a CID foal with bone marrow cells resulted in a graft-versus-host reaction (Ardans *et al.,* 1977). It seems that histocompatible bone marrow transplants should be made in CID foals, for a success would suggest that stem cells are capable of reconstitution in this disease of foals and that the thymic remnant is functioning adequately. It is unlikely that the CID foal represents a useful model for bone marrow transplantation in humans because the disparity between our knowledge of human and horse histocompatibility systems is so great.

11. Availability of Affected Foals

Combined immunodeficiency occurs at a surprisingly high prevalence in Arabian and part-Arabian foals. Studies based on nonbiased sampling procedures have reported CID foal prevalences in the United States of 2.3% (Poppie and McGuire, 1977) and 2.7% (A. E. McChesney, personal communication). In Australia a prevalence of 2–3% has been reported (Studdert, 1978). An estimate of the number of CID carriers in the breeding population was made from the prevalence of CID foals and considering an autosomal recessive mode of inheritance. The carrier percentage estimate was 28% (Poppie and McGuire, 1977), which is extremely high for a fatal genetic disease. It is difficult to determine the reason for such a high frequency of heterozygotes, although it may be that the CID trait is closely linked to the desirable traits that are being selected. Alternatively, a founder effect is possible, as the numbers of Arabian horses have undergone marked increases in the recent past.

We have accumulated a herd of 33 CID carrier mares and 2 carrier stallions for research purposes at Washington State University.

12. Discussion of Future Research Possibilities

The CID syndrome in foals can appropriately be referred to as severe and it represents, with the exception of children, the only occurrence of severe CID. Because a significant number of children with severe CID lack a histocompatible bone marrow donor and reliable alternative forms of therapy are unavailable, research on treatment methods seems imperative. An important potential use of CID foals is to discover and evaluate treatment procedures.

Knowledge of the biochemical basis of the different forms of CID in children may result in new forms of therapy such as enzyme replacement and genetic engineering. Foals with CID have normal adenosine deaminase and nucleoside phosphorylase, but obviously have a biochemical defect. Affected foals are plentiful in number and can provide large amounts of tissue and cells for study. We feel studies on the biochemistry of CID are the second important potential use for CID foals.

Although there are a number of other research possibilities, we shall mention only one. The CID foals offer a unique and defined system to study the role of the immune respone in infectious disease. For instance, a nononcogenic retrovirus infection (equine infectious anemia) persists in horses and causes anemia, hepatitis, and glomerulitis, all apparently immunologically mediated. Infection of CID foals with this virus would provide unequivocal evidence as to the role of the immune response in the persistence of the virus and the induction of lesions.

ACKNOWLEDGMENTS

This work was supported by grants from the National Institutes of Health (HD 08886), Morris Animal Foundation, and Arabian Horse Registry of America.

References

Ammann, A. J., Wara, D. W., Salmon, S., and Perkins, H., 1973, Thymus transplantation: Permanent reconstitution of cellular immunity in a patient with sex-linked combined immunodeficiency, *N. Engl. J. Med.* **289**:5.

Ardans, A. A., Trommershausen-Smith, A., Osburn, B. I., Mayhew, I. G., Trees, C., Park, M. I., Sawyer, M., and Stabenfeldt, G. H., 1977, Immunotherapy in two foals with combined immunodeficiency, resulting in graft-versus-host reaction, *J. Am. Med. Assoc.* **170**:167.

Banks, K. L., and McGuire, T. C., 1975, Surface receptors on neutrophils and monocytes from immunodeficient and normal horses, *Immunology* **28**:581.

Banks, K. L., McGuire, T. C., and Jerrells, T. R., 1976, Absence of B lymphocytes in a horse with primary agammaglobulinemia, *Clin. Immunol. Immunopathol.* **5**:282.

Borzy, M. S., Schulte-Wissermann, H., Gilbert, E., Horowitz, S. D., Pellett, J., and Hong, R., 1979, Thymic morphology in immunodeficiency diseases: Results of thymic biopsies, *Clin. Immunol. Immunopathol.* **12**:31.

Buckley, R. H., 1977, Immunoreconstitution, *Pediatr. Clin. North Am.* **24**:313.

Buening, G. M., Perryman, L. E., and McGuire, T. C., 1978, Immunoglobulins and secretory component in the external secretions of foals with combined immunodeficiency, *Infect. Immun.* **19**:695.

Castles, J. J., Gershwin, M. E., Saito, W., Ardans, A., and Osburn, B., 1977, The activity of purine salvage pathway enzymes in murine and horse models of congenital and acquired dysimmunity, *Dev. Comp. Immunol.* **1**:165.

Clark, E. G., Turner, A. S., Boysen, B. G., and Rouse, B. T., 1978, Listeriosis in an Arabian foal with combined immunodeficiency, *J. Am. Vet. Med. Assoc.* **172**:363.

Clayton, F. W., 1976, Combined (B & T-lymphocyte) immunodeficiency in an Arabian foal, *Can. Vet. J.* **17**:26.

Deem, D. A., Traver, D. S., Thacker, H. L., and Perryman, L. E., 1979, Agammaglobulinemia in a horse, *J. Am. Vet. Med. Assoc.* **175**:469.

Giblett, E. R., Anderson, J. E., Cohen, F., Pollara, B., and Meuwissen, H. J., 1972, Adenosine-deaminase deficiency in two patients with severely impaired cellular immunity, *Lancet* **2**:1067.

Helms, C. M., and Allen, P. Z., 1971, A comparative immunological examination of some immunoglobulins of several equine species, *Comp. Biochem. Physiol.* **38**:439.

Henry, J. N., and Gagnon, A. N., 1976, Adenovirus pneumonia in an Arabian foal. Case report, *Can. Vet. J.* **17**:220.

Hirschhorn, R., and Martin D. W., 1978, Enzyme defects in immunodeficiency disease, *Springer Seminars Immunopathol.* **1**:299.

Hodgin, E. C., McGuire, T. C., Perryman, L. E., and Grant, B. D., 1978, Evaluation of delayed hypersensitivity responses in normal horses and immunodeficient foals, *Am. J. Vet. Res.* **39**:1161.

Hong, R., Santosham, M., Schulte-Wissermann, H., Horowitz, S., Hsu, S. H., and Winkelstein, J. A., 1976, Reconstitution of B and T lymphocyte function in severe combined immunodeficiency disease after transplantation with thymic epithelium, *Lancet* **2**:1270.

Johnson, K. G., and Hutchins, D. R., 1967, Suspected adenovirus bronchitis in Arabian foals, *Aust. Vet. J.* **43**:600.

Lew, A. M., Hosking, C. S., and Studdert, M. J., 1980, Immunologic aspects of combined immunodeficiency disease in Arabian Foals, *Am. J. Vet. Res.* **41**:1161.

McChesney, A. E., England, J. J., Adcock, J. L., Stackhouse, L. L., and Chow, T. L., 1970, Adenoviral infection in suckling Arabian foals, *Pathol. Vet.* **7**:547.

McChesney, A. E., England, J. J., Whiteman, C. E., Adcock, J. L., Rich, L. J., and Chow, T. L., 1974, Experimental transmission of equine adenovirus in Arabian and non-Arabian foals, *Am. J. Vet. Res.* **35**:1015.

McGuire, T. C., and Crawford, T. B., 1973, Passive immunity in the foal: Measurement of immunoglobulin classes and specific antibody, *Am. J. Vet. Res.* **34**:1299.

McGuire, T. C., and Poppie, M. J., 1973, Hypogammaglobulinemia and thymic hypoplasia in horses: A primary combined immunodeficiency disorder, *Infect. Immun.* **8**:272.

McGuire, T. C., Poppie, M. J., and Banks, K. L., 1974, Combined (B- and T-lymphocyte) immunodeficiency: A fatal genetic disease in Arabian foals, *J. Am. Vet. Med. Assoc.* **164**:70.

McGuire, T. C., Banks, K. L., and Poppie, M. J., 1975a, Combined immunodeficiency in horses: Characterization of the lymphocyte defect, *Clin. Immunol. Immunopathol.* **3**:555.

McGuire, T. C., Poppie, M. J., and Banks, K. L., 1975b, Hypogammaglobulinemia predisposing to infection in foals, *J. Am. Vet. Med. Assoc.* **166**:71.

McGuire, T. C., Banks, K. L., and Davis, W. C., 1976a, Alterations of the thymus and other lymphoid tissue in young horses with combined immunodeficiency, *Am. J. Pathol.* **84**:39.

McGuire, T. C., Banks, K. L., Evans, D. R., and Poppie, M. J., 1976b, Agammaglobulinemia in a horse with evidence of functional T lymphocytes, *Am. J. Vet. Res.* **37**:41.

McGuire, T. C., Pollara, B., Moore, J. J., and Poppie, M. J., 1976c, Evaluation of adenosine deaminase and other purine salvage pathway enzymes in horses with combined immunodeficiency, *Infect. Immun.* **13**:995.

McGuire, T. C., Crawford, T. B., Hallowell, A. L., and Macomber, L. E., 1977, Failure of colostral immunoglobulin transfer as an explanation for most infections and deaths of neonatal foals, *J. Am. Vet. Med. Assoc.* **170**:1302.

Magnuson, N. S., and Perryman, L. E., 1979, *In vitro* effects of adenosine on lymphocytes and erythrocytes from horses with combined immunodeficiency, *J. Clin. Invest.* **64**:89.

Pahwa, R., Pahwa, S., O'Reilly, R., and Good, R. A., 1978, Treatment of the immunodeficiency diseases. Progress toward replacement therapy emphasizing cellular and macromolecular engineering, *Springer Seminars Immunopathol.* **1**:355.

Perryman, L. E., 1979, Primary and secondary immune deficiencies of domestic animals, *Adv. Vet. Sci. Comp. Med.* **23**:23.

Perryman, L. E., 1981, Use of fetal tissues for immunoreconstitution in horses with severe combined immunodeficiency, in: *Fetal Liver Transplantation* (G. Lucarelli, T. Fleidner, and R. P. Gale, eds.), Excerpta Medica, Amsterdam, in press.

Perryman, L. E., and Liu, I. K. M., 1980, Graft-versus-host reactions in foals with combined immunodeficiency, *Am. J. Vet. Res.* **41**:187.

Perryman, L. E., and McGuire, T. C., 1978, Mixed lymphocyte culture responses in combined immunodeficiency of horses, *Transplantation* **25:**50.

Perryman, L. E., and McGuire, T. C., 1980, Evaluation of immune system: Failures in horses and ponies, *J. Am. Vet. Med. Assoc.* **176:**1374.

Perryman, L. E., and Torbeck, R. L., 1980, Combined immunodeficiency of Arabian horses: Confirmation of an autosomal recessive mode of inheritance, *J. Am. Vet. Med. Assoc.* **176:**1250.

Perryman, L. E., McGuire, T. C., and Hilbert, B. J., 1977, Selective immunoglobulin M deficiency in foals, *J. Am. Vet. Med. Assoc.* **170:**212.

Perryman, L. E., McGuire, T. C., and Crawford, T. B., 1978, Maintenance of foals with combined immunodeficiency: Causes and control of secondary infections, *Am. J. Vet. Res.* **39:**1043.

Perryman, L. E., Magnuson, N. S., and Torbeck, R. L., 1979a, Combined immunodeficiency in horses, in: *Birth Defects Symposium IX. Inborn Errors of Specific Immunity* (B. Pollara, R. J. Pickering, H. J. Meuwissen, and I. H. Portar, eds.), pp. 391–401, Academic Press, New York.

Perryman, L. E., Buening, G. M., McGuire, T. C., Torbeck, R. L., Poppie, M. J., and Sale, G. E., 1979b, Fetal tissue transplantation for immunotherapy of combined immunodeficiency in horses, *Clin. Immunol. Immunopathol.* **12:**238.

Perryman, L. E., McGuire, T. C., and Torbeck, R. L., 1980, Ontogeny of lymphocyte function in the fetal horse, *Am. J. Vet. Res.* **41:**1197.

Pickering, R. J., Pollara, B., and Meuwissen, H. J., 1974, Workshop in severe combined immunological deficiency disease and adenosine deaminase deficiency, *Clin. Immunol. Immunopathol.* **3:**301.

Polmar, S. H., Stern, R. C., Schwartz, A. L., Wetzler, E. M., Chase, P. A., and Hirschhorn, R., 1976, Enzyme replacement therapy for adenosine deaminase deficiency and severe combined immunodeficiency, *N. Engl. J. Med.* **295:**1337.

Poppie, M. J., and McGuire, T. C., 1976, Combined immunodeficiency with a failure of colostral immunoglobulin transfer in foals, *Vet. Rec.* **99:**44.

Poppie, M. J., and McGuire, T. C., 1977, Combined immunodeficiency in foals of Arabian breeding: Evaluation of mode of inheritance and estimation of prevalence of affected foals and carrier mares and stallions, *J. Am. Vet. Med. Assoc.* **170:**31.

Radwan, A. I., and Crawford, T. B., 1974, The mechanisms of neutralization of sensitized equine arteritis virus by complement components, *J. Gen. Virol.* **25:**229.

Rejnek, J., Prokesova, L., Sterzl, J., and Matousek, V., 1973, The presence of IgG and IgM in full term horse umbilical cord sera, *Immunochemistry* **10:**397.

Snyder, S. P., England, J. J., and McChesney, A. E., 1978, Cryptosporidiosis in immunodeficient Arabian foals, *Vet. Pathol.* **15:**12.

Splitter, G. A., Incefy, G. S., Dardanne, M., Iwata, T., and McGuire, T. C., 1979a, Arabian horses with severe combined immunodeficiency: Evaluation of functional thymic hormones, *Dev. Comp. Immunol.* **3:**359.

Splitter, G. A., Incefy, G. S., Iwata, T., and McGuire, T. C., 1979b, Evaluation of functional thymic hormones in Arabian horses with severe combined immunodeficiency, *Clin. Exp. Immunol.* **38:**37.

Studdert, M. J., 1978, Primary, severe, combined immunodeficiency disease of Arabian foals, *Aust. Vet. J.* **5:**411.

Thompson, D. B., Studdert, M. J., Beiharz, R. G., and Littlejohns, I. R., 1975, Inheritance of a lethal immunodeficiency disease of Arabian foals, *Aust. Vet. J.* **51:**109.

Whitwell, K. E., 1978, Combined immunodeficiency in Arabian foals, *Vet. Rec.* **103:**568.

Winchester, R. J., Fu, S. M., Hoffman, T., and Kunkel, H. G., 1975, IgG on lymphocyte surfaces: Technical problems and the significance of a third cell population, *J. Immunol.* **114:**1210.

PART VIII
DEFECTS IN EFFECTOR PATHWAYS

9
Complement Deficiencies of Laboratory Animals

CARL H. HAMMER, THELMA GAITHER, AND MICHAEL M. FRANK

1. Introduction

Complement is a collective term used to designate a series of plasma glycoproteins whose activation is responsible for many features of the inflammatory response as well as for a number of aspects of host defense. The complement proteins interact in a precise sequence of activation steps that lead to the formation of a series of complement protein cleavage fragments (Fig. 1). At several points in the cascade the interaction of these cleavage fragments leads to the assembly of complex enzymes that function to continue the reaction sequence. Many of the smaller complement-derived cleavage fragments formed during activation have important biological properties mediating one or another feature of the inflammatory process. A unique aspect of the later steps in the complement cascade is that the activated terminal proteins become inserted into the lipid bilayer of many biological membranes, thereby forming holes in the membrane and interrupting its biological integrity.

Two major mechanisms for activation of the complement cascade are currently recognized. One phylogenetically older and perhaps more primitive mechanism, termed the alternative complement pathway, can be activated by a number of biochemical substances. Some, with precisely defined structures such as certain repeating polysaccharide units, can be found in certain bacterial or fungal cell walls. This pathway does not have an absolute requirement for antibody. The second pathway, which is thought to be phylogenetically newer, is activated by antigen–antibody complexes containing antibody of certain classes and subclasses (IgM and IgG subclasses 1, 2, and 3 in man). This interaction leads to the

CARL H. HAMMER, THELMA GAITHER, AND MICHAEL M. FRANK • Laboratory of Clinical Investigation, National Institute of Allergy and Infectious Diseases, National Institutes of Health, Bethesda, Maryland 20205.

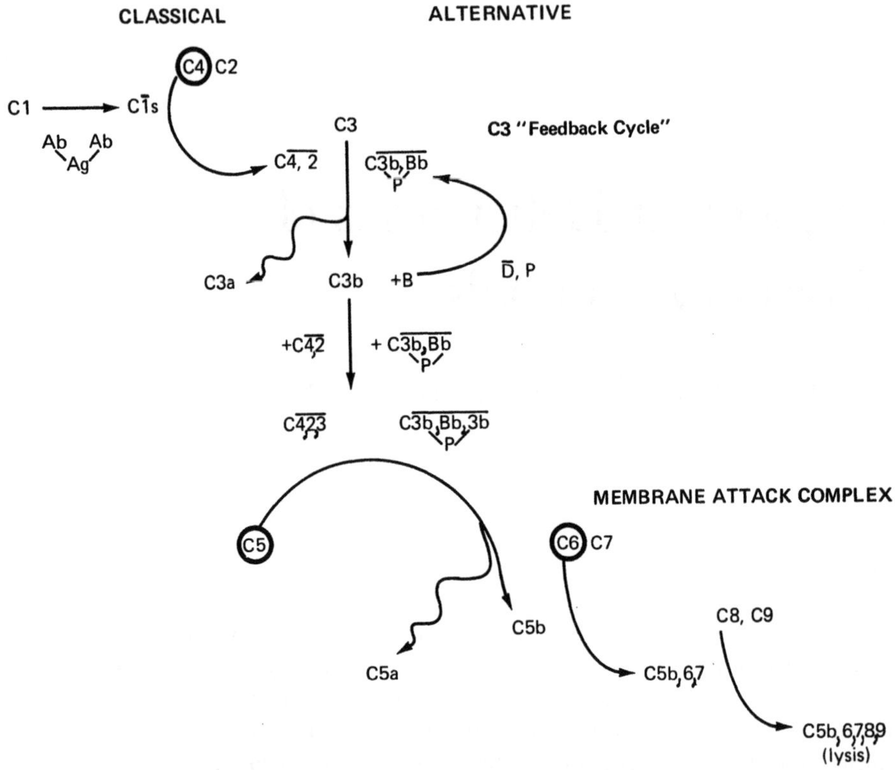

FIGURE 1. The complement cascade. Shown are both classical and alternative pathways as well as the terminal steps in the sequence, termed the membrane attack complex, and the C3 feedback cycle. In the classical pathway antigen combined with 2 molecules of IgG antibody interacts with C1. This interaction leads to activation of the subunit C1s to the enzymatically active $C\bar{1}s$. This in turn cleaves C2 and C4 to form a new enzyme $\overline{C4,2}$, which cleaves C3 into C3a and C3b. $\overline{C4,2}$ in the presence of C3b forms a new enzyme capable of coordinating with and cleaving C5. One C5 fragment, C5b, in turn reacts with C6, C7, C8, and C9 to cause lysis. In the alternative pathway an enzyme analogous to $\overline{C4,2}$ is formed from C3b and the $\bar{D}$ cleavage fragment of B, Bb, designated $\overline{C3bBb}$. This enzyme is stabilized by properdin (P). It is capable of further C3 cleavage. The C3b resulting from this cleavage can feedback to form additional $\overline{C3bBb}$. The most well-studied deficiencies are circled.

rapid and efficient activation of the so-called early-acting components of the classical pathway. Both activation sequences lead to the generation of an enzyme capable of cleaving and activating C3, a key component of both pathways. The C3 is cleaved into two fragments termed C3a and C3b. The former can cause marked vasodilation via histamine release from mast cells. The latter plays a major role in the process of opsonization. The immediate interaction of C3b with the surface of a suitable acceptor particle leads to coating of the particle with C3b. This C3b is capable of interacting with cells with C3b receptors. These receptors are present on the surface of most phagocytic cells as well as other cells such as B lymphocytes, etc. The interaction of the receptors with particle-bound C3b leads to adherence of the particle to the phagocytic cell membrane, thereby facilitating phagocytosis. The interaction of C3b with the B-lymphocyte C3b receptor triggers these cells to release lymphokines, soluble substances that regulate the state of activation and the metabolic state of cells of the immune system. In this way complement activation acts as a bridge between the humoral and the cellular immune systems.

Activation of complement components generates a number of fragments that possess important biological activities. Thus, the smaller fragments generated by C4, C3, and C5 cleavage act as "anaphylatoxins," permeability factors that cause mast cells to degranulate and release their mediators by noncytotoxic mechanisms. Several factors, most particularly C5a, act as chemotactic factors and induce the directed migration of phagocytic cells like neutrophils and macrophages toward the phagocytic stimulus. A fragment of C3, C3e, is also reported to induce leukocytosis via leukocyte release from the bone marrow (Ghebrehiwet and Müller-Eberhard, 1979). Thus, activation of complement can produce a number of effects that in concert act to promote host defense. The generation of small activation fragments mediates many features of inflammation. The opsonization of particles aids phagocytosis. The action of the complete sequence may induce lysis of the offending microbe.

Much of our present information about the function of complement and its activity in both host protection and the induction of pathologic states comes from the careful evaluation of genetically controlled defects in the complement cascade that have been discovered in animals—so-called experiments of nature. Because of the precise biochemical specificity of the complement protein interactions, a defect in synthesis of a protein will usually interrupt the cascade of reactions at the point of the defect. Evaluation of the biologic functions of complement in animals with precisely defined defects allows one to gain insight into complement function both normally and in disease states. In this chapter we review the well-studied defects known to exist in animals. We correlate this information with our current understanding of complement component biochemistry and the function of the components in biological systems. We also review the available information about the genetic control of the defect, our current understanding of component synthesis, and similar defects as defined in man (Agnello, 1978). It should be understood that much of our understanding of the defects that have been defined in man is based on these studies of component defects and their effects in animals.

2. The Well-Studied Deficiency States

2.1. C4 Deficiency in Guinea Pigs

2.1.1. History and Genetics

The first report of a complement defect in a laboratory animal dates back to 1919 when H. D. Moore described a guinea pig strain whose serum was unable to lyse antibody-sensitized horse erythrocytes. It was further shown that serum from these animals was significantly depressed in the ability to opsonize bacteria for ingestion by guinea pig leukocytes and that complement-deficient animals were more susceptible to killing by injection of *B. cholerae suis* than were normal guinea pigs. Moore implicated an autosomal recessive transmittance of the complement deficiency trait, but the data concerning the inheritance of the defect were not available until 1923 when R. R. Hyde reported the experimental results of F. A. Rich that clearly showed a simple Mendelian recessive inheritance. Hyde observed that incubation of yeast with normal guinea pig serum, which at that time was know to remove the activity of the classical third component of complement (later shown to consist of C3 through C9), produced a reagent no longer able to correct the defect in the deficient serum. Heated serum (56°C, 30 min) still retained the capacity to correct the defect. He believed that the serum was deficient in "C3." It was not until 1926 that C4, another heat-stable component of the complement system, was discovered. In light of our present knowledge of C1 and C2 sensitivity to heat inactivation as compared to the thermal stability of C4, as well as C3–9, it is remotely possible that the complement deficiency resided in the fourth component. However, Hyde's deficient strain of guinea pigs subsequently died out and an exact determination of their true defect is no longer possible.

A deficiency of the fourth component of complement was discovered in an outbred closed colony of guinea pigs at the NIH (termed NIH multipurpose) (Ellman *et al.*, 1970). In experiments designed for entirely different purposes, one guinea pig from the NIH closed colony was noted to respond to the injection of opsonized bacteria with the development of a precipitating antibody that would react with the serum of several inbred strains of guinea pigs. Analysis of this antiserum demonstrated its capacity to precipitate partially purified C4 on Ouchterlony analysis and to agglutinate sheep erythrocytes containing hemolysin and complement components C1 and C4 ($EAC\overline{1}$,4b)* but not cells lacking C4 ($EAC\overline{1}$). Analysis of the deficient serum failed to detect functional C4 or total hemolytic complement activity, all other component activities being present. On screening, 2% of the closed colony of animals were shown to be lacking C4 antigen and were also totally deficient in C4 functional activity. The possibility that an inhibitor of C4 was present in the serum of the deficient guinea pigs was ruled out, for the addition of either purified C4 or normal guinea pig serum could restore functional activity to deficient serum.

Mating studies with C4-deficient animals gave results clearly demonstrating that the C4 deficiency is transmitted as a simple autosomal trait. When a C4-deficient male animal was mated with a normal guinea pig, all four F_1 progeny (two male, two female) were judged to have intermediate functional and antigenic levels of C4. Matings of animals that were both lacking in C4 resulted in progeny totally lacking in C4 function and antigenic protein. In a reported backcross of an F_1 female with a deficient male parent, three offspring were produced. One, a female, demonstrated reduced C4 function and antigen levels while the two males had no detectable levels of C4. A series of other matings all demonstrated simple Mendelian genetics. This deficiency state in animals resembles that reported in man. In studies of an 18-year-old girl and in another family a 7-year-old boy, the genetic defect was shown to be transmitted as an autosomal recessive trait and members of both families had decreased levels of serum C4 (Hauptman *et al.*, 1974; Ochs *et al.*, 1977; Schaller *et al.*, 1977).

Shevach *et al.* (1976) were able to demonstrate in mating studies with C4 heterozygous and normal guinea pigs that the gene controlling C4 synthesis is closely linked to the

*The symbols E, A, and C represent respectively a sheep erythrocyte, one molecule of IgM or two molecules of IgG antibody, and complement molecules, the individual complement components being designated by number, i.e., C1, C2, etc. The letters a and b are used to designate fragments of the complement molecules. The bar over certain complement components indicates enzyme activity.

so-called *GPLA,* the major histocompatibility locus of the guinea pig. Furthermore, the C4-deficient guinea pigs appeared to have the same haplotype as inbred strain-13 animals, as cytotoxic alloantiserum could not be raised by reciprocal immunization of the two strains. This was in contrast to strain-2 guinea pigs, which clearly differed at this locus. Additional support for their gentic identity at this locus was demonstrated by the ability of inbred C4-deficient animals bearing the B1 antigen to accept a syngeneic strain-13 sarcoma. Recently, genetic polymorphism of guinea pig complement protein C4 was shown, which was electrophoretically detectable only in plasma samples (Bitter-Suermann *et al.,* 1977). Three different guinea pig C4 variants were observed, designated C4-F, C4-S, and C4-S1. From breeding experiments their mode of inheritance was shown to be under the control of three codominant alleles of a single autosomal locus. It was also shown that factor B polymorphism could be identified as two protein variants designated Bf-S and Bf-F. Studies of inheritance of Bf types showed that these allotypes were controlled by two alleles at a single locus. Individual factor B and C4 allotypes were compared within a given sample. The results of 175 such analyses revealed close linkage between the C4 and the factor B loci; Bf-S and C4-F types and Bf-F and C4-S or C4-S1 types are always associated. As strain-13 guinea pigs show a C4-FF electrophoretic pattern and these animals have the same haplotype as C4-deficient guinea pigs (Shevach *et al.,* 1976), which show Bf-SS phenotype consistently, the authors' speculation that the C4 deficiency arose from a mutation of a C4-F gene is reasonable. Subsequently these authors presented inheritance data for C4, factor B, and GPLA allotypes that linked both the Bf and the C4 loci to the *GPLA* (Krönke *et al.,* 1977). Linkage of genes controlling synthesis of human C4 and the major histocompatibility locus of man *(HLA)* has also been demonstrated in family studies of C4 deficiency (Rittner *et al.,* 1975; Ochs *et al.,* 1977) and by segregation studies of structural genetic polymorphisms of C4 in normal families (Teisberg *et al.,* 1976, 1977; O'Neill *et al.,* 1978). In the mouse, it was shown that a serum substance, Ss protein a product of the mouse major histocompatibility complex *(H-2)* (Schreffler and Owen, 1963), is probably the fourth component of mouse complement (Meo *et al.,* 1975; Carroll and Capra, 1978; Parker *et al.,* 1979).

2.1.2. Biochemistry and Synthesis

The fourth component of complement was discovered in 1926 by Gordon and coworkers. These authors found that ammonia inactivated a principle (the fourth component) different from the other three known components of complement (C1, C2, "C3") and this was termed C4. However, it was not until 1963 that Müller-Eberhard and Biro physically isolated and characterized the fourth component of complement from human serum as a distinct protein, beta 1E globulin (β1E). The assignment of C4 activity to the β1E molecule was demonstrated by the coincidental electrophoretic mobility, sedimentation, and chromatographic properties of the two entities. Further evidence of their identity was shown by the ability of sheep erythrocytes sensitized with antibody and carrying the first component of complement to adsorb β1E in a diisopropyl fluorophosphate (DFP)-inhibitable and dose-dependent reaction. Subsequent lysis of EAC$\overline{1}$, β1E by a C4-deficient serum could be blocked by antibody specific for β1E.

When subjected to physicochemical characterization, purified human C4 (mol. wt. 209,000) was shown to consist of three unequal, covalently linked, polypeptide chains, α,

β, and γ, with apparent molecular weights of 93,000, 78,000, and 33,000, respectively (Schreiber and Müller-Eberhard, 1974). Subsequent reports have confirmed these values and further characterize the fourth component as a glycoprotein containing as much as 8% carbohydrate associated with the α and β chains and possibly the γ chain (Bolotin *et al.*, 1977; Gigli *et al.*, 1977; Hall and Colten, 1977a). The molecular size and chain structure of purified guinea pig C4 have been shown to bc similar to those of human C4 (Hall and Colten, 1977a).

The C4 molecule plays a critical role in the assembly of two important enzymes of the classical pathway of complement activation, C3-convertase ($\overline{\text{C4b,2a}}$) and C5-convertase ($\overline{\text{C4b,2a,3b}}$). These complex proteases are responsible for cleavage of their natural substrates C3 and C5 with the attendant production of inflammatory products C3a and C5a, respectively. In addition, C5 cleavage results in the generation of C5b, a large split fragment capable of binding C6 and initiating formation of the membrane attack complex C5b–9 (Müller-Eberhard, 1975; Mayer, 1978; Mayer *et al.*, 1979).

The enzymatic nature of activated C$\overline{1}$ was demonstrated by several laboratories in the early 1950s by its capacity to hydrolyze synthetic esters, inactivate C4 and C2 functional activity, and its susceptibility to the serine esterase inhibitor DFP (Levine, 1955; Becker, 1956a,b; Lepow *et al.*, 1956). The later observation that this protease was responsible for cleavage and binding of complement components C4 and C2 was initially hypothesized by Mayer *et al.* (1962) for guinea pig C2. It was later substantiated in a series of studies employing an antibody to C2 that was capable of blocking the $\overline{\text{C4b,2a}}$ enzyme (Sitomer *et al.*, 1966; Stroud *et al.*, 1966). The assembly of the human C3-convertase counterpart with highly purified components C4 and C2 was demonstrated by Müller-Eberhard and colleagues. Activation and binding of C4 by C$\overline{1}$ to a cell-surface receptor (Müller-Eberhard and Biro, 1963) were later shown to result from the cleavage of C4 to C4b, a 202,000-dalton fragment with transient binding capacity for the activating surface (Budzko and Müller-Eberhard, 1970; Patrick *et al.*, 1970). The formation of C4b initiates formation of the C3-cleaving enzyme. A small, biologically active peptide, C4a, of 6000 daltons obtained from the α chain of C4, is released into the fluid phase on C4 cleavage. The capacity of fluid-phase C4b or cell-bound C4b to bind C2, a reversible step potentiated by Mg^{2+} ions (Sitomer *et al.*, 1966; Müller-Eberhard *et al.*, 1967), is a prerequisite for efficient cleavage of C2 into two large fragments with the subsequent generation of protease activity. One product, C2a (76,000 daltons), combines with C4b (Stroud *et al.*, 1966; Polley and Müller-Eberhard, 1968) to generate the complex $\overline{\text{C4b,2a}}$ that has C3-cleaving capacity (Polley and Müller-Eberhard, 1968; Shin and Mayer, 1968).

Cleavage of C3 by the $\overline{\text{C4b,2a}}$ enzyme generates a new enzyme $\overline{\text{C4b,2a,3b}}$, or C5-convertase, formed by association of $\overline{\text{C4b,2a}}$ with its parent split product, C3b, bound on the cell surface near the enzymic site (Cooper and Müller-Eberhard, 1970; Shin *et al.*, 1971a,b). Work from several laboratories suggests that the C2a subunit of both enzymes contains the active site for their respective substrates (Goldlust *et al.*, 1974; Cooper, 1975; Medicus *et al.*, 1976a). It may be inferred from the above findings that a principal role of the C4b subunit within the enzymic complex is to subserve the stabilization of the C2a fragment in a conformation capable of preserving the proteolytic site. This concept is supported by the known capacity of activated C$\overline{1}$ to cleave C2 in the fluid phase whether or not C4b is present, but only in the presence of C4b (i.e., $\overline{\text{C4b,2a}}$) does it generate C3-cleaving activity (Polley and Müller-Eberhard, 1968).

Regulation of the capacity of cell-bound C4b to participate in generation of C3-convertase activity with all attendant biological consequences has only just recently begun to be understood. Several laboratories have shown that this control is effected by two serum proteins, one being an enzyme designated C3bINA known also to participate in C3b inactivation and a newly discovered cofactor called C4b-binding protein (Shiraishi and Stroud, 1975; Nagasawa *et al.*, 1976; Fugita and Nussenzweig, 1979).

The site(s) of C4 biosynthesis in both guinea pigs and humans has been under investigation for the past 25 years. Using immunochemical methods to detect C4 production by incorporation of labeled amino acids or by sensitive functional C4 plaque assays, a number of laboratories have shown that the macrophage or macrophagelike cell from various tissues is the primary site of C4 biosynthesis (Colten, 1976). Early studies reported by Thorbecke and co-workers have shown that ^{14}C-labeled amino acids could be incorporated into protein immunochemically identical with C4 by a variety of primate cell types during culture. These authors found that peritoneal macrophages, fibroblasts, kidney and liver cultures, thymoma tissue, and an epitheliallike cell line were capable of C4 production (Stecher and Thorbecke, 1967a,b; Stecher *et al.*, 1967). In contrast to Thorbecke *et al.* (1965) who could not demonstrate C4 synthesis in lymphoid tissue, Lai A Fat and Furth (1975) were able to show incorporation of ^{14}C-labeled amino acids into C4 produced in human spleen and some lymph node cultures. *In vitro* production of functionally active guinea pig C4 from bone marrow, spleen, and liver slice cultures was first reported by Siboo and Vas (1965) and also from liver slices later by Jenson *et al.* (1971).

Synthesis of C4 by tissues of the reticuloendothelial system of the guinea pig has been investigated in detail over the last 10 years by Burkholder and co-workers. Using a modified Jerne technique, these investigators were able to identify the cell type(s) involved in *in vitro* C4 production by the formation of hemolytic plaques. Their initial investigations identified macrophages from peritoneal exudates, bronchopulmonary washings, and suspensions of cells prepared from liver as the site(s) of C4 synthesis (Littleton *et al.*, 1970). Subsequently, by the use of density-gradient centrifugation a small population of cells from the spleen and liver characterized by their low density were isolated and shown to synthesize C4. These were large mononuclear cells and were restricted to the region of the gradient occupied by phagocytic cells (Ilgen and Burkholder, 1974a; Ilgen *et al.*, 1974). In these earlier reports, either puromycin or cycloheximide was shown to inhibit the C4-dependent plaque formation and thus the presence of C4 was concluded to result as a consequence of *in vitro* synthesis rather than release of preformed protein. A later study reported that these C4-synthesizing cells could also be inhibited by small quantities of specific antibody to C4 (Ilgen and Burkholder, 1974b). This finding was interpreted to indicate that a membrane determinant is present on C4-synthesizing cells which is similar to C4 and which may regulate C4 synthesis. More recent work has been directed at elucidating the ultrastructure of plaque-forming cells (primarily peritoneal and alveolar macrophages) by the use of immunolabeling techniques and electron microscopy (Barber *et al.*, 1976; Barber and Burkholder, 1978). In studies of C4-producing alveolar cells, immunolabeling allowed the visualization of C4 in the endoplasmic reticulum (ER), perinuclear space and Golgi complex of cells, the increased amounts of ER being consistent with a cell actively engaged in protein synthesis.

The earlier evidence that the guinea pig macrophage or macrophagelike cell was responsible for C4 production was confirmed by Colten (1972a). He showed that human fetal liver cells (abundant in large mononuclear types) actively synthesize C2 and C4. With a plaque assay using rat EDTA-serum as a source of late components to improve the sensitivity of detection of C2 and C4, Wyatt *et al.* (1972) identified guinea pig peritoneal cells capable of producing both C2 and C4 simultaneously.

The discovery of C4 deficiency in guinea pigs presented an ideal model system in which the activities of the alternative and classical pathways of complement activation may be separated and studied both *in vitro* and *in vivo*. Additionally, specific information concerning the genetic control and biosynthesis of this well-characterized complement component could be obtained by the appropriate use of modern tissue culture techniques and sensitive functional assays for complement activity.

The early studies of Frank *et al.* (1971) on the complement component levels in homozygous C4-deficient guinea pigs showed that the serum from these animals contained only about one-half the amount of C2 and as little as 10–20% the amount of C1 as was present in normal guinea pig serum. The C3–9 titers of C4-deficient animals were normal. Previous studies of the biosynthesis of both C4 and C2 by macrophage or macrophagelike cells of guinea pigs described in the preceding section suggested that study of this cell type would be of interest in C4-deficient animals. The results of *in vitro* biosynthetic experiments designed to analyze the synthetic capability of cells from homozygous and heterozygous C4-deficient animals showed good correlations between synthetic rates and the C4 and C2 levels of serum from the respective animals. Thus, cells isolated from homozygous deficient animals produced no C4 *in vitro* and C4 production was less than normal in the heterozygote. Likewise, C2 synthesis was only 40% of normal with cells from the C4-deficient animal (Colten and Frank, 1972). In subsequent studies, hybrids of C4-deficient peritoneal exudate (PE) cells were formed *in vitro* by fusion with a cell line of human origin (HeLa). As earlier shown for C4-deficient PE cells, HeLa cells alone produced no detectable C4 but some resulting hybrid cells synthesized functional C4 of *human* origin (Colten and Parkman, 1972). It was further shown that the C4-deficient guinea pigs produced a 45,000-dalton, heat-stable soluble factor that induced C4 production in the human cell (Colten, 1972b, 1976). While this regulator of C4 synthesis was produced by normal cells, significantly more could be recovered from C4-deficient cells, presumably indicating a quantitative difference in production. One interpretation presented by the authors for the production of human C4 by these hybrids is that C4 deficiency arises from a C4 structural gene defect. A regulator gene may control and augment the expression of the gene product. The C4-deficient cell might overproduce this factor, but in the absence of a structural C4 gene, could not make C4 protein. This regulator may induce C4 synthesis in the hybrid cell which has the information for C4 human synthesis in the genome.

Recent evidence obtained from a cell-free biosynthetic system suggests that guinea pig C4 is synthesized as a single-stranded precursor, pro-C4 (200,000 daltons), that subsequent to release from the polysome is converted to three-chain C4 (Hall and Colten, 1977b). Pro-C4 is not detected in C4-deficient cells although some antigenic C4 fragments are synthesized (Hall and Colten, 1977b). The detection of intracellular pro-C4 in short-term cultures of guinea pig liver and the report that 5% of the total C4 in normal human serum may represent the human analog of pro-C4 (Gigli, 1978) lead one to speculate that conversion occurs only during transit from the cell, possibly as a consequence of glycosylation. The finding that cell-free synthesized or intracellular precursor C4 is smaller than the C4 that is secreted in culture may be accounted for by the presence of carbohydrate in the latter protein as is found associated with human C4.

2.1.3. Biological Consequences

The C4-deficient guinea pigs were discovered at a time when a renewal of interest occurred in the possible existence of an alternative pathway of complement activation. Experiments by Gewurz *et al.* (1968a) had suggested that incubation of fresh normal serum with endotoxic lipopolysaccharide produced inactivation of the later-acting complement components, C3 through C9, without significant consumption of C1, C4, and C2. An alternative pathway of activation bypassing utilization of C1, C4, and C2 but responsible for C3–9 consumption was also suggested in several other studies (Marcus *et al.,* 1970; Sandberg *et al.,* 1970), but conclusive evidence for absolute blockade of the early classical components was not demonstrated. Initial studies by Frank *et al.* (1971) utilizing sera from C4-deficient guinea pigs, entirely lacking in native C4 function or antigen, showed that *E. coli* endotoxin and antigen–antibody complexes of HSA-rabbit anti-HSA were able to fix or activate the late-acting complement components. Endotoxin consumed C3–9 in C4-deficient serum almost as well as in normal guinea pig serum, while activation with complexes was only about half as efficient. The authors concluded that an alternative pathway for terminal-component activation must be present in the sera of deficient animals. Extensive investigation of complement-dependent processes in the C4-deficient guinea pigs and in their serum has identified those functions that require an intact classical pathway. Many immunologic parameters studied were found to be normal in C4-deficient animals while a number of functions were shown to be impaired or absent. It was found that *in vivo* immune effector responses such as passive cutaneous anaphylaxis elicited by DNP-bovine serum albumin injection, contact and delayed hypersensitivity, and the cellular exudative response to a foreign body were normal (Ellman *et al.,* 1971). Passive cutaneous anaphylaxis and delayed hypersensitivity responses are not known to require complement. The normal Arthus-type reactions that were also observed were indicative of the involvement of an alternative pathway of activation. This reaction is believed to result from the complement-dependent accumulation of neutrophils in the walls of venules as a result of C3–9 activation, the deposition of C3, and the concomitant generation of chemotactic factors (Cochrane, 1968; Müller-Eberhard, 1968). It was found that like C4-sufficient animals, C4-deficient animals were able to deposit C3b at sites of induced direct Arthus reactions (Frank *et al.,* 1973).

The normal Arthus response of C4-deficient guinea pigs prompted a more thorough examination of the nature, origin, and kinetics of chemotactic factor generation in the serum of both normal and deficient animals. While the alternate pathway was capable of mediating generation of chemotactic factors by endotoxin or immune complexes, it was clearly demonstrated that the presence of an intact classical pathway was required for rapid generation of chemotactic activity initiated by immune complexes (Clark *et al.,* 1973). The factors responsible for polymorphonuclear chemotactic activity were shown to be products of C5 and not C3 cleavage. Similar findings were obtained in studies of chemotactic factor generation in normal human serum and human serum deficient in C2 (Gallin *et al.,* 1975). The authors found rapid production of chemotactic factor (C5a) when initiated by antibody and classical pathway components but only after a latent period when mediated by the alternative pathway. In contrast, others have found that serum from a patient with systemic lupus erythematosus and an inherited C4 deficiency was impaired in its ability to generate chemotactic activity under selected conditions (i.e., low serum concentration or short activation time) (Clark and Klebanoff, 1978). The extent of the defect was also modulated by thc type of activating agent employed; zymosan generation of chemotactic activity was proportional to serum concentrations, while endotoxin produced no activity during a 60-min incubation. The addition of purified C4 to the zymosan-activated serum restored the activity to only 60% of normal while little restoration was seen for endotoxin activation. The failure of C4 to correct completely the impaired chemotactic activity suggests that an additional defect besides C4 deficiency exists in this patient's serum.

C4-deficient animals were more susceptible to the lethal effects of endotoxin injection than normal guinea pigs and they survived about as well as cobra venom factor (CVF) (C3–9) depleted normal animals (May *et al.,* 1972a). C4-deficient CVF-treated animals fared more poorly. It was postulated that classical pathway activation contributes to endotoxin detoxification and requires in addition participation of C3 and/or other terminal components (C5–9) for full protection. Other *in vivo* experiments concerning the biological effects of endotoxin administration on platelets and blood coagulation in C4-deficient animals were able to dissect the roles of the alternative and classical pathways. Thus, while C4-sufficient guinea pigs experienced profound thrombocytopenia within minutes of endotoxin challenge, the platelet levels of C4-deficient animals remained normal (Kane *et al.,* 1973). Susceptibility to endotoxin-induced hypercoagulation was also found lacking in C4-deficient guinea pigs, although these animals had normal clotting capability. Both thrombocytopenia and hypercoagulability were shown to require activation of C3–9 as evidenced in experiments with CVF-treated normal animals. In coagulation studies in C4-deficient guinea pigs, Dodds *et al.* (1977) reported that intrinsic coagulation is slower than in normal controls. It was reported that prothrombin consumption times and factor VIII activity were also low. The combination of all three deficiencies was speculated to be the cause of the coagulation abnormality of C4-deficient animals.

The immune adherence reaction (IA) is defined as the agglutination of human erythrocytes in the presence of immune complex or cell-bound C3b. The similar binding properties to macrophages and neutrophils suggest that C3b is responsible for promotion of phagocytosis and resistance to infection. The existence of two pathways for generation of the opsonic fragment provided studies to assess the contribution of each in the production of IA (May *et al.,* 1972b). It was found that optimally sensitized red cell membranes became IA positive in normal but not C4-deficient serum, thus requiring the classial pathway. However, endotoxin or soluble immune complexes could generate reactivity equally well in both sera, implicating the alternative pathway.

Further studies demonstrated that the Forssman reaction, which causes death in normal guinea pigs upon injection of rabbit anti-Forssman antibody, did not occur in C4-deficient guinea pigs. Restoration with normal guinea pig serum as a source of C4 allowed the reaction to proceed once sufficient time had elapsed to allow C4 to diffuse into tissues. Here too, the classical pathway is responsible for biological effects presumably initiated by bound antigen–antibody complexes (May and Frank, 1972). The ineffectiveness of the alternative complement pathway in mediating *in vitro* antibody-dependent cytolytic or IA reactions by antimembrane antibodies (May *et al.,* 1972c) is compatible with the *in vivo* findings that C4-deficient animals are resistant to the induction of Forssman shock. Examination of C4-deficient guinea pig serum for lysis of a complement-sensitive strain of *Escherichia coli* showed a marked defect in the kinetics of killing but not in the overall extent (Root *et al.,* 1972). C4-deficient serum was also shown to opsonize a number of different microorgansms, but close examination using a kinetic assay established a phagocytic defect. From these studies it was concluded that for full expression of bacterial opsonization and killing an intact classical pathway is required. Others measuring opsonization of zymosan particles by C4-deficient guinea pig serum showed impaired opsonic function in dilute (5%) C4-deficient serum but not in 10% serum (Clark and Klebanoff, 1978). The kinetics of opsonization in this system also showed a defect early in the time course, although the endpoint in many of these reactions was normal.

In experiments conceptually similar to those of Root *et al.* (1972), a number of pathogenic vectors have since been investigated utilizing C4-deficient animals as a probe for studying the requirement of complement and its pathways of activation. Thus, in *Cryptococcus neoformans* infection both the classical and the alternative pathways were found necessary for optimal kinetics of opsonization, although the alternative pathway alone was sufficient for opsonization (Diamond *et al.,* 1974). In complement consumption assays with *Fusobacterium polymorphum* cell wall, an exclusive requirement of the alternative pathway was indicated (Hawley and Falkler, 1977). Studies by Wikel (1979) also demonstrated that only the alternative pathway of complement activation is important in development to resistance to the ixodid tick, *Dermacentor andersoni.* Protection from *Candida albicans* infection in nonimmune guinea pigs was also shown to require an intact alternative pathway including the terminal components of complement (Gelfand *et al.,* 1978). In contrast, *in vivo* studies by Strunk *et al.* (1977) of guinea pig infection with herpes simplex virus suggested that neither the alternative, classical, nor common pathways (C3–9) were important for virus clearance.

In the first studies performed with C4-deficient guinea pigs, clearance of isologous erythrocytes sensitized *in vitro* with specific antibody was shown to be impaired (Ellman *et al.,* 1971). By infusion of normal guinea pig serum the clearance defect was temporarily corrected. Subsequently, the clearance from the circulation by the reticuloendothelial system of ^{51}Cr-labeled erythrocytes coated with IgG and IgM complement-fixing antibodies has been studied in depth (Frank, 1977). These studies emphasize the importance of complement in mediating the clearance of antibody-coated erythrocytes. They demonstrate that complement activation greatly augments the clearance of IgG-coated erythrocytes and show that complement activation is an absolute requirement for the clearance of IgM-sensitized cells. In both of these situations activation of the classical rather than the alternative pathway is responsible for the complement-mediated effect.

Much investigation in the past decade has been directed at uncovering a possible role for complement (C3) in the induction of the allergic response (Pepys, 1976). The linkage of complement components C2, factor B, and C4 to the *MHC* of both the human and the guinea pig systems (Section 2.1.1) has elicited much speculation regarding their genetic association and participation in the afferent phase of antibody production. The results of studies by Ellman *et al.* (1971) of antibody synthesis in C4-deficient as compared to normal animals suggested a mild but statistically significant defect. C4-deficient and heterozygous guinea pigs immunized with DNP-bovine gamma globulin produced equivalent amounts of antibody 2 weeks after immunization. However, the response to limiting amounts of both ovalbumin and bovine serum albumin by C4-deficient animals was half that of normal guinea pigs. In contrast to these results, studies by Jackson *et al.* (1979) in a boy with inherited C4-deficiency showed that the patient's responses to one specific T-cell-dependent antigen, bacteriophage ϕX174, was clearly abnormal. The primary IgM response was significantly depressed and no IgM to IgG transition or amplification in the secondary response was observed. Because of these observations, the response of guinea pigs to immunization with the antigen ϕX174 was examined. C4-deficient guinea pigs showed similar markedly depressed responses to this immunogen and a total lack of development of immunologic memory that was corrected by administration of the deficient component (Ochs *et al.,* 1978). More recently these studies have been extended to include C3 and C7 (human) and C6 (rabbit) deficiency with similar results. The defect was reported to be more severe in deficiencies of the early components C3 and C4 (Ochs *et al.,* 1979). Ferone *et al.* (1976) reported inhibition of both a mixed lymphocyte reaction and the proliferative response to phytohemagglutinin mitogen by antisera to human C4; the lymphocytes were previously shown to express C4 antigen on their surface. Similar studies were performed in the guinea pig, assessing inibition of mixed leukocyte reactions between strain-13 and strain -2 guinea pigs and inhibition of proliferative responses by mitogens and soluble protein antigens by antiserum to guinea pig C4 (Burger and Shevach, 1979). No inhibition was found with normal cells and no deficiency was found in similar studies with lymphocytes from C4-deficient guinea pigs. The authors concluded that no functional relationship was demonstrable between C4 and the *MHC* in the afferent limb of the immune response in the guinea pig, in contrast to that shown for humans.

2.2. C5 Deficiency in Mice

2.2.1. History and Genetics

In 1962 Rosenberg and Tachibana utilized a sensitive hemolytic test system to study the levels of complement activity in a number of inbred strains of mice. Two strains, DBA/2 and B10.D2 (old), were found to be lacking in hemolytic complement function. The results of appropriate genetic crosses subsequently established that the hemolytic complement defect was inherited as a single Mendelian recessive trait. The allele for this locus was designated Hc^{o} (Herzenberg *et al.,* 1963; Tachibana *et al.,* 1963). Immunization of complement-deficient mice with the globulin fraction of complement-positive mice produced antisera recognizing a beta globulin present only in the sera of complement-positive strains (Erickson *et al.,* 1964). Genetic experiments showed no recombination between the beta globulin antigen locus and the hemolytic complement locus, suggesting the identity of the two loci. Heterologous antisera to the beta globulin also reacted only with the sera of complement-positive mice and it was concluded that the Hc^{o} allele produced no product (beta globulin) or one that is immunochemically distinct.

Independent observations by Cinader and Dubiski reported in 1963 identified a new allotypic specificity or polymorphic serum protein designated MuB1 present in the sera of particular inbred strains of mice. As in the experiments by Erickson *et al.* (1964), the antigen MuB1 was detected in the serum of mice by immunochemical analysis using antisera obtained by immunization of deficient inbred mice with sufficient inbred mouse serum. It was also detected by use of antiserum produced in rabbits to MuB1 specificities (Cinader and Dubiski, 1964). The mouse antiserum was shown to react with the serum of other rodents as well as many other species of mammals including man. This observation as well as the inability of rabbit antibodies to MuB1 to detect MuB1 specificities in MuB1-deficient mice suggested that the latter did not contain an antigen corresponding to MuB1 (Cinader *et al.*, 1964). Immunologic tests for the presence or absence of MuB1 antigen in hybrids of inbred strains showed that, as in the *Hc* system, the MuB1 deficiency is transmitted as an autosomal recessive trait. The absence of MuB1 was shown to correlate with the absence of hemolytic complement, Hc, and thus MuB1 was postulated to be the missing complement component, beta globulin, described by Erickson *et al.* (1964).

Prior to the discovery of mice with a specific complement component deficiency, it was generally recognized that mice have low serum hemolytic complement activity in the assays typically employed. It was suggested by some investigators that this might result from the presence of a complement-inhibiting factor (Borsos and Cooper, 1961). Subsequent findings by Terry *et al.* (1964) demonstrated no deficiencies in mouse serum C1, C4, or C2 activity in a number of inbred strains including those deficient in total complement. They found, however, a deficiency in activity of the classical third component (C3–9) which correlated with the total complement levels in deficient animals. They were unable to rule out the presence of a complement inhibitor in deficient sera. A sex-dependent control of complement concentration within inbred strains was shown; both total complement and C3 complex activity were lower in females than males. This finding was also reported by Cinader *et al.* (1964) and by Dubiski and Cinader (1966) for MuB1 antigen concentration. In the latter case, adult female mice had only about two-thirds the concentration of MuB1 antigen found in adult males. As Linscott and Cochrane (1964) demonstrated that both B10.D2 mice (old) lacking complement activity and B10.D2 (new) sufficient in complement activity contained beta 1C globulin, the apparent complement deficiency in "old line" mice was not attributable to the absence of the third component of complement, C3. In 1965, a known protein contaminant of β1C, β1F glycoprotein, was characterized as the fifth component of complement (Nilsson and Müller-Eberhard, 1965). An immunological relationship between MuB1 and β1F (C5) was demonstrated using specific mouse antiserum for these two proteins (Nilsson and Müller-Eberhard, 1967). Highly purified human C5 was able to reconstitute the complement-deficient mouse serum. Thus MuB1 (Hc) represents the fifth component of mouse complement (C5).

The *Hc* or *C5* gene was shown not to be closely linked to the *H-2* locus (Herzenberg *et al.*, 1963). In the first reported case of total C5 deficiency in man (Rosenfeld *et al.*, 1976a,b), genetic linkage of the *C5* gene with the *HLA* was also excluded (Rosenfeld *et al.*, 1977). More recently Weitkamp *et al.* (1978) in screening black and white Americans for polymorphic forms of C5 by immunofixation electrophoresis were unable to demonstrate a single qualitative variation, similar to the findings of Vaz-Guedes *et al.* (1978) in a study of over 2000 human sera. However, one black individual had decreased levels of both C5 functional activity as well as antigen. A study of 20 members of this family confirmed the presence of a C5-deficiency gene in a second black family. No homozygous recessive individuals were identified. There was no linkage of the quantitative *C5* variation with any markers studied including *C3* or *HLA*.

2.2.2. Biochemistry and Synthesis

Nilsson and Müller-Eberhard (1965) isolated and characterized a thermolabile β1F globulin from human serum as the fifth component of complement. In studies of guinea pig serum, the identity of a fifth component in the classical pathway was shown by Nelson *et al.* (1966). Later, Nilsson *et al.* (1972) developed a modified technique for more efficient recovery of the biologically active component although no information regarding the product's specific functional activity was presented. Approximately 20% of the immunochemically detectable C5 was recoverable. Recently, Tack *et al.* (1979) developed a preparative method of C5 purification from human plasma that yields 30% recovery of immunochemically detectable C5 with full functional activity. A method for purification of guinea pig C5 was developed by Cook *et al.* (1971).

Human C5 is a glycoprotein containing about 20% carbohydrate and consists of two unequal subunits linked by both noncovalent and disulfide bonds (Nilsson *et al.*, 1972, 1975; Müller-Eberhard, 1975; Minta and Man, 1977; Tack *et al.*, 1979). The molecular weight of the intact molecule is approximately 190,000–220,000, the α and β chains being 115,000–140,000 and 75,000–83,000, respectively.

C5 binding and cleavage are dependent upon formation of either one of two distinct enzymes, the classical, $\overline{\text{C4b,2a,3a}}$ (Shin *et al.*, 1968; Cooper and Müller-Eberhard, 1970; Shin *et al.*, 1971a,b), and the alternative, $\overline{\text{P,C3b}_2\text{,Bb}}$ (Brade *et al.*, 1973; Daha *et al.*, 1976; Medicus *et al.*, 1976b), C5-convertases. These enzymes result from prior binding and cleavage of C3 by C3-cleaving enzymes $\overline{\text{C4b,2a}}$ and $\overline{\text{P,C3b,Bb}}$, respectively, with subsequent incorporation (association) of the C3b fragment. Cleavage of C5 by its natural proteases results in the formation of two unequal, biologically active fragments: C5a, mol. wt. 10,000–15,000 (Cochrane and Müller-Eberhard, 1968; Shin *et al.*, 1968; Nilsson *et al.*, 1975), and C5b, mol. wt. 180,000–210,000 (Nilsson *et al.*, 1975; Podack *et al.*, 1978a; Yamamoto and Gewurz, 1978). The smaller C5a fragment is a source of neutrophil (Shin *et al.*, 1968; Ward and Newman, 1969; Snyderman *et al.*, 1970) and tumor-cell (Orr *et al.*, 1978) chemotactic factors and anaphylatoxin (Cochrane and Müller-Eberhard, 1968; Shin *et al.*, 1968). The C5a molecule has been shown to be derived from the amino-terminal region of the α chain of intact C5 (Nilsson *et al.*, 1975; Tack *et al.*, 1979). It has recently been shown that biological activity could, in fact, be obtained from the isolated α chain by tryptic digestion (Phan and Ward, 1979). The large fragment, C5b, has been shown to participate in opsonization and phagocytosis of red cells by polymorphonuclear leukocytes (Segerling and Opferkuch, 1977) and to block C3b receptors on Raji cells, also suggesting the presence of C5/C5b receptors on lymphoid cells (Landen and Dierich, 1979). Most recent studies of the C5b fragment, however, have been directed at elucidating its interaction with the terminal components of complement C6–9 and the assembly of the cytolytic C5b–9 membrane attack complex (Kolb *et al.*, 1972; Mayer, 1972; Mayer *et al.*, 1979).

Subsequent to isolation of C5 as a distinct serum protein, a number of related studies by different laboratories were undertaken to elaborate the complex role of C5 in immune cytolysis. Cooper and Müller-Eberhard (1970) demonstrated physical uptake of human C5 by cell-bound $\overline{\text{C4b,2a,3b}}$. While the efficiency of uptake onto the cell was low (4%), bound C5 was very efficient in producing lysis. The binding efficiency of the cellular intermediates EAC$\overline{1}$,$\overline{\text{4,2,3b}}$ for C5 was significantly improved in the presence of C6 and C7 and produced a themostable intermediate EAC1–7. In the absence of C6 and C7, the EAC1–5 intermediate was very labile and lost its hemolytic site at a rate proportional to the temperature of incubation. Recent findings demonstrated a weak interaction between cell-bound C3b and fluid-phase C5 of about $5 \times 10^5\ M^{-1}$ at physiological ionic strength (Isenman *et al.,* 1980). Results by Shin *et al.* (1971b) were similar with regard to the lability of the C5 functional site. They also showed that consumption of C5 by EAC$\overline{1,4,2,3b}$ and formation of C5b sites varied with the concentration of reactants, temperature, pH, and ionic strength, results compatible with the enzymatic cleavage of the C5 molecule.

Interaction of C5b with C6 has been shown to increase the functional stability of the labile C5b site (Hadding *et al.,* 1970; Shin *et al.,* 1971b) such that C5b,6 complex generated by the "reactive lysis" method (Lachmann and Thompson, 1970; Lachmann *et al.,* 1970a; Thompson and Lachmann, 1970; Podack *et al.,* 1978a; Yamamoto and Gewurz, 1978) or by elution from EAC1–6 (Goldlust *et al.,* 1974) retains its cytolytic capacity. Further reaction of C5b,6 complex with C7 generates a trimolecular complex C5b,6,7 (Götze and Müller-Eberhard, 1970; Arroyave and Müller-Eberhard, 1973; Podack *et al.,* 1978b). This complex in its nascent form has the transient capacity to bind to cell membranes and represents the initial stage in the assembly of the C5b–9 macromolecular membrane-active complex (MAC).

Identification of Hc (MuB1) as the fifth component of complement afforded an opportunity to study fetal synthesis and maternal–fetal transfer of complement in a genetically deficient strain of mice (Tachibana and Rosenberg, 1966). Double-diffusion precipitin and hemolytic complement tests were used to assay for the presence of C5 in the tissue fluid, in the sera of embryos, and in the corresponding placentas at various stages of development as well as in the sera of the offspring. It was shown that C5 is synthesized in the fetus at 10 to 12 days of gestation and that passive transfer of C5 from mother to fetus does not occur. Bone marrow transplantation was used by Phillips *et al.* (1969) to confer C5 activity that persisted up to 40 days Hc^{o} mice. *In vitro* tissue culture studies were able to show incorporation of ^{14}C-labeled amino acids into C5 protein. Spleen and parenchymal hepatoma cells were shown to produce the complement component.

Others have used inactivated Sendai virus to fuse complement-sufficient cells with deficient cells to produce hybrids that would not only synthesize C5 *in vitro* but be capable of the restoration of complement activity *in vivo* (Levy and Ladda, 1971; Levy *et al.,* 1973). Hybrids of chicken erythrocytes or normal kidney cells with the C5-deficient macrophage produced C5 *in vitro* and were shown to produce detectable C5 antigen as well as to restore C5 function *in vivo.* Only sera from restored animals were able to generate chemotactic activity, a biologic property of activated C5. Cultures of chicken spleen, kidney, and liver cells failed to elaborate a factor that would restore C5 function in deficient mouse serum, nor did homogenates of these cells. By immunodiffusion, no detectable C5 antigen could be demonstrated. Results with the mouse–chicken hybrids suggested that the presence of *mouse* C5 antigen and functional activity in the recipients of these cells was due to the expression of the C5-deficient mouse structural gene directed by regulatory factors from the chick nucleus.

Primary rat liver cell cultures, rich in parenchymal cells but depleted in macrophages, as well as clonal strains of rat hepatoma, synthesize functional C5 (Strunk *et al.,* 1975; Colten, 1976).

In man, *in vitro* C5 synthesis was shown by production of the biologically active component in fetal and postnatal liver, fetal lung, fetal spleen, and fetal intestine (Colten, 1973). In another study *in vitro* incorporation of ^{14}C-labeled amino acids into C5 protein occurred in fetal tissue cultures of liver, spleen, placenta, peritoneal and bone marrow cells, thymus, and colon (Kohler, 1973).

There have been several recent reports of attempts to define the genetic lesion in C5-deficient mice. Using an *in vivo* labeling technique Patel and Minta (1979) were able to incorporate ^{14}C-labeled amino acids into C5 protein immunochemically recoverable from both the serum and the supernatants of liver homogenates from normal AKR/J mice. Upon sodium dodecyl sulfate polyacrylamide gel electrophoresis (SDS-PAGE) analysis, they showed that C5 obtained from the liver was recovered as a nonreducible, single protein chain of 200,000 daltons, or pro-C5. In contrast, C5 recovered from the plasma consisted of two chains of unequal size, an α chain of 115,000 daltons and a β chain of 82,000 daltons. Thus, it is structurally similar to both human and guinea pig C5. When the serum and liver homogenates of AKR/J C5-deficient mice were analyzed for the presence of C5, none was found, suggesting a complete lack of C5 synthesis in this strain. Concurrently, Ooi and Colten (1979) have reported somewhat different findings in studies of C5 synthesis in five different C5-deficient mouse strains. *In vitro* tissue culture of peritoneal macrophages from normal mice incorporated [^{35}S]methionine into immunoprecipitable C5 of both extracellular and intracellular origins. Analysis of the reduced and alkylated protein showed a single polypeptide of 220,000 daltons for C5 of extracellular origin, and 200,000 daltons for the intracellularly synthesized, but presumably nonglycosylated component. In five different C5-deficient cell cultures, no immunoprecipitable C5 was detected extracellularly; in contrast to the results of Patel and Minta (1979), however, pro-C5 of 200,000 daltons was identified in the intracellular compartment. These data were interpreted to suggest that mutant, C5-deficient mice synthesize pro-C5 but that the product is not secreted, possibly due to the absence of a glycosylase or transferase enzyme. Alternatively, the pro-C5 may lack the required site for attachment of carbohydrate. The fact that different techniques were used by these two authors and that the latter studies used a more sensitive technique to measure incorporation into C5 may account for these differences.

2.2.3. Biological Consequences

The availability of two coisogenic lines of mice, one possessing hemolytic complement [B10.D2 (new)] and the other deficient [B10.D2 (old)], has prompted many investigations of the role of complement in the process of graft rejection and in the immunosuppressive effects of antilymphocyte serum. Phillips *et al.* (1968) showed that while limited numbers of sarcoma I tumor cells (5×10^5) injected intraperitoneally are rejected by both complement-deficient and -sufficient presensitized strains, only the B10.D2 (new) was capable of rejecting large numbers of cells (5×10^7). Skin allografts from complement-deficient A/WySn mice, strongly histoincompatible to B10.D2 with respect to H-2, were shown by Weitzel and Rother (1970) to have prolonged survival in the B10.D2 (old) mice. Conflicting evidence for the participation of complement in graft rejection was presented by Salerno (1969). His results showed that spleen cells transplanted from preimmunized BALB/c (compatible at the *H-2* locus) or C57BL/6 (incompatible at the *H-2* locus) into recipient

B10.D2 (new) and B10.D2 (old) mice survived equally well and thus were not affected by the absence of C5.

Similar results were obtained in skin homografts of CF-1 and C57BL/6J on B10.D2 (new) and B10.D2 (old). In both recipient lines of mice, rejection occurred at the same rate (Caren and Rosenberg, 1965). Cinader *et al.* (1971) showed that antilymphocyte serum (ALS)-treated strains of C5-deficient mice were not as capable of rejecting skin allografts as were normal mice, implicating complement in the allograft rejection process. Similar results using antithymocyte serum (ATS) were obtained by Weitzel and Rother (1970). In their experiments, ATS injection prior to transplantation resulted in additional prolongation of graft survival past the period found in C5-deficient mice not treated with ATS. However, accelerated rejection of the graft occurred if ATS was used following grafting.

Other experiments by Cinader *et al.* (1971) examined the effect of ALS on the humoral antibody response in C5-deficient and normal mice. They found that *in vivo* ALS immunosuppression of antibody to sheep red blood cells was identical in the presence or absence of an intact complement system. Similar findings were demonstrated by Barth and Carroll (1970) using ALS in DBA/2, AKR/N, and B10.D2 (old) mice, all of which are complement deficient (C5), and BALB/c as the normal control. Thus a positive role for the terminal components of complement in elimination of antigen-sensitive cells by cytolysis is unlikely. Presumably complement, however, could play a role through opsonization via C3b and could facilitate elimination of lymphocytes by phagocytosis.

Conflicting observations such as those described above for graft rejection are not unique to this one area of investigation utilizing complement-deficient and -sufficient mice. Nephritis in mice, both that experimentally induced by the injection of nephrotoxic serum and that which spontaneously develops (Howie and Helyer, 1965) in the NZB (C5-deficient) strain, was studied with regard to the role that complement plays in immune tissue damage. Unanue *et al.* (1967) studied experimentally induced nephritis, a biphasic disease, in normal and complement-deficient mice and could show no difference in the severity of the disease induced in complement-positive or -negative strains. Experiments by Lanier *et al.* (1971) were designed to study the development of spontaneous nephritis in (NZB × NZW)F_2 hybrids, which express either normal C5 levels, intermediate levels, or no C5. They found that the presence of C5 did not contribute to the development of nephritis and that it may even have some beneficial effect. In contrast, Lindberg and Rosenberg (1968), utilizing hybrids of the coisogenic strains B10.D2 (old) and B10.D2 (new), found significant complement-related differences in urinary protein excretion and survival following induction of experimental nephritis. In these experiments, C5-sufficient mice were at greater risk, implicating the terminal-complement components in immune tissue injury. Opposite conclusions were drawn by Nakamura and Weigle (1968), who induced thyroiditis in A/JAX, B10.D2 (old), and DBA/2J, all of which lack complement, and in normal mice. Their data demonstrated that complement-deficient mice were more susceptible to development of thyroiditis. Passive cutaneous anaphylaxis (PCA) was investigated in both C5-deficient and normal mice to demonstrate a possible role for complement (Ben-Efraim and Cinader, 1964). Sequential presentation of antigen followed immediately by antibody to complement-positive and -negative mice demonstrated no difference in the extent of the PCA response. However, marked differences were observed between the two types of mice when antigen was presented several hours after antibody. The occurrence of two phases of the cutaneous reaction was observed, one of which appeared to require an intact complement system.

B10.D2 (old) and B10.D2 (new) were studied for the development of the Arthus reaction by sensitization with hemocyanin in complete Freund's adjuvant and a subsequent challenge with the antigen. Similar Arthus reactions were produced in mice of both strains. Thus this reaction does not require the terminal components of complement (Crisler and Frank, 1965).

Reports from a number of laboratories have implicated a role for complement (C3 in particular) in induction of the immune response (see Section 2.1.3; Pepys, 1976). Recently Martinelli *et al.* (1978) have studied AKR/J $C5^-$, DBA/2J $C5^-$, and DBA/1J $C5^+$ mice in their ability to respond to the T-dependent immunogen sheep erythrocytes (SRBC). Cobra venom factor (CVF) treatment of normal mice prior to immunization depresses the humoral antibody response. It was observed that, when subjected to SRBC immunization, CVF-treated $C5^+$ and $C5^-$ mice showed essentially similar suppression of both IgG and IgM responses. Their ability to respond to SRBC in the absence of CVF treatment was also shown to be equal. Thus, development of an immune response to SRBC as well as CVF suppression of antibody production in the mouse does not appear to require C5 or other terminal components.

Studies by Goldman *et al.* (1977, 1979) have shown that administration of anti-C5 antibody raised in C5-deficient mice or administration of homozygous C5-deficient parental lymphoid cells to newborn heterozygous F_1 hybrids induces chronic suppression of C5 production. The suppression of C5 by anti-C5 was dependent on the source of antibody and the strain combination. In experiments with parental lymphoid cells, suppression was independent of C4, C2, or C6 and thus was not due to complement activation.

The requirement of humoral opsonins for the phagocytosis of bacteria has been known for many years (Halpern *et al.*, 1957). Prior to the discovery and maintenance of complement-deficient animal colonies, antigen–antibody complexes were administered to "decomplement" animals in an effort to determine the role of complement *in vivo* in various immunologic reactions (Biozzi and Stiffel, 1962; Spiegelberg *et al.*, 1963). However, it was suspected that antibodies and other serum factors were also affected by this treatment. The availability of mice deficient in C5 led to numerous experiments designed to clarify the role of the late-acting complement components in resistance to various infections.

In 1964, Stiffel and co-workers reported studies of the phagocytosis of *Salmonella typhimurium* in two strains of complement-deficient mice. Their results indicated that bacterial phagocytosis was unrelated to serum complement titer. When resistance to the common mouse pathogen, *Corynebacterium kutscheri*, was compared in C5-deficient and normal mice, it was shown that normal mice fared only slightly better than their C5-deficient counterparts (Caren and Rosenberg, 1966). These findings were supported by Glynn and Medhurst (1967), who reported that although the absence of complement had no effect on the phagocytosis of *E. coli* by the reticuloendothelial system, the killing of an organism sensitive to complement was impaired in the C5-deficient animals.

Johnston *et al.* (1969) reported an extensive investigation of the phagocytosis of pneumococci (*D. pneumoniae*, type II) utilizing a spectrophotometric method for measuring phagocytosis. By employing sera deficient in C2, C3, C5, or C6 and by the sequential fixation of purified C1, C4, C2, and C3 to the bacterial surface, the importance of the early- and late-acting components was examined. Johnston *et al.* concluded that C1–3, but not the later components, C5–9, were necessary for optimal phagocytosis of pneumococci *in vitro.*

To the contrary, Shin and co-workers (1969) demonstrated that C5 played a definite, though lesser role than C3 in the defense against pneumococcal infection induced by encapsulated or unencapsulated strains of bacteria. The participation of C5 in antipneumococcal defense was demonstrated *in vivo* as well as *in vitro.* C5 was also reported to act as an opsonin for the encapsulated strain of *Staphylococcus aureus* in studies of intraperitoneal and subcutaneous infections induced in normal and C5-deficient mice (Easmon and Glynn, 1976). In other studies, C5-deficient mice were found to be more susceptible to *Listeria monocytogenes* infection. As resistance to *Listeria* is thought to be predominantly cell mediated, the role of C5 in the early attraction of neutrophils to the inflammatory site and/or its ability to interact with activated T or B cells may contribute to the impaired resistance to the organisms in C5-deficient mice (Lawrence and Schell, 1978).

A limited number of studies have been conducted to determine the contribution of complement in host resistance to parasite infections. In studies utilizing C5-deficient mice as a model, the complement deficiency seemed to have little or no effect on the course of parasitemia. The contribution of complement components C5–9 in the control of *Toxoplasma gondii* was found to be quantitatively insignificant (Araujo *et al.*, 1975). Similarly, complement-mediated lysis was reported not to be involved in the elimination of *Trypanosoma musculi* (Jarvinen and Dalmasso, 1977).

The role of complement during experimental *C. albicans* infection in mice was examined by Morelli and Rosenberg. The results of *in vitro* studies indicated that phagocytosis of *Candida* by mouse peripheral blood leukocytes was enhanced by normal mouse complement, even in the absence of specific antibody. Intracellular killing of *Candida* by peritoneal cells, however, was independent of antibody and did not require a complete complement system (Morelli and Rosenberg, 1971a). *In vivo* studies by these investigators also provided evidence of a role for complement in defense against *Candida* in that normal animals survived longer than C5-deficient mice when injected with the organisms (Morelli and Rosenberg, 1971b).

Studies have also been undertaken to determine the effect of complement on *Mycoplasma canis* and *M. pulmonis.* Tachibana *et al.* (1970) found that *in vitro* C5-deficient mouse serum was ineffective in killing *M. Canis,* while normal mouse serum, activated by antibody, was mycoplasmacidal. Keystone *et al.* (1978) studied the degree of arthritis induced in normal and C5-deficient mice inoculated parenterally with *M. pulmonis* organisms. Their results indicated that C5 is important in the elimination of *M. pulmonis* in that C5-deficient mice developed a more persistent and severe arthritis and larger numbers of organisms could be isolated from the joints of the deficient animals.

The role of complement in recovery from viral infections has also been investigated (Hicks *et al.*, 1978). Mice were infected with an H3N2 influenza A recombinant (MRC-9). Infected C5-deficient mice were more susceptible to influenza pneumonia and experienced a higher incidence of pulmonary consolidation and mortality as compared to normal mice. As neutralizing antibody responses were the same in the two types of mice, the delayed viral clearance in C5-deficient mice could be attributed to the lack of an intact complement system.

The importance of C5 as a mediator of the inflammatory response has been the subject of many investigations in which C5-deficient mice have been used as an experimental animal model. The accumulation of polymorphonuclear (PMN) leukocytes, which is viewed as an important mechanism for preventing the dissemination of infectious agents, has been studied *in vivo* and *in vitro*. Snyderman and co-workers (1971) presented data suggesting that *in vivo*, C5 is important in the early accumulation of PMN leukocytes in response to an inflammatory stimulus. When endotoxin was injected into the peritoneal cavity of C5-deficient mice, there was little chemotactic activity generated and a significant increase in PMN leukocytes did not occur until 12 hr after injection. In normal mice the generation of chemotactic activity was observed, followed by an accumulation of PMN leukocytes as early as 3 hr after injection. Similar observations were made by Sveen in 1978, who evaluated PMN leukocyte migration in an *in vivo* chamber as well as in the peritoneal cavities of normal and C5-deficient mice after injection with endotoxin. These findings conflicted with an earlier report by Stecher and Sorkin (1969), who observed no differences in the generation of chemotactic factors in response to endotoxin and antigen–antibody complexes by C5-deficient mouse serum and C6-deficient rabbit serum when compared with normal sera. The different conclusions drawn from these *in vitro* studies of C5-deficient serum may reflect changes in experimental conditions that would be difficult to identify and control. On the other hand, chemotactic responses in C6-deficient rabbits may be attributed to C5a, which is known to be chemotactic.

2.3. C6 Deficiency in Rabbits

2.3.1. History and Genetics

In 1961 a strain of rabbits deficient in lytic complement activity was described by Rother and Rother in Freiburg, Germany. Initial studies of the serum of these rabbits showed that it was incapable of lysing antibody-sensitized sheep erythrocytes. It was shown that the rabbits lack activity of what was then termed total C3 (later shown to comprise C3–9). Further studies demonstrated that the defect was due to a deficiency in the sixth component of complement, C6 (Rother *et al.*, 1966). A second, unrelated strain of rabbits deficient in C6 was developed in Mexico by Biro and Garcia in 1965. Later, Lachmann described a third strain of rabbits with a similar defect that was discovered by chance in Cambridge, England (Lachmann, 1970).

Each of the strains described as deficient in C6 had little or no measurable serum C6 activity in hemolytic assays. Hemolytic activity of the serum was fully reconstituted by adding purified human or rabbit C6 (Rother *et al.*, 1966) or purified guinea pig C6 (Nelson and Biro, 1968). Furthermore, upon injection with purified C6, the deficient animals produced antibody to C6 that was capable of inactivating the C6 in normal rabbit serum (Rother *et al.*, 1966). Lachmann provided evidence that the deficient Cambridge and Mexican strains did not possess an inactive analog of C6 by demonstrating that antisera to rabbit C6 raised in a heterologous species, the chicken, as well as that raised in rabbits did not cross-react with any of the C6-deficient sera tested. The deficiency in rabbits appears to be due to a genetically controlled defect in C6 synthesis that is apparently fairly common, considering the discoveries of three unrelated, widely separated, deficient strains.

Anticomplementary activity was not associated with the C6 deficiency in the Freiburg strain (Rother *et al.,* 1966). However, Nelson and Biro (1968) isolated an exceedingly heat-labile C6 inactivator from the Mexican strain. The C6 inactivator, which was also present at lower levels in normal rabbit serum, reacted specifically with the cellular intermediate EAC$\overline{1}$,$\overline{4,2,3}$,5b,6 to destroy its hemolytic activity.

C6 deficiency in rabbits was shown to be transmitted as a single, autosomal recessive trait. Lachmann studied over 200 offspring, covering three generations of a "Himalayan" buck crossed originally with five does. Heterozygous animals were found to have a mean serum C6 level 40% that of the mean level in normals. Genetic studies indicated that the C6 deficiency was not linked to the gamma globulin allotype or the hemoglobin Val/Leu structure (Lachmann, 1970). A C6 defect with apparently similar inheritance has been identified in man. In studies of these deficient humans no genetic relationship between C6 deficiency and the *HLA* system was shown in two separate family studies (Mittal *et al.,* 1976; Hobart *et al.,* 1977a). Moreover, there was no suggestion of linkage between *C6* and 33 other genetic marker loci, including *HLA,* in studies of Norwegian pedigrees conducted by Olving *et al.* (1979). These studies stand in marked contrast to the report of Raum *et al.* (1976), which found an association between C6 deficiency and *HLA* (*AW 24* haplotype), suggesting an error in the initial report. On the other hand, there is evidence that in humans, *C6* and *C7* are closely linked (Hobart *et al.,* 1977a). In fact, coexistent partial deficiencies in C6 and C7 have been demonstrated (Glass *et al.,* 1978). In most cases, however, human C6 deficiency is not associated with reduced C7 (Leddy *et al.,* 1974; Lim *et al.,* 1976; Glass *et al.,* 1978). Similarly, in C6-deficient rabbits, C7 activity was found to be normal (Nelson and Biro, 1968). Recently, the simultaneous occurrence of hereditary C6 and C2 deficiency was noted in a French Canadian family in which three members were deficient in both C2 and C6. The C6 deficiency did not show *HLA* linkage, while the C2 deficiency segregated with markers in the *HLA* region, indicating that the C6 and C2 deficiencies represent two separate and unrelated hereditary deficiencies (Delage *et al.,* 1979). There is also evidence that in humans the heterozygous genetic deficiency for C6 may be due to a silent or null gene at the same locus as the normal *C6* variant gene (Hobart *et al.,* 1977a; Glass *et al.,* 1978). To date, no such evidence has been established as the basis for C6 deficiency in rabbits.

2.3.2. Biochemistry and Synthesis

C6 is one of the five components of the complement system that comprise what has been described as the membrane attack complex (MAC) (Section 2.2.2).

Several studies have shown that C6 and C7 are structurally similar and their genesis from a common molecular ancestor has been suggested. Nilsson first described C6 and C7 as slow-migrating beta globulins with similar sedimentation rates, between 5 and 6 S. However, the components that he isolated from the euglobulin fraction of human serum were contaminated with other globulins (Nilsson, 1967). The characterization of a homogeneous preparation of C6 was first reported by Arroyave and Müller-Eberhard in 1971, who isolated C6 from human, rabbit, and guinea pig serum. The molecular weight of the beta globulin was 95,000 by polyacrylamide gel electrophoresis and 125,000 by molecular sieve chromatography and sucrose density-gradient ultracentrifugation. However, the recovery of functional C6 was very poor; only 1/20th of the C6 protein recovered appeared to have activity. In 1976, Podack and co-workers reported a procedure for isolating C6 and C7 without the loss of hemolytic activity. Using SDS-PAGE analysis, the molecular weights of C6 and C7 were similar, 111,000 and 102,000, respectively. Their electrophoretic mobilities were the same (-2.6×10^{-5} cm^2 sec^{-1} V^{-1}) and both behaved as single polypeptide chains under reducing conditions on SDS-PAGE (Podack *et al.*, 1976).

Studies further characterizing C6 and C7 were reported by Podack and co-workers in 1979. The components were purified using insolubilized antibodies in affinity chromatography, a procedure that was reported to yield highly purified C6 and C7 with full hemolytic activity. Similarities in amino acid and carbohydrate content of the two proteins were reported. Analogous fragmentation was obtained with short-term trypsinization, implying that the tertiary structures of C6 and C7 are similar. Both C6 and C7 were polymorphic, as shown by isoelectric focusing, with isoelectric points ranging from pH 6.15 to 6.7. The UV spectra were also similar, with extinction coefficients of 1.71 and 1.92 cm^2 mg^{-1} for C6 and C7, respectively.

The first reported attempt to identify tissues responsible for C6 synthesis was undertaken by Rother and co-workers (1968). Rabbit tissues that were analyzed for C6 production included liver, spleen, peritoneal macrophages, and kidney. Production of C6 activity was found in liver cultures and possibly in macrophages and splenic tissue cultures. The hypothesis that the liver is the primary *in vivo* site of C6 synthesis was tested by examining the allotype of C6 produced by a liver transplant recipient whose C6 phenotype (A) was different from that of the organ donor (phenotype B). The recipient's C6 phenotype changed from A to B within 10 days of the transplant and remained so during 17 weeks of the study, demonstrating that the liver is the predominant site of C6 synthesis in man (Hobart *et al.*, 1977b).

2.3.3. Biological Consequences

C6-deficient rabbits appear to withstand common laboratory infections as well as normal rabbits, which is due in part to their normal production of antibody. Furthermore, serological reactions requiring the participation of the first four components of complement are also normal in C6-deficient serum. This includes immune adherence, phagocytosis by PMN leukocytes, and immune clearance of bacteria (Rother and Till, 1974). However, there is evidence of impairment in one component of their immune defense mechanism. The deficient serum lacks the ability to kill gram-negative bacteria, even after their sensitization with antibody (Rother *et al.*, 1964a). An impairment in passive but not active Arthus reactivity (Rother *et al.*, 1964b) and suppression of the delayed hypersensitivity reaction to tuberculin (K. Rother *et al.*, 1967a) have also been reported. C6-deficient rabbits also failed to reject homografts more frequently than normal rabbits (Volk *et al.*, 1964; U. Rother *et al.*, 1967). On the other hand, after injection with anti-rabbit kidney antiserum, C6-deficient rabbits developed nephrotoxic nephritis as severe as that of controls, suggesting that C6 is not essential to the development of this experimental disease (K. Rother *et al.*, 1967b). Somewhat contradictory findings were reported by Biro (1966) in studies of the Mexican strain of rabbits; he concluded that while C6 does enhance the Arthus reactions, it was impossible to show its absolute requirement in either the active or the passive Arthus reaction. He also found that homograft rejection and delayed hypersensitivity were both normal in these animals. Apparently, further studies are needed to clarify the role of C6 in hypersensitivity.

Ward *et al.* (1965) reported that C6-deficient rabbit serum was defective in generating leukocytic chemotactic activity *in vitro*. However, Stecher and Sorkin (1969) and Snyderman *et al.* (1970) reported normal chemotactic activity in C6-deficient serum in the presence of either endotoxin or immune complexes. The role of the trimolecular complex, C5b,6,7, in chemotaxis, described by Ward *et al.* (1966) and Lachmann *et al.* (1970b), further implicated C6 as a participant in this biological activity. The discrepancies in these findings may be partly explained by the fact that other potent chemotactic factors such as C5a are known to be generated by antigen–antibody complexes (Shin *et al.*, 1968; Fernandez *et al.*, 1978).

In studies reported by Johnson and Ward (1971) it was suggested that C6 is necessary for rabbits to survive injection with endotoxin and C6 is required for endotoxin detoxification. However, this situation is quite complex and precisely the opposite findings have been reported by several different groups. Using low-toxicity endotoxin, Fong *et al.* (1974) found the survival rate of injected C6-deficient rabbits to be high. They also showed that C6 deficiency is associated with a qualitative and quantitative protection against the endotoxin-induced local Shwartzman reaction. On the other hand, C6 deficiency was reported not to prevent intravascular clotting and the generalized Shwartzman reaction induced by endotoxin in cortisone-treated rabbits (Stafford *et al.*, 1976).

In 1971 Zimmerman and co-workers reported a clotting defect in C6-deficient rabbit plasma in which whole blood clotting time was prolonged and prothrombin consumption was reduced when compared to normals. Both defects were corrected by adding physiological amounts of purified C6 to the deficient blood. Although true for C6-deficient rabbits, studies of C6-deficient human serum (Heusinkveld *et al.*, 1974) found no clear evidence of a reduction in prothrombin consumption, suggesting that not all findings in animal models will prove directly relevant to the clinical state in man. Abnormal coagulation in C6-deficient rabbits has been attributed to the failure of C6-deficient plasma to release platelet factors. Direct evidence of a platelet-related defect in these animals was shown by Siraganian in 1972. It was demonstrated that, unlike normal rabbit plasma, C6-deficient plasma expressed no platelet-release reaction when treated with complement-activating factors, such as zymosan, inulin, or antigen–antibody complexes. Furthermore, platelet factor 3 (PF 3) activation was shown to be greatly impaired in C6-deficient rabbits injected with endotoxin (Brown and Lachmann, 1973). Whereas all of the normal rabbits died when injected with a single 6-mg dose of endotoxin, some of the C6-deficient animals survived. Their survival was correlated with the absence of PF 3 activity and with the recovery of their platelets. In other studies of platelet aggregation and secretion, it was shown that the aggregation response to zymosan is biphasic and that the second phase, involving the release of platelet constituents, does not occur in C6-deficient plasma. The primary phase of aggregation, which does occur in C6-deficient plasma, is thought to result from direct interaction of zymosan and platelets, while secondary aggregation occurs only after platelet release (Christian and Gordon, 1975).

C6 deficiency appears to protect against lethal endotoxin shock in some cases, primarily by limiting the extent of platelet damage and the subsequent release of platelet factors that trigger disseminated intravascular coagulation.

3. Other Deficiency States

3.1. C1 Deficiency in Chickens

In 1974, Gabrielsen and co-workers reported finding a significant reduction in hemolytic C1 activity in hypogammaglobulinemic chickens that had been bursectomized by administration of cyclophosphamide. The results obtained from this experimental model support earlier observations of deficient hemolytic C1 and lowered circulating C1q in patients with the X-linked recessive form of agammaglobulinemia (Jonsen and Kass, 1957; Müller-Eberhard and Kunkel, 1961; Gewurz *et al.*, 1968b; Kohler and Müller-Eberhard, 1969). Patients with thymic alymphoplasia are reported to have very low levels of C1. A direct relationship between concentrations of serum C1q and IgG, but not IgA or IgM, was reported by Kohler and Müller-Eberhard in 1969. In other studies using radiolabeled C1q, these investigators found no evidence of decreased C1q synthesis in the agammaglobulinemic patients they studied. However, they concluded that the radiolabeled C1q became complexed with IgG and was promptly removed from the circulation of hypogammaglobulinemic patients (1972). This concept was further supported by the work of Atkinson *et al.* (1978). They described a paient with agammaglobulinemia and low C1q who was infused intravenously with an experimental C1q-free gamma globulin preparation. Immediately following the infusion of gammaglobulin, C1q levels were at the preinfusion value; however, there was a striking rise in C1q over the next 24 hr. Recently, Loos and co-workers (1980) described a 10-year-old male with a complete deficiency in C1q who had C1r and C1s levels three to four times higher than normal. However, the IgG level in this patient was not reported.

3.2. C4 Deficiency in Rats

The discovery of a single-component deficiency (C4) in a strain of Wistar rats was first reported by Arroyave and co-workers in 1974. A more detailed description of this genetic deficiency was given in 1977 (Arroyave *et al.*, 1977). Hemolytic activity was fully reconstituted by adding purified human C4 and there was no evidence of C4 inhibitory activity in the deficient serum. The deficiency was reported to be transmitted as an autosomal recessive trait. Thus far, extensive analysis of the biological capacities of the strain has not been performed.

3.3. C6 Deficiency in Hamsters

A deficiency in C6 in some strains of golden Syrian hamsters with a high incidence of proliferative enteritis was reported by Yang *et al.* (1974). The deficient serum was not anticomplementary and had the normal capacity to induce immune adherence and phagocytosis. C6 deficiency was confirmed by titration of isolated complement components of normal and deficient hamster sera. Here again an extensive evaluation of this strain is not available.

4. Summary

The term complement refers to a series of plasma glycoproteins that are responsible for many features of the inflammatory response and for a number of aspects of host defense against invading pathogens. Two converging pathways of complement activation have been described. The classical pathway is activated by antigen–antibody complexes. The alternative pathway can be activated in antibody-free serum. Both pathways lead to activation of the later-acting components in the complement cascade, C3–9.

Much of our current understanding about the interaction of the components and their biological function has come from studies of animals deficient in one or another of the complement proteins. Genetically controlled protein deficiencies have been described in the early-acting component C4 (guinea pig) and in two proteins of the late-acting component group, C5 (mice) and C6 (rabbit). Less well studied defects in C1 (chickens), C4 (rats), and C6 (hamsters) have also been described. This review examines in detail the biochemical and physiological consequences of these deficiency states.

ACKNOWLEDGMENT

Sincere thanks go to Mrs. Cindy Earp for her efforts in typing the manuscript.

References

Agnello, V., 1978, Complement deficiency states, *Medicine* **57**:1.

Araujo, F. G., Rosenberg, L. T., and Remington, J. S., 1975, Experimental *Toxoplasma gondii* infection in mice: The role of the fifth component of complement, *Proc. Soc. Exp. Biol. Med.* **149**:800.

Arroyave, C. M., and Müller-Eberhard, H. J., 1971, Isolation of the sixth component of complement from human serum, *Immunochemistry* **8**:995.

Arroyave, C. M., and Müller-Eberhard, H. J., 1973, Interactions between human C5, C6 and C7 and their functional significance in complement dependent cytolysis, *J. Immunol.* **111**:536.

Arroyave, C. M., Levy, R. M., and Johnson, J. S., 1974, Genetic deficiency of the fourth component of complement in Wistar rats, *Fed. Proc.* **33**:795.

Arroyave, C. M., Levy, R. M., and Johnson, J. S., 1977, Genetic deficiency of the fourth component of complement (C4) in Wistar rats, *Immunology* **33**:453.

Atkinson, J. P., Fisher, R., Reinhart, R., and Frank, M. M., 1978, Reduced concentration of the first component of complement in hypogammaglobulinemia: Correction by infusion of gammaglobulin, *Clin. Immunol. Immunopathol.* **9**:350.

Barber, T. A., and Burkholder, P. M., 1978, Enumeration and ultrastructure of C4-producing free alveolar cells from guinea pig lung, *J. Immunol.* **120**:716.

Barber, T. A., Burkholder, P. M., and Ilgen, C. L., 1976, Production of the fourth component of guinea pig complement (C4) by peritoneal macrophages: Immunolabeling and ultrastructure, *J. Reticuloendothelial Soc.* **20**:267.

Barth, R. F., and Carroll, G. F., 1970, Immunosuppressive effects of antilymphocyte serum in complement-deficient mice: Evidence that immune cytolysis is not essential for ALS activity, *J. Immunol.* **104**:522.

Becker, E. L., 1956a, Concerning the mechanism of complement action. I. Inhibition of complement activity by diisopropylfluorophosphate, *J. Immunol.* **77**:462.

Becker, E. L., 1956b, Concerning the mechanism of complement action. II. The nature of the first component of guinea pig complement, *J. Immunol.* **77**:469.

Ben-Efraim, S., and Cinader, B., 1964, The role of complement in the passive cutaneous reaction of mice, *J. Exp. Med.* **120**:925.
Biozzi, G., and Stiffel, C., 1962, Role of normal and immune opsonins in the phagocytosis of bacteria and erythrocytes by the reticulo-endothelial cells, in: *2nd International Symposium on Immunopathology* (P. Grabar and P. Miescher, eds.), p. 249, Schwabe, Basel.
Biro, C. E., 1966, The role of the sixth component of complement in some types of hypersensitivity, *Immunology* **10**:563.
Biro, C. E., and Garcia, G., 1965, The antigenicity of aggregated and aggregate-free human gammaglobuilin for rabbits, *Immunology* **8**:411.
Bitter-Suermann, D., Krönke, M., Brade, V., and Hadding, V., 1977, Inherited polymorphism of guinea pig factor B and C4: Evidence for genetic linkage between C4 and Bf loci, *J. Immunol.* **118**:1822.
Bokisch, V. A., and Sobel, A. T., 1974, Receptor for the fourth component of complement on human B lymphocytes and cultured human lymphoblastoid cells, *J. Exp. Med.* **140**:1336.
Bolotin, C., Morris, S., Tack, B., and Prahl, J., 1977, Purification and structural analysis of the fourth component of human complement, *Biochemistry* **16**:2008.
Borsos, T., and Cooper, M., 1961, On the hemolytic activity of mouse complement, *Proc. Soc. Exp. Biol. Med.* **107**:227.
Brade, V., Lee, G. D., Nicholson, A., Shin, H. S., and Mayer, M. M., 1973, The reaction of zymosan with the properdin system in normal and C4 deficient guinea pig serum, *J. Immunol.* **111**:1389.
Brown, D. L., and Lachmann, P. J., 1973, The behavior of complement and platelets in lethal endotoxin shock in rabbits, *Int. Arch. Allergy* **45**:193.
Budzko, D. B., and Müller-Eberhard, H. J., 1970, Cleavage of the fourth component of human complement (C4) by C1 esterase: Isolation and characterization of the low molecular weight product, *Immunochemistry* **7**:227.
Burger, C., and Shevach, E. M., 1979, Evaluation of the role of C4 in the cellular immune response *in vitro*, *J. Immunol.* **122**:2388.
Caren, L. D., and Rosenberg, L. T., 1965, Complement in skin grafting in mice, *Immunology* **9**:359.
Caren, L. D., and Rosenberg, L. T., 1966, The role of complement in resistance to endogenous and exogenous infection with a common mouse pathogen, *Corynebacterium kutscheri*, *J. Exp. Med.* **124**:689.
Carroll M. C., and Capra, J. D., 1978, Studies on the murine Ss protein: Demonstration that the Ss protein is functionally the fourth component of complement, *Proc. Natl. Acad. Sci. USA* **75**:2424.
Christian, F. A., and Gordon, J. L., 1975, Platelet function in C6-deficient rabbits: Aggregation and secretion induced by collagen and zymosan, *Immunology* **29**:131.
Cinader, B., and Dubiski, S., 1963, An alpha-globulin allotype in the mouse (MuB1), *Nature (London)* **200**:781.
Cinader, B., and Dubiski, S., 1964, Effect of autologous protein on the specificity of the antibody response: Mouse and rabbit antibody to MuB1, *Nature (London)* **202**:102.
Cinader, B., Dubiski, S., and Wardlow, A. C., 1964, Distribution, inheritance and properties of an antigen, MuB1, and its relation to hemolytic complement, *J. Exp. Med.* **120**:879.
Cinader, B., Jeejeebhoy, H. F., Koh, S. W., and Rabbat, A. G., 1971, Immunosuppressive and graft-rejecting antibodies in heterologous antilymphocytic serum, *J. Exp. Med.* **133**:81.
Clark, R. A., and Klebanoff, S. J., 1978, Role of the classical and alternative complement pathways in chemotaxis and opsonization: Studies of human serum deficient in C4, *J. Immunol.* **120**:1102.
Clark, R. A., Frank, M. M., and Kimball, H. R., 1973, Generation of chemotactic factor in guinea pig serum via activation of the classical and alternate complement pathways, *Clin. Immunol. Immunopathol.* **1**:414.
Cochrane, C. G., 1968, Immunologic tissue injury mediated by neutrophilic leucocytes, *Adv. Immunol.* **9**:97.
Cochrane, C. G., and Müller-Eberhard, H. J., 1968, The derivation of two distinct anaphylatoxin activities from the third and fifth components of human complement, *J. Exp. Med.* **127**:371.
Colten, H. R., 1972a, Ontogeny of the human complement system: *In vitro* biosynthesis of individual complement components by fetal tissues, *J. Clin. Invest.* **51**:725.
Colten, H. R., 1972b, *In vitro* systhesis of a regulator of mammalian gene expression, *Proc. Natl. Acad. Sci. USA* **69**:2233.
Colten, H. R., 1973, Biosynthesis of the fifth component of complement (C5) by human fetal tissues, *Clin. Immunol. Immunopathol.* **1**:346.
Colten, H. R., 1976, Biosynthesis of complement, *Adv. Immunol.* **22**:67.

Colten, H. R., and Frank, M. M., 1972, Biosynthesis of the second (C2) and fourth (C4) components of complement *in vitro* by tissues isolated from guinea pigs with genetically determined C4 deficiency, *Immunology* **22:**991.

Colten, H. R., and Parkman, R., 1972, Biosynthesis of C4 (fourth component of complement) by hybrids of C4-deficient guinea pig cells and HeLa cells. *Science* **176:**1029.

Cook, C. T., Shin, H. S., Mayer, M. M., and Laudenslayer, K. A., 1971, The fifth component of the guinea pig complement system. I. Purification and characterization, *J. Immunol.* **106:**467.

Cooper, N. R., 1969, Immune adherence by the fourth component of complement, *Science* **165:**396.

Cooper, N. R., 1975, Enzymatic activity of the second component of complement, *Biochemistry* **14:**4245.

Cooper, N. R., and Müller-Eberhard, H. J., 1970, The reaction mechanism of human C5 in immune hemolysis, *J. Exp. Med.* **132:**775.

Crisler, C., and Frank, M. M., 1965, Skin graft rejection and the Arthus reaction in mice deficient in the third component (C3) of complement, *Fed. Proc.* **24:**447.

Daha, M. R., Fearon, D. T., and Austen, K. F., 1976, C3 requirements for formation of alternative pathway C5 convertase, *J. Immunol.* **117:**630.

Delage, J. M., Lehner-Netsch, G., LaFleur, R., and Simard, J., 1979, Simultaneous occurrence of hereditary C6 and C2 deficiency in a French-Canadian family, *Immunology* **37:**419.

Diamond, R. D., May, J. E., Kane, M. A., Frank, M. M., and Bennett, J. E., 1974, The role of the classical and alternate complement pathways in host defenses against *Cryptococcus neoformans* infection, *J. Immunol.* **112:**2260.

Dodds, W. J., Raymond, S. L., Moynihan, A. C., Pickering, R. J., and Amiraian, K., 1977, Coagulation studies in C4-deficient guinea pigs, *Proc. Soc. Exp. Biol. Med.* **154:**382.

Dubiski, S., and Cinader, B., 1966, Gene dosage effect of serum concentration of a complement component, MuB1, *Proc. Soc. Exp. Biol. Med.* **122:**775.

Easmon, C. S. F., and Glynn, A. A., 1976, Comparison of subcutaneous and intraperitoneal staphylococcal infections in normal and complement deficient mice, *Infect. Immun.* **13:**399.

Ellman, L., Green, I., and Frank, M., 1970, Genetically controlled deficiency of the fourth component of complement in the guinea pig, *Science* **170:**74.

Ellman, L., Green, I., Judge, F., and Frank, M. M., 1971, *In vivo* studies in C4-deficient guinea pigs, *J. Exp. Med.* **134:**162.

Erickson, R. P., Tachibana, D. K., Herzenberg, L. A., and Rosenberg, L. T., 1964, A single gene controlling hemolytic complement and a serum antigen in the mouse, *J. Immunol.* **92:**611.

Fernandez, H. N., Henson, P. M., Otani, A., and Hugli, T. E., 1978, Chemotactic response to human C3a and C5a anaphylatoxins, *J. Immunol.* **120:**109.

Ferrone, S., Pellegrino, M. A., and Cooper, N. R., 1976, Expression of C4 on human lymphoid cells and possible involvement in immune recognition phenomena, *Science* **193:**53.

Fong, J. S. C., Rother, K. O., and Good, R. A., 1974, The role of complement in the localized Shwartzman reaction: Studies with congenitally C6-deficient rabbits, *Clin. Exp. Immunol.* **16:**77.

Frank, M. M., 1977, The effect of IgM and IgG complement fixing antibodies and of complement on erythrocyte survival, in: *The Nature and Significance of Complement Activation* (W. Pollack, P. Mollison, and A. M. Reiss, eds.), pp. 13–27, Ortho Research Institute of Medical Sciences, Raritan, N.J.

Frank, M. M., May, J., Gaither, T., and Ellman, L., 1971, *In vitro* studies of complement function in sera of C4-deficient guinea pigs, *J. Exp. Med.* **134:**176.

Frank, M. M., Ellman, L., Green, I., and Cochrane, C. G., 1973, Site of deposition of C3 in Arthus reactions of C4 deficient guinea pigs, *J. Immunol.* **110:**1447.

Fujita, T., and Nussenzweig, V., 1979, The role of C4-binding protein and β1H in proteolysis of C4b and C3b, *J. Exp. Med.* **150:**267.

Gabrielsen, A. E., Linna, T. J., Weitekamp, D. P., and Pickering, R. J., 1974, Reduced haemolytic C1 activity in serum of hypogammaglobulinaemic chickens, *Immunology* **27:**463.

Gallin, J. I., Clark, R. A., and Frank, M. M., 1975, Kinetic analysis of chemotactic factor generation in human serum via activation of the classical and alternate complement pathways, *Clin. Immunol. Immunopathol.* **3:**334.

Gelfand, J. A., Hurley, D. L., Fauci, A. S., and Frank, M. M., 1978, Role of complement in host defense against experimental disseminated candidiasis, *J. Infect. Dis.* **138:**9.

Gewurz, H., Shin, H. S., and Mergenhagen, S. E., 1968a, Interactions of the complement system with endotoxin lipopolysaccharide: Consumption of each of the six terminal complement components, *J. Exp. Med.* **128**:1049.

Gewurz, H., Pickering, R. J., Christian, C. L., Snyderman, R., Mergenhagen, S. E., and Good, R. A., 1968b, Decreased C1q protein concentration and agglutinating activity in agammaglobulinemia syndromes: An inborn error reflected in the complement system, *Clin. Exp. Immunol.* **3**:437.

Ghebrehiwet, B., and Müller-Eberhard, H. J., 1979, C3e: An acidic fragment of human C3 with leucocytosis-inducing activity. *J. Immunol.* **123**:616.

Gigli, I., 1978, A single chain precursor of C4 in human serum, *Nature (London)* **272**:836.

Gigli, I., Zabern, I. V., and Porter, R. R., 1977, The isolation and structure of C4, the fourth component of human complement, *Biochem. J.* **165**:439.

Glass, D., Raum, D., Balavitch, D., Kagan, E., Rabson, A., Schur, P. H., and Alper, C. A., 1978, Inherited deficiency of the sixth component of complement: A silent or null gene, *J. Immunol.* **120**:538.

Glynn, A. A., and Medhurst, F. A., 1967, Possible extracellular and intracellular bactericidal actions of mouse complement, *Nature (London)* **213**:608.

Goldlust, M. B., Shin, H. S., Hammer, C. H., and Mayer, M. M., 1974, Studies of complement complex C5b,6 eluted from EAC16: Reaction of C5b,6 with EAC4b,3b and evidence on the role of C2a and C3b in the activation of C5, *J. Immunol.* **113**:998.

Goldman, J. N., and Goldman, M. B., 1977, Antibody induced suppression of the fifth component of complement in mice, *J. Immunol.* **120**:400.

Goldman, M. B., French, D. L., and Goldman, J. N., 1979, Cell-mediated suppression of the fifth component of complement in mice, *J. Immunol.* **122**:246.

Gordon, J., Whitehead, H. R., and Wormall, A., 1926, CXXIX. The action of ammonia on complement. The fourth component. *Biochem. J.* **20**:1028.

Götze, O., and Müller-Eberhard, H. J., 1970, Lysis of erythrocytes by complement in the absence of antibody, *J. Exp. Med.* **132**:898.

Hadding, V., Bitter-Suermann, D., and Wellensick, H. J., 1970, Independent and consecutive action of C5, C6 and C7 in immune hemolysis. II. Formation and decay of the intermediate complexes EAC1-5 and EAC1-6, *Immunochemistry* **7**:967.

Hall, R. E., and Colten, H. R., 1977a, Molecular size and subunit structure of the fourth component of guinea pig complement, *J. Immunol.* **118**:1903.

Hall, R. E., and Colten, H. R., 1977b, Cell-free synthesis of the fourth component of guinea pig complement (C4): Identification of a precursor of serum C4 (pro-C4), *Proc. Natl. Acad. Sci. USA* **74**:1707.

Halpern, B. N., Biozzi, G., Benacerraf, B., and Stiffel, C., 1957, Phagocytosis of foreign red cells by the reticulo-endothelial system, *Am. J. Physiol.* **189**:520.

Hauptman, G., Grosshans, E., and Heid, E., 1974, Lupus érythémateux aigus et déficits héréditaires en complement: A propos d'un cas par deficit complement C4, *Ann. Dermatol. Syphiligr.* **101**:479.

Hawley, C. E., and Falkler, W. A., Jr., 1977, Anticomplementary activity of *Fusobacterium polymorphum* in normal and C4-deficient sources of guinea pig complement, *Infect. Immun.* **18**:124.

Herzenberg, L. A., Tachibana, D. K., Herzenberg, L. A., and Rosenberg, L. T., 1963, A gene locus concerned with hemolytic complement in *Mus musculus, Genetics* **48**:711.

Heusinkveld, R. S., Leddy, J. P., Klemperer, M. R., and Breckenridge, R. T., 1974, Hereditary deficiency of the sixth component of complement in man. II. Studies of hemostasis, *J. Clin. Invest.* **53**:554.

Hicks, J. T., Ennis, F. A., Kim, E., and Verbonitz, M., 1978, The importance of an intact complement pathway in recovery from a primary viral infection: Influenza in decomplemented and in C5-deficient mice, *J. Immunol.* **121**:1437.

Hobart, M. J., Cook, P. J. L., and Lachmann, P. J., 1977a, Linkage studies with C6, *J. Immunogenet.* **4**:423.

Hobart, M. J., Lachmann, P. J., and Calne, R. Y., 1977b, C6: Synthesis by the liver *in vivo, J. Exp. Med.* **146**:629.

Howie, J. B., and Helyer, B. J., 1965, Autoimmune disease in mice, *Ann. N.Y. Acad. Sci.* **124**:167.

Hyde, R. R., 1923, Complement-deficient guinea pig serum, *J. Immunol.* **8**:267.

Ilgen, C. L., and Burkholder, P. M., 1974a, Isolation of C4-synthesizing cells from guinea pig liver by Ficoll density gradient centrifugation, *Immunology* **26**:197.

Ilgen, C. L., and Burkholder, P. M., 1974b, Effect of anti-C4 on cellular synthesis of C4 by guinea pig peritoneal macrophages, pulmonary macrophages, liver or spleen cells, *J. Reticuloendothelial Soc.* **15**:185.

Ilgen, C. L., Bossen, E. H., Rowlands, D. T., Jr., and Burkholder, P. M., 1974, Isolation and characterization of C4-synthesizing cells from guinea pig spleen, *Immunology* **26:**659.

Isenman, D. E., Podack, E. R., and Cooper, N. R., 1980, The interaction of C5 with C3b in the free solution: A sufficient condition for cleavage by a fluid phase C3/C5 convertase, *J. Immunol.* **124:**326.

Jackson, C. G., Ochs, H. D., and Wedgwood, R. J., 1979, Immune response of a patient with deficiency of the fourth component of complement and systemic lupus erythematosus, *N. Engl. J. Med.* **300:**1124.

Jarvinen, J. A., and Dalmasso, A. P., 1977, *Trypanosoma musculi* infections in normocomplementemic, C5-deficient and C3-depleted mice, *Infect. Immun.* **16:**557.

Jenson, J. A., Garces, M. C., and Iglesias, E., 1971, Specific inactivation of the fourth complement of component. I. *In vivo* studies, *Infect. Immun.* **4:**12.

Johnson, K. J., and Ward, P. A., 1971, Protective function of C6 in rabbits treated with bacterial endotoxin, *J. Immunol.* **106:**1125.

Johnston, R. B., Jr., Klemperer, M. R., Alper, C. A., and Rosen, F. S., 1969, The enhancement of bacterial phagocytosis by serum, the role of complement components and two cofactors, *J. Exp. Med.* **129:**1275.

Jonsen, J., and Kass, E., 1957, Agammaglobulinemia—Report of components of complement in a case, *Acta Pathol. Microbiol. Scand.* **41:**393.

Kane, M. A., May, J. E., and Frank, M. M., 1973, Interactions of the classical and alternate complement pathways with endotoxin lipopolysaccharide: Effect on platelets and blood coagulation, *J. Clin. Invest.* **52:**370.

Keystone, E., Taylor-Robinson, D., Pope, C., Taylor, G., and Furr, P., 1978, Effect of inherited deficiency of the fifth component of complement on arthritis induced in mice by *Mycoplasma pulmonis, Arthritis Rheum.* **21:**792.

Kohler, P. F., 1973, Maturation of the human complement system. 1. Onset time and sites of fetal C1q, C4, C3 and C5 synthesis, *J. Clin. Invest.* **52:**671.

Kohler, P. F., and Müller-Eberhard, H. J., 1969, Complement-immunoglobulin relation: Deficiency of C1q associated with impaired immunoglobulin G synthesis, *Science* **163:**474.

Kohler, P. F., and Müller-Eberhard, H. J., 1972, Metabolism of human C1q, studies in hypogammaglobulinemia, myeloma, and systemic lupus erythematosus, *J. Clin. Invest.* **51:**868.

Kolb, W. P., Haxby, J. A., Arroyave, C. M., and Müller-Eberhard, H. J., 1972, Molecular analysis of the membrane attack mechanism of complement, *J. Exp. Med.* **135:**549.

Krönke, M., Hadding, U., Geczy, A. F., DeWeck, A. L., and Bitter-Suermann, D., 1977, Linkage of guinea pig Bf and C4 to the GPLA, *J. Immunol.* **119:**2016.

Lachmann, P. J., 1970, C6-deficiency in rabbits, in: *Protides of the Biological Fluids* (H. Peeters, ed.), pp. 301–309, Pergamon Press, Oxford.

Lachmann, P. J., and Thompson, R. A., 1970, Reactive lysis: The complement-mediated lysis of unsensitized cells. II. The characterization of activated reactor as $\overline{C56}$ and the participation of C8 and C9, *J. Exp. Med.* **131:**643.

Lachmann, P. J., Munn, E. A., and Weissmann, G., 1970a, Complement mediated lysis of liposomes produced by the reactive lysis procedure, *Immunology* **19:**983.

Lachmann, P. J., Kay, A. B., and Thompson, R. A., 1970b, The chemotactic activity for neutrophil and eosinophil leukocytes of the trimolecular complex of the fifth, sixth and seventh components of human complement ($\overline{C567}$) prepared in free solution by the reactive lysis procedure, *Immunology* **19:**895.

Lai A Fat, R. F. M., and van Furth, R. 1975, *In vitro* synthesis of some complement components (C1q, C3 and C4) by lymphoid tissues and circulating leukocytes in man, *Immunology* **28:**359.

Landen, B., and Dierich, M. P., 1979, Identity of C3- and C5-receptors on lymphoid cells, *J. Immunol.* **122:**1015.

Lanier, B. G., McDuffie, F. C., and Holley, K. E., 1971, Role of C5 in the nephritis of NZB/W mice, *J. Immunol.* **106:**740.

Lawrence, D. A., and Schell, R. F., 1978, Susceptibility of C5-deficient mice to listeriosis: Modulation by concanavalin A, *Cell. Immunol.* **39:**336.

Leddy, J. P., Frank, M. M., Gaither, T., Baum, J., and Klemperer, M. R., 1974, Hereditary deficiency of the sixth component of complement in man. I. Immunochemical, biologic, and family studies, *J. Clin. Invest.* **53:**544.

Lepow, I. H., Ratnoff, O. D., and Pillemer, L., 1956, Elution of an esterase from antigen-antibody aggregates treated with human complement, *Proc. Soc. Exp. Biol. Med.* **92:**111.

Levine, L., 1955, Inhibition of immune hemolysis by diisopropylfluorophosphate, *Biochim. Biophys. Acta* **18**:283.

Levy, N. L., and Ladda, R. L., 1971, Restoration of haemolytic complement activity in C5-deficient mice by gene complementation in hybrid cells, *Nature New Biol.* **229**:51.

Levy, N. L., Snyderman, R., Ladda, R. L., and Lieberman, R., 1973, Cytogenetic engineering *in vivo:* Restoration of biologic complement activity to C5-deficient mice by intravenous inoculation of hybrid cells, *Proc. Natl. Acad. Sci. USA* **70**:3125.

Lim, D., Gewurz, A., Lint, T. F., Chaze, M., Sephein, B., and Gewurz, H., 1976, Absence of the sixth component of complement in a patient with repeated episodes of meningococcal meningitis, *J. Pediatr.* **89**:42.

Lindberg, L. H., and Rosenberg, L. T., 1968, Nephrotoxic serum nephritis in mice with a genetic deficiency in complement, *J. Immunol.* **100**:34.

Linscott, W. D., and Cochrane, C. G., 1964, Guinea pig β1C-globulin: Its relationship to the third component of complement and its alteration following interaction with immune complexes, *J. Immunol.* **93**:972.

Littleton, C., Keisler, D., and Burkholder, P. M., 1970, Cellular basis for synthesis of the fourth component of guinea pig complement as determined by a haemolytic plaque technique, *Immunology* **18**:693.

Loos, M., Laurell, A. B., Sjöholm, A. G., Mårtensson, U., and Berkel, A. I., 1980, Immunochemical and functional analysis of a complete C1q deficiency in man: Evidence that C1r and C1s are in the native form, and that they reassociate with purified C1q to form macromolecular C1, *J. Immunol.* **124**:59.

Marcus, R. L., Shin, H. S., and Mayer, M. M., 1970, An alternate complement pathway: Evidence for C3 convertase activity other than $C\overline{4,2}$ on serum treated endotoxin lipopolysaccharide, *Fed. Proc.* **29**:304.

Martinelli, G. P., Matsuda, T., Waks, H. S., and Osler, A. G., 1978, Studies on immunosuppression by cobra venom factor III. On early responses to sheep erythrocytes in C5-deficient mice, *J. Immunol.* **121**:2052.

May, J. E., and Frank, M. M., 1972, Complement-mediated tissue damage: Contribution of the classical and alternate complement pathways in the Forssman reaction, *J. Immunol.* **108**:1517.

May, J. E., Kane, M. A., and Frank, M. M., 1972a, Host defense against bacterial endotoxemia—Contribution of the early and late components of complement to detoxification, *J. Immunol.* **109**:893.

May, J. E., Kane, M. A., and Frank, M. M., 1972b, Immune adherence by the alternate complement pathway, *Proc. Soc. Exp. Biol. Med.* **141**:287.

May, J. E., Green, I., and Frank, M. M., 1972c, The alternate complement pathway in cell damage: Antibody-mediated cytolysis of erythrocytes and nucleated cells, *J. Immunol.* **109**:595.

Mayer, M. M., 1972, Mechanism of cytolysis by complement, *Proc. Natl. Acad. Sci. USA* **69**:2954.

Mayer, M. M., 1978, Complement, past and present, in: *The Harvey Lectures* (Series 72, 1976–1977), pp. 139–193, Academic Press, New York.

Mayer, M. M., Asher, E. T., and Borsos, T., 1962, Inhibition of guinea pig C′2 by rabbit anti-C′2, *Fed. Proc.* **21**:17.

Mayer, M. M., Hammer, C. H., Michaels, D. W., and Shin, M. L., 1979, Immunologically mediated membrane damage: The mechanism of complement action and the similarity of lymphocyte-mediated cytotoxicity, *Immunochemistry* **15**:813.

Medicus, R. G., Götze, O., and Müller-Eberhard, H. J., 1976a, The serine protease nature of the C3 and C5 convertases of the classical and alternative complement pathways, *Scand. J. Immunol.* **5**:1049.

Medicus, R. G., Götze, O., and Müller-Eberhard, H. J., 1976b, Alternative pathway of complement: Recruitment of precursor properdin by the labile C3/C5 convertase and the potentiation of the pathway, *J. Exp. Med.* **144**:1076.

Meo, T., Krasteff, T., and Shreffler, D. C., 1975, Immunochemical characterization of murine H-2 controlled Ss (serum substance) protein through identification of its human homologue as the fourth component of complement, *Proc. Natl. Acad. Sci. USA* **72**:4536.

Minta, J. O., and Man, D. P., 1977, Cleavage of human C5 by trypsin: Characterization of the digestion products by gel electrophoresis, *J. Immunol.* **119**:1597.

Mittal, K. K., Wolski, D. P., Lim, D., Gewurz, A., and Schmid, F. R., 1976, Genetic independence between the HLA system and deficiency in the first and sixth components of complement, *Tissue Antigens* **7**:97.

Moore, H. D., 1919, Complementary and opsonic functions in their relation to immunity. A study of the serum of guinea pigs naturally deficient in complement, *J. Immunol.* **4**:425.

Morelli, R., and Rosenberg, L. T., 1971a, The role of complement in the phagocytosis of *Candida albicans* by mouse peripheral blood leukocytes, *J. Immunol.* **107**:476.

Morelli, R., and Rosenberg, L. T., 1971b, Role of complement during experimental *Candida* infection in mice, *Infect. Immun.* **3**:521.

Müller-Eberhard, H. J., 1968, Chemistry and reaction mechanisms of complement, *Adv. Immunol.* **3**:1.

Müller-Eberhard, H. J., 1975, Complement, *Annu. Rev. Biochem.* **44**:697.

Müller-Eberhard, H. J., and Biro, C. E., 1963, Isolation and description of the fourth component of human complement, *J. Exp. Med.* **118**:447.

Müller-Eberhard, H. J., and Kunkel, H. G., 1961, Isolation of a thermolabile serum protein which precipitates γ-globulin aggregates and participates in immune hemolysis, *Proc. Soc. Exp. Biol. Med.* **106**:29.

Müller-Eberhard, H. J., Polley, M. J., and Calcott, M. A., 1967, Formation and functional significance of a molecular complex derived from the second and the fourth component of complement, *J. Exp. Med.* **125**:359.

Nagasawa, S., Shiraishi, S., and Stroud, R M., 1976, The molecular structure of C4 and its cleavage products, *J. Immunol.* **116**:1743.

Nakamura, R. M., and Weigle, W. O., 1968, Experimental thyroiditis in complement intact and deficient mice following injections of heterologous thyroglobulins without adjuvant, *Proc. Soc. Exp. Biol. Med.* **129**:412.

Nelson, R. A., Jr., Jensen, J., Gigli, I., and Tamura, N., 1966, Methods for the separation, purification and measurement of the nine components of hemolytic complement in guinea-pig serum, *Immunochemistry* **3**:111.

Nelson, R. A., Jr., and Biro, C. E., 1968, Complement components of a haemolytically deficient strain of rabbits, *Immunology* **14**:525.

Nilsson, U. R., 1967, Separation and partial purification of the sixth, seventh and eighth components of human haemolytic complement, *Acta Pathol. Microbiol. Scand.* **70**:469.

Nilsson, U. R., and Müller-Eberhard, H. J., 1965, Isolation of β1F-globulin from human serum and its characterization as the fifth component of complement, *J. Exp. Med.* **122**:277.

Nilsson, U. R., and Müller-Eberhard, H. J., 1967, Deficiency of the fifth component of complement in mice with an inherited complement defect, *J. Exp. Med.* **124**:1.

Nilsson, U. R., Tomar, R. H., and Taylor, F. B., Jr., 1972, Additional studies on human C5: Development of a modified purification method and characterization of the purified product by polyacrylamide gel electrophoresis, *Immunochemistry* **9**:709.

Nilsson, U. R., Mandle, R. J., Jr., and McConnell-Mapes, J. A., 1975, Human C3 and C5: Subunit structure and modifications by trypsin and $\overline{C43}$–$\overline{C423}$, *J. Immunol.* **114**:815.

Ochs, H. D., Rosenfeld, S. I., Thomas, E. D., Giblett, E. R., Alper, C. A., Dupont, B., Schaller, J. G., Gilliland, B. C., and Hansen, J. A., 1977, Linkage between the fourth component of complement and the major histocompatibility complex, *N. Engl. J. Med.* **296**:470.

Ochs, H. D., Jackson, C. G., Heller, S. R., and Wedgwood, R. J., 1978, Defective antibody response to a T-dependent antigen in C4 deficient guinea pigs and its correction by addition of C4, *Fed. Proc.* **37**:1477.

Ochs, H. D., Heller, S. R., Jackson, C. G., Osofsky, S. G., Rabson, A. R., and Wedgwood, R. J., 1979, Complement and the antibody response, *Clin. Res.* **27**:322A.

Olving, J. H., Olaisen, B., Gedde-Dahl, T., Jr., and Teisberg, P., 1979, Genetic linkage relations of the sixth component of complement (C6), *Humangenetik* **46**:181.

O'Neill, G. J., Yang, S. Y., and Dupont, B., 1978, Two HLA-linked loci controlling the fourth component of human complement, *Proc. Natl. Acad. Sci. USA* **75**:5165.

Ooi, Y. M., and Colten, H. R., 1979, Genetic defect in secretion of complement C5 in mice, *Nature (London)* **282**:207.

Orr, W., Varani, J., and Ward, P. A., 1978, Characteristics of the chemotactic response of neoplastic cells to a factor derived from the fifth component of complement, *Am. J. Pathol.* **93**:405.

Parker, K. L., Roos, M. H., and Schreffler, D. C., 1979, Structural characterization of the murine fourth component of complement and sex-limited protein and their precursors: Evidence for two loci in the S region of the H-2 complex, *Proc. Natl. Acad. Sci. USA* **76**:5853.

Patel, F., and Minta, J. O., 1979, Biosynthesis of a single chain pro-C5 by normal mouse liver mRNA: Analysis of the molecular basis of C5 deficiency in AKR/J mice, *J. Immunol.* **123**:2408.

Patrick, R. A., Taubman, S. B., and Lepow, I. H., 1970, Cleavage of the fourth component of human complement (C4) by activated C1s, *Immunochemistry* **7**:217.

Pepys, M. B., 1976, Role of complement in the induction of immunological responses, *Transplant. Rev.* **32**:93.

Phan, S. H., and Ward, P. A., 1979, Generation of biologic activity from the purified α-chain of C5, *J. Immunol.* **123**:2735.

Phillips, M. E., Rother, V., and Rother, K., 1968, Serum complement in the rejection of sarcoma I ascites tumor grafts, *J. Immunol.* **100**:493.

Phillips, M. E., Rother, V. A., Rother, K. O., and Thorbecke, G. J., 1969, Studies on serum proteins of chimeras. III. Detection of donor-type C′5 in allogeneic and congenic post-irradiation chimeras, *Immunology* **17**:315.

Podack, E. R., Kolb, W. P., and Müller-Eberhard, H. J., 1976, Purification of the sixth and seventh component of human complement without loss of hemolytic activity, *J. Immunol.* **116**:263.

Podack, E. R., Kolb, W. P., and Müller-Eberhard, H. J., 1978a, The C5b-6 complex: Formation, isolation, and inhibition of its activity by lipoprotein and the S-protein of human serum, *J. Immunol.* **120**:1841.

Podack, E. R., Biesecker, G., Kolb, W. P., and Müller-Eberhard, H. J., 1978b, The C5b-6 complex: Reaction with C7, C8, C9, *J. Immunol.* **121**:484.

Podack, E. R., Kolb, W. P., Esser, A. F., and Müller-Eberhard, H. J., 1979, Structural similarities between C6 and C7 of human complement, *J. Immunol.* **123**:1071.

Polley, M. J., and Müller-Eberhard, H. J., 1968, The second component of human complement: Its isolation, fragmentation by C′1 esterase and incorporation into C′3 convertase, *J. Exp. Med.* **128**:533.

Raum, D., Glass, D., Carpenter, C. B., and Schur, P. H., 1976, C6 deficiency–HLA association, *Fed. Proc.* **35**:655.

Rittner, C., Hauptmann, G., Grosse-Wilde, H., Grosshans, E., Tongio, M. M., and Mayer, S., 1975, Linkage between HL-A "major histocompatibility complex" and genes controlling the syntehsis of the fourth component of complement, in: *Histocompatibility Testing* (F. Kissmeyer-Nielsen, ed.), pp. 945–954, Munksgaard, Copenhagen.

Root, R. K., Ellman, L., and Frank, M. M., 1972, Bacterial and opsonic properties of C4-deficient guinea pig serum, *J. Immunol.* **109**:477.

Rosenberg, L. T., and Tachibana, D. K., 1962, Activity of mouse complement, *J. Immunol.* **89**:861.

Rosenfeld, S. I., Kelly, M. E., and Leddy, J. P., 1976a, Hereditary deficiency of the fifth component of complement in man. I. Clinical, immunochemical, and family studies, *J. Clin. Invest.* **57**:1626.

Rosenfeld, S. I., Baum, J., Steigbigel, R. T., and Leddy, J. P., 1976b, Hereditary deficiency of the fifth component of complement in man. II. Biological properties of C5-deficient human serum, *J. Clin. Invest.* **57**:1635.

Rosenfeld, S. I., Weitkamp, L. R., and Ward, F., 1977, Hereditary C5 deficiency in man: Genetic linkage studies, *J. Immunol.* **119**:604.

Rother, K., and Till, G., 1974, *In vivo*-Störungen des komplementsystems, in: *Komplement, Biochemie and Pathologie* (K. Rother, ed.), p. 238, Steinkopff, Darmstadt.

Rother, K., Rother, U., Petersen, K. F., Gemsa, D., and Mitze, F., 1964a, Immune bactericidal activity of complement: Separation and description of intermediate steps, *J. Immunol.* **93**:319.

Rother, K., Rother, U., and Schindera, F., 1964b, Passive Arthus-reaktion bei komplementdefektin kaninchen, *Z. Immunitaetsforsch.* **126**:473.

Rother, K., Rother, U., Müller-Eberhard, H. J., and Nilsson, U., 1966, Deficiency of the sixth component of complement in rabbits with an inherited complement defect, *J. Exp. Med.* **124**:773.

Rother, K., McCluskey, R. T., and Rother, U., 1967a, Tuberculin hypersensitivity in C6 deficient rabbits, *Fed. Proc.* **26**:787.

Rother, K., Rother, U., Vassalli, P., and McCluskey, R. T., 1967b, Nephrotoxic serum nephritis in C6-deficient rabbits. I. Study of the second phase of the disease, *J. Immunol.* **98**:965.

Rother, U., and Rother, K., 1961, Uber einen angeborenen komplementdefekt bei kaninchen, *Z. Immunitaetsforsch.* **121**:224.

Rother, U., Ballantyne, D. L., Cohen, C., and Rother, K., 1967, Allograft rejection in C6 defective rabbits, *J. Exp. Med.* **126**:565.

Rother, U., Thorbecke, G. J., Stecher-Levin, V. J., Hurlimann, J., and Rother, K., 1968, Formation of C′6 by rabbit liver tissue *in vitro, Immunology* **14**:649.

Salerno, A., 1969, Survival of transplanted spleen cells in allogeneic mice. The effects of strength of histoincompatibility and of C′ deficiency, *Transplantation* **8**:189.

Sandberg, A. L., Osler, A. G., Shin, H. S., and Oliveira, B., 1970, The biological activities of guinea pig antibodies: Modes of complement interaction with γ_1 and γ_2 immunoglobulins, *J. Immunol.* **104**:329.

Schaller, J. G., Gilliland, B. G., Ochs, H. D., Leddy, J. P., Agodoa, L. C. Y., and Rosenfeld, S. I., 1977, Severe systemic lupus erythematosus with nephritis in a boy with deficiency of the fourth component of complement, *Arthritis Rheum.* **20**:1519.

Schreffler, D. C., and Owen, R. D., 1963, A serologically detected variant in mouse serum: Inheritance and association with the histocompatibility-2 locus, *Genetics* **48**:9.

Schreiber, R. D., and Müller-Eberhard, H. J., 1974, Fourth component of human complement: Description of a three polypeptide chain structure, *J. Exp. Med.* **140:**1324.

Segerling, M., and Opferkuch, W., 1977, C5-dependent enhancement of rosette formation and phagocytosis by human polymorphonuclear leucocytes, *Monogr. Allergy* **12:**101.

Shevach, E. M., Frank, M. M., and Green, I., 1976, Linkage of the gene controlling synthesis of the fourth component of complement to the major histocompatibility complex of the guinea pig, *Immunogenetics* **3:**595.

Shin, H. S., and Mayer, M. M. 1968, The third component of the guinea pig complement system. II. Kinetic study of the reaction of EAC$\overline{4,2}$a with guinea pig C′3. Enzymatic nature of C3 consumption, multi-phase character of fixation, and hemolytic titration of C′3, *Biochemistry* **7:**2997.

Shin, H. S., Snyderman, R., Friedman, E., Mellors, A., and Mayer, M. M., 1968, Chemotactic and anaphylatoxic fragment cleaved from the fifth component of guinea pig complement, *Science* **162:**361.

Shin, H. S., Smith, M. R., and Wood, W. B., Jr., 1969, Heat labile opsonins to pneumococcus. II. Involvement of C3 and C5, *J. Exp. Med.* **130:**1229.

Shin, H. S., Pickering, R. J., and Mayer, M. M., 1971a, The fifth component of the guinea pit complement system. II. Mechanism of SAC$\overline{1,4,2,3}$,5b formation and C5 consumption by EAC$\overline{1,4,2,3}$, *J. Immunol.* **106:**473.

Shin, H. S., Pickering, R. J., and Mayer, M. M., 1971b, The fifth component of the guinea pig complement system. III. Dissociation and transfer of C5b, and the probable site of C5b fixation, *J. Immunol.* **106:**480.

Shiraishi, S., and Stroud, R. M., 1975, Cleavage products of C4b produced by enzymes in human serum, *Immunochemistry* **12:**935.

Siboo, R., and Vas, S. I., 1965, Studies on *in vitro* antibody production. III. Production of complement. *Can. J. Microbiol.* **11:**415.

Siriganian, R. P., 1972, Platelet requirement in the interaction of the complement and clotting systems, *Nature New Biol.* **239:**208.

Sitomer, G., Stroud, R. M., and Mayer, M. M., 1966, Reversible adsorption of C′2 by EAC′4: Role of Mg^{2+}, enumeration of competent SAC′4, two-step nature of C′2a fixation and estimation of its efficiency, *Immunochemistry* **3:**57.

Snyderman, R., Phillips, J., and Mergenhagen, S. E., 1970, Polymorphonuclear leukocyte chemotactic activity in rabbit serum and guinea pig serum treated with immune complexes: Evidence for C5a as the major chemotactic factor, *Infect. Immun.* **1:**521.

Snyderman, R., Phillips, J. K., and Mergenhagen, S. E., 1971, Biological activity of complement *in vivo:* Role of C5 in the accumulation of polymorphonuclear leukocytes in inflammatory exudates, *J. Exp. Med.* **134:**1131.

Spiegelberg, H., Miescher, P., and Benacerraf, B., 1963, Studies on the role of complement in the immune clearance of *Escherichia coli* and rat erythrocytes by the reticuloendothelial system in mice, *J. Immunol.* **90:**751.

Stafford, B. T., Rapaport, S. I., and Shen, S. M. C., 1976, Effects of endotoxin in cortisone-treated rabbits with a hereditary deficiency of the sixth component of complement (C6 deficiency), *Thromb. Haemostas.* **35:**460.

Stecher, V. J., and Sorkin, E., 1969, Studies on chemotaxis. XII. Generation of chemotactic activity for polymorphonuclear leucocytes in sera with complement deficiencies, *Immunology* **16:**231.

Stecher, V. J., and Thorbecke, G. J., 1967a, Sites of synthesis of serum proteins. I. Serum proteins produced by macrophages *in vitro, J. Immunol.* **99:**643.

Stecher, V. J., and Thorbecke, G. J., 1967b, Sites of synthesis of serum proteins. III. Production of β1C, β1E, and transferrin by primate and rodent cell lines, *J. Immunol.* **99:**660.

Stecher, V. J., Morse, J. H., and Thorbecke, G. J., 1967, Sites of production of primate serum proteins associated with the complement system, *Proc. Soc. Exp. Biol. Med.* **124:**433.

Stiffel, C., Biozzi, G., Mouton, D., Bouthillier, Y., and Decreusefond, C., 1964, Studies on phagocytosis of bacteria by the reticuloendothelial system in a strain of mice lacking hemolytic complement, *J. Immunol.* **93:**246.

Stroud, R. M., Mayer, M. M., Miller, J. A., and McKenzie, A. T., 1966, C′2a^{d}, an inactive derivative of C′2 released during decay of EAC′4,2a, *Immunochemistry* **3:**163.

Strunk, R. C., Tashjian, A. H., and Colten, H. R., 1975, Complement biosynthesis *in vitro* by rat hepatoma cell strains, *J. Immunol.* **114:**331.

Strunk, R. C., John, T. J., and Sieber, O. F., 1977, Herpes simplex virus infections in guinea pigs deficient in the fourth component of complement, *Infect. Immun.* **15:**165.

Sveen, K., 1978, The importance of C5 and the role of the alternative complement pathway in leukocyte chemotaxis induced *in vivo* and *in vitro* by *Bacteroides fragilis* lipopolysaccharide, *Acta Pathol. Microbiol. Scand.* **86**:93.

Tachibana, D. K., and Rosenberg, L. T., 1966, Fetal synthesis of Hc′, a component of mouse complement, *J. Immunol.* **97**:213.

Tachibana, D. K., Ulrich, M., and Rosenberg, L. T., 1963, The inheritance of hemolytic complement activity in CF-1 mice, *J. Immunol.* **91**:230.

Tachibana, D. K., Hayflick, L., and Rosenberg, L. T., 1970, Effect of complement and genetically defective mouse complement on *Mycoplasma canis, J. Infect. Dis.* **121**:541.

Tack, B. F., Morris, S. C., and Prahl, J. W., 1979, Fifth component of human complement: Purification from plasma and polypeptide chain structure, *Biochemistry* **18**:1490.

Teisberg, P., Åkesson, I., Olaisen, B., Gedde-Dahl, T., Jr., and Thorsby, E., 1976, Genetic polymorphism of C4 in man and localization of a structural C4 locus to the HLA gene complex of chromosome 6, *Nature (London)* **264**:253.

Teisberg, P., Olaisen, B., Jonassen, R., Gedde-Dahl, T., Jr., and Thorsby, E., 1977, The genetic polymorphism of the fourth component of human complement: Methodological aspects and a presentation of linkage and association data relevant to its localization in the HLA region, *J. Exp. Med.* **146**:1380.

Terry, W. D., Borsos, T., and Rapp, H. J., 1964, Differences in serum complement activity among inbred strains of mice, *J. Immunol.* **92**:576.

Thompson, R. A., and Lachmann, P. J., 1970, Reactive lysis: The complement-mediated lysis of unsensitized cells. I. The characterization of the indicator factor and its identification as C7, *J. Exp. Med.* **131**:629.

Thorbecke, G. J., Hochwald, G. M., Furth, R. van, Müller-Eberhard, H. J., and Jacobson, E. B., 1965, Problems in determining the sites of synthesis of complement components in: *Ciba Foundation Symposium on Complement* (G. E. W. Wolstenholme and J. Knight, eds.), p. 99, Churchill, London.

Unanue, E. R., Mardiney, M. R., Jr., and Dixon, F. J., 1967, Nephrotoxic serum nephritis in complement intact and deficient mice, *J. Immunol.* **98**:609.

Vaz-Guedes, M. A., Hobart, M. J., and Lachmann, P. J., 1978, Absence of variation in human C5, *J. Immunogenet.* **5**:279.

Volk, H., Mauersberger, D., Rother, K., and Rother, U., 1964, Prolonged survival of skin homografts in rabbits defective in the third component of complement, *Ann. N.Y. Acad. Sci.* **120**:26.

Ward, P. A., and Newman, L. J., 1969, A neutrophil chemotactic factor from human C5, *J. Immunol.* **102**:93.

Ward, P. A., Cochrane, C. G., and Müller-Eberhard, H. J., 1965, The role of serum complement in chemotaxis of leukocytes *in vitro, J. Exp. Med.* **122**:327.

Ward, P. A., Cochrane, C. G., and Müller-Eberhard, H. J., 1966, Further studies on the chemotactic factor of complement and its formation *in vivo, Immunology* **11**:141.

Weitkamp, L. R., Rosenfeld, S., and Johnston, E., 1978, Complement C5: Immunofixation electrophoresis, quantitative variants, and nonlinkage to HLA, *Cytogenet. Cell Genet.* **22**:651.

Weitzel, H. K., and Rother, K., 1970, Studies on the role of serum complement in allograft rejection and in immunosuppression by antithymocyte serum (ATS), *Eur. Surg. Res.* **2**:310.

Wikel, S. K., 1979, Acquired resistance to ticks: Expression of resistance by C4-deficient guinea pigs, *Am. J. Trop. Med. Hyg.* **28**(3):586.

Wyatt, H. V., Colten, H. R., and Borsos, T., 1972, Production of the second (C2) and fourth (C4) components of guinea pig complement by single peritoneal cells: Evidence that one cell may produce both components, *J. Immunol.* **108**:1609.

Yamamoto, K., and Gewurz, H., 1978, The complex of C5b and C6: Isolation, characterization, and identification of a modified form of C5b consisting of three polypeptide chains, *J. Immunol.* **120**:2008.

Yang, S. Y., Jensen, R., Folke, L., Good, R. A., and Day, N. K., 1974, Complement deficiency in hamsters, *Fed. Proc.* **33**:795.

Zimmerman, T. S., Arroyave, C. M., and Müller-Eberhard, H. J., 1971, A blood coagulation abnormality in rabbits deficient in the sixth component of complement (C6) and its correction by purified C6, *J. Exp. Med.* **134**:1591.

10

Defects in Murine Responsiveness to Bacterial Lipopolysaccharide

The C3H/HeJ and C57BL/10ScCr Strains

ROBERT J. SCIBIENSKI

1. Introduction

The term lipopolysaccharide (LPS) by common usage has come to apply almost exclusively to the complex glycolipid component of the outer membrane of gram-negative bacteria. This component constitutes the endotoxic principle of these organisms and has been found to exhibit a broad variety of biological activities in higher organisms. In view of the fact that many of these activities impinge in one way or another on immunologic phenomena, it is not surprising that much of the work with LPS has been in this field. However, several of the biological effects of LPS are of broader interest to cell biologists and membrane biologists in general and some informative work on nonimmunologic aspects of these compounds has also been reported.

Research into the mechanism of action of LPS has been stimulated by the existence of inbred strains of mice that are uniquely insensitive to the majority of its biological effects. There were reports of inherited resistance to LPS as early as 1940 (Hill *et al.*, 1940), but it was not until a decade ago, when the immunologic properties of LPS were beginning to be recognized, that the genetic aspects of LPS activity began to receive serious attention. Since that time a considerable amount of information on the subject has accumulated and it is these studies that form the basis for this chapter. However, in order to orient the reader who perhaps may not be totally conversant with LPS biology, I have chosen to include some general background information on LPS and its various activities.

ROBERT J. SCIBIENSKI • Department of Medical Microbiology, School of Medicine, University of California, Davis, California 95616.

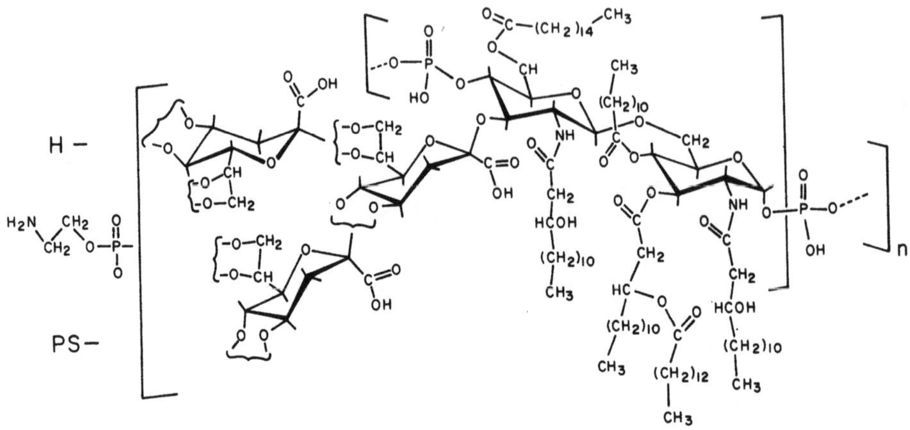

FIGURE 1. Proposed structure of bacterial lipopolysaccharide. The arrangement of the fatty acids is hypothetical, the exact arrangements being unknown and possibly variable. The exact arrangement of the KDO trisaccharide and constituents is also not known. PS–, polysaccharide side chain containing the "O" antigenic specificity of the molecule. See text for details and references.

2. Origin and Structure of LPS

LPS comprises a major component of the outer membrane of gram-negative bacteria. It is located in the exterior region of this membrane and its polysaccharide components constitute the major or "O" antigens of the bacterium. Its function in the bacterial cell is not known. LPS is routinely extracted from whole bacterial cells under rather harsh conditions involving treatment with concentrated organic solvents. The two most commonly used methods utilize either trichloroacetic acid (Boivin *et al.,* 1933) or hot, aqueous, 45% phenol (Westphal and Jann, 1965). More recently a third procedure utilizing aqueous butanol has been introduced (Morrison and Leive, 1975). It is now clear that lipopolysaccharides extracted by these different methods differ significantly in their properties, a point that will be dealt with later.

As pointed out, LPS is a complex glycolipid. For most lipopolysaccharides the basic structural unit consists of a β-1, 6-linked glucosamine dimer from which extend a minimum of five fatty acids (in both amide and ester linkages) and one glycosidically linked polysaccharide chain (Fig. 1). The glucosamine dimers invariable have phosphate groups at the 1 and 4′ positions and are thought to exist as pyrophosphate-linked polymers in the native form. However, the extent of and variability in the degree of polymerization are unknown. The glucosamine dimer with attached fatty acids is commonly referred to as lipid A.

The lipids of LPS consist of C12, C14, and C16 straight-chain fatty acids in ester linkage and one of a variety of hydroxy fatty acids in either amide or ester linkage (Rietschel *et al.,* 1972, 1977; Lüderitz, 1977). In some cases a second hydroxy fatty acid is thought to be esterified to the hydroxyl group of the first (Rietschel *et al.,* 1972, 1977). The polysaccharide "side-chain" component of LPS consists of three residues of 3-deoxy-D-mannoctulosonate (ketodeoxyoctonate, KDO) in branched linkage, followed by several heptoses and hexoses, and finally a repeating oligosaccharide unit, which may be either linear or branched, consisting of a variety of different hexoses (Lüderitz *et al.,* 1971). The exact linkage arrangements in the KDO "core" oligosaccharide have not been established. One or more ethanolamine residues may be linked to the KDO and heptose residues through phosphodiester bridges. (For a more thorough treatment of LPS structure the following references should be consulted: Lüderitz *et al.,* 1971; Lüderitz, 1977; Rietschel *et al.,* 1977.)

In addition to LPS, two other components that exhibit some activities in common with LPS can be extracted from gram-negative bacteria. Both of these components are lipoproteins in nature. One, in fact, coextracts with LPS when butanol or TCA is utilized as the deproteinizing agent (Skidmore *et al.,* 1975b; Sultzer and Goodman, 1976). This component is usually referred to as endotoxin protein (EP) or lipid-associated protein (LAP). The other protein component appears to normally reside at the inner surface of the exterior membrane and is covalently linked to the peptidoglycan of the cell wall (Braun, 1977).

3. Biological Activities of LPS

As pointed out, LPS exhibits a broad variety of biological activities in higher organisms. These activities can be loosely organized into five major categories based predominately on the target tissue (in some cases known, in others inferred) associated with the

TABLE I
Biological Activities of Lipopolysaccharide

Group	Activity	Target tissue
I	Toxicity	Multiple
II	Macrophage activation Tumor necrosis Enhancement of nonspecific resistance to infection Serum amyloid A induction Glucocorticoid antagonizing-factor induction Immunologic adjuvanticity	Reticuloendothelial
III	Mitogenesis Polyclonal activation Immunologic adjuvanticity Interferon production Immunogenicity	Lymphoid
IV	Activation of cells of nonlymphoreticular origin	Fibroblast, platelet, other?
V	Complement activation Serum arylesterase activity Leukocyte extravasation	Extracellular fluids (e.g., serum, lymph)

activity. These activities are presented in Table I. Among these, toxicity was perhaps the earliest recognized activity but remains poorly understood even today. The toxic effect most probably results from a combination of several of the other effects of LPS and the target tissues are clearly multiple. Among the more obvious elements of the toxicity are fluid loss, leukocyte extravasation, shock, and a febrile response (Berry, 1971; Gewurz *et al.,* 1971; Hinshaw, 1971; Snell, 1971; Galanos, 1975; Galanos *et al.,* 1977a; Lüderitz, 1977).

The second category includes those activities that are most probably associated with cells of the reticuloendothelial system. In some cases the association is direct and obvious, as with macrophage activation. Thus, LPS has been shown to enhance glucose utilization (Rosenstreich *et al.,* 1977; Ryan *et al.,* 1979), production of both lymphocyte-activating factor (Gery *et al.,* 1972; Rosenstreich *et al.,* 1978) and colony-stimulating factor (Chervenick, 1972; Quesenberry *et al.,* 1972), and prostaglandin synthesis (Rosenstreich *et al.,* 1977, 1978; Kurland *et al.,* 1979) by macrophages. Macrophages have also been shown to participate in the production of glucocorticoid-antagonizing factor (Berry *et al.,* 1977) and serum amyloid A induction (Sipe *et al.,* 1979) following LPS challenge. Furthermore, although it has not to my knowledge been directly demonstrated, the immunologic adjuvant activities (Claman, 1963; Louis *et al.,* 1973; Armerding and Katz, 1974; Newburger *et al.,* 1974; Skidmore *et al.,* 1975a; Talcott *et al.,* 1975; Scibienski and Gershwin, 1977; McGhee *et al.,* 1979a) and enhancement on nonspecific resistance (Shear, 1944; Parr *et al.,* 1973; Galanos *et al.,* 1977a,b) associated with LPS are likely, at least in part, to derive from its macrophage-stimulating activities.

The third category consists of those activities associated with stimulation of the lymphoid tissues. Among these is perhaps the most widely recognized immunological activity of LPS, that of B-cell mitogenicity and polyclonal activation (Andersson *et al.,* 1972a; Greaves and Janossy, 1972; Peavy *et al.,* 1974). However, several studies have suggested that not all B cells respond to LPS when assessed by these criteria (Andersson *et al.,* 1972b, 1977). Furthermore, although LPS does not appear to stimulate T cells directly, there have been reports of LPS enhancement of T-cell responses to other mitogens (Forbes *et al.,* 1975). Whether such enhancement obtains through a direct effect on T cells or indirectly through macrophages or other accessory cells is not known. The same can be said of the adjuvant properties of LPS, which have been shown to involve a T-cell component in a number of cases (Armerding and Katz, 1974; Newburger *et al.,* 1974; Scibienski and Gershwin, 1977). Regarding induction of interferon synthesis, I have included it under this category because it is a known product of activated lymphocytes (Green and Cooperband, 1971; Merigan, 1971) and because its exact source in LPS-challenged animals has not to my knowledge been established. However, it seems likely that interferon might arise from a variety of different tissues following LPS exposure. The final activity of this category is, of course, the immunogenicity of the "O" antigen component of LPS.

The fourth category involves activation of measurable functions in cells of other than lymphoreticular origin. This category includes induction of glucose utilization and DNA synthesis in fibroblasts (Vaheri *et al.,* 1973) and induction of mediator release from platelets (Des Prez *et al.,* 1961). Whether LPS can be shown to affect other cells of nonlymphoreticular origin has not, to my knowledge, been assessed.

The fifth and final category includes LPS activities where the most immediate effect of the LPS is on extracellular fluid systems such as the complement cascade. Indeed, LPS has been shown to activate complement through both the classical and the alternate pathways (Morrison and Kline, 1977). A second serum-associated activity reportedly activated by LPS is an arylesterase enzymatic activity that may play a role in LPS detoxification (Bøg-Hanson *et al.*, 1979). I have also included leukocyte extravasation under this category because it seems, at least in part, to result from complement activation (Sultzer and Goodman, 1977).

4. Structure–Function Correlation of LPS Biological Activities

As is evident from Fig. 1, LPS has an exceedingly complex chemical structure that is still not completely resolved. One must therefore proceed with caution in attempting to assign various of the biological functions to the various structural components. Nevertheless a considerable amount of work has been carried out in this area with some success. The two most commonly used approaches have been alkaline hydrolysis of the native molecule and derivation of the LPS from mutants that are defective at some step in the synthesis. Both of these approaches have led to the conclusion that the majority of the biological activities of LPS reside in the lipid elements of the structure. Thus, alkaline hydrolysis is selective in that the majority of the carbohydrate remains unaffected whereas the ester- and amide-bonded components are cleaved. Such treatment abrogates all biological activities (Sultzer and Goodman, 1977) except activation of the alternate pathway of complement (Morrison and Kline, 1977) and certain of the adjuvant properties of the molecule (Nowotny, 1977).

The critical role of the lipid component has been confirmed by studies with LPS extracted from mutants lacking the oligosaccharide repeating units of the "O" antigen component. Some of these mutants are known to lack everything but lipid A and the KDO trisaccharide. Nevertheless all such carbohydrate-deficient lipopolysaccharides exhibit biological activity equivalent to the native molecule (Lüderitz *et al.*, 1971; Sultzer, 1971).

As pointed out above, alkaline-hydrolyzed LPS has been shown to be capable of activating the alternate pathway of complement (Morrison and Kline, 1977). The carbohydrate component of LPS has also been shown to play a role in LPS adjuvanticity. In one study the purified oligosaccharide component enhanced the antibody response of mice to SRBC (Nowotny, 1977). In another study the carbohydrate component was shown to be required for enhancement of the primary splenic antibody response to a protein antigen coupled to the LPS (Scibienski, 1980). However, in a followup study, the absence of the carbohydrate had only a minimal depressive effect on the antibody response of draining lymph nodes following footpad immunization (R. J. Scibienski, unpublished data). Therefore the exact role of the carbohydrate in the LPS adjuvant effect in this system remains unclear.

5. Genetically Determined Nonresponsiveness to LPS

Inherited resistance to LPS toxicity was originally reported four decades ago (Hill *et al.*, 1940). However, essentially no work was done on the phenomenon until the middle 1960s when several investigators again reported strain differences in the sensitivity of mice to endotoxin (Heppner and Weiss, 1965; Sacquet and Charlier, 1965; Sultzer, 1968, 1969).

The papers of Sultzer were particularly significant in that they not only documented the extreme resistance of the C3H/HeJ strain to LPS toxicity, but also for the first time described a difference between C3H/HeJ mice and mice of other strains in terms of the nontoxic effects of LPS. In that particular case it was shown that LPS-associated leukocyte extravasation patterns in C3H/HeJ animals were significantly different from those of several other strains (Sultzer, 1968, 1969). Since that time numerous studies have been carried out and it is now well established that C3H/HeJ mice are nonresponsive to LPS with respect to the great majority of its biological activities. Thus, in addition to leukocyte extravasation (Sultzer, 1968, 1969, 1971), C3H/HeJ mice are nonresponsive to LPS at the level of polyclonal B-cell activation (Watson and Riblet, 1974; Coutinho *et al.*, 1975a; Sultzer, 1976), adjuvanticity (Skidmore *et al.*, 1975a; Talcott *et al.*, 1975), macrophage activation (Rosenstreich *et al.*, 1978; Kurland *et al.*, 1979; Ryan *et al.*, 1979), fibroblast activation (Ryan and McAdam, 1977), tumor necrosis (Nowotny, 1977; Männel *et al.*, 1979), enhancement of nonspecific resistance to bacterial infection (Chedid *et al.*, 1976; Parant *et al.*, 1977; Galelli *et al.*, 1979), induction of interferon (Apte *et al.*, 1977), serum amyloid A (McAdam and Sipe, 1976), glucocorticoid-antagonizing factor (Berry *et al.*, 1977), and even stimulation of certain T-cell-associated responses (Koenig *et al.*, 1977; McGhee *et al.*, 1979a). In contrast, the B-cell DNA synthetic response of C3H/HeJ mice to a number of other B-cell mitogens including dextran sulfate, purified protein derivative (PPD), α-bromo cyclic AMP, polyriboinosinic acid-polyribocytidylic acid [poly(I·C)], bacterial lipoprotein, and endotoxin protein, is intact (Sultzer and Nilsson, 1972; Coutinho *et al.*, 1975a; Sultzer and Goodman, 1976; Braun, 1977; Rosenstreich and Blumenthal, 1977; McAdam and Ryan, 1978).

Much of the work that has been done on LPS responsiveness has, of necessity, been phenomenological. However, the availability of LPS-nonresponder mice has stimulated several areas of experimentation that in sum have increased our understanding of the mechanisms underlying the various LPS activities. These are reviewed and analyzed in the following sections.

5.1. Genetics and Strain Distribution of LPS Nonresponsiveness

A considerable amount of work has been devoted to defining the genetics of the LPS nonresponse trait. Breeding experiments have established that LPS response capability is most likely associated with a single genetic locus. Thus, in numerous experiments involving F_2 and backcross generations derived from responder × nonresponder F_1 animals, no segregation of any of the biological activities of LPS has ever been observed (Rosenstreich *et al.*, 1978; Watson *et al.*, 1978a). In fact, there have been only two cases where breeding experiments suggested that more than one locus might be involved. In one case, the degree of variance in the leukocyte extravasation response to LPS of the F_2 generation of an original C3H/HeJ × A/HeJ cross was consistent with polygenic control (Sultzer, 1972). However, in view of the fact that the leukocyte extravasation response probably involves complement activation in addition to direct stimulation of cellular function (Sultzer and Goodman, 1977), such a result might be expected even if the cellular response to LPS were regulated by a single locus. The second study suggesting possible regulation of the LPS response by multiple genes involved interbreeding experiments with a large number of

C3H sublines that could be grouped into high, intermediate, and low responders based on B-cell activation (Glode and Rosenstreich, 1976). In these experiments, crosses between high (C3H/St) and intermediate (C3H/HeN) responder yielded only high responders whereas crosses between high and low responders (C3H/HeJ) yielded intermediate responders. In contrast, crosses between intermediate and low strains again yielded an intermediate response (relative to the parental strains). These data were interpreted to indicate either multigenic control or multiple alleles wherein the "high" allele was dominant over the intermediate while the "low" allele was codominant with either the high or the intermediate (Glode and Rosenstreich, 1976). However, this interpretation rests on the assumption that the response levels observed in the parental strains were not subject to other, non-LPS-related factors. It seems conceivable that the high and intermediate strains are, in fact, genetically identical vis à vis LPS response capability, but that additional factors are at play in the intermediate strains that limit DNA synthetic responses in general. These authors themselves commented on the fact that the responses they obtained were quite variable even between animals of a given strain. In retrospect, it may have been enlightening had these investigators also assessed the responsiveness of their animals to other B-cell mitogens.

In numerous other studies of LPS responsiveness, single-locus genetics has been routinely observed. Thus, low-responder × high-responder crosses yield either intermediate- or high-responder offspring, backcrosses yield the expected 50:50 distribution, and F_2 generations exhibit the expected 1:2:1 distribution (Watson and Riblet, 1974; Coutinho *et al.*, 1975b; Talcott *et al.*, 1975; Apte *et al.*, 1977; Parant *et al.*, 1977; Watson *et al.*, 1977a,b; Rosenstreich *et al.*, 1978; McGhee *et al.*, 1979b). Whether one observes high or intermediate responsiveness in F_1 and F_2 generations would seem to be dependent on the activity being assessed and/or the responder strain utilized. For example, in a study of B-cell mitogenesis, Watson and co-workers found that crossing C3H/HeJ with CWB (C3H.SW-*Ig-lb*) mice yielded only high-responder offspring, whereas when the C3H/HeJ strain was crossed with a number of other responder strains the F_1 offspring were intermediate in responsiveness between the two parental strains (Watson and Riblet, 1974, 1975; Watson *et al.*, 1977a). The reason that F_1 animals derived from crosses with the CWB strain are high responders is at present unknown. However, in other studies the response level of F_1 offspring of responder × nonresponder crosses has been found to be dose related. Thus, McGhee and co-workers found that LPS-associated adjuvanticity, interferon induction, and lethality were all of intermediate levels in responder × nonresponder F_1 animals but that at high levels of LPS (100 μg) the interferon responses of the responder and F_1 animals were equivalent (McGhee *et al.*, 1979b). A similar lack of difference in the interferon response to high doses of LPS of responder and F_1 animals was observed by Apte and co-workers (Apte *et al.*, 1977). Concentration effects have also been observed in the case of certain macrophage responses to LPS (McAdam and Sipe, 1976; Rosenstreich *et al.*, 1978). It has even been reported that C3H/HeJ animals themselves will respond partially to higher doses of LPS (Coutinho, 1976). Furthermore, adrenalectomy renders responder and F_1 animals equally sensitive to the toxic effects of LPS (Chedid *et al.*, 1976; Parant *et al.*, 1977), and bacterial agents such as *Bordetella* and *Mycobacterium* have been reported to enhance the responsiveness of nonresponder mice to LPS (Sultzer and Goodman, 1977; Vogel *et al.*, 1979b). In view of these considerations, it seems reasonable to conclude that the LPS-responder and -nonresponder alleles are probably codominantly expressed.

In addition to the C3H/HeJ strain, LPS nonresponsiveness has also been documented in the C57BL/10ScCr strain (Coutinho *et al.,* 1977a; Coutinho and Meo, 1978; McAdam and Ryan, 1978) and was subsequently shown to be present in the C57BL/10ScN strain from which the B10 ScCr originated (Vogel *et al.,* 1979a). These strains of mice are nonresponsive as assessed by B-cell activation, macrophage stimulation, and inducton of serum amyloid A (Coutinho *et al.,* 1977a; McAdam and Ryan, 1978; Vogel *et al.* 1979a). Crosses between C57BL/10ScCr and C3H/HeJ animals fail to yield complementation (Coutinho and Meo, 1978), suggesting that the defect is in the same gene in both strains. However, as pointed out by these authors, the history of the two strains and the nature of the B10Sc trait make it unlikely that both strains express the same mutation. Thus, the sublines that eventually segregated out to yield the strains diverged long before the C3H sublines, only one of which expresses the defect, began to diverge (Klein, 1975). Furthermore, crosses between B10Sc and responder B10Sn mice yielded only the high-responder phenotype in the F_1 offspring as assessed by B-cell activation. No intermediates were observed over an extended LPS dose range (Coutinho and Meo, 1978). In contrast, as pointed out above, crosses involving C3H/HeJ animals generally yield F_1 offspring of intermediate responsiveness when assessed by this method. It therefore seems likely that the two strains carry independent mutations in the same gene.

The genetic analysis of the LPS nonresponsiveness of C3H/HeJ mice has been extended to analysis of linkage with other genetic markers of the murine system. Coutinho and co-workers reported that LPS responsiveness segregated independently of both the *Mls* and the *Mod-1* loci (Coutinho *et al.,* 1977b). Watson and co-workers assessed the H_2, *Ig*, and *Mup* (mouse urinary protein) loci for linkage to LPS responsiveness and found concordant inheritance with the *Mup* locus in 13 of 14 backcross lines, thus establishing that the *LPS* locus is on chromosome 4 of the mouse (Watson *et al.,* 1977b). Subsequent studies by Watson and co-workers placed the *LPS* gene between *Mup-1* and polysyndactyly *(PS)* (Watson *et al.,* 1978a). A similar linkage for the *LPS* gene of the C57BL/10Sc strain has been reported (Coutinho and Meo, 1978).

5.2. Biological Activities Affected by the *LPS* Gene

The existence of mice genetically nonresponsive to LPS has stimulated a number of studies that have furthered our understanding of the mechanisms of action of LPS. Coutinho and co-workers have utilized these animals to prepare and characterize an antiserum that appears to react with a cell-surface component relevant to LPS activity. Thus, a rabbit antiserum was prepared against C3H/Tif splenic B cells and subsequently absorbed extensively with C3H/HeJ tissues. This antiserum was shown to react with a fraction of C3H/Tif B cells corresponding to that fraction of cells that were LPS responsive, and furthermore to react almost exclusively with the LPS-responsive B-cell subset (Forni and Coutinho, 1978). A study of the strain distribution of this marker revealed that, in addition to the C3H/HeJ strain, it was also absent from C57BL/10ScCr animals. [It is perhaps of interest to note that expression of this marker was also very low in the CBA/N strain (Forni and Coutinho, 1978). CBA/N mice have been shown to have a B-cell developmental defect that partially affects their LPS responsiveness (Scher *et al.,* 1975; Rosenstreich *et al.,* 1978).] In subsequent studies, reactivity with this antiserum was shown in a series of genetic analyses to be concordantly expressed with LPS responsiveness (Coutinho *et al.*, 1978; Forni and Coutinho, 1978). In these experiments the observation was also made that responder × nonresponder F_1 offspring expressed the marker in only half as many cells as the responder parent, raising the possibility of allelic exclusion of the *LPS* gene (Coutinho *et al.*, 1978). These authors further mentioned that their unpublished experiments were consistent with this possibility, as the frequency of LPS-responsive cells in F_1 animals seemed to be only half as high as in the responder parents.

In the same series of experiments these investigators further demonstrated that the IgG fraction of this antiserum was capable of stimulating DNA synthesis in LPS-responder spleen cells in a dose-dependent manner. They also found that LPS, but not other B-cell mitogens including bacterial lipoprotein (see later), was capable of competing with the antiserum for binding to the B-cell surface. However, this latter finding is rather difficult to interpret, for other investigators have reported no difference in LPS binding to spleen cells from responder and nonresponder strains (Watson and Riblet, 1975; Kabir and Rosenstreich, 1977). Nevertheless, the availability of an antiserum with these characteristics should do much toward enhancing our understanding of the mechanisms of action of LPS.

The above-mentioned studies by Coutinho and co-workers were predominantly concerned with B-cell responsiveness to LPS. Other studies have utilized LPS-nonresponder mice to better define the LPS responsiveness of other cell types. Thus, a number of reports have dealt with the T-cell-associated activities of LPS. C3H/HeJ mice have been reported to be defective in LPS enhancement of T-cell stimulation by lectins (Rosenstreich *et al.*, 1977) and LPS-induced phenotypic conversion of T-cell precursors (Koenig *et al.*, 1977). A role for T cells in B-cell stimulation by LPS also has been suggested by studies in the C3H/HeJ strain, albeit inconsistently. Thus, although some investigators have reported that addition of responder T cells and/or macrophages to responder B cells has no effect on the response of the latter to LPS (Watson and Riblet, 1975; Coutinho, 1976; Rosenstreich *et al.*, 1977; Watson *et al.*, 1977a), others have reported that the response of B cells to LPS can be potentiated by such additions (McGhee *et al.*, 1979a). The latter investigators further reported that the minimal response of C3H/HeJ B cells to LPS could be similarly potentiated by addition of C3H/HeN responder T cells, but that the C3H/HeN B-cell response could not be enhanced by addition of C3H/HeJ T cells. These latter studies therefore implicate a direct action of LPS on T cells, as any potentiation that resulted from non-LPS-associated T-cell effects would be expected to be intact in the C3H/HeJ animals. This latter finding is consistent with other studies that have demonstrated deficient responsiveness to LPS in congenitally athymic nude mice (Kagnoff *et al.*, 1974; Berry *et al.*, 1977).

The C3H/HeJ strain has also been instrumental in delineating the mechanism of the serum amyloid A (SAA) response. SAA is an acute-phase serum protein that undergoes a very striking increase in concentration following LPS stimulation of responsive mice (McAdam and Sipe, 1976). This response is not observed in C3H/HeJ or C57BL/10Sc animals (McAdam and Ryan, 1978; Watson *et al.*, 1978b). It has been shown that the increase in SAA is itself a secondary effect of an SAA-stimulating factor (SAASF) that is the primary product of the LPS stimulation. Thus, although C3H/HeJ animals do not respond directly to LPS with an increase in SAA, they do mount such a response when exposed to the serum of LPS-challenged C3H/HeN responder mice (Sipe *et al.*, 1979). Radiation chimeras produced by repopulating irradiated C3H/HeJ recipients with C3H/

HeN bone marrow do respond to LPS with an increase in SAA (Sipe *et al.*, 1979), suggesting that the stimulating factor is a product of lymphoid or reticuloendothelial elements. Furthermore, incubation of thioglycollate-induced peritoneal macrophages with LPS results in SAASF induction, whereas LPS stimulation of spleen cells, which contained less than 10% macrophages, did not result in induction of SAASF activity, leading to the conclusion that the factor is in fact derived from macrophages (Sipe *et al.*, 1979; Selinger *et al.*, 1980). More recently it has been shown the SAASF derived from LPS-stimulated responder macrophages will induce SAA synthesis by cultured hepatocytes from nonresponder animals, suggesting that the liver may be the site of action of the stimulating factor (Selinger *et al.*, 1980).

Radiation chimeras derived from responder–nonresponder combinations have also proven useful in delineating certain of the other *in vivo* biological activities of LPS. Thus, C3H/HeJ → C3H/HeN chimeras are resistant (Rosenstreich *et al.*, 1977; Michalek *et al.*, 1980). This suggests that the toxicity of LPS derives predominantly from its activities on the lymphoid and reticuloendothelial elements. These elements probably also underlie the tumor necrosis response to LPS. Thus, neither C3H/HeN nor C3H/HeJ fibrosarcomas could be necrotized by LPS when grafted into C3H/HeJ recipients, whereas tumors of both origins were necrotized by LPS when grafted into C3H/HeN recipients (Männel *et al.*, 1979). Furthermore, C3H/HeJ → C3H/HeN chimeras exhibited no tumor necrotic response to LPS whereas C3H/HeN → C3H/HeJ chimeras did exhibit such a response (Männel *et al.*, 1979). These findings suggest that the tumor-necrotizing activity of LPS does not result from a direct action on the tumor itself nor on the vasculature of the host but, rather, is a secondary effect resulting from stimulation of lymphoid and/or reticuloendothelial components. A similar conclusion can be drawn concerning induction of nonspecific resistance to bacterial infection, for chimeras, when assessed for this activity, also respond according to bone marrow type (Galelli *et al.*, 1979).

6. Bacterial Lipoprotein Mitogens and Their Effects in LPS-Nonresponder Mice

As pointed out earlier, two other biologically active components can be extracted from the exterior layers of gram-negative bacteria. One of these coextracts with LPS when mild procedures such as those involving butanol or TCA deproteinization are utilized (Skidmore *et al.*, 1975b). This component, which has been obtained in purified form (Morrison *et al.*, 1976b; Sultzer and Goodman, 1976) has been variously termed endotoxin protein or lipid-associated protein. The other component, termed bacterial lipoprotein, resides on the inner face of the outer membrane and is covalently linked to the peptidoglycan of the cell wall (Braun, 1977). Both of these components are mitogenic for murine B lymphocytes, including those of LPS-nonresponder strains (Melchers *et al.*, 1975; Skidmore *et al.*, 1975b; Morrison *et al.*, 1976; Sultzer and Goodman, 1976; Forni and Coutinho, 1978; Goodman *et al.*, 1978). However, endotoxin protein does not exhibit the total spectrum of LPS activities when administered to C3H/HeJ mice. Thus, it does not act as an adjuvant in these animals nor does it act to circumvent T-cell tolerance (Goodman *et al.*, 1978), both of which activities have been demonstrated for LPS in responder strains (Chiller and Weigle, 1973). Interestingly, however, when endotoxin protein is administered to C3H/HeJ animals along with a deaggregated immunoglobulin tolerogen, induction of tolerance is inhibited (Goodman *et al.,* 1978), an effect similar to that seen with LPS in responder mice (Louis *et al.,* 1973). These findings suggest that the adjuvanticity and tolerance-circumventing properties of LPS involve effects distinct from its B-cell-activation properties.

A number of studies have been carried out that assessed the effect of LPS on the response of C3H/HeJ B cells to other B-cell mitogens. As pointed out, the C3H/HeJ and C57BL/10Sc strains are uniquely nonresponsive to LPS and respond quite well to other B-cell mitogens, including bacterial lipoprotein and endotoxin protein. It has been found that whereas LPS has no effect on the response of B cells from these strains to PPD or dextran sulfate, the response to poly(I·C) and to endotoxin protein is inhibited by LPS (Skidmore *et al.,* 1975b; Glode *et al.,* 1976; Rosenstreich *et al.,* 1977; Sultzer and Goodman, 1977; Watson *et al.,* 1977a). These findings therefore suggest that LPS, endotoxin protein, and poly(I·C) all act on a subpopulation of B cells distinct from those stimulated by PPD and dextran sulfate, or, if common B-cell subpopulations are involved, that LPS, endotoxin protein, and poly(I·C) all act through a common pathway distinct from that utilized by other mitogens. There are contrary indications, however, in that poly(I·C) and LPS responses have been reported to be additive in responder strains (Watson *et al.,* 1977a), suggesting that each is capable of stimulating a subpopulation of B cells that is not sensitive to the other. It should also be remembered that endotoxin protein does not inhibit Coutinho's antiserum against the putative LPS "receptor." Furthermore, LPS has been reported to have an immunosuppressive property and this property is evident in nonresponder strains (Haas *et al.,* 1978), raising the possibility that the above-mentioned suppression may obtain through a pathway other than those involved in LPS-activating effects. However, these considerations notwithstanding, the available data strongly suggest a concordance in the mechanism(s) of B-cell activation by LPS, endotoxin protein, and poly-(I·C).

7. Speculations on the Mechanism of Action of LPS

Certain conclusions can be drawn from the data reviewed here. To begin with, it is clear from the genetic evidence that LPS responsiveness is accounted for by a small number of genetic elements, most probably one. Second, the C3H/HeJ and C57BL/10Sc strains are not totally nonresponsive to LPS, as partial responses can be observed either in the presence of other stimulants or at high concentrations of LPS. Third, LPS has multicellular effects, being able to directly stimulate B cells, macrophages, fibroblasts, and probably T cells also, and all of these functions are deficient in nonresponder strains. Fourth, with respect to B-cell activation, LPS is suppressive in nonresponder strains. Finally, there are no detectable differences in the binding of LPS to spleen cells from responder and nonresponder strains, and treatment of the cell surface with proteases or neuraminidase has no effect on LPS binding (Kabir and Rosenstreich, 1977).

Based on the genetic data and on the serological studies of Coutinho and co-workers, it is tempting to conclude that LPS acts through a single cell-surface component that has been altered by the LPS mutation. It is clear that said component need not be a specific receptor in the classical sense of the term, as LPS binds equally well to responder and nonresponder cells. Whether such a putative component is nevertheless capable of specifically interacting with LPS would seem to be exceedingly difficult to determine given currently available technology. However, in view of the multiple membrane effects that LPS might cxert, it seems a bit premature to conclude that LPS mediates its effects through specific interactions. Indeed, the only datum that in any way argues for such an occurrence is the finding of Coutinho and co-workers that LPS, but not other B-cell mitogens, competes with their antiresponder antisera in binding studies (Coutinho *et al.*, 1978). However, this finding does not establish a molecular association between LPS and its putative receptor, much less some specific form of interaction. Moreover, the facts that such widely diverse cell types as lymphocytes, macrophages, and fibroblasts are all responsive to LPS and that such responsiveness is in all cases associated with a single genetic locus argue strongly that we are dealing with a membrane component that has a similar function in different cell types. This component would thus be somewhat analogous to the nucleotide cyclase systems that can be activated by different receptor systems depending on the cellular type in which they are located. Whether we are in fact dealing with a cyclic nucleotide pathway is not known. Indeed, little is known about the role of cyclic nucleotides in lymphocyte activation in general (Parker, 1978). Moreover, even if this analogy is an accurate one, it still does not tell us what the nature of the interaction between LPS and this component might be. It seems equally likely that LPS might directly interact with such a component or bring about its effects by altering the environment in which the component resides. For example, several enzymes have been shown to be sensitive to the lipid composition and biophysical state of the membrane in which they are located (Cronan and Vagelos, 1972; Baldassare *et al.*, 1977), and it seems likely that many other membrane-bound enzyme systems might be similarly affected. In this regard, studies have been carried out that suggest tht LPS can profoundly affect the membranes of eukaryotic cells. Thus, in a series of experiments with the polysaccharide-deficient LPS extracted from *E. coli* strain mR595, Brailovsky and co-workers (1977) found that the amount of LPS bound to rat fibroblasts had a considerable effect on the properties of the fibroblast membranes. When lower concentrations of LPS were bound to the cell, addition of antibodies specific for mR595 LPS followed by anitimmunoglobulin antibodies led to capping of the resulting complexes. In contrast, when higher concentrations of LPS were bound to the cells, addition of the double-antibody system had no effect, indicating that the LPS was immobilized. More surprisingly, at the higher LPS concentration the cell-surface concanavalin A receptors were also found to be immobilized and the ability of the lectin to agglutinate the cells was found to be interfered with, although there was no inhibition of lectin binding to the cell surface. Finally, the capping phenomenon observed at the lower concentration of LPS could be inhibited by colchicine and cytochalasin B (Brailovsky *et al.*, 1977). As concanavalin A is capable of binding to both glycoproteins and glycolipids, these results suggest that high concentrations of LPS are capable of a general inhibition of lateral membrane diffusion. Furthermore, the colchicine/cytochalasin effects strongly suggest that the LPS interacts with one or more components that span the membrane to interact with the cytoskeleton. Based on these findings it seems clear that LPS might interact with a number of different membrane components in eukaryotic cells. Additionally, although biophysical studies have not been carried out in eukaryotic cells, such studies in gram-negative prokaryotic cells suggest that the character of their LPS can affect the fluidity of the membrane in which it is located. Thus, mutants deficient in the carbohydrate component of LPS have a more fluid outer membrane than do wild-type organisms that have a complete polysaccharide component (Rottem and Leive, 1977). This suggests that inhibition of the lateral diffusion of LPS through noncovalent interactions among its polysaccharide side chains can significantly affect the microviscosity of the membrane in which the LPS is embedded. It seems reasonable to assume that similar phenomena might occur in eukaryotic membranes and such membrane effects could conceivably be the basis for the cell-activating properties of LPS.

If LPS does in fact exert its effects through alteration of membrane biophysical parameters, then a second group of enzymes must be considered as potential targets of the LPS mutation. These are the enzymes involved in synthesis and assembly of the membrane lipids. Thus, an alteration in the character of the membrane itself could cause LPS to exert a different membrane effect than it normally does, thereby resulting in a deficient response. Considering the complexity of eukaryotic membrane systems, the probability that such a mutation would independently arise in the same gene on two different occasions seems rather low. Nevertheless this possibility cannot be ruled out on the basis of the presently available evidence.

To summarize, several different cell types are capable of responding to LPS with a broad spectrum of activities. All of these activities involve activation pathways that include at least one common element. On two different occasions, mutations have arisen at the genetic locus that codes for this element, leading to LPS nonresponsiveness. The fact that mutations at a single locus affect the response of widely divergent cell types to LPS suggests that the element involved is either a component of a second messenger pathway or a component involved in membrane genesis, the product of which is critical to LPS-induced membrane alterations that in turn lead to cell activation. The available data do not rule out either alternative but would seem to support the former. This is particularly true in the case of the serological studies of Coutinho and co-workers. Further characterization of their antiserum and production of similar antisera, perhaps utilizing monoclonal antibody techniques, should contribute significantly to our understanding of the mechanism of action of LPS on eukaryotic cells and perhaps a better understanding of the normal response mechanisms of these cells.

ACKNOWLEDGMENTS

I would like to thank Drs. A. Coutinho, K. P. W. J. McAdam, D. L. Rosenstreich, W. O. Weigle, and S. N. Vogel for their willingness to share pertinent unpublished findings with me.

References

Andersson, J., Sjöberg, O., and Möller, G., 1972a, Mitogens as probes for immunocyte activation and cellular cooperation, *Transplant. Rev.* **11**:131.

Andersson, J., Sjöberg, O., and Möller, G., 1972b, Induction of immunoglobulin and antibody synthesis *in vitro* by lipopolysaccharide, *Eur. J. Immunol.* **2**:349.

Andersson, J., Coutinho, A., Lernhardt, W., and Melchers, F., 1977, Clonal growth and maturation to immunoglobulin secretion *in vitro* of every growth inducible B lymphocyte, *Cell* **10**:27.

Apte, R. N., Ascher, O., and Pluznik, D. H., 1977, Genetic analysis of generation of serum interferon by bacterial lipopolysaccharide, *J. Immunol.* **119**:1898.

Armerding, D., and Katz, D. H., 1974, Activation of T and B lymphocytes *in vitro*. I. Regulatory influence of bacterial lipopolysaccharide (LPS) on specific T cell helper functions, *J. Exp. Med.* **139**:24.

Baldassare, J. J., Brenckle, G. M., Hoffman, M., and Silbert, D. F., 1977, Modification of membrane lipid. Functional properties of membrane in relation to fatty acid structure, *J. Biol. Chem.* **252**:8797.

Berry, L. J., 1971, Metabolic effects of bacterial endotoxins, in: *Microbial Toxins* (S. Kadis, G. Weinbaum, and S. J. Ajl, eds.), Vol. V, pp. 165–204, Academic Press, New York.

Berry, L. J., Moore, R. N., Goodrum, K. J., and Couch, R. E., 1977, Cellular requirements for enzyme inhibition by endotoxin in mice, in: *Microbiology, 1977* (D. Schlessinger, ed.), pp. 321–325, Am. Soc. Microbiol., Washington, D.C.

Bøg-Hansen, T. C., Hertel-Wulff, B., and Back, U., 1979, Lipoprotein changes induced by bacterial lipopolysaccharides in "non-responder" C3H/HeJ mice, *Febs Lett.* **104**:441.

Boivin, A., Mesrobeanu, I., and Mesrobeanu, L., 1933, Preparation of the specific polysaccharides of bacteria, *C.R. Soc. Biol.* **113**:490.

Brailovsky, C. A., Nigam, V. N., and Pesant, S., 1977, Cell surface mobility of a bacterial R-form glycolipid, mR595, after binding to rat cells, and its effect on the mobility of concanavalin A receptors, *Exp. Cell Res.* **109**:389.

Braun, V., 1977, Lipoprotein from the outer membrane of *Escherichia coli* as antigen, immunogen, and mitogen, in: *Microbiology, 1977* (D. Schlessinger, ed.), pp. 257–261, Am. Soc. Microbiol., Washington, D.C.

Chedid, L., Parant, M., Damais, C., Parant, F., Juy, D., and Galelli, A., 1976, Failure of endotoxin to increase nonspecific resistance to infection of lipopolysaccharide low responder mice, *Infect. Immun.* **13**:722.

Chervenick, P. A., 1972, Effect of endotoxin and postendotoxin plasma on *in vitro* granulopoiesis, *J. Clin. Med.* **79**:1014.

Chiller, J. M., and Weigle, W. O., 1973, Termination of tolerance to human gamma-globulin in mice by antigen and bacterial lipopolysaccharide, *J. Exp. Med.* **137**:740.

Claman, H. N., 1963, Tolerance to a protein antigen in adult mice and the effect of non-specific factors, *J. Immunol.* **91**:833.

Coutinho, A., 1976, Genetic control of B cell responses. II. Identification of the spleen defect in C3H/HeJ mice, *Scand. J. Immunol.* **5**:129.

Coutinho, A., and Meo, T., 1978, Genetic basis for unresponsiveness to lipopolysaccharide in C57BL/10Cr mice, *Immunogenetics* **7**:17.

Coutinho, A., Gronowiz, E., and Sultzer, B. M., 1975a, Genetical control of B cell responses. I. Selective unresponsiveness to lipopolysaccharides, *Scand. J. Immunol. 4:139.*

Coutinho, A., Moller, G., and Gronowiz, E., 1975b, Genetical control of B cell responses. IV. Inheritance of the unresponsiveness to lipopolysaccharide, *J. Exp. Med.* **142**:253.

Coutinho, A., Forni, L., Melchers, F., and Watanabe, T., 1977a, Genetic defect in responsiveness to the B cell mitogen lipopolysaccharide, *Eur. J. Immunol.* **7**:325.

Coutinho, A., Meo, T., and Watanabe, T., 1977b, Independent segregation of two functional markers expressed on the same B-cell subset in the mouse: The MLS determinants and LPS receptors, *Scand. J. Immunol.* **6**:1005.

Coutinho, A., Forni, L., and Watanabe, T., 1978, Genetic and functional characterization of an antiserum to the lipid A-specific triggering receptor on murine B lymphocytes, *Eur. J. Immunol.* **8**:63.

Cronan, J. E., Jr., and Vagelos, P. R., 1972, Metabolism and function of the membrane phospholipids of *Escherichia coli, Biochim. Biophys. Acta* **265**:25.

Des Prez, R. M., Horowitz, H. I., and Hook, E. W., 1961, Effects of bacterial endotoxin on rabbit platelets. I. Platelet aggregation and release of platelet factors *in vitro, J. Exp. Med.* **114**:857.

Forbes, J. T., Nakao, Y., and Smith, R. T., 1975, T mitogens trigger LPS responsiveness in mouse thymus cells, *J. Immunol.* **114**:1004.

Forni, L., and Coutinho, A., 1978, An antiserum which recognizes lipopolysaccharide reactive B cells in the mouse, *Eur. J. Immunol.* **8**:56.

Galanos, C., 1975, Physical state and biological activity of lipopolysaccharides. Toxicity and immunogenicity of the lipid A component, *Z. Immunitaetsforsch.* **149**:214.

Galanos, C., Freundenberg, M., Hase, L., Jay, F., and Ruschmann, E., 1977a, Biological properties and immunological properties of lipid A, in: *Microbioloby, 1977* (D. Schlessinger, ed.), pp. 269–276, Am. Soc. Microbiol., Washington, D.C.

Galanos, C., Lüderitz, O., Reitschel, E. T., and Westphal, O., 1977b, Newer aspects of the chemistry and biology of bacterial lipopolysaccharides, with special reference to their lipid A component, in: *International Review of Biochemistry; Volume 14, Biochemistry of Lipids II* (T. W. Goodwin, ed.), pp. 239–335, University Park Press, Baltimore.

Galelli, A., Le Garrec, Y., and Chedid, L., 1979, Transfer by bone marrow cells of increased natural resistance to *Klebsiella pneumoniae* induced by lipopolysaccharide in genetically deficient C3H/He mice, *Infect. Immun.* **23**:232.

Gery, I., Gershon, R., and Waksman, B. H., 1972, Potentiation of the T-lymphocyte response to mitogens. I. The responding cell, *J. Exp. Med.* **136**:128.

Gewurz, H., Snyderman, R., Mergenhagen, S. E., and Shin, H. S., 1971, Effects of endotoxic lipopolysaccharides on the complement system, in: *Microbial Toxins* (S. Kadis, G. Weinbaum, and S. J. Ajl, eds.), Vol. V, pp. 127–146, Academic Press, New York.

Glode, L. M., and Rosenstreich, D. L., 1976, Genetic control of B cell activation by bacterial lipopolysaccharide is mediated by multiple distinct genes or alleles, *J. Immunol.* **117**:2061.

Glode, L. M., Scher, I., Osborne, B., and Rosenstreich, D. L., 1976, Cellular mechanism of endotoxin unresponsiveness in C3H/HeJ mice, *J. Immunol.* **116**:454.

Goodman, M. J., Parks, D. E., and Weigle, W. O., 1978, Immunologic responsiveness of the C3H/HeJ mouse: Differential ability of butanol-extracted lipopolysaccharide (LPS) to evoke LPS-mediated effects, *J. Exp. Med.* **147**:800.

Greaves, M. F., and Janossy, G., 1972, Elicitation of selective T and B lymphocyte responses by cell surface binding ligands, *Transplant. Rev.* **11**:87.

Green, L. A., and Cooperband, S. R., 1971, Production and assay of human lymphocyte interferon induced by anitgens, in: *In Vitro Methods in Cell-Mediated Immunity* (B. R. Bloom and P. R. Glade, eds.), pp. 501–514, Academic Press, New York.

Haas, G. P., Johnson, A. G., and Nowotny, A., 1978, Suppression of the immune response in C3H/HeJ mice by protein-free lipopolysaccharides, *J. Exp. Med.* **148**:1081.

Heppner, G., and Weiss, D. W., 1965, High susceptibility of strain A mice to endotoxin and endotoxin-red blood cell mixtures, *J. Bacteriol.* **90**:696.

Hill, A. B., Hatswell, J. M., and Topley, W. W. C., 1940, The inheritance of resistance, demonstrated by the development of a strain of mice resistant to experimental inoculation with a bacterial endotoxin, *J. Hyg.* **40**:538.

Hinshaw, L. B., 1971, Release of vasoactive agents and the vascular effects of endotoxin, in: *Microbial Toxins* (S. Kadis, G. Weinbaum, and S. J. Ajl, eds.), Vol. V, pp. 209–273, Academic Press, New York.

Kabir, S., and Rosenstreich, D. L., 1977, Binding of bacterial endotoxin to murine spleen lymphocytes, *Infect. Immun.* **15**:156.

Kagnoff, M. F., Billings, P., and Cohn, M., 1974, Functional characteristics of Peyer's patch lymphoid cells. II. Lipopolysaccharide is thymus dependent, *J. Exp. Med.* **139**:407.

Klein, J., 1975, *Biology of the Mouse Histocompatibility Complex,* Springer-Verlag, New York.

Koenig, S., Hoffmann, M. K., and Thomas, L., 1977, Induction of phenotypic lymphocyte differentiation in LPS unresponsive mice by an LPS induced serum factor and by a lipid A associated protein, *J. Immunol.* **118**:1910.

Kurland, J. I., Pelvs, L. M., Ralph, P., Bockman, R. L., and Moore, M. A. S., 1979, Induction of prostaglandin synthesis in normal and neoplastic macrophages: Role for colony-stimulating factor(s) distinct from effects on myeloid progenitor cell proliferation, *Proc. Natl. Acad. Sci. USA* **76**:2326.

Louis, J. A. Chiller, J. M., and Weigle, W. O., 1973, The ability of bacterial lipopolysaccharide to modulate the induction of unresponsiveness to a state of immunity. Cellular parameters, *J. Exp. Med.* **148**:1481.

Lüderitz, O., 1977, Endotoxins and other cell wall components of Gram-negative bacteria and their biological activities, in: *Microbiology, 1977* (D. Schlessinger, ed.), pp. 239–246, Am. Soc. Microbiol., Washington, D.C.

Lüderitz, O., Westphal, O., Staub, A. M., and Nikaido, H., 1971, Isolation and chemical and immunological characterization of bacterial lipopolysaccharides, in: *Microbial Toxins* (S. Kadis, G. Weinbaum, and S. J. Ajl, eds.), Vol. IV, pp. 145–224, Academic Press, New York.

McAdam, K. P. W. J., and Ryan, J. L., 1978, C57BL/101/CR mice: Nonresponders to activation by the lipid A moiety of bacterial lipopolysaccharide, *J. Immunol.* **120**:249.

McAdam, K. P. W. J., and Sipe, J. D., 1976, Murine model for human secondary amyloidosis: Genetic variability of the acute-phase serum protein SAA response to endotoxins and casein, *J. Exp. Med.* **144:**1121.

McGhee, J. R., Farrar, J. J., Michalek, S. M., Mergenhagen, S. E., and Rosenstreich, D. L., 1979a, Cellular requirements for lipopolysaccharide adjuvanticity. A role for both T lymphocytes and macrophages for *in vitro* responses to particulate antigens, *J. Exp. Med.* **149:**793.

McGhee, J. R., Michalek, S. M., Moore, R. N., Mergenhagen, S. E., and Rosenstreich, D. L., 1979b, Genetic control of *in vivo* sensitivity to lipopolysaccharide: Evidence for codominant inheritance, *J. Immunol.* **122:**2052.

Männel, D. N., Rosenstreich, D. L., and Mergenhagen, S. E., 1979, Mechanism of lipopolysaccharide-induced tumor necrosis: Requirement for lipopolysaccharide-sensitive lymphoreticular cells, *Infect. Immun.* **24:**573.

Melchers, F., Braun, V., and Galanos, C., 1975, The lipoprotein of the outer membrane of *Escherichia coli:* A B-lymphocyte mitogen, *J. Exp. Med.* **142:**473.

Merigan, T. C., 1971, A plaque inhibition assay for human interferon employing human neonate skin fibroblast monolayers and bovine vesicular stomatitis virus, in: *In Vitro Methods in Cell-Mediated Immunity* (B. R. Bloom and P. R. Glade, eds.), pp. 489–500, Academic Press, New York.

Michalek, S. M., Moore, R. N., McGhee, J. R., Rosenstreich, D. L., and Mergenhagen, S. E., 1980, The primary role of lymphoreticular cells in the mediation of host responses to bacterial endotoxin, *J. Infect. Dis.* **141:**55.

Morrison, D. C., and Kline, L. ., 1977, Activation of the classical and properdin pathways of complement by bacterial lipopolysaccharides, *J. Immunol.* **118:**362.

Morrison, D., and Leive, L., 1975, Fractions of lipopolysaccharide from *Escherichia coli* 0111:B4 prepared by two extraction procedures, *J. Biol. Chem.* **250:**2911.

Morrison, D. C., Betz, S. J., and Jacobs, D. M., 1976, Isolation of a lipid A bound polypeptide responsible for "LPS-initiated" mitogenesis of C3H/HeJ spleen cells, *J. Exp. Med.* **144:**840.

Newberger, P. E., Hamaoka, T., and Katz, D. H., 1974, Potentiation of helper T cell function in IgE antibody responses by bacterial lipopolysaccharide (LPS), *J. Immunol.* **113:**824.

Nowotny, A., 1977, Relation of structure to function in bacterial endotoxins, in: *Microbiology, 1977* (D. Schlessinger, ed.), pp. 247–252, Am. Soc. Microbiol., Washington, D.C.

Parant, M., Parant, F., and Chedid, L., 1977, Inheritance of lipopolysaccharide-enhanced nonspecific resistance to infection and of susceptibility to endotoxic shock in lipopolysaccharide low-responder mice, *Infect. Immun.* **16:**432.

Parker, C. W., 1978, cAMP and lectin induced mitogenesis in lymphocytes: Possible implications for neoplastic cell growth, *Adv. Cyclic Nucleotide Res.* **9:**647.

Parr, S. Wheeler, E., and Alexander, P., 1973, Similarities of the antitumor actions of endotoxin, lipid A and double stranded RNA, *Br. J. Cancer* **27:**370.

Peavy, D. L., Adler, W. H., Shands, J. W., and Smith, R. T., 1974, Selective effects of mitogens on subpopulations of mouse lymphoid cells, *Cell. Immunol.* **11:**86.

Quesenberry, P., Morley, A., Stohlman, F., Rickard, K., Howard, D., and Smith, M., 1972, Effect of endotoxins on granulopoiesis and colony-stimulating factors, *N. Engl. Med.* **286:**227.

Rietschel, E. T., Gottert, H., Lüderitz, O., and Westphal, O., 1972, Nature and linkages of the fatty acids present in the lipid A component of *Salmonella* lipopolysaccharides, *Eur. J. Biochem.* **28:**166.

Rietschel, E. T., Hase, S., King, M., Redmond, J., and Lehmann, V., 1977, Chemical structure of lipid A, in: *Microbiology, 1977* (D. Schlessinger, ed.), pp. 262–268, Am. Soc. Microbiol., Washington, D.C.

Rosenstreich, D. L., and Blumenthal, R., 1977, Ionophorous activity and murine B lymphocyte mitogens, *J. Immunol.* **118:**129.

Rosenstreich, D. L., Glode, L. M., Wahl, L. M., Sandberg, A. L., and Mergenhagen, S. E., 1977, Analysis of the cellular defects of endotoxin unresponsive C3H/HeJ mice, in: *Microbiology, 1977* (D. Schlessinger, ed.), pp. 314–320, Am. Soc. Microbiol., Washington, D.C.

Rosenstreich, D. L., Vogel, S. N., Jacques, A. R., Wahl, L. M., and Oppenheim, J. J., 1978, Macrophage sensitivity to endotoxin: Genetic control by a single codominant gene, *J. Immunol.* **121:**1664.

Rottem, S., and Leive, L., 1977, Effect of variations in lipopolysaccharide on the fluidity of the outer membrane of *Escherichia coli, J. Biol. Chem.* **252:**2077.

Ryan, J. L., and McAdam, K. P. W. J., 1977, Genetic non-responsiveness of murine fibroblasts to bacterial endotoxin, *Nature (London)* **269:**153.

Ryan, J. L., Glode, L. M., and Rosenstreich, D. L., 1979, Lack of responsiveness of C3H/HeJ macrophages to lipopolysaccharide: The cellular basis of LPS stimulated metabolism, *J. Immunol.* **122:**932.

Sacquet, E., and Charlier, H., 1965, Sensibilité de diverses races de souris à l'action litale de l'endotoxine de *Salmonella typhosa*. Influence de la flore microbienne associée, *Ann. Inst. Pasteur Paris* **108**:353.

Scher, I., Ahmed, A., Sharrow, S. O., Steinberg, A. D., and Paul, W. E., 1975, Genetic control of B-lymphocyte function in the CBA/N mouse strain: A model for examining the mechanism of B lymphocyte activation, in: *Role of Mitogens in Immunobiology* (J. J. Openheim and D. Rosenstreich, eds.), pp. 325–335, Academic Press, New York.

Scibienski, R. J., 1980, Immunological properties of protein-lipopolysaccharide complexes. III. Role of carbohydrate in the LPS adjuvant effect, *Mol. Immunol.*, **17**:21.

Scibienski, R. J., and Gershwin, M. E., 1977, Immunologic properties of protein–lipopolysaccharide complexes. I. Antibody response of normal, thymectomized, and nude mice to a lysozyme–lipopolysaccharide complex, *J. Immunol.* **119**:504.

Selinger, M. J., McAdam, K. P. W. J., Kaplan, M. M., Sipe, J. D., Rosenstreich, D. L., and Vogel, S. N., 1980, *in vitro* production of SAA by mouse hepatocytes, in: *Proceedings of the Symposium on Amyloidosis*, (G. G. Glennes, ed.) Excerpta Medica, Amsterdam.

Shear, M. J., 1944, Chemical treatment of tumors. IX. Reaction of mice with primary subcutaneous tumors to injection of a hemorrhage producing bacterial polysaccharide, *J. Natl. Cancer Inst.* **4**:461.

Sipe, J. D., Vogel, S. N., Ryan, J. L., McAdam, K. P. W. J., and Rosenstreich, D. L., 1979, Detection of a mediator derived from endotoxin-stimulated macrophages that induces the acute phase serum amyloid A response in mice, *J. Exp. Med.* **150**:597.

Skidmore, B. J., Chiller, J. M., Morrison, D. C., and Weigle, W. O., 1975a, Immunologic properties of bacterial lipopolysaccharide (LPS): Correlation between the mitogenic, adjuvant, and immunogenic activities, *J. Immunol.* **114**:770.

Skidmore, B. J., Morrison, D. C., Chiller, J. M., and Weigle, W. O., 1975b, Immunologic properties of bacterial lipopolysaccharide (LPS). II. The unresponsiveness of C3H/HeJ mouse spleen cells to LPS induced mitogenesis is dependent on the method used to extract the LPS, *J. Exp. Med.* **142**:1488.

Snell, E. S., 1971, Endotoxin and the pathogenesis of fever, in: *Microbial Toxins* (S. Kadis, G. Weinbaum, and S. J. Ajl, eds.), Vol. V, pp. 277–335, Academic Press, New York.

Sultzer, B. M., 1968, Genetic control of leukocyte responses to endotoxin, *Nature (London)* **219**:1253.

Sultzer, B. M., 1969, Genetic factors in leukocyte responses to endotoxin: Further studies in mice, *J. Immunol.* **103**:32.

Sultzer, B. M., 1971, Chemical modification of endotoxin and inactivation of its biological activities, in: *Microbial Toxins* (S. Kadis, G. Weinbaum, and S. J. Ajl, eds.), Vol. V, pp. 91–126, Academic Press, New York.

Sultzer, B. M., 1972, Genetic control of host responses to endotoxin, *Infect. Immun.* **5**:107.

Sultzer, B. M., 1976, Genetic analysis of lymphocyte activation by lipopolysaccharide endotoxin, *Infect. Immun.* **13**:1579.

Sultzer, B. M., and Goodman, G. W., 1976, Endotoxin protein: A B cell mitogen and polyclonal activator of C3H/HeJ lymphocytes, *J. Exp. Med.* **144**:821.

Sultzer, B. M., and Goodman, G. W., 1977, Characteristics of endotoxin-resistant low-responder mice, in: *Microbiology, 1977* (D. Schlessinger, ed.), pp. 304–309, Am. Soc. Microbiol., Washington, D.C.

Sultzer, B. M., and Nilsson, B. S., 1972, PPD tuberculin—A B cell mitogen, *Nature New Biol.* **240**:198.

Talcott, J. A., Ness, D. B., and Grumet, F. C., 1975, Adjuvant effect of LPS on *in vivo* antibody response: Genetic defect in C3H/HeJ mice, *Immunogenetics* **2**:507.

Vaheri, A., Rouslahti, E., Sarvas, M., and Nurminen, M., 1973, Mitogenic effect by lipopolysaccharide and pokeweed lectin on density-inhibited chick embryo fibroblasts, *J. Exp. Med.* **138**:1356.

Vogel, S. N., Hansen, C. T., and Rosenstreich, D. L., 1979a, Characterization of a congenitally LPS-resistant, athymic mouse strain, *J. Immunol.* **122**:619.

Vogel, S. N., Moore, R. N., Sipe, J. D., and Rosenstreich, D. L., 1979b, Partial reversal of LPS unresponsiveness in BCG-infected C3H/HeJ mice, *Fed. Proc.* **38**:1001.

Watson, J., and Riblet, R., 1974, Genetic control of responses to bacterial lipopolysaccharides in mice. I. Evidence for a single gene that influences mitogenic and immunogenic responses to lipopolysaccharides, *J. Exp. Med.* **140**:1147.

Watson, J., and Riblet, R., 1975, Genetic control of responses to bacterial lipopolysaccharides in mice. II. A gene that influences a membrane component involved in the activation of bone-marrow derived lymphocytes by lipopolysaccharide, *J. Immunol.* **114**:1462.

Watson, J., Kelly, K., and Largen, M., 1977a, Genetic and cellular aspects of host response to endotoxin, in: *Microbiology, 1977* (D. Schlessinger, ed.), pp. 298–303, Am. Soc. Microbiol., Washington, D.C.

Watson, J., Riblet, R., and Taylor, B. A., 1977b, The response of recombinant inbred strains of mice to bacterial lipopolysaccharides, *J. Immunol.* **118**:2088.

Watson, J., Kelly, K., Largen, M., and Taylor, B., 1978a, The genetic mapping of a defective LPS response gene in C3H/HeJ mice, *J. Immunol.* **120**:422.

Watson, J., Largen, M., and McAdam, K. P. W. J., 1978b, Genetic control of endotoxic responses in mice, *J. Exp. Med.* **147**:39.

Westphal, O., and Jann, K., 1965, Bacterial lipopolysaccharide extractions with phenol-water and further application of the procedure in: *Methods in Carbohydrate Chemistry* (R. L. Whistler, ed.), Vol. 5, p. 83, Academic Press, New York.

Part IX

Modification of Immune Responsiveness in Gnotobiotic Environments

11
Host Defense Mechanisms in Gnotobiotic Animals

PATRICIA M. BEALMEAR

1. Introduction

A question that is often raised in immunology is, what constitutes a true primary immune response? In work with laboratory animals, a reaction to an antigen to which the animal has not previously had any deliberate exposure is generally considered to be a primary response. However, during a normal life span animals encounter a large number of antigens in the environment that might cross-react with antigens used in investigative work. Thus the true primary nature of an immune response to the deliberate antigenic exposure might sometimes be questionable. As the investigation of the immune response becomes more complex, there is increasing evidence that the so-called "normal" animal is an unsatisfactory host in which to study the *true* primary immune response (Kim *et al.*, 1965; Springer, 1966). Since the pioneer work of Mishell and Dutton in 1967, many investigators have been studying the primary response *in vitro* in cells taken from supposedly immunologically virgin animals in an effort to understand the sequence of events in primary antibody induction. These responses were induced by using natural antigens such as bacteriophages and various animal erythrocytes. The detection of background numbers of plaque-forming cells reactive against animal erythrocytes and "natural" antibodies reactive against bacteriophages and other antigens in the animals used, made investigators question whether these *in vitro* responses were true primary responses or only early secondary responses. The germfree animal has provided an invaluable system in which the true primary response to antigen can be studied.

During the past fifty years a methodology for complete environmental control has been developed by which animals can be produced and maintained free of any demonstrable horizontally transmitted microbial or parasitic flora; however, the vertically transmitted

PATRICIA M. BEALMEAR • Departments of Biological Resources and Dermatology, Roswell Park Memorial Institute, Buffalo, New York 14263.

leukemia virus and possibly other viruses have not been eliminated (De Harven, 1964; Pollard *et al.*, 1964; Pollard and Matsuzawa, 1964, 1966; Salomon, 1968). These animals are referred to as germfree, axenic, or gnotobiotic* animals as distinguished from specific-pathogen-free (SPF) and conventional animals (Trexler, 1963, 1979).

The history of gnotobiology dates back to Pasteur (1885), who concluded that life could not exist without microbes, following his unsuccessful experimentation with sterile plants in an attempt to resolve questions about nitrogen fixation. The initial experiments were conducted by Nuttall and Thierfelder in Berlin in 1895. After 2 years of work, they concluded that life without bacteria was possible, but did not pursue the matter further. Following Nuttall and Thierfelder, Schottelius (1899) worked for over a decade trying to raise germfree chickens. By 1914 the early period of germfree methodology ended following demonstrations by Cohendy, Kuster, and Guyenot that germfree life was possible with chickens, goats, and flies, respectively (Cohendy and Wollman, 1914; Luckey, 1968). The second period of animal gnotobiology occurred from 1915 to 1934, during which time Glimstedt in Sweden, Reyniers in America, Balzam in Poland, and Kimura, Naito, and Kobayashi in Japan were concerned with the problems involved in rearing germfree animals. These pioneers paved the way for the third period of gnotobiologic revival from 1935 to 1954, in which biologists produced colonies of germfree animals, made practical isolators, and studied the details of germfree life. The current period began in 1955 and was marked by the acceptance of this new research tool and its application to the study of infectious disease, surgery, and animal production. The immunologists have perhaps gained the most from this unique animal model, because it has enabled them to study the immune response per se in the absence of the complications produced by microorganisms and their products.

Germfree animals were originally derived from healthy conventional mothers by cesarian delivery, at term, directly into sterile isolator systems (Reyniers, 1946, 1957; Gustafsson, 1948; Trexler and Reynolds, 1957). The newly derived animals were hand-fed a steam-sterilized diet by intubation until weaning age and then fed a steam-sterilized diet *ad libitum*. They were propagated and maintained under conditions controlled for microflora, diet, temperature, humidity, light, and noise (Pollard and Nordin, 1971). They reproduced by natural means and served as foster mothers for new strains of animals that were later introduced into the germfree environment by cesarian derivation (Pollard, 1964). The list of animals that have been produced in a germfree environment is long and includes rats, mice, guinea pigs, rabbits, cats, dogs, monkeys, sheep, cattle, pigs, chickens, and even insects. A few human babies suspected of having immune defects have also been cesarian derived and introduced into a germfree environment for limited periods of time. A detailed review of each of these species is beyond the scope of this work, and thus only those species that have contributed the most to the advancement of immunological knowledge will be

***Gnotobiotic:* A word derived from the Greek *gnotos* and *biota* meaning "known flora and fauna." *Gnotobiote* (gnotobiotic animal): One of an animal stock or strain derived by aseptic cesarean section (or sterile hatching of eggs) which are reared and continuously maintained with germfree techniques under isolator conditions and in which the composition of any associated fauna and flora, if present, is fully defined by accepted current methodology. *Germfree animal* (axenic animal): A gnotobiote which is free from all demonstrable associated forms of life including bacteria, viruses, fungi, protozoa and other saprophytic or parasitic forms." (Source: *Gnotobiotes,* Institute of Laboratory Animal Resources, International Standard Book No. 0-309-01858-7, National Academy of Science, Washington, D.C.)

considered, e.g., the colostrum-deprived piglet and mice will be covered in depth; rats, guinea pigs, rabbits, primates, dogs, and chickens will receive only brief consideration.

2. General Characteristics of the Germfree Animal

2.1. Lymphatic Tissue in Germfree Animals

2.1.1. Lymphoid Organs

The morphological and functional development of the lymphatic and hematopoietic tissues has been investigated in several species of germfree animals, i.e., mice, rats, guinea pigs, rabbits, piglets, and chickens (Glimstedt, 1936; Tajima, 1955; Miyakawa *et al.,* 1957; Gordon, 1959; Phillips *et al.,* 1959; Thorbecke, 1959; Gordon and Wostmann, 1960; Bealmear and Wilson, 1966; Pollard, 1967a; Šterzl and Silverstein, 1967; Bauer, 1968; Pleasants, 1968; Kruml *et al.,* 1969; Chakhava, 1973; Rolls *et al.,* 1975; F. Kovářů *et al.,* 1979). In general, the development of the lymphatic system in secondary lymphatic organs of the germfree animal is retarded, if these organs are in contact with the gastrointestinal tract (Gordon, 1959; Štěpánková *et al.,* 1979; Sterzl and Kovářů, 1978). There is little difference between the lymphatic tissues of germfree and conventional animals when these organs develop at sites remote from contact with the microflora (Gordon, 1959; Uno *et al.,* 1968).

The germfree animal appears normal on autopsy, but when organ weights are measured, conspicuous differences are observed between germfree and conventional animals. Various parts of the digestive tract, i.e., the small intestine, and the mesenteric lymph node have lower weights in the germfree group (Gordon, 1959). The heart, brain, and endocrine glands also show similar differences between germfree and conventional animals, with the exception of the adrenal glands of the germfree rat, which appear heavier. This decreased weight in certain organs of germfree animals can be accounted for by decreases in water content; lymphocytes, plasma cells, and macrophages of the lamina propria and submucosa; and the amount of connective tissue present (Gordon, 1959; Gordon and Wostmann, 1960). These differences are observed in germfree animals irrespective of species and sex, and they disappear within 2–3 weeks after introduction of a normal microflora. It can be concluded that the morphology of germfree animals is quantitatively, but not qualitatively, different from that of conventional control animals (Thorbecke, 1959). When immunized with specific antigens, they are capable of forming secondary nodules in their lymphoid organs, plasma cells, circulating antibody, and interferon (Wagner, 1955, 1959, 1961; Miyakawa, 1959; Wagner and Wostmann, 1959, 1961; Olson and Wostmann, 1966a,b; Bealmear and Wilson, 1967a,c; Horowitz and Bauer, 1967; Sell, 1967; Bauer, 1968; Thonard *et al.,* 1968; Trávníček *et al.,* 1968; Ohwaki *et al.,* 1977).

The lymph nodes of germfree mice resemble those of neonatal animals; they are smaller in volume and lower in activity with few, if any, germinal centers in evidence (Thorbecke, 1959; Gordon and Wostmann, 1960; Pollard, 1967a). These same observations apply to splenic tissue and other disseminated lymphatic tissues. The small germinal centers that are sometimes observed probably reflect "a) 'normal' baseline activity; b) responses to antigenic components in the diet; or c) the response of the host to viral agents now known to be indigenous to the mice" (Pollard, 1967a,b). Leukemia virus has been found in all strains of germfree mice examined (Pollard, 1964; Kajima and Pollard, 1965). These small germinal centers can even be found in some germfree mice fed a low-molecular-weight, antigen-free diet (Pleasants *et al.*, 1964).

The lymph nodes and spleens of germfree rats are more inactive than those of mice (Krumbhaar, 1948; Gordon and Wostmann, 1960; Pollard, 1967a; Iljma and Yamane, 1968), and they probably provide a more uniform baseline of activity than that observed in mice. No viral particles were found in the rats using either tissue culture or electron microscopic procedures (Pollard and Kajima, 1966, 1970).

When comparing germfree and conventional piglets and rabbits, F. Kovářů *et al.* (1975, 1979) found a persistence of extramedullary hematopoiesis, which appeared in the secondary lymphatic organs and also in the thymus. This involves mainly erythropoietic rather than myeloid activity.

Hashimoto and Hashimoto (1968) made a comparative study of the bones and bone marrow of germfree and conventional Fischer rats. They found less hematopoietic activity in germfree than in conventional rats, with granulopoietic activity especially low. Transformation to fatty marrow was observed sooner in germfree rats. The epiphyseal cartilage was less thick, and lance-shaped trabeculae, osteoblasts, and osteoclasts in the metaphysis were fewer in number in germfree rats.

2.1.2. Stem Cells

2.1.2a. CFU-s. Techniques based on colony formation either *in vivo* with spleen colonies (CFU-s) or *in vitro* in semisolid agar (CFU-c) have provided information about cellular organization and communication within the hematopoietic system. The spleen colony technique of Till and McCulloch (1961) provided the first tool for determining stem-cell numbers *in vivo.* In an effort to determine why lethally X-irradiated germfree mice that had received either syngeneic or allogeneic bone marrow survived longer than their conventional counterparts, an explanation was sought in terms of the number of stem cells found in the bone marrow of germfree and conventional donors (Connell and Wilson, 1965; Wilson and Bealmear, 1967). The equal numbers of spleen colonies found in germfree and conventional mice demonstrated that the presence of bacteria and/or their products does not inhibit the formation of colonies or restrict stem cells. As the number of spleen colonies is the same when the same number of marrow cells is given from germfree and conventional donors, it appears that the number of stem cells in a given quantity of marrow is also the same. The peripheral blood counts during recovery are also not significantly different in germfree and conventional bone marrow chimeras (Connell and Wilson, 1965; Wilson and Bealmear, 1967).

2.1.2b. CFU-c. At present there are *in vitro* clonal assay systems in which early progenitor cells of the granulocytic and monocytic series (CFU-c), and erythroid and megakaryoctic series undergo differentiation and proliferation, when the appropriate stimulating factors are available (Moore, 1979). Pluripotent stem cells differentiate and proliferate *in vivo* and *in vitro* under the influence of regulatory factors produced by the stromal cells of the hematopoietic inductive microenvironment, which is composed of macrophages, endothelial cells, and fat-containing cells. The macrophage appears to play a unique regulatory role by elaborating mediators that influence both the immune response and hematopoiesis (Moore, 1979).

Using the semisolid agar culture system (Bradley and Metcalf, 1966), it was shown that certain bone marrow cells will proliferate *in vitro* and form colonies of granulocytes and mononuclear cells named CFU-c. This *in vitro* colony formation is dependent upon the presence of a factor called colony-stimulating factor (CSF) (Metcalf and Foster, 1967; Robinson *et al.,* 1967). The sera of many normal conventional mice contain detectable levels of CSF. This factor is probably protein in nature with a molecular weight of approximately 50,000 (Stanley *et al.,* 1968).

Metcalf *et al.* (1967) showed that CSF levels were exceptionally low in germfree mice. In the AKR germfree mice with lymphoid leukemia, however, CSF levels were high. These data suggested that infections of various types might result in elevations of serum CSF. In Swiss mice infected with LDH-elevating virus, CSF serum levels were elevated in acute stages of the disease. CSF levels were also high in sera of human patients with acute infections (Foster *et al.,* 1968) and mononucleosis (Metcalf and Wahren, 1968).

In a study with germfree and conventional Swiss mice, Metcalf and Foster (1969) demonstrated that bone marrow colonies *in vitro* were very similar for both germfree and conventional bone marrow. There was a linear relationship between the number of colonies developing and the dose of CSF; however, there was a tendency for greater uniformity in the incidence of CFU-c in germfree mice than in conventional mice. Serum levels of CSF were very low in germfree mice (Metcalf and Foster, 1969). Byrne *et al.* (1979) characterized the granulocyte-macrophage progenitor cells of germfree mice by buoyant density analysis.

Morley *et al.* (1972) investigated the role of the gut microflora in stimulating CSF in the serum by comparing CSF levels after X irradiation in germfree and conventional CFW mice. In germfree mice there was no detectable rise of CSF activity after X irradiation, whereas there was a clear-cut elevation at variable times after X irradiation in conventional mice. These results suggested that bacteria are important in the generation of CSF. As injection of endotoxin produces high levels of CSF, it is hypothesized that radiation damages the gut of conventional animals, which results in bacteremia and endotoxemia. Both conventional and germfree CFW mice responded to endotoxin with similar increases in serum CSF before and after X irradiation (Quesenberry *et al.,* 1974a,b). Chang and Pollard (1973) confirmed the observations on serum CSF levels in irradiated germfree CFW mice, and showed that gnotobiotic CFW mice monoassociated with gram-negative *Escherichia coli* developed CSF elevations, while those monoassociated with gram-positive *Streptococcus faecalis* did not. Limulus reactive material, presumably endotoxin, was demonstrated in the plasma of lethally X-irradiated conventional CFW mice, but not in germfree mice (Quesenberry *et al.,* 1978).

Mahmood and Robinson (1978) and Robinson *et al.,* (1977) suggested that neutrophils may modulate their own production by inactivation of microorganisms and their products, thereby negating their stimulatory effect on CSF production. Joshi *et al.* (1979) measured changes in serum CSF levels, bone marrow colony-forming cells (CFU-c), absolute granulocyte counts, and bacterial levels in germfree mice after introduction of bacteria into their environment. These studies showed that a marked rise in CSF production occurred immediately after the germfree mice were moved into a conventional environment. The rise in CSF was correlated with the rise in stool bacterial levels and it was followed by a rise in the absolute granulocyte count. As the absolute granulocyte count rose, both CSF production and stool bacterial levels decreased. These studies are compatible with the hypothesis that CSF production is modulated by the levels of microorganisms and their products and that neutrophils serve as a negative feedback in the scheme of granulocyte regulation by inactivation of the stimulatory effect of microorganisms (Joshi *et al.,* 1979).

The exact role that CSF plays *in vivo* on leukopoiesis remains obscure, because of its short lifetime *in vivo* and the lack of large pools of a purified CSF preparation (Mirand and Hoffman, 1973). Various maneuvers can affect CSF levels in germfree mice. Six days after unilateral ureteral ligation, CSF levels are 4 times higher than in nonoperated germfree mice. In other germfree mice infected with polycythemic Friend virus (FV-P), unilateral ureteral ligation raises serum CSF markedly to levels 6 times those in nonoperated, noninfected controls. In mice subjected to unilateral ureteral ligature, the neutrophil count in peripheral blood increased by a factor of 36 with FV-P infection. Mirand and Hoffman (1973) concluded that the data support the hypothesis that CSF functions *in vivo* as a specific leukopoietin and that FV-P may be responsible for altering some endocrine regulatory mechanism involving leukopoietin, thus leading to a marked granulocytopoiesis.

2.1.3. Host Defense Mechanisms

2.1.3a. Cell-Mediated Immunity. Germfree mice are capable of a typical delayed-type hypersensitivity (DTH) response as demonstrated by their response to *Corynebacterium parvum* (Scott *et al.,* 1978). Wolberg *et al.* (1977) speculated that it was probably impossible to get a true primary immune response to *C. parvum* in conventional mice due to preexisting immunity generated by cross-reactivity with bacteria. Scott *et al.* (1978), therefore, examined the responsiveness of germfree mice to *C. parvum.* Because a component of the stimulatory effects of *C. parvum* is thought to be the result of a specific immune response, these effects might be reduced in germfree mice. Contrary to expectations, *C. parvum* was found to be fully capable of lymphoreticular stimulation in germfree mice as evidenced by splenomegaly. The development of cell-mediated immunity to *C. parvum* in germfree mice was similarly unimpaired; this aspect of *C. parvum* immunity is implicated in its antitumor activity (Bomford and Christie, 1975; Christie and Bomford, 1975; Tuttle and North, 1975). These data show that the stimulatory effects of *C. parvum* may be fully established in the absence of previous sensitization. Fray *et al.* (1978) also described the *in vitro* tumor inhibitory capacity of *C. parvum*-activated peritoneal cells from germfree mice as similar to that of *C. parvum*-activated cells from conventional mice.

MacDonald and Carter (1978) found that germfree mice, sensitized to picryl chloride by painting on the abdomen and challenge on the left ear at day 6, showed the same level of contact sensitivity as that shown by conventional mice. The DTH lesions in germfree and conventional mice were indistinguishable histologically. Germfree mice also exhibited contact sensitivity to oxazolone, which was in direct contrast to the impaired DTH in germfree guinea pigs reported by Lev and Battisto in 1970. A diminished level of DTH to tubercle bacilli in the germfree mouse was also observed (Ueda *et al.,* 1975).

Smith *et al.* (1972) demonstrated that germfree mice reject skin allografts at the same rate as conventional mice. Germfree mice were also capable of rejecting allogeneic tumor transplants at a normal rate (Bealmear, 1965; Bealmear and Wilson, 1967b).

The dissimilarity between the germfree mouse and the germfree guinea pig in their response to allergenic haptens may be explained in terms of the cells that are responding in each case. The contact sensitivity lesion in the germfree mouse consists of lymphocytes, neutrophils, and macrophages, whereas in the guinea pig much attention has been focused on the role of the basophil in contact sensitivity. Richerson *et al.* (1969) coined the term "cutaneous basophil hypersensitivity" to characterize the contact sensitivity lesion in the guinea pig. Germfree guinea pigs also accept allografts and permit their growth (Miyakawa *et al.,* 1958). As germfree mice display a normal contact sensitivity in the ear, but show a reduced response in the footpad after immunization, the nature of the antigen and its route of administration may be important factors in determining whether or not an animal can express DTH (MacDonald and Carter, 1978).

To add to the controversy concerning the ability of the germfree guinea pig to develop cell-mediated immune responses, Green and Balish (1977) tested the responses of randomly bred strain-2 germfree and conventional guinea pigs to *Trichophyton mentagrophytes* (American Type Culture Collection 18748). All infected guinea pigs elicited positive DTH skin tests following primary infection, and spleen and lymph node lymphocyte cultures underwent blastogenesis to dermatophyte antigens. Following the secondary infection, which was shorter in duration than the primary infection, delayed hypersensitivity skin tests and blastogenesis responses were still positive to dermatophyte antigens. These results suggested that the cell-mediated immune potential of the germfree guinea pig is not impaired, and can be activated with a pathogenic microorganism.

A DTH reaction can also be elicited by an injection of 10^8 sheep red blood cells (SRBC) into the rear footpad of conventional mice previously immunized with small doses of SRBC (MacDonald and Carter, 1979). Germfree mice produced no DTH to the same doses of SRBC when immunized by the i.v. route, and only weak reactions when immunized by the s.c. (footpad) route. Varying the dose of SRBC did not elicit a response in germfree mice. The transfer of lymphocytes from conventional mice, however, did confer on germfree mice the capacity to express DTH to SRBC. If germfree mice were monoassociated with a gram-negative microflora, they could generate a normal DTH response to SRBC.

Germfree mice were capable of blastogenesis responses to the T-cell mitogens phytohemagglutinin (PHA) and concanavalin A (Con A) and to specific alloantigens in the mixed lymphocyte reaction (MLR) (Beardsley, 1976; Bealmear and Hale, 1977; Bealmear *et al.,* 1979). The magnitude of these responses was dramatically influenced by the diet; mice maintained on a diet of 6% vegetable fat had significantly higher responses in all of the above assays than mice maintained on a diet of 11% fat of both animal and vegetable origin (Bealmear *et al.,* 1979). When the diet was changed from 11% to 6% fat, blastogenesis responses returned to normal after approximately 2 weeks on the lower fat diet.

Human dialyzable transfer factor (TF) was found capable of inducing *in vivo* (skin test) and *in vitro* (blastogenesis) cellular immune responses in gnotobiotic nonhuman primates (Eichberg *et al.,* 1976a). These animals were germfree or gnotobiotic, and therefore had not been skin tested previously. These data supported the hypothesis that induction of cell-mediated immunity by *in vivo* treatment with TF does not require previous exposure to specific and/or cross-reacting antigens, and this induction may be due to an antigen-specific informational effect of TF. Due to lack of antigenic stimulus, it is possible that germfree primates have more mature uncommitted mononuclear cells capable of responding to the effects of TF. Animals reared in a germfree environment are still exposed to antigens in the diet. Sensitization to these killed microbes in the diet may account for the relatively normal numbers of circulating lymphocytes and the responses to TF. If germfree primates were raised on a completely antigen-free, chemically defined diet (Greenstein *et al.,* 1957; Pleasants *et al.,* 1968), and therefore not exposed to antigen, they could serve as ideal models for resolving questions concerning the role of antigenic priming and other questions relevant to the mechanisms of TF activity (Eichberg *et al.,* 1976a).

2.1.3b. Humoral Immunity. Assuming that immunoglobulin secretion originates only upon antigenic stimulation, the serum protein pattern of the germfree animal seems to reflect the low level of antigenic stimulation. Gamma globulins are almost always low in concentration but never entirely absent (Wostmann, 1961; Sell, 1964a,b, 1967). Some controversy exists about the γM concentration in mouse serum; Arnason *et al.* (1964) showed normal levels, whereas Fahey and Sell (1965) found this fraction to be below detectable limits. Differences in strain, diet, and husbandry may contribute to this controversy about serum protein levels (Asofsky *et al.,* 1968).

The germfree rat demonstrates a serum gamma globulin pattern that appears to be independent of the diet. Immunoelectrophoresis shows that all proteins present in the "gamma range" of the conventional animal are present, but in lower concentrations, in the germfree rat (Wostmann, 1968). In the case of the germfree mouse, guinea pig (Wostmann, 1961; Sell, 1964a,b), and rabbit (Wostmann, 1961), the serum protein pattern is markedly affected by the diet. In germfree rabbits and guinea pigs the gamma globulin concentration is substantially increased by incorporation of bovine milk proteins into the diet (Wostmann, 1961). In guinea pigs the 7 S γ1 fraction appears to be especially diet dependent (Sell, 1964a).

The fractional rate of catabolism of gamma globulin in germfree chickens and guinea pigs appears comparable to that found in conventional animals (Sell, 1964a; Wostmann and Olson, 1964). The half-lives of 7–8 days in guinea pigs and 4.2 days in chickens reflect the turnover of the 7 S gamma globulins. Fahey and Sell (1965) demonstrated in mice that 7 S gamma globulin turnover depends on the concentration of the 7 S class of gamma globulins. Sell and Fahey (1964) determined the total gamma globulin and observed the turnover of labeled homologous and heterologous gamma globulin in germfree and conventional mice. The rate of synthesis of gamma globulin in conventional mice was approximately 50 times greater than that of germfree mice; the greatly reduced serum concentrations of gamma globulin in germfree mice reflected their lower rate of synthesis. The ability to catabolize heterologous gamma globulin was found to be essentially the same in germfree and conventional mice.

Upon exposure to a normal microflora, the ex-germfree rat demonstrates an early increase in alpha globulins, which presumably reflects the formation of an acute-phase protein (Wostmann and Gordon, 1960; Balish *et al.,* 1973). Beta globulins increase shortly after the alpha globulins, but the gamma globulins remain at the low germfree level for at least 2 weeks.

Asofsky and Hylton (1967) observed no detectable IgA synthesis in germfree mice. Immunoglobulin synthesis was investigated by incubating various tissues with ^{14}C-labeled lysine and isoleucine, and then examining the tissue supernatants for proteins co-precipitating with specific antiglobulin antisera and unlabeled "carrier" immunoglobulins by radioimmunoelectrophoretic analysis. They made the following observations: "(1) Serum, colostrum, and intestinal washings from conventionally reared (CV) mice contained IgA; this protein was undetectable in similar material from germfree (GF) mice. (2) Almost all lymphoid tissues from adult CV mice synthesized all immunoglobulins including IgA; those from GF mice formed only IgM and 7 S $\gamma 1$. (3) Lactating mammary tissue from both CV and GF mice produced about equal amounts of protein co-precipitating with serum IgA." In the case of the GF tissue this protein was of lower molecular weight than IgA, but similar to that of the T chain as judged by sedimentation in the ultracentrifuge. Ileum and jejunum from CV mice, but not from GF mice, synthesized IgA. Moreau *et al.* (1977) induced IgA plasmocytes in germfree C3H mice by monoassociation with gram-negative bacteria. Hashimoto *et al.* (1973) induced IgA production in germfree ICR mice with a mouse adenovirus.

When immunized with specific antigen, the germfree animal can produce specific antibody to the antigen (Olson and Wostmann, 1966a,b; Bealmear and Wilson, 1967a); the response is usually equal to and sometimes even higher than that of the conventional control animals. There is, however, a slight delay of about 24 hr before antibody can be detected in the germfree animal. Ohwaki *et al.* (1977) compared the responses of germfree and conventional ICR mice to SRBC, dinitrophenyl-lysine-Ficoll (DNP-Lys-Ficoll), and dinitrophenylated bovine serum albumin (DNP-BSA). IgM plaque-forming cells (PFC) of both germfree and conventional mice increased exponentially during the first 3 weeks after birth and then reached a plateau. The response of germfree mice was always slightly lower than that of conventional mice, but the difference was not significant. No detectable IgG PFC were found in either germfree or conventional mice during the first week of life, but their numbers increased abruptly and reached a plateau during the third week of life. In contrast to IgM PFC, IgG PFC of germfree mice were fewer than in conventional mice by a factor of about 10. In the case of DNP-Lys-Ficoll, a thymus-independent (TI) antigen, no significant difference was found between germfree and conventional mice in both IgM and IgG PFC. On the other hand, DNP-BSA, a thymus-dependent (TD) antigen, produced significant differences in both IgM and IgG DNP-PFC, i.e., conventional mice produced about 10 times more IgM and IgG PFC than germfree mice. After secondary immunization with DNP-BSA, a rapid development of IgG PFC occurred in both germfree and conventional mice, but the response in germfree mice was significantly less than in conventional mice.

Six-week-old normal germfree and conventional mice were examined for the ratio of Ig-bearing and Thy-1.2-bearing cells in their spleens by an indirect immunofluorescence technique. A statistically significant difference was observed in the percentage of Thy-1.2-bearing cells but not in Ig-bearing cells between germfree and conventional mice. The total numbers of nucleated cells were similar, but germfree mice appeared to have smaller numbers of Thy-1.2-bearing cells than conventional mice. Ohwaki *et al.* (1977) concluded that the spleen of germfree mice may contain fewer helper T cells. The reason for this lower responsiveness to TD antigens in the germfree mice is not quite clear, but it is possible that bacterial products from the gut flora may influence cell differentiation. Sheid *et al.* (1973) demonstrated that bacterial lipopolysaccharides can stimulate the differentiation of T cells *in vitro*.

The serum gamma globulin levels of conventional guinea pigs fall during the first 4 weeks of life due to catabolism of maternal gamma globulin transferred via the placenta

(Sell, 1967). The levels then rise sharply from 4 to 12 weeks of age so that from 3 months to 1 year of age the total serum gamma globulin concentration ranges between 500 and 1000 mg/100 ml. In contrast, the germfree guinea pig continues to catabolize maternal gamma globulin until about 6 weeks of age, at which time a constant level of approximately 100 mg/100 ml is established and maintained for at least 1 year of life. Studies by Sell (1967) of the specific antibody, total serum gamma globulin, specific immunoglobulins, and histologic responses of germfree and conventional guinea pigs to immunization with soluble bovine gamma globulin (BGG), alum-precipitated BGG, and BGG in complete Freund's adjuvant indicated that the germfree guinea pig has the same potential to respond to these antigens as the conventional guinea pig.

Olson and Wostmann (1966b) attempted to determine the quantity of potential antibody-producing cells in cervical and mesenteric lymph nodes of germfree and conventional mice. The lymph nodes of germfree animals contained as little as 1/12th the number of blast and potential antibody-producing cells as their conventional counterparts. When stimulated with 7 S human gamma globulin (HGG) i.p., potential antibody-producing cells showed a greater increase in germfree mice after the challenge with antigen in comparison to conventional controls (Olson and Wostmann, 1966a). This finding suggests that germfree hosts may have a greater increase in competent cells after challenge because of their uncommitted status, which is attributable to lack of previous antigenic stimulation (Gordon and Pesti, 1971).

Germfree mice respond normally to the B-cell mitogens pokeweed (PWM) and lipopolysaccharide (LPS) (Bealmear and Hale, 1977; Bealmear *et al.,* 1979). The response of germfree mice is slightly lower than that of conventional mice, but this is somewhat dependent upon the antigen load and the fat content of the diet. Kiyono *et al.* (1980) measured several responses to bacterial endotoxin (LPS) in germfree (GF), *E. coli* monoassociated (Mono), and conventional (CV) mice, including immunogenicity, mitogenicity, and lethality. GF mice exhibited significant antibody responses to LPS at low concentrations and over a broad dosage range; however, the development of their responses differed from those of CV mice. CV mice exhibited peak responses at 14 days after LPS administration, and their titers decreased over the next 2 weeks. GF mice had moderate responses by 2 weeks and attained their highest titers at 4 weeks; peak GF titers were higher than peak CV titers. When TNP-LPS immunogen was administered by either the intraperitoneal or the intragastric route, the highest numbers of splenic anti-TNP-LPS PFC occurred in GF mice, whereas the Mono animals exhibited intermediate responsiveness. This pattern was dose dependent, for GF mice responded at TNP-LPS concentrations that were minimally immunogenic for Mono mice and nonimmunogenic for CV mice. These differences were also observed in anti-TNP responses *in vitro.* No differences were noted in the mitogenic responses to PHA, Con A, PWM, and LPS in these three groups. Both GF and Mono mice, however, yielded higher responses to LPS than were observed in CV mice. Mono mice were intermediate in their sensitivity to LPS-induced lethality. GF mice were approximately fivefold more resistant than CV mice. Kiyono *et al.* (1980) concluded that GF mice exhibit greater *in vitro* mitogenic and immunogenic responses and are more refractory to endotoxin than CV mice. McGhee *et al.* (1980) also examined the *in vitro* immune responsiveness of cells from GF mice by using both lectins (PHA and Con A) and endotoxin. Splenocytes from GF and CV mice manifested similar responses to the T-lymphocyte mitogens PHA and Con A, but GF mice had two- to threefold greater responses to LPS than CV. Lower responses to LPS were observed after conventionalization of GF mice with indigenous BALB/c flora. Similar results were observed when a second LPS effect was measured, using anti-TNP PFC responses after TNP-LPS immunization. To determine the nature of the cell(s) involved in this suppression, macrophages (Mϕ) and purified lymphocytes from GF and CV spleens, and various cell combinations were tested for their mitogenic and immunogenic responsiveness to LPS. No apparent differences in mitogenic responses or anti-TNP PFC could be detected between GF and CV Mϕ when tested with T and B cells from GF mice. Lower responses were always seen with CV T and B cells and either GF or CV Mϕ, suggesting a lymphocyte origin for suppression. Purified T cells from either CV Peyer's patches or spleen significantly reduced mitogenic and TNP-LPS immune responses in GF cultures. The presence of T-suppressor cells was further supported by the observation that GF and nude mice exhibited higher responses to TNP-LPS. McGhee *et al.* (1980) suggested that the normal gram-negative microflora of the gut induces the production, possibly via lipid A, of a T-cell population that regulates B-cell responses to bacterial endotoxin.

2.1.3c. Macrophages. There are no significant differences between germfree and conventional mice in the number, distribution, or phagocytic ability of macrophages in their spleens and lymph nodes (Bauer *et al.*, 1963, 1964). When phagocytic indices were determined by carbon clearance on three strains of germfree mice and their conventional controls (Doll, 1962), no significant differences were found in carbon clearance when clearance rates were correlated to relative combined weights of liver and spleen. Heise and Myrvik (1966) suggested the functional similarity of macrophages of germfree and conventional animals from their finding of similar concentrations of lysozyme, cathepsin, and acid phosphatase in peritoneal macrophages from germfree and conventional rats. These enzymes were found in lower concentrations when taken from alveolar macrophages of germfree animals. Perkins *et al.* (1967) found fewer phagocytic cells in germfree than conventional mice following i.p. injection of a sterile stimulant such as gelatin, although the phagocytic capacity of individual cells for opsonized SRBC *in vitro* was the same for cells from both germfree and conventional animals.

When microorganisms rather than carbon are used to measure phagocytic ability, conventional macrophages are consistently more active than germfree. MacDonald *et al.* (1978) reported that no difference could be detected in the rate of blood clearance of *Listeria* between germfree and conventional mice, but there was considerably less killing of *Listeria* in the spleens and livers of the germfree mice than the conventional mice. Using an *in vitro* assay, peritoneal macrophages from germfree mice also demonstrated less anti-*Listeria* activity than those of conventional mice (MacDonald *et al.*, 1978). Podoprigora and Zaitsev (1979) found decreased phagocytosis of *E. coli* 055 in 3- to 4-month-old germfree rats. On the contrary, macrophages of 2.5-year-old germfree rats showed a 2- to 2.5-fold higher phagocytic index for *S. albus* than conventional rats (Podoprigora and Zaitsev, 1979). They also reported the number of Kupffer cells in the livers of germfree rats to be only half that of conventional rats. However, there was a progressive decrease of phagocytic activity in old conventional rats that was not observed in old germfree rats (Podoprigora and Zaitsev, 1979).

Woodward (1978) reported that the cytology of the lavage-recoverable peritoneal cell population from germfree mice was similar to that of their conventional controls. There were no differences in total counts, differential counts, or [^{3}H]thymidine labeling indices of their peritoneal leukocytes. In germfree nude mice, however, peritoneal macrophages were not nonspecifically cytotoxic to tumor cells *in vitro* (Meltzer, 1976). Meltzer's data suggested that environmental stimuli, rather than thymic deficiency, induced activated tumoricidal macrophages in nude mice.

Steinman *et al.* (1974) reported the presence of a small population of an apparently novel cell type in peripheral lymphoid organs, which adheres to glass and plastic surfaces. These unique cells are called "dendritic" cells, and they do not have the functional properties of lymphocytes, macrophages, or nonphagocytic "reticular" cells. They are derived from a proliferating pool located in bone marrow and from the glass-nonadherent cells of the spleen. They did not observe an increase in splenic dendritic cell numbers following immunization, but this probably reflected a high level of ongoing antigenic stimulation from antigens entering via the gut. The number of these dendritic cells must be influenced by the microflora, however, because both the total number and the concentrations increased (13- and 2-fold, respectively) in germfree mouse spleens 2 weeks after these mice were placed in a conventional environment. Several properties of these dendritic cells distinguished them from mononuclear phagocytes. Dendritic cells were not found in newborn mice, and their concentration in both spleen and mesenteric lymph nodes did not reach adult levels until 3–4 weeks of age. These cells almost completely disappeared from adherent populations following administration of steroids or following exposure to ionizing radiation. Splenic dendritic cells originate from precursors both in the bone marrow and in the spleen itself, probably the red pulp. Dendritic cells in the spleen and lymph nodes undergo slight of moderate increases in numbers during a primary immune response.

2.1.3d. Properdin and Complement. Wagner (1959) observed that the levels of properdin, complement, and complement components in the sera of germfree rats were at the

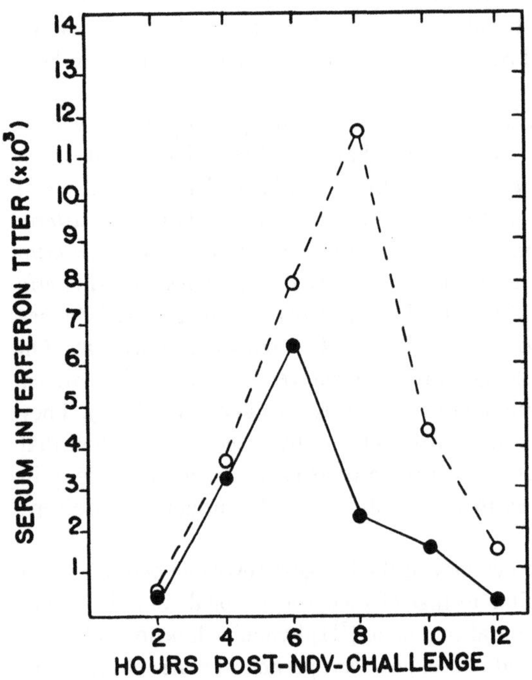

FIGURE 1. Serum interferon levels induced by Newcastle disease virus in germfree (○) and conventional (●) CFW mice. [Reprinted with permission from Havell *et al.* (1970).]

lowest limits of the conventional normal range. It was noted, however, that the sera of the germfree rats demonstrated positive agglutination titers with *M. epidermidis* (an organism usually found in the diet prior to sterilization), and thus these rats could not be considered to be immunologically naive. These observations on properdin were later confirmed by Gustafsson and Laurell (1960). Newton *et al.* (1960) compared complement levels of germfree and conventional guinea pigs and found results comparable to those of Wagner (1959) in the rats. They also found an age-conditioned increment of complement levels in both groups. Šterzl (1963) confirmed that levels of complement in germfree animals develop independently after birth following antigenic stimulation from the diet.

2.1.3e. Interferon. The major site of interferon production is believed to be the reticuloendothilial system (DeSomer and Billiau, 1966); this system is relatively underdeveloped (Thorbecke and Benacerraf, 1959) in germfree animals. Comparative studies of interferon in germfree and conventional animals have shown differences in the kinetics as well as the magnitude of the response (Considine and Starr, 1967; Starr and Considine, 1968; Havell *et al.*, 1970). When germfree and conventional CFW mice were challenged i.v. with an optimum interferon-inducing dose of Newcastle disease virus (8×10^8 PFU), the response of the conventional mice reached a peak of about 6000 interferon units/ml 6 hr after challenge (Havell *et al.*, 1970). In contrast, the germfree mice reached a peak response of 12,000 units/ml 12 hr later, after the curve for conventional mice had declined sharply (Fig. 1). When challenged with Newcastle disease virus, gnotobiotic mice monoassociated with a gram-negative organism gave results comparable to those of conventional mice (Fig. 2). Mice monoassociated with a gram-positive organism gave results comparable to those of germfree mice. There was no significant difference in interferon production between

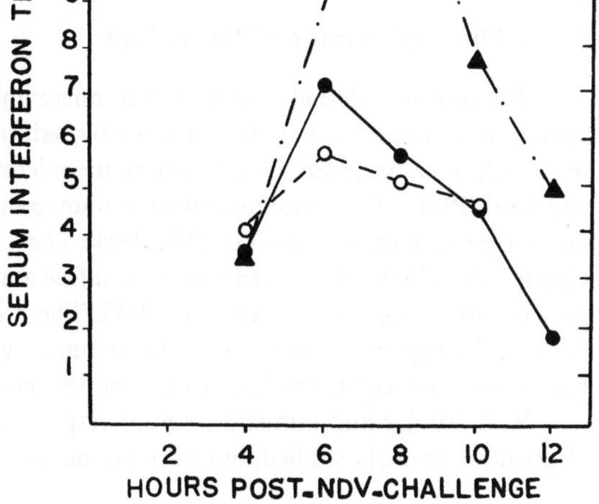

FIGURE 2. Serum interferon levels induced by Newcastle disease virus, in three groups of ex-germfree mice, each associated with a single bacterial species. (●) *Proteus mirablis;* (○) *Salmonella typhimurium;* (▲) *Streptococcus faecalis.* [Reprinted with permission from Havell *et al.* (1970).]

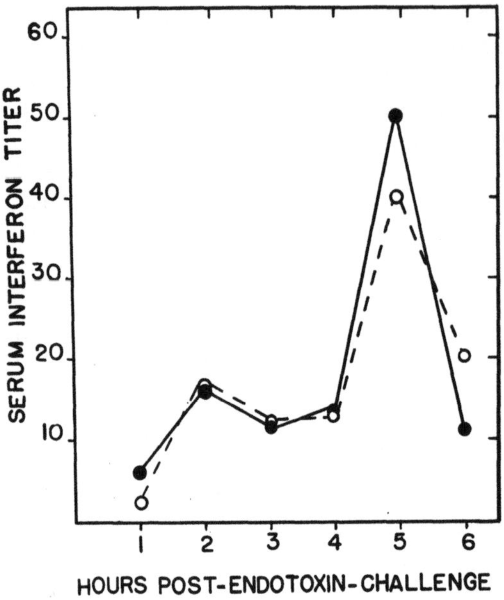

FIGURE 3. Serum interferon levels induced by endotoxin in germfree (○) and conventional (●) CFW mice. [Reprinted with permission from Havell *et al.* (1970).]

germfree and conventional mice challenged with i.v. administration of 250 μg of endotoxin (Fig. 3).

Two possible explanations have been offered for the apparent increased responsiveness of germfree mice to the induction of interferon by virus (Havell *et al.,* 1970): (1) there is a true increased interferon production (Ho, 1964; Stinebring and Youngner, 1964) or (2) there is decreased catabolism of interferon. The constant endogenous stimulation of conventional animals with endotoxin may result in relative hyporeactivity. As there is an increased half-life of antibody in germfree chickens because of reduced catabolism (Wostmann and Olson, 1969), it is possible that a similar mechanism might be operating in germfree mice in the case of interferon, which could explain their higher responsiveness to interferon induction by virus.

2.1.4. Diet and Immune Mechanisms

The reticuloendothelial system of germfree animals is less well developed than that of conventional animals. It is characterized by reduced weight in those tissues that are normally subject to antigenic stimulation by the microflora of the gut (Gordon, 1959; Gordon and Pesti, 1971). The concentration of immunocompetent cells is lower in those tissues that are in close contact with the gut (Thorbecke *et al.,* 1957; Olson and Wostmann, 1966a,b). Gamma globulin levels are reduced in germfree animals, but these appear to be diet dependent in certain species (Wostmann, 1961). There is a total absence of the IgG line in the immunoelectrophoretic analysis of the serum of young adult germfree mice reared on a low-molecular-weight, "antigen-free" diet developed by Pleasants *et al.* (1964).

Both cellular and humoral immune responses are quite adequate in all of the species of germfree animals studied, but different authors report different primary response patterns, which may or may not differ from those of their conventional counterparts (Kim *et al.*, 1965; Olson and Wostmann, 1966; Sell, 1967). In all of these experiments, dietary stimulation appears to be a major variable.

In studies on the biochemistry of the antimicrobial defense mechanism, adult germfree rats were monoassociated with *Salmonella typhimurium*. In the early studies the animals were maintained on diet L 462 (Wostmann, 1959). When these animals were monoassociated, a weight loss of 25–30% occurred, and approximately one-fourth of the animals died within the first 2 weeks. In later experiments (Kellogg and Wostmann, 1967), diets 5010 C and L 485 were used, with much less weight loss and no deaths. Body weights and the number of deaths were recorded before and after monoassociation. The data showed that while in both experiments the animals maintained on diet L 462 demonstrated severe weight loss and death, monoassociated rats fed either the 5010 C or the L 485 diet showed little weight loss and no mortality. Before monoassociation, diet L 462 had appeared superior to diets 5010 C and L 485 as judged by effects on body weight. These experiments suggested that diet exerts a pronounced influence on the resistance to pathogenic microorganisms.

In order to further clarify the role of diet on the immune mechanisms of germfree animals, germfree CFW mice were fed four different solid diets: L 462, L 474 E_{12} (Wostmann and Kellogg, 1967), 5010 C, and L 485. The animals were sacrificed at 65 days of age and the total serum protein, leukocyte count, body weight, weight and histology of the mesenteric lymph nodes and liver, bone marrow cellularity, and anti-SRBC plaque-forming cells (PFC background) in the spleen were determined. These values were compared to comparable germfree mice reared on the low-molecular-weight, "antigen-free" diet, and to conventional mice reared on diet L 485 (Wostmann *et al.,* 1969).

Part of these data are shown in Table I. There is no difference in total serum protein in any of the groups; weights of the mesenteric lymph nodes of all germfree groups are comparable, but they are lower than those found in the conventional mice. Background anti-SRBC PFC values were also similar in all germfree groups and lower than those in the conventional group. The major differences among the germfree groups were found in their leukocyte patterns. Total leukocyte counts in germfree mice fed diets L 462 and L 474 E_{12} were comparable and approximately twice as high as in animals maintained on diets 5010 C and L 485. The latter values were again approximately 2–3 times higher than those found in the mice fed the water-soluble formula. Lymphocyte values also followed this general pattern. Total granulocyte counts in animals reared on diet L 462 were statistically higher than in the animals reared on the 5010 C or L 485 diet; counts in mice fed diet L 474 E_{12} were higher, but not significantly higher. Total granulocyte counts in mice maintained on the water soluble formula tended to be even lower than those in animals reared on solid diets 5010 C and L 474 E_{12}. Comparison with conventional animals reared on diet L 485 showed that the leukocyte pattern is affected far more by the diet than by the microbial status.

Few of the above differences were reflected by the other elements of the reticuloendothelial system (RES). The only significant difference noted in the bone marrow among the four solid-diet groups was a significantly higher percentage of granulocyte precursors in germfree mice reared on diet L 474 E_{12}. Granulocyte levels on this diet approached the values found in the conventional animals. Histologically the lymph nodes of all four groups were quite comparable, with dense primary nodules predominating and an occasional

TABLE I

Results Obtained from Germfree and Conventional Mice Maintained on Solid and Water-Soluble, Low-Molecular-Weight Diets[a]

	Germfree										Conventional
	L 474 E_{12}		L 462		5010 C		L 485		H_2O-I	H_2O-II	L 485
	(9)[b]	(10)	(10)	(10)	(10)	(10)	(10)	(10)	(4)	(4)	(20)
Body weight (g)	26.8	21.2	28.8	21.7	28.3	22.9	26.8	22.4	—	21.3	24.7
	±0.5[c]	±0.7	±0.8	±0.3	±0.9	±0.5	±0.4	±0.7	—	±1.0	±0.8
Serum protein (%)	6.1	5.8	6.1	6.0	5.9	5.8	6.0	5.8	5.9	6.2	6.0
	±0.1	±0.1	±0.1	±0.2	±0.1	±0.1	±0.1	±0.1	±0.3	±0.1	±0.147
WBC × 10^3											
Total	18.4	19.0	20.0	18.5	8.6	11.0	8.4	8.2	4.6	2.2	9.9
	±1.6	±1.3	±1.7	±1.7	±0.4	±0.6	±0.7	±0.8	±0.7	±0.2	±0.11
Agranulocytes	15.3	16.8	17.0	16.7	7.0	9.7	6.0	7.3	3.1	1.8	7.5
	±1.6	±1.3	±1.5	±1.6	±0.6	±0.6	±0.5	±0.7	±0.4	±0.2	±0.2
Granulocytes	3.1	2.2	3.0	1.8	1.6	1.3	2.4	0.9	1.5	0.4	2.4
	±0.3	±0.4	±0.4	±0.2	±0.3	±0.3	±0.5	±0.1	±0.4	±0.1	±0.2
Weight (mg/100 ml)											
Mesenteric node	47	69	38	82	45	73	45	67	—	77	—
	±3	±7	±4	±5	±5	±5	±6	±6		±3	
			42 +3[d]								134 + 11[d]
PFC/spleen	29	—	33	—	24	—	32	—	—	32	91
	±6	—	±5	—	±4	—	±7	—	—	±7	±22[e]

[a]Reprinted with permission from Wostmann *et al.* (1969).
[b]Number of animals in parentheses.
[c]Mean ± S.E.
[d]Data from earlier experimental series (2).
[e]Data from earlier experimental series (15).

inactive center visible in the nodule. Some plasma cells were present in the inactive centers, but few if any mitotic figures were in evidence. There were large areas of white pulp in all the spleens examined, but no active centers, few plasma cells, and few mitotic figures. There was, however, a pronounced increase in Kupffer cells in the livers of mice fed diet L 474 E_{12}, L 462, or the water-soluble diet. These data indicate that diet may be a major factor in the functional development of the RES (Wostmann *et al.*, 1969). When the components of the various solid diets were compared, both diets L 462 and L 474 E_{12} were found to contain substantial amounts of casein and other bovine milk products. Incorporation of commercially produced casein into the diet of germfree animals may lead to extensive stimulation of the RES by nonviable microbial forms and possibly by the protein itself (Sell, 1964b). In the case of the germfree rats monoassociated with *S. typhimurium*, the rearing on diets 5010 C and L 485 appeared to be advantageous in comparison with the results obtained with diet L 462, indicating that dietary factors other than casein may have a strong influence on antimicrobial defense mechanisms.

When the serum proteins were examined in this study, the serum of mice reared on the four solid diets contained normal amounts of the electrophoretically discernible protein fractions, with the exception of gamma globulin, which occurred at a much lower concentration (Wostmann *et al.*, 1970). Germfree mice reared on the water-soluble diet showed a similar pattern with normal levels of total serum protein and albumin, but gamma globulins were mostly below the detectable range. Body weights of the mice fed the water-soluble diet were approximately 20% lower than those of the mice fed the solid diets (22 vs. 27 g).

Immunoelectrophoresis and further quantitation of the immunoglobulins in the sera of germfree mice on solid diets have generally indicated that all globulins usually regarded as immune globulins are present, although in reduced amounts (Arnason *et al.*, 1964; Fahey and Sell, 1965). The patterns obtained with sera from mice on the four solid diets confirmed the earlier observations. The IgG fraction was clearly shown and IgA was also indicated. The typical IgM line was absent. Sera from mice reared on the water-soluble diet, however, lacked a sharply delineated IgG fraction, although the presence of some IgA was suggested. In the absence of a demonstrable IgG line, two faint arcs became visible. The faster moving one may well be a vestige of IgG2, while the slower globulin fraction appears in the position where Hurlemann *et al.* (1966) indicate C-reactive protein (CRP). Conventional mouse serum demonstrated the usual spectrum of immune globulins (Fig. 4). Electrophoretically determined gamma globulin levels in young adult mice will vary from 15 to more than 50% of conventional levels, and appear to be strongly diet dependent (Wostmann *et al.*, 1971). The feeding of a Millipore-filtered, water-soluble diet drastically reduces gamma globulins in the germfree mouse. However, in the immunoelectrophoretic patterns of the sera of older animals, well-defined IgG lines are again visible (Fig. 5), indicating either a slowly developing specific response or a possible further evolution of an originally undetected random gamma globulin production. When similar water-soluble diets were not only Millipore-filtered but also prefiltered through a Diaflo membrane, which retains all materials of molecular weight greater than 10,000, the picture changes even more drastically (Fig. 6). Exposure of the inbred germfree C3H/He mice to this diet for periods up to 1 year did not lead to the detection of IgG or IgA in the serum, and only occasionally to possible traces of IgM. In many of these immunoelectrophoretic patterns an arc could be seen in the IgG1 range. However, this arc did not develop when specific anti-

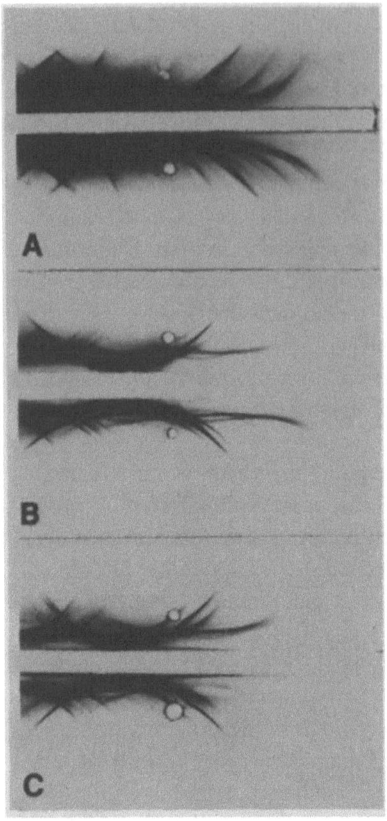

FIGURE 4. Cationic range of immunoelectrophoretic patterns indicating influence of diet on germfree murine gamma globulin fractions. Top well received 5 μl and bottom well 10 μl of antigen on each plate except for (C), in which 10 μl of each representative serum was used. (A,B) Conventional and germfree mice, respectively, fed L 485 commercial-type solid diet. (C) The upper well received serum from a germfree mouse fed a solid diet based on rice starch and casein; the lower well received serum from a germfree mouse fed a semisolid diet of comparable composition, but with an amino acid solution substituted for casein. [Reprinted with permission from Wostmann *et al.* (1970).]

IgG1 serum was used, and it was thought to represent CRP. These data indicate that the immune globulins demonstrable in the sera of germfree mice fed various types of solid diets are the result of dietary stimulation. Combined with older data on the effect of dietary components on gamma globulin levels of germfree rabbits and guinea pigs (Wostmann, 1961), these observations appear to establish diet as a potential major source of stimulation of immune mechanisms leading to humoral antibody responses.

Another factor that may profoundly influence the immune status of the individual may be the dietary lipids. Advanced cancer is often associated with impaired immunocompetence. Several possible mechanisms may be responsible for this impairment of immune competence, but a conceivable mechanism might depend on the correlation between the derangement in lipid metabolism observed in cancer patients and their common reduction in immunocompetence. A relationship between serum lipoprotein levels and cancer has been noted by several investigators. As early as 1959 Barclay and his associates reported significant decreases of high-density lipoproteins (HDL) and an increase of low-density lipoproteins (LDL) in cases of advanced breast cancer and correlated the degree of alteration with the extent of the disease. Studies on other forms of cancer strengthened these findings and demonstrated that many forms of cancer are accompanied by decreased HDL. The histological type of cancer does not play a critical role in this decrease. Serum lipoproteins are complexes of proteins and lipids differing in size and composition. Their concentration in plasma is known to be altered by changes in food intake, hormonal factors, and pathophysiological events. The cause of the reported changes in serum lipoproteins in cancer patients, however, requires further clarification. Since it has been observed that patients with high levels of LDL (β-lipoproteins) and low levels of HDL (α-lipoproteins) have an impaired immune response when their lymphocytes are tested *in vitro,* Bealmear

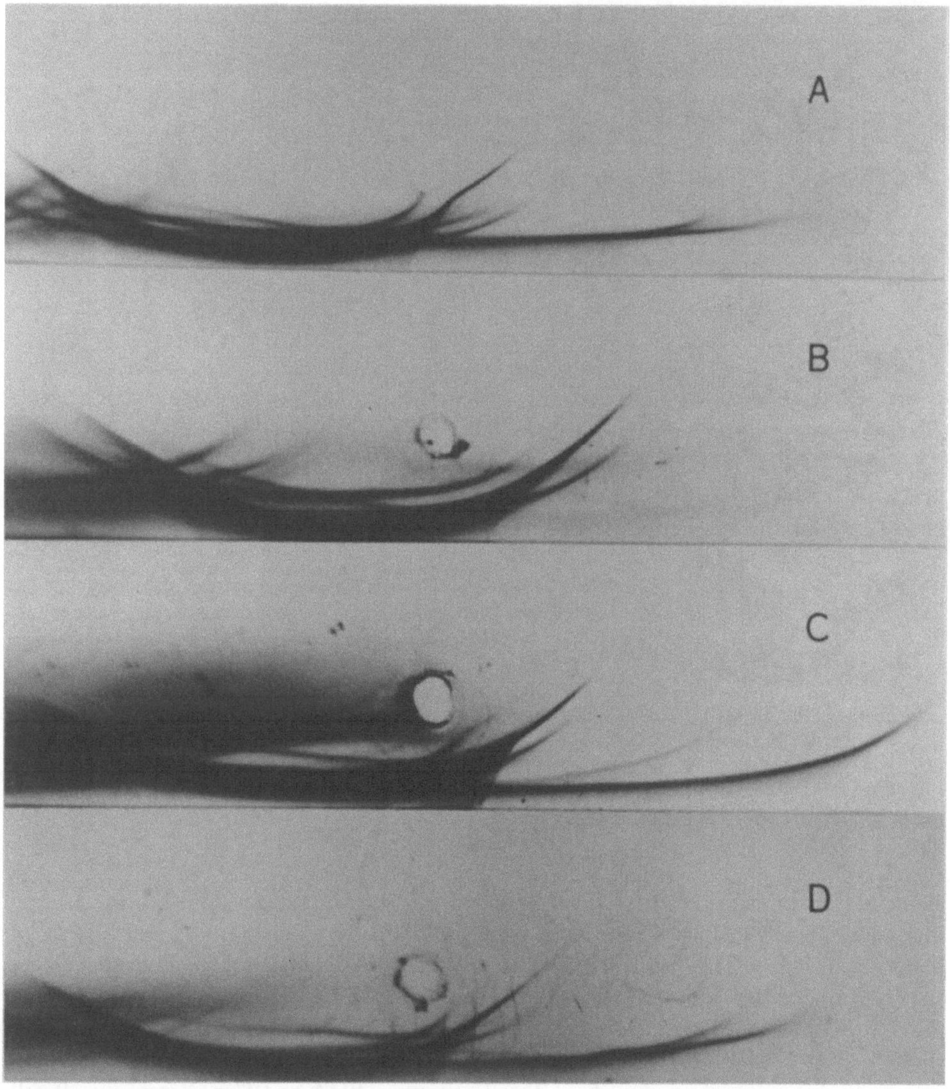

FIGURE 5. Influence of diet on gamma globulins of germfree CFW mice. (A) Heat-sterilized solid diet L 485; (B) low-molecular-weight, chemically defined, water-soluble diet sterilized by Millipore filtration; (C) conventional mouse fed diet L 485. (A–C) Age approximately 75 days. (D) Same as (B), but at age 330 days. [Reprinted with permission from Wostmann *et al.* (1971).]

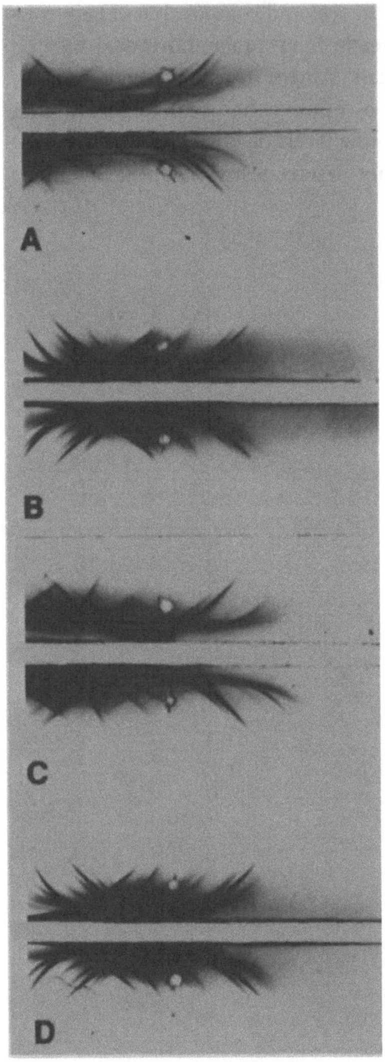

FIGURE 6. Gamma globulins in germfree C3H mice maintained on Diaflo-prefiltered, low-molecular-weight, water-soluble diet. (A) Immunoelectrophoretic patterns of 14-month-old mice; (B) patterns of 5-month-old mice on day of SRBC treatment; (C) patterns of 5-month-old mice 6 days after SRBC treatment; (D, top) pattern of 8-month-old mice maintained on L 485 commercial-type solid diet and (bottom) same animals after 4 months of Diaflow-prefiltered diet. Note the absence of gamma globulins in (D). [Reprinted with permission from Wostmann *et al.* (1971).]

et al. (1979) investigated the relationship between dietary fat, serum LDL levels, and immunocompetence. Two Teklad diets were used in these studies, L 485 with 6% fat of plant origin and S 2335 with 11% fat of both plant and animal origin. Both diets were sterilized under the same conditions. C3H/He and DBA/2 germfree and conventional mice were maintained on these diets and their immune responses tested at various ages ranging from 3 weeks to 1 year. Spleen cell suspensions taken from these mice were used in *in vitro* blastogenesis assays. Two T-cell mitogens, PHA and Con A, and two B-cell mitogens, PWM and LPS, were used. The response to alloantigens was measured by the one-way mixed lymphocyte reaction (MLR). Results of a 3-year study with animals on these diets indicated that animals on the high-fat diet had significantly lower responses in terms of both cell-mediated and humoral immunity. Animals on the high-fat diet were capable of a normal response after they had been returned to the 6%-fat diet for approximately 2 weeks. Serum obtained from mice on the high-fat diet also had an inhibitory effect on *in vitro* lymphocyte responses; this suggested that the lipoproteins might be exerting this suppressive effect. These observations are strengthened by those of Curtiss *et al.* (1980), who noted that when LDL were administered i.v. to C3H mice 24 hr before antigen, there was a profound reduction in the primary response to SRBC as measured by PFC and serum hemagglutination titers. These findings have relevance in the possible manipulation of the immune system of cancer patients by diet.

2.2. Radiobiology

Radiation studies have utilized germfree animals to assess the effect of microorganisms on the response of mammals to whole-body ionizing radiation, the recovery phenomena, bone marrow therapy, and the action of protective agents. These studies have demonstrated that bacteria and other agents contribute to the pathologic state by altering the normal anatomic and physiologic status of the host directly and through the pharmacologic action of products of microbial origin.

2.2.1. Radiation Pathology

The cause of death following acute whole-body irradiation is dose dependent (Alexander, 1957; Grosch, 1965; Gilbert, 1960). Death is believed to result from infection at the lower end of the lethal range; at higher doses, hematopoietic failure; at still higher doses (1000–10,000 rads), intestinal damage characterized by denudation of the intestinal lining with a survival time of 3 to 5 days (Quastler, 1956). Germfree animals have a higher survival percentage in the midlethal dose range and live longer than their conventional counterparts in the supralethal dose range (McLaughlin *et al.,* 1964; Walburg *et al.,* 1964; Horowitz and Bauer, 1968). The LD_{50} X-ray dose (Fig. 7) is about 50 R higher for germ-

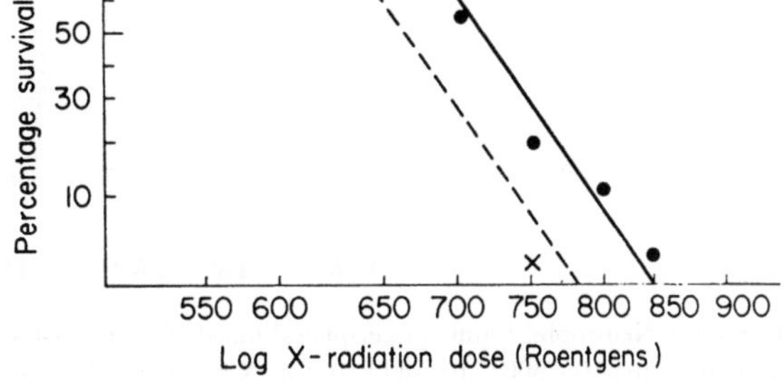

FIGURE 7. Dose–response curves of axenic and conventional mice exposed to X radiation. [Reprinted with permission from Wilson (1963).]

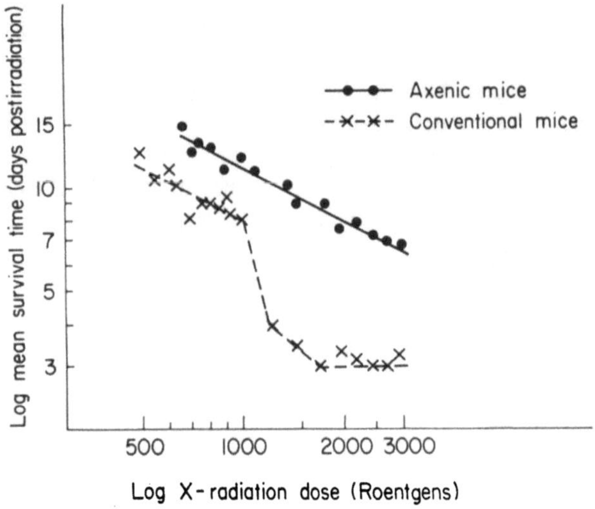

FIGURE 8. Survival times of axenic and conventional mice after exposure to a single large dose of X radiation. [Reprinted with permission from Wilson (1963).]

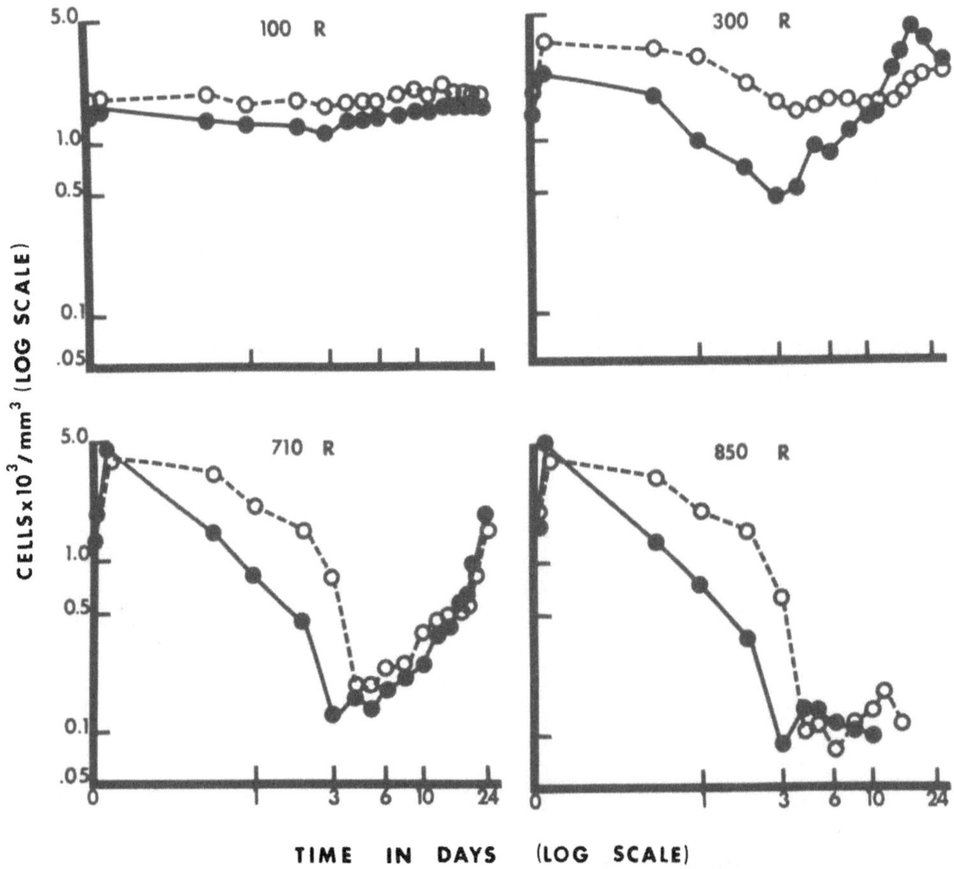

FIGURE 9. Neutrophil counts in peripheral blood of axenic (--O--) and conventional (—●—) Swiss-Webster mice following 100, 300, 710, and 850 R of X radiation. [Wilson, Matsuzawa, Connell, and Bealmear, unpublished observations. Published with permission.]

free mice than for conventional mice (Wilson, 1963). A dose–response curve for Swiss–Webster germfree and conventional mice is shown in Fig. 8. In order to separate the damage to the hematopoietic system due directly to irradiation from that due to postirradiation infection and an altered relationship to intestinal microorganisms, the peripheral blood changes in germfree and conventional Swiss–Webster mice were compared for 24 days after whole-body X irradiation (Wilson *et al.,* 1964c). This study revealed less severe or less rapid depression of neutrophils and thrombocytes in germfree mice exposed to midlethal and supralethal X-ray doses (Figs. 9, 10). It therefore appeared that the presence of bacteria in irradiated animals contributed in part to damage to the bone marrow as evidenced by a decrease in the peripheral blood count of cells originating in myeloid tissue, but the presence of bacteria did not alter the effect on lymphoid tissue (Fig. 11).

The erythrocyte and reticulocyte levels, however, of both germfree and conventional mice (Figs. 12, 13) were significantly depressed at the midlethal and supralethal X-ray doses (Wilson *et al.,* 1964c).

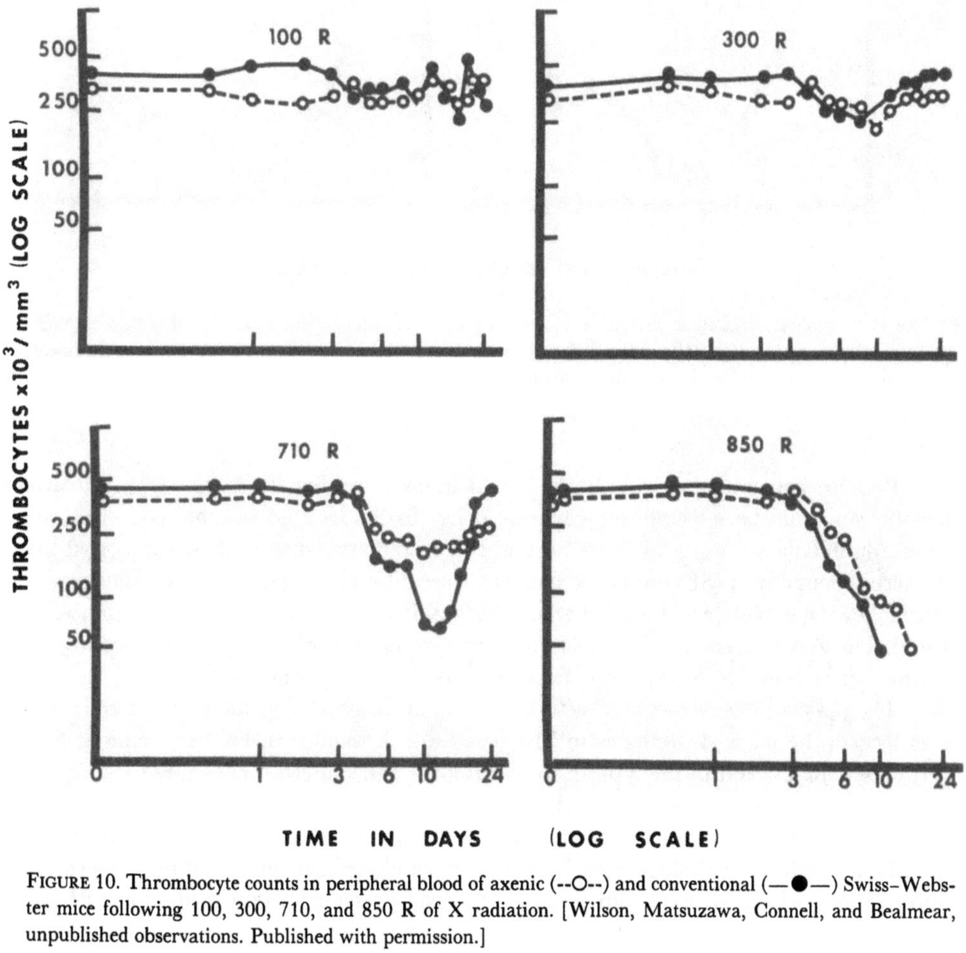

FIGURE 10. Thrombocyte counts in peripheral blood of axenic (--O--) and conventional (—●—) Swiss–Webster mice following 100, 300, 710, and 850 R of X radiation. [Wilson, Matsuzawa, Connell, and Bealmear, unpublished observations. Published with permission.]

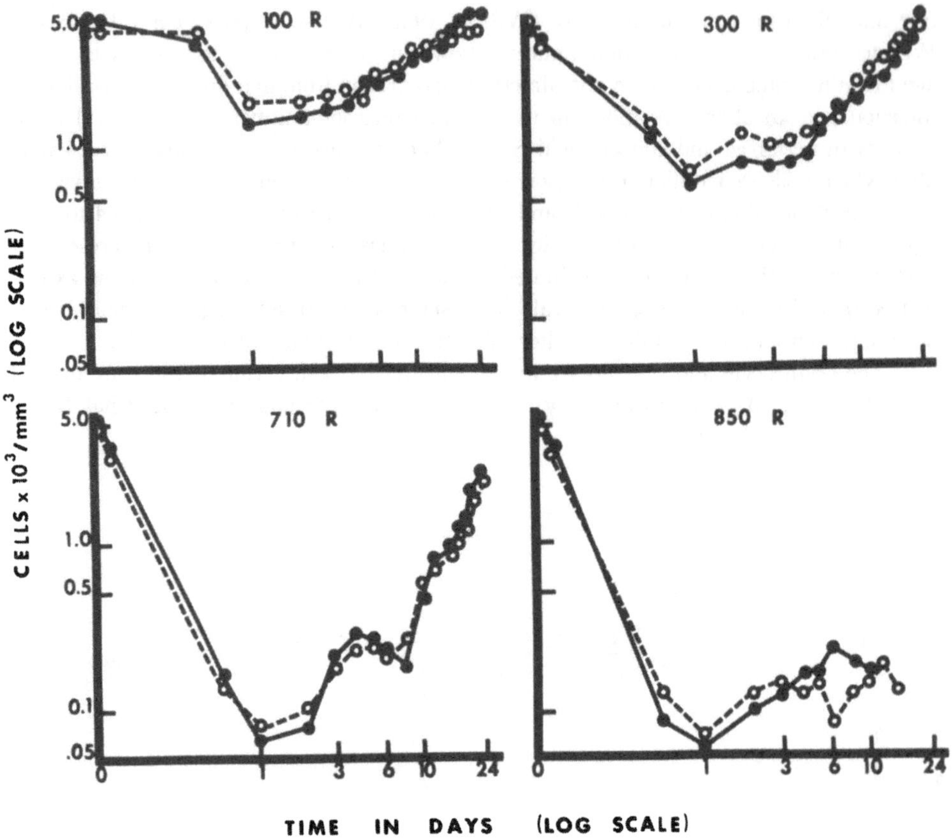

FIGURE 11. Lymphocyte counts in peripheral blood of axenic (--O--) and conventional (—●—) Swiss-Webster mice following 100, 300, 710, and 850 R of X radiation. [Wilson, Matsuzawa, Connell, and Bealmear, unpublished observations. Published with permission.]

Because hematopoiesis is arrested in whole-body irradiation, a disturbance in iron metabolism could be expected as a sequel to irradiation. Secondary pathology resulting from iron imbalance may play an important role in radiation disease. It was reported that the serum bound iron (SBI) response patterns were different in germfree and conventional mice following exposure to 850 R of whole-body X irradiation (Barry, 1967; Wilson *et al.*, 1969). The hyperferremia in X-ray-treated germfree mice (Fig. 14) was sharply contrasted by the hypoferremia of X-ray-treated conventional mice (Fig. 15).

The possible involvement of bacterial endotoxin in provoking the hypoferremic state was seen in the marked decreases in SBI levels that it induces in both conventional and germfree animals, and in the typical "conventional" SBI pattern it causes when administered to X-irradiated germfree mice (Fig. 14).

More explicit proof for the involvement of endotoxin and other vasoactive substances in radiation-induced pathology came from data on irradiated gnotobiotic animals. Germfree mice monoassociated with a gram-positive organism presented a postirradiation SBI pat-

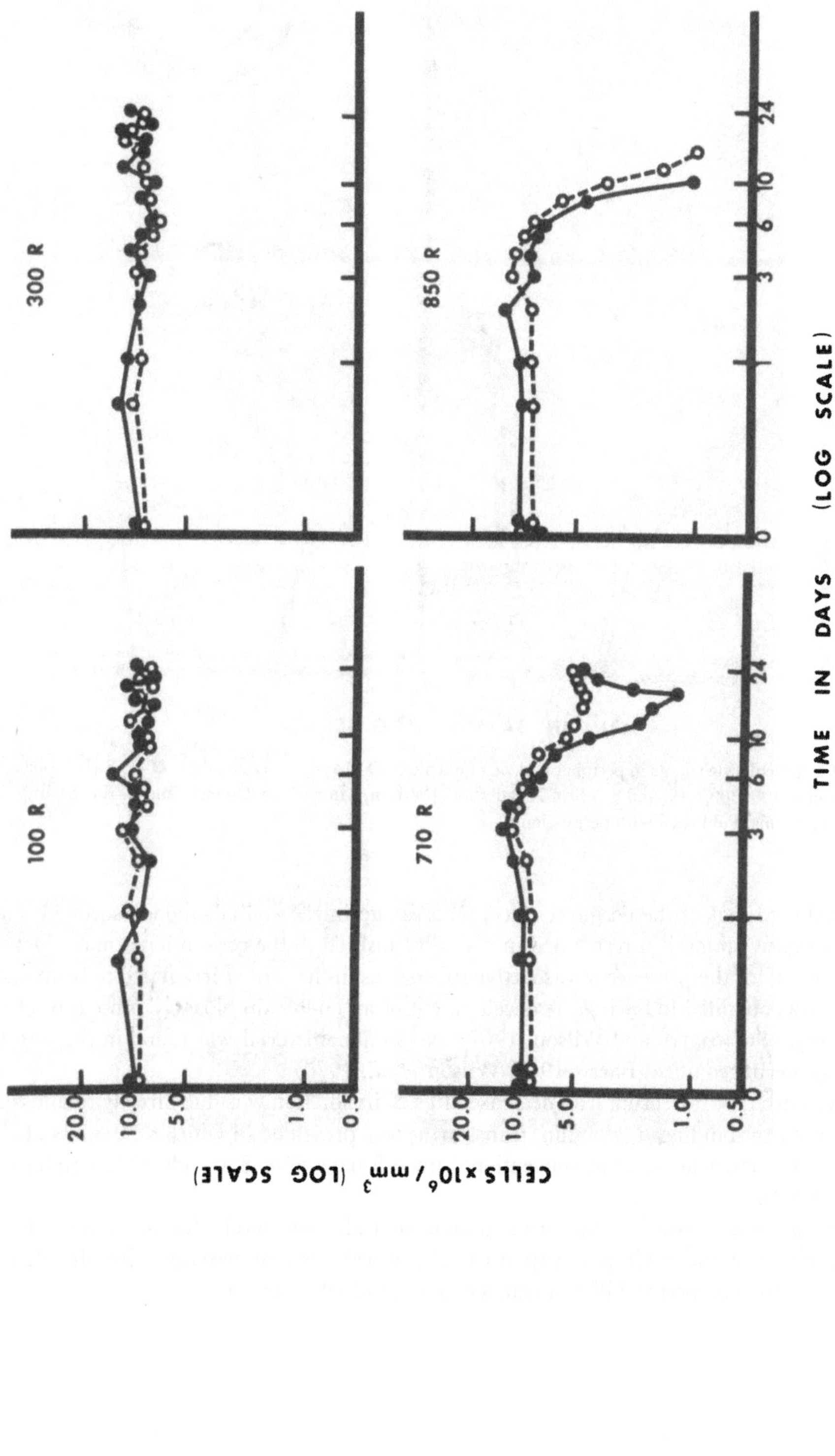

FIGURE 12. Erythrocyte counts in peripheral blood of axenic (--O--) and conventional (—●—) Swiss–Webster mice following 100, 300, 710, and 850 R of X radiation. [Wilson, Matsuzawa, Connell, and Bealmear, unpublished observations. Published with permission.]

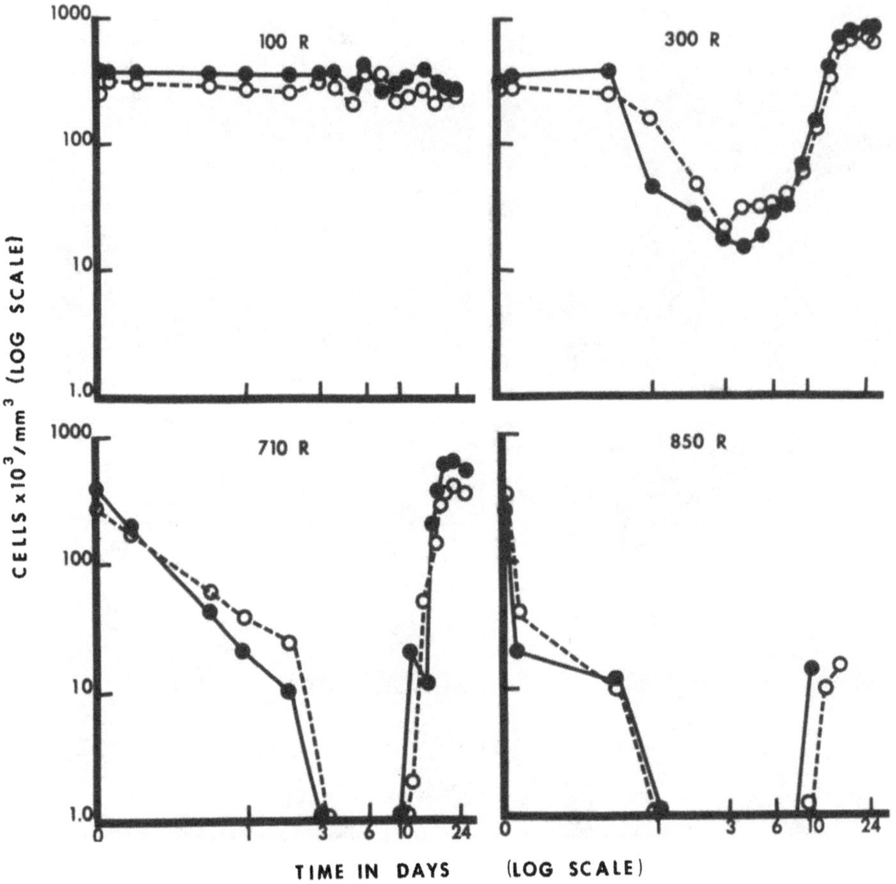

FIGURE 13. Reticulocyte counts in peripheral blood of axenic (--O--) and conventional (—●—) Swiss-Webster mice following 100, 300, 710, and 850 R of X radiation. [Wilson, Matsuzawa, Connell, and Bealmear, unpublished observations. Published with permission.]

tern similar to that of the germfree mice, whereas germfree animals monoassociated with a gram-negative organism responded in a similar fashion to the conventional mice. Definitive evidence for the presence of vasoactive substances in the sera of irradiated conventional animals was obtained following the development of an endotoxin bioassay procedure (Fig. 16) (Barry, 1967; Barry and Wilson, 1967). No similar material was found in the sera of irradiated germfree mice (Barry, 1967; Wilson *et al.,* 1970).

Evidence that the serum iron decrease after X irradiation was due directly to a loss of the specific iron-binding β_1 globulin, transferrin, was presented in which a 50% loss of the protein was shown to occur in conventional mice 5 days after 850 rads of X irradiation (Barry, 1967).

The gross decrease in total serum protein and albumin levels that accompanied the administration of endotoxin gave support to the hypothesis that accompanying the alteration in the vascular permeability, a leakage of serum had occurred.

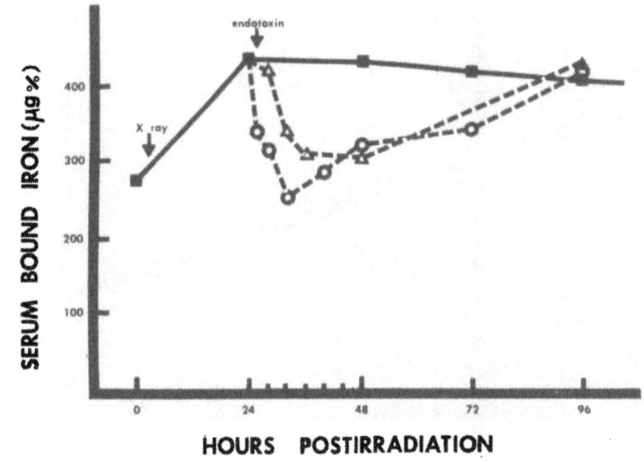

FIGURE 14. Serum bound iron levels of germfree CFW mice following 850 R of X radiation, receiving 10 µg *E. coli* endotoxin (—O—), 10 µg *S. typhosa* endotoxin (—Δ—), or no endotoxin (—■—). [Reprinted with permission from Wilson *et al.* (1970).]

The results of the study were consistent with the following proposed mechanism of action for the induction of hypoferremia in X-irradiated conventional animals: (1) X irradiation produces conditions that favor the absorption of endotoxins; (2) bacterial products that are vasoactive substances cause changes in the permeability of the vessel membranes, either directly or indirectly through other agents; (3) this increased permeability allows serum proteins to leak out into surrounding tissue spaces, resulting in a decrease in the transferrin level and an ultimate hypoferremia (Barry, 1967; Wilson *et al.,* 1969, 1970). Additional experiments using ^{59}Fe-labeled transferrin and radioautographs of the spleen, lymph nodes, and liver at the time of maximum hypoferremia demonstrated that the transferrin was indeed lost to the tissues (Bealmear and Wilson, 1968).

In the "gut death" range, germfree mice lived twice as long as conventional mice given the same dose of irradiation (Fig. 8). Death in this dose range is characterized by denu-

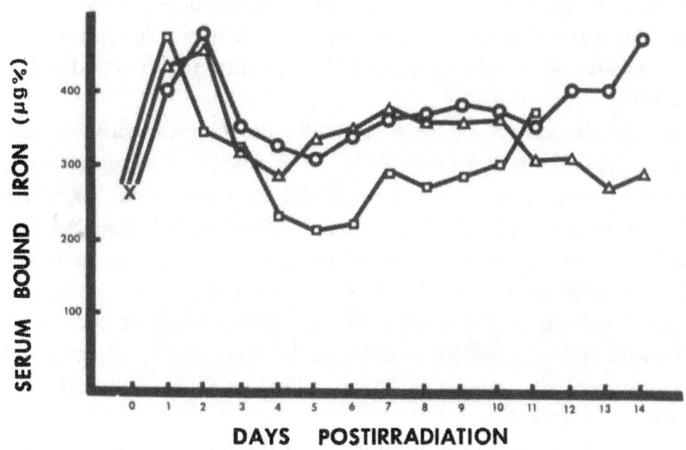

FIGURE 15. Serum bound iron levels of conventional CFW mice exposed to 450 (—Δ—), 660 (—O—), or 850 (—□—) R of X radiation. [Reprinted with permission from Wilson *et al.* (1969).]

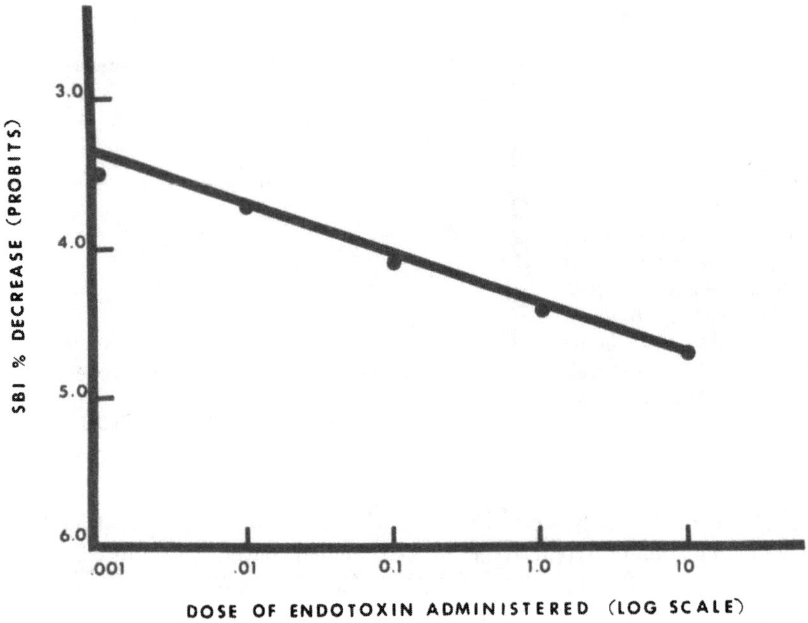

FIGURE 16. *E. coli* endotoxin dose–response curve. Percent serum bound iron decrease (probit transformation) vs. endotoxin dose (μg). [Reprinted with permission from Wilson *et al.* (1970).]

dation of the intestinal lining and a survival time of 3–5 days (Quastler, 1956). Abrams *et al.* (1963) have observed that the cell renewal rate of the intestinal mucosa of germfree mice is about half that of conventional mice. It was reasonable to assume that the rate of denudation is a function of the rate of cell renewal and that the germfree animal with a lower renewal rate would live longer than conventional mice when both are exposed to the same dose of radiation in the supralethal range. A series of experiments was designed to test the hypothesis that the survival time of an animal exposed to whole-body irradiation in the "gut death" range is proportional to the life span of its intestinal mucosal cells (Matsuzawa and Wilson, 1965).

Following a whole-body X-ray dose of 3000 rads, the mean survival time of conventional CFW mice was 3.5 days; germfree mice, 7.3 days; and mice monoassociated with *Pseudomonas aeruginosa, Streptococcus faecalis,* and *Escherichia coli,* 5.4, 6.9, and 6.2 days, respectively. In ex-germfree mice the mean survival time was 4.6 days; the survival time of ex-germfree mice shortened immediately after contamination, and remained somewhat longer than that of conventional mice (Wilson *et al.,* 1968).

The population of cells in the intestinal crypts of both germfree and conventional mice was rapidly reduced after irradiation; in general, the depopulation of epithelial cells in the crypts of both germfree and conventional mice followed the same pattern but was delayed about 1 day in the germfree mice.

As ionizing radiation blocks cell production in the mucosa, the mucosal lining of the intestine should be denuded after the last cells to migrate from the crypts onto the villi reach the tip of the villi. By this reasoning, the rate of cell depletion is a function of the time for the epithelial cells produced in the crypts to traverse the villi and disappear, i.e., the life span of the cells of the intestinal mucosa. In germfree mice the time for the cells to migrate to the tips of the villi is 4.3 days and the mean survival time is 7.3 days. The migration time for conventional mice is only 2.1 days, and the mean survival time is 3.5 days. In both kinds of mice the migration time is about 60% of the mean survival time. The difference between the life span of the cells was probably effected by a decrease in cell population.

It was hypothesized by Abrams and co-workers (1963) and Matsuzawa and Wilson (1965) that either the intestinal flora is responsible for acceleration of cell turnover in the conventional mouse or the absence of the flora retards the process in germfree mice. Lesher *et al.* (1964) hypothesized that the decreased cell production and proliferation are due to a reduced metabolic activity or nutritional deficiency.

Under normal conditions, the growth of bacteria in the small intestine is limited and concentrations are low (Dixon, 1960). Dixon and Paulley (1963) found no difference in small intestine morphology unless abnormal strains of bacteria were allowed to accumulate. As bacteria alone do not seem to account for this difference in cell turnover, it was hypothesized that there is a biochemical abnormality in the germfree animal resulting from the lack of several known strains of bacteria that are capable of hydrolyzing conjugated bile acids.

Bile acids, products of cholesterol metabolism, are unconjugated when first formed but they are conjugated immediately with either taurine or glycine before they leave the liver (Norman and Sjovall, 1958). In conventional rats, cholic acid (unconjugated) turnover in either form is 2.2 days while in germfree or antibiotic-treated rats it takes 11.2 days (Gustafsson *et al.,* 1957). When the conjugated acid reaches the conventional cecum, it is hydrolyzed to its unconjugated or free form by bacterial enzymes (Gustafsson *et al.,* 1966; Hill and Drasar, 1967). Norman and Grubb (1956) showed that *Clostridia, Enterococci,* and some *Bacteroides* were capable of hydrolyzing conjugated bile acids and that *Lactobacillus, Flavobacteria, Escherichia coli, Proteus vulgaris, Staphylococcus aureus,* and *Bacillus subtilis* were incapable of hydrolyzing conjugated bile acids. A cell-free extract has been isolated from *Clostridium perfringens* that is capable of hydrolyzing bile acid conjugates *in vitro* (Nair *et al.,* 1965). Further bacterial degradation of cholic acid yields deoxycholic acid (Lindstedt and Samuelsson, 1959). The unconjugated bile acid produced by hydrolysis is partly excreted and the rest is recycled. As it travels through the small intestine it is absorbed (Dietschy *et al.,* 1966). Because unconjugated bile acids are not found either in the excrement or in the enterohepatic system of germfree animals, intestinal enzymes may be ruled out as being responsible for hydrolyzing conjugated acids.

Fry and Staffeldt (1964) have shown the deleterious effects of sodium deoxycholate on mucosal cells of mice; their studies would seem to support the hypothesis that the unconjugated form is toxic to the cell, as it disrupts the cell when it is absorbed. Because the cell turnover rate of the germfree intestinal mucosa is slower than that of the conventional mucosa, the possibility exists that the higher concentration of conjugated bile acid in the germfree animal is the inhibitory factor that decreases the cell turnover rate by interference with normal cell metabolism.

An explanation of the difference in survival time of germfree and conventional mice irradiated in the "gut death" range was sought in terms of the concentrations of conjugated and unconjugated bile acids present in the small intestine of germfree and conventional mice, and the effect of these bile acids, e.g., cholic acid, which bacteria produce from the conjugated bile acids, e.g., taurocholic acid, released into the intestine, accelerate cellular turnover on the villus, thereby shortening the life span of the cells. Two groups of germfree mice were maintained on a low-fat diet for 7 months after weaning. One group received a supplement of 0.15% cholic acid and the other received 0.15% sodium taurocholate. The migration of the mucosal cells on the villi of these animals was studied by tritiated thymidine labeling of crypt cells in mitosis, sacrificing mice daily after labeling, and following the progress of labeled cells on the villi histologically. The life span of the mucosal cells of those receiving cholic acid was 2 days, comparable to the conventional state. Taurocholate did not alter the migration rate. Furthermore, it took 4 days for the label to disappear from the mucosa of the cholic acid group versus 7 days in the taurocholate group. These data seemed to support the hypothesis (Ranken *et al.*, 1969, 1971).

2.2.2. Radiation Treatment

The germfree animal provides a model system for studying the radiation syndrome in the absence of complicating infections. If the mechanisms of radiation damage per se can be separated from the effects of microorganisms and their products, then it may become feasible to treat the syndrome. Bone marrow transplants have been the most successful treatment of heavily irradiated and immunosuppressed leukemia patients, the hematopoietic radiation syndrome, combined immune deficiency disease, and aplastic anemia. To date, limited success has been achieved with bone marrow transplantation in man, and transplantation trials have often been complicated by infection.

2.2.2a. Survival after Bone Marrow Transplantation. Development of "secondary disease" or graft-versus-host disease (GVHD) is the usual consequence of treatment of immunosuppressed or X-irradiated mice with allogeneic bone marrow. Mice that survive after allogeneic bone marrow treatment manifest a wasting syndrome at 20 to 50 days postirradiation. This syndrome is characterized by weight loss, skin lesions, diarrhea, ataxic gait, epilation, kyphotic posture, and lymphoid aplasia. Most mice die within 20–40 days postirradiation with severe symptoms. Although it is impossible to extrapolate data directly from mouse to man, the mouse system nevertheless still has an important role to play in the resolution of many transplantation problems in general and in bone marrow transplantation in particular. This is especially true if the complications of infection can be divorced from GVHD per se through the use of germfree animals.

Lethally irradiated germfree mice treated with allogeneic bone marrow do not develop typical secondary disease (Connell and Wilson, 1965) (Table II). Details of these experiments were also reported by Wilson and Bealmear (1967). Both germfree and conventional mice of the CFW and C3H strains were exposed to a range of normally supralethal doses of whole-body X irradiation, treated with syngeneic and allogeneic bone marrow, and their responses compared. The $LD_{50(30)}$ for the untreated germfree mice is 705 rads; for conventional mice, it is 660 rads. Allogeneic bone marrow treatment raised these values to 770–890 rads for conventional mice and 1280–1650 rads for germfree mice.

The recovery of conventional mice after 800 rads and syngeneic or allogeneic bone marrow and germfree mice after 1000 rads and the same marrow treatment was studied by using peripheral blood counts, organ weights, and total and differential marrow counts as indices of recovery. With the exception of a slight leukocytopenia in conventional allo-

TABLE II
Relative Occurrence of Secondary Disease Symptoms in Germfree and Conventional Mice[a]

Symptoms associated with secondary disease	Conventional mice: Relative occurrence	Conventional mice: Time of appearance after X irradiation (days)	Germfree mice: Relative occurrence	Germfree mice: Time of appearance after X irradiation (days)
Diarrhea	Frequent	20–40	Slight	60–90
Gross skin lesions	Moderate	40–60	None	
Extreme depilation	Slight	60–90	Frequent after 1200 R or more	60–90
Kyphotic posture	Moderate	20–60	Very slight	60–90
Depressed lymphocyte count	Frequent and extreme	20–40	Frequent (not great depression)	20–40
Weight loss	Frequent	20–40	Frequent after 1200–1600 R	30–40
Mortality	Frequent	20–50	Seldom	60–100

[a]Reprinted with permission from Connell and Wilson (1965).

geneic chimeras, there were no discernible differences in the marrow and spleen repopulation processes in any of these groups.

In subsequent studies, the pathology of secondary disease in radiation chimeras was thoroughly investigated (Jones, 1970; Jones *et al.,* 1971). Adult C3H/He germfree mice (11 weeks of age) of both sexes were exposed to 1000 rads of X irradiation; conventional mice were exposed to 825 rads. Twenty-four hours after X irradiation, the mice received 10^7 allogeneic DBA/2 bone marrow cells intravenously. Pairs of animals were sacrificed at regular intervals for the initial 30-day observation period and at weekly intervals thereafter until 2 years posttransplant. Peripheral blood and bone marrow counts, body weights, and the liver, thymus, and spleen weights were evaluated. Tissues were fixed for histological examination. Parallel experiments with syngeneic (C3H/He to C3H/He) and xenogeneic (Fischer 344 rats to C3H/He) were also conducted.

A transitory splenomegaly appeared on day 6 posttransplant in both germfree and conventional mice. In the germfree mice spleen weights returned to normal on day 14 and remained at normal levels throughout the observation period. After an initial depression in leukocyte counts, an increase was observed by day 8. Hematocrit values returned to normal by day 12. A persistent increase in both number and size of Peyer's patches was observed by day 28. Recovery patterns were inconsistent in conventional mice.

Throughout a 120-day period, 100% of the germfree syngeneic chimeras, 98% of the allogeneic chimeras, and 3% of the xenogeneic chimeras survived (Jones *et al.,* 1971). None of the conventional allogeneic or xenogeneic chimeras survived the 120-day period. Alkaline phosphatase (AP) staining of the granulocytes in the xenogeneic chimeras (rat granulocytes are AP positive) and the support of donor-type skin in the allogeneic chimeras suggested that survivors were truly chimeric animals. Both allogeneic and syngeneic chimeras were immunized with SRBC and their responses measured by the Jerne–Nordin hemolytic plaque assay. By use of the appropriate isoantisera, it was determined conclusively that the cells responsible for the response to SRBC were always of donor origin and that the survivors were truly chimeric animals (Bealmear *et al.,* 1973). To date, germfree allogeneic radiation chimeras have survived in the germfree state until 21 months posttransplant (the normal life span of these mice) without gross symptoms of secondary disease. One xenogeneic chimera survived 13 months posttransplant in the germfree state.

While germfree xenogeneic chimeras presented both severe gross and histological evidence of secondary disease (Fig. 17), the presence of external symptoms among the germfree allogeneic chimeras was rare. However, there was evidence to support the presence of GVH-caused histopathology among both the germfree and the conventional allogeneic chimeras (Figs. 17, 18). Examples of degeneration were found in all tissues examined—lymph nodes, spleen, thymus, small intestine, liver, lung, and kidney—although the greatest atrophy was observed in the lymphoid tissues (Jones, 1970; Bealmear *et al.,* 1973). With hematoxylin-eosin stained sections, the GVHD reaction was not detectable until about 90 days posttransplant in the germfree allogeneic chimeras; therefore, in an attempt to find the large pyroninophilic cells associated with GVHD during the early days postgrafting, the stain was changed to methyl green and pyronin. These large pyroninophilic cells appeared as early as day 4 posttransplant, and their numbers increased with time (Bealmear *et al.,* 1973). This indicated that GVHD begins immediately after transplant, but is usually not fatal until the animal has lived out three-fourths of its normal life span. Peripheral blood lymphocytes reflected early proliferation of lymphocytes in the tissues and a decline as secondary disease arose. Values of bone marrow lymphocytes were above normal.

A major point yet to be established is the adaptability of the surviving germfree chimera to the conventional state. The feasibility of that transition lies in whether the donor line of lymphoid cells has become tolerant toward the antigenicity of the host tissue, or whether in the absence of tolerance, tissue destruction continues with time and will result in major dysfunction of the chimeric animal.

When a group of allogeneic chimeras were removed from germfree isolators and placed in the conventional environment to be used for skin graft experiments, about 50% of the chimeras died soon after conventionalization, but the other 50% survived, acquired a normal microflora, and adjusted to the conventional state. A gradual acquisition of a microflora seems to be important in this situation.

As 50% of the conventionalized allogeneic chimeras did not survive, the immunological capability or limitation of the germfree allogeneic chimera undergoing GVHD was questioned. The indirect test for chimerism mentioned above by means of challenge with skin grafts from the donor strain of mice and from a third-party strain revealed that the allogeneic chimeras were tolerant to donor-strain skin and rejected third-party skin (Jones, 1970). This indicated that they were immunologically competent, at least for skin graft rejection.

Both allogeneic and syngeneic chimeras were immunized with SRBC and their responses measured by the hemolytic plaque assay. The magnitude of the response was reduced in the allogeneic chimeras as compared to the syngeneic chimeras (Loughman, 1972; Bealmear *et al.,* 1973; Loughman *et al.,* 1973). An *in vitro* immunization and challenge of spleen cells from allogeneic and syngeneic chimeras gave comparable results; the response of the allogeneic chimeras was greatly reduced.

When immunized with the *Salmonella* H antigen flagella or polymerized flagellin (POL), which is a T-cell-independent antigen, the syngeneic and allogeneic chimeras responded comparably (Bealmear *et al.,* 1973). When tested for cell-mediated immunity or

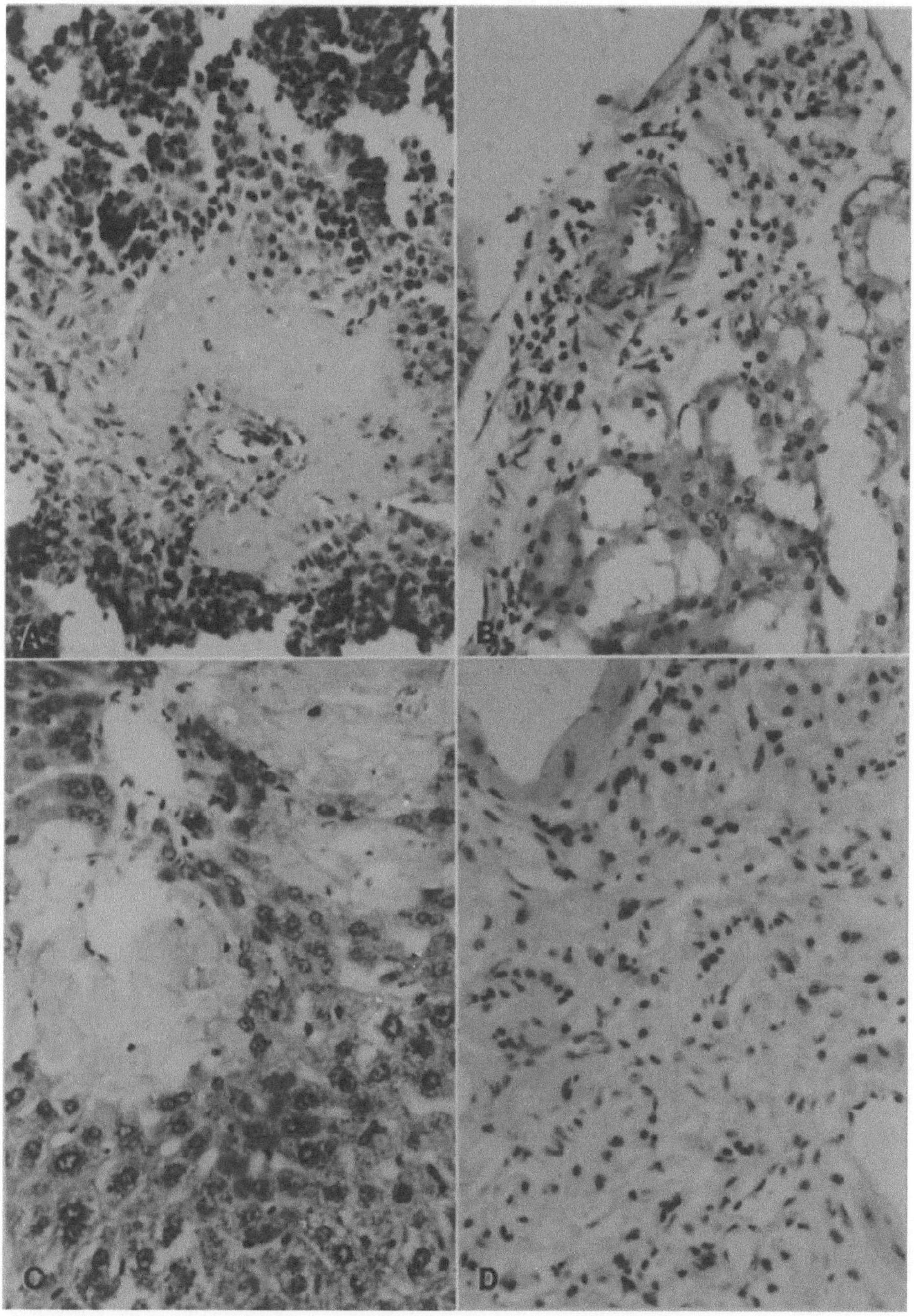

FIGURE 17. Histopathology of graft-versus-host disease in xenogeneic and allogeneic germfree mouse radiation chimeras. (A) Perivascular degeneration of spleen at 30 days after a xenogeneic transplant of rat bone marrow to a C3H/He germfree mouse. (B) Perivascular infiltration and degeneration of the kidney at 30 days after xenogeneic transplant. (C) Aseptic necrosis of liver in the periportal region at 93 days after an allogeneic transplant of DBA/2 bone marrow to a C3H/He mouse. (D) Severe lung degeneration near the bronchus at 389 days after allogeneic transplant. (A–D, × 800). [Reprinted with permission from Jones (1970).]

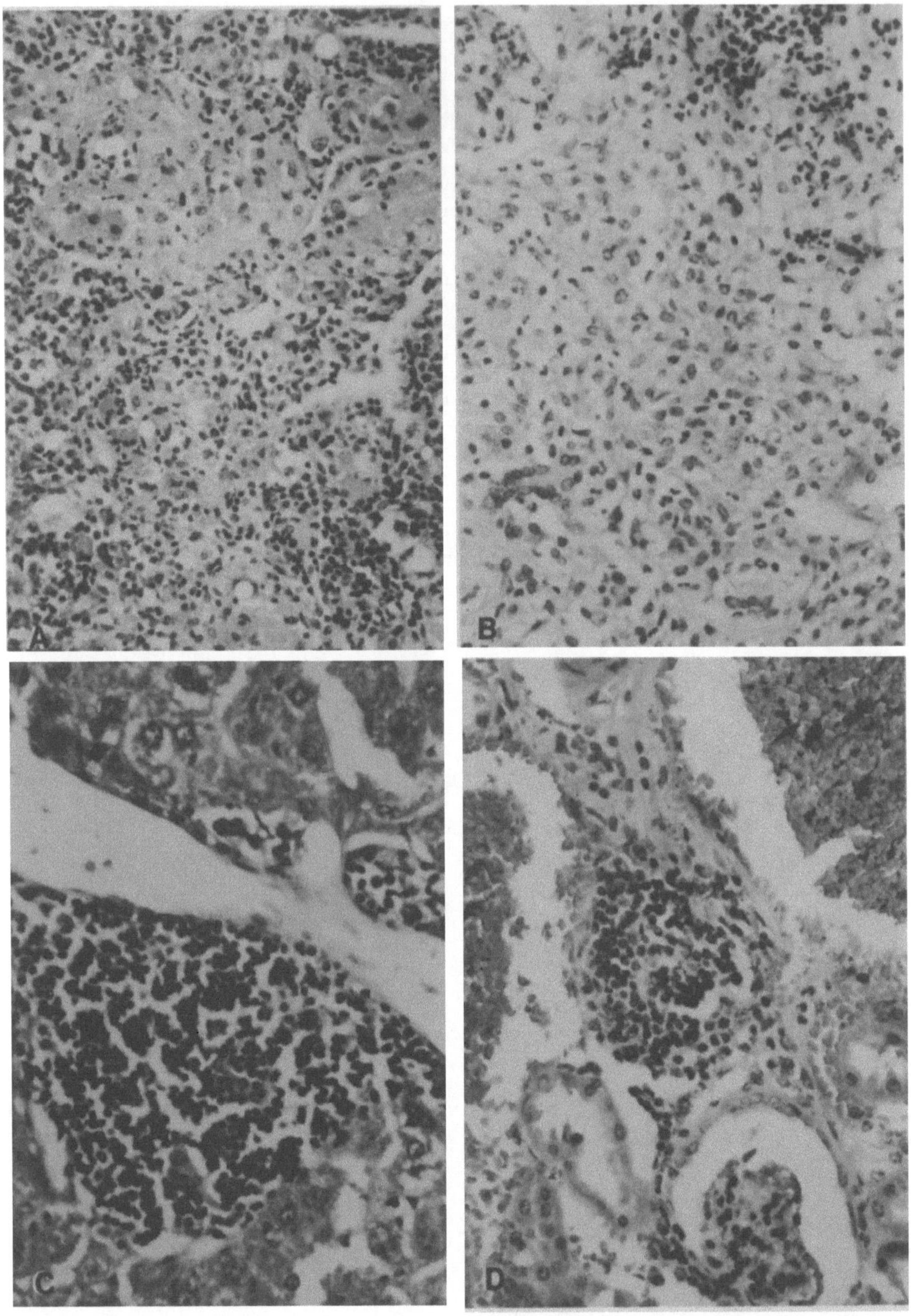

FIGURE 18. Histopathology of graft-versus-host disease in allogeneic conventional mouse radiation chimeras. (A) Atrophy of lymph node of C3H/He conventional mouse 27 days after bone marrow transplant from a DBA/2 conventional mouse (× 640). (B) Thymic degeneration at 27 days after allogeneic transplant (× 800). (C) Perivascular infiltration of the liver with mononuclear cells at 140 days after transplant of allogeneic bone marrow (× 800). (D) Perivascular infiltration and degeneration in the kidney at 140 days after allogeneic bone marrow transplant (× 800). [Reprinted with permission from Jones (1970).]

the T-cell response to third-party tumor cells as measured by the chromium release assay, they did not respond (Bealmear *et al.,* 1973; Dauphinee, 1973).

It was concluded that the absence of a microflora delays, but does not completely avert the fatal effects of GVHD; the pathology of secondary disease, however, is mitigated. The germfree allogeneic radiation chimera has a greatly reduced or absent T-cell response as demonstrated by its lack of response to third-party tumor cells. The responses, which are both helper T cell and B cell mediated (e.g., response to SRBC), are reduced in the chimera, but not absent. The B-cell response of the allogeneic chimera is normal. Though the thymus of the allogeneic chimera appears to be morphologically normal, it would seem that the stroma of the thymus and the cells that reconstitute it must be of the same *H-2* locus for the T cells to function normally. These data corroborate those of Griscelli and Durandy (1976), who reported that patients with severe combined immune deficiency disease (SCID) transplanted with HLA-identical fetal thymus demonstrated T-cell–B-cell cooperation between the T cells of the donor and the histocompatible B cells of the recipient; no such T-cell–B-cell cooperation was observed when the fetal thymus donor was histoincompatible. McCombs *et al.* (1976) indicated that the possible histocompatibility requirement between lymphocyte and macrophage necessary for proliferation induced by soluble antigen is not a requirement for the proliferative response to mitogen and alloantigen; they did not, however, demonstrate any need for histocompatibility between cells and thymic stroma because their work was limited to *in vitro* assays with single-cell suspensions.

There is also a possibility of an increase in suppressor-cell activity in these allogeneic chimeric animals. Isa *et al.* (1976) observed an increase in suppressor T cells among cytotoxic lymphocytes taken from C57BL/6J mice bearing the teratocarcinoma 402AX. In this case the presence of alloantigens from the tumor seemed to stimulate the production of suppressor T lymphocytes. Rich and Rich (1976) also found that alloantigen-activated spleen-localizing thymocytes (T cells) suppressed mixed lymphocyte reactions 40–95%. Further evidence is necessary in order to decide whether the apparent increase in suppressor activity is really the result of the presence of alloantigens.

Though the germfree state did not eliminate GVHD in allogeneic chimeras, it did permit these animals to live out their normal life spans without obvious symptoms of the disease. Thus the germfree state enables investigators to have a prolonged time period in which to study GVHD and hopefully to develop methods for its further amelioration.

2.2.2b. Pretreatment of Bone Marrow Inoculum with Anti-Thy-1.2 (θ) Serum. As T cells are the major effectors of GVH disease, it was of interest to investigate the possible amelioration of this disease by eliminating immunocompetent T cells in the bone marrow by pretreatment with anti-θ serum, which is specific for T-cell-surface antigens. It was hypothesized that the elimination of the competent T cells at the time of inoculation would prevent antigen processing and presentation to the responding cells. When new T cells would differentiate from the stem cells inoculated, they would differentiate in the thymic environment of the host and would perhaps not recognize the host as foreign and would thus become tolerant. In a sense this would recreate the prenatal situation. A most important and significant finding in these experiments is the complete absence of any evidence of GVHD either grossly or histologically in these pretreated chimeras (Bealmear, 1975). The details of these experiments are given below.

C3H/He germfree mice 11 weeks of age were exposed to 1050 rads from a cobalt-sixty source. The recipients were inoculated (i.v.) with 10^7 bone marrow cells from syngeneic donor (C3H/He) of allogeneic donor (DBA/2) mice. Both strains of bone marrow cells were incubated with anti-θ (anti-Thy-1.2) serum and complement for 1 hr at room temperature before i.v. inoculation. Untreated bone marrow cells from both strains were used as controls. Chimeras were sampled each month, and the following parameters were tested: survival, hematology, histology of thc lymphoid organs, and immunological competence. Tests for immunological competence were performed with the mitogens PHA, Con A, PWM, and LPS, both *in vivo* and *in vitro* immunizations with SRBC followed by the hemolytic plaque assay, mixed lymphocyte cultures (MLC) assay, cell-mediated lympholysis (CML) assay, and the footpad response against methylated BSA.

At 15 months posttransplantation, 94% of the syngeneic chimeras were alive, 88% of the allogeneic, and 86% of the allogeneic and syngeneic chimeras that received cells pretreated with anti-θ serum. Nine allogeneic chimeras survived with a normal life span of 24 months. No large group of anti-θ-treated chimeras was kept beyond 15 months, because they were used for other tests. A few anti-θ-treated chimeras survived to 21 months posttransplantation (the normal life span of these mice).

Responses to PWM and LPS, which are B-cell mitogens, were normal or above in both syngeneic and allogeneic cell-recipient groups until 10 months posttransplant, when the responses decreased to a stimulation index of less than two. This may be due to accelerated aging by the exposure to radiation. PHA responses were normal or above in both syngeneic groups, but were suppressed in the allogeneic groups after 8 months posttransplantation. It was postulated that this is the result of suppressor-cell activity, as the Con A responses of the suppressed animals were elevated, and suppressor cells are responsive to Con A (Bealmear, 1975).

The presence of alloantigens may stimulate the production of suppressor cells as in the case of allogeneic tumor-sensitized mice (Argyris and Horowitz, 1976). Whether these suppressor cells are splenic macrophages like those reported by Argyris and Horowitz or are suppressor T cells remains to be more fully investigated; in the rat system, Veit and Feldman (1976) demonstrated that the suppressor elements in spleens of tumor-bearing animals were macrophages. In a GVH reaction in which a regional depression in T-cell function was demonstrated, Lapp *et al.* (1975) described the emergence of two distinct forms of suppressor cells, specific T-cell and nonspecific non-T-cell suppressors. Miller *et al.* (1975) found cells capable of inhibiting the effector-cell destruction of ^{51}Cr-labeled target cells that appeared during a GVH reaction in mice. These suppressor cells appeared to have little or no θ antigen and interacted with cytotoxic effector cells through cell contact or short-range contact. The major target for suppressor T cells appears to be other T cells, with B cells representing a less likely or minor target, at least the B cells that respond to thymus-independent antigens (Gershon, 1975). This corroborates the data in which there was T-cell suppression but no apparent B-cell dysfunction.

Responses to both *in vivo* and *in vitro* immunizations with SRBC ranged from normal to elevated in both syngeneic groups and were suppressed in both allogeneic groups. It was again postulated that this reduced response was due to an imbalance between the helper T cells and the suppressor cells. The apparently suppressed cells gave a significant response to Con A. The responses in the MLC and CML assays followed the same pattern as seen for the T-cell mitogens and the response to SRBC with T cells suppressed.

It should be emphasized that no histopathological changes were detected in the allogeneic chimeras that received cells pretreated with anti-θ serum. This was in direct contrast to the extensive histopathology observed in regular allogeneic chimeras.

As GVHD is the major problem in the practical application of bone marrow transplantation in medicine, the resolution of this problem can only be achieved through thorough research of the essential elements of GVHD. Germfree animals offer distinct experimental advantages for delineation of the pathogenesis of GVHD in the absence of the complications introduced by microorganisms and their products. Control of GVHD would facilitate the use of bone marrow transplantation in man in the management of immune deficiency disease, aplastic anemia, immunosuppressed leukemic patients, and patients accidentally exposed to large doses of whole-body irradiation. The preliminary observations on the use of anti-θ serum treatment of the bone marrow inoculum appear to represent a feasible approach to the problems associated with allogeneic bone marrow transplantation. The main significance of this research lies in the potential use of bone marrow transplantation in the management of these diseases.

2.3. Aging and Carcinogenesis

The value of germfree animals for the study of aging and carcinogenesis has long been recognized (Gordon and Pesti, 1971). With the germfree animal, a clear differentiation can be made between the possible role of microbial factors on the host and other nonmicrobial environmental agents (Gordon *et al.,* 1966).

2.3.1. Survival and Spontaneous Tumors

In a study of 300 germfree and 300 conventional Swiss–Webster mice, the mean ages of the mice at natural death were: germfree males 723 ± 19 days, females, 681 ± 12; conventional males, 480 ± 10, females, 516 ± 10 (Gordon *et al.,* 1966). In another study (Walburg and Cosgrove, 1967, Walburg, 1973), 50 germfree and 50 conventional ICR mice were maintained until natural death, at which time the mean ages of the mice were: germfree males, 556 ± 43 days, females, 535 ± 46; conventional males, 536 ± 41, females, 547 ± 45. In the first trimester of life, no significant difference in survival was observed in the two groups. In the middle trimester, the germfree mice had an increased survival rate. There was a 40% cumulative mortality for germfree males and females at about 580 days, but the same mortality level was reached by the conventional mice at 410 days. Among the germfree mice from both studies, the most prevalent lesion at natural death was associated with the enlargement of the cecum. Cecal volvuli were frequently observed, and occasionally the gallbladder was distended. None of these lesions were observed in the conventional controls. Neoplastic diseases were evenly distributed between germfree and conventional mice in both studies. In old germfree rats that were sacrificed in a healthy state, no lesions were observed, with the exception of a few benign tumors (Pollard and Kajima, 1970). The apparent survival similarity for female and male germfree mice is contrasted to the higher survival of the female conventional mice, which is probably related to the greater resistance to infection observed in females (Wheater and Hurst, 1962). During 6 years of observation, seven germfree Lobund mice of the BALB/c, C3H/f, and Swiss–Webster strains developed spontaneous breast adenocarcinomas, and B-type virus particles were detected in two of them (Kajima and Pollard, 1965). Nielsen *et al.* (1967, 1973) observed germfree and conventional CFW mice for a 5-year period for spontaneous neoplasms. At 18–24 months of age, 29% of the conventional mice had leukemia and 23% had other tumors. In contrast, germfree and ex-germfree mice had only 0.06 and 6.2% malignancies, respectively. Electron microscopic evaluations of tissues demonstrated an increase in C-type particles in the conventional mice.

Germfree rats, free of detectable viruses, have been the source of valuable data on spontaneous and chemically induced tumors (Pollard *et al.*, 1964; Pollard and Kajima, 1967, 1970; Pollard *et al.*, 1968; Zacharia and Pollard, 1969; Miyakawa *et al.*, 1973; Sacksteder *et al.*, 1973). When the lymph nodes and spleens of tumor-bearing rats were compared to those of disease-free controls, significant changes were observed in the lymphoid tissues from the tumor-bearing animals: "(1) the lymph nodes contained increased numbers of plasma cells, but were free of germinal zones. (2) The red pulp of the spleens contained increased numbers of large and small lymphoid cells, but were free of plasma cells and germinal zones. Exposures of conventional rats to irrelevant antigenic stimuli in the environment causes the tumor-related changes to be "masked" by development of germinal zones and plasma cells in their lymph nodes and spleens. The tissue changes thus manifested by animals with growing tumors may represent a defective immunological response to tumor antigens" (Pollard *et al.*, 1968).

Spontaneous neoplastic tranformation of germfree rat embryos was observed *in vitro* with cells taken from germfree Wistar embryos. They showed morphologic conversion to epitheliumlike cells on the 8th passage, and by the 19th passage *in vitro*, the converted cells induced progressively growing local fibrosarcomas in newborn rats. These tumors, however, could not be transplanted to other rats. The inoculated rats of three strains, the normal and the transformed cells in culture, and the tumors were all free of detectable virus particles and viral antigens (Sharon and Pollard, 1969). These data indicate that transformation of cells *in vitro* and subsequent tumor development in transplanted hosts can occur without the involvement of microorganisms or their products. Germfree Wistar rats live longer than their conventional counterparts, and as they survive beyond 30 months of age, increasing numbers of spontaneous tumors are observed. The majority of these spontaneous tumors are benign, but a few rats have developed adenocarcinomas of the prostate gland, which were followed by metastasis to the lungs and other visceral organs. An occasional thymoma or pituitary tumor can also be found (Pollard and Kajima, 1970; Pollard, 1976). The two studies on old germfree Wistar rats were conducted in 1962 and 1969. The incidence of spontaneous breast tumors and of thymomas was reduced in 1969, and all tumors in the 1969 study were benign; the only relevant change in the 1969 group was a new improved diet.

2.3.2. Chemical Carcinogenesis

Germfree animals of many species are as susceptible to chemical carcinogenesis as their conventional counterparts. The incidence of tumors and the time of development are comparable in germfree and conventional animals. Transplantable fibrosarcomas have been induced by s.c. inoculation of 3-methylcholanthrene (MCA) in oil into adult germfree Swiss–Webster, Ha/ICR, CFW, and C3H strains of mice, and Fischer and Wistar rats

(Pollard *et al.,* 1964; Bealmear, 1965; Bealmear and Wilson, 1967b). Reyniers and Sacksteder (1959) also induced fibrosarcomas in germfree Bantam and Leghorn chickens with MCA. Pollard (1963) induced the appearance of mammary tumors in female germfree rats following intragastric administration of 20 mg of 7,12-dimethylbenzanthracene (DMBA).

Pollard and Salomon (1963) demonstrated that germfree newborn mice are also susceptible to chemical carcinogenesis. Lung adenomas appeared in Swiss mice following a single neonatal s.c. inoculation of MCA.

Strain variations in susceptibility to oncogenic chemicals have been observed in germfree and conventional rats (Pollard and Luckert, 1977). Wistar rats were resistant and Sprague–Dawley rats were susceptible to induction of mammary and intestinal adenocarcinomas by DMBA and 1,2-dimethylhydrazine (DMH), respectively. Leukemia could not be induced in germfree rats by whole-body X irradiation, but Gross, passage A, leukemia virus from mice will induce leukemia in rats when deliberately inoculated during the neonatal period (Pollard and Kajima, 1966).

The effect of the intestinal microflora on colon carcinogenesis by DMH, azoxymethane (AOM), and *N*-methyl-*N*′-nitro-*N*′-nitrosoguanidine (MNNG) was studied in Fischer rats (Reddy, 1979). The germfree status reduced, but did not eliminate, the colon tumors induced by repeated s.c. or intrarectal injections of DMH. The number of colonic tumors was higher in germfree than in conventional rats exposed to AOM or MNNG. Shih *et al.* (1975) also demonstrated that the presence of a microflora may be protective against some carcinogens. When they injected Sprague–Dawley germfree and conventional rats with MNNG and *N*-methyl-*N*-nitrosourea (MNU), the results with both compounds showed that germfree rats were much more susceptible to the carcinogenic and toxic effects of these carcinogens than were conventional rats. The mechanism of this protective role of the microflora might be either in stimulating host immunity, e.g., by endotoxin stimulation, or by deactivating or degrading the carcinogen. When the effect of both dietary fat and the intestinal microflora were considered in colon and breast carcinogenesis (Reddy *et al.,* 1979), the incidence of intestinal tumors following 3,2′-dimethyl-4-aminobiphenyl (DMAB) administration was lower in germfree than in conventional rats. The mammary tumor incidence in germfree female rats was also lower than in conventional female rats (Reddy *et al.,* 1979). These data suggest that the gut microflora may play a modifying role, either directly or indirectly, in colon carcinogenesis and possibly in mammary carcinogenesis evoked by DMAB. Reddy *et al.* (1977) have also seen a strong association between colon cancer and fecal excretion of bile acids in man. Their data indicate that primary bile acids, when further modified to secondary bile acids by intestinal bacteria, exert a strong promoting effect in colon carcinogenesis. Salomon (1968) concluded that the big difference in the response to carcinogens in germfree and conventional animals occurs primarily with the oral carcinogens, where bacterial action in the gut may convert potentially carcinogenic agents to active forms; e.g., cycasin, a methylazoxymethanol glycoside derived from the nut *Cycas circinalis* L., produces neoplasms of the kidney, intestines, and liver and is highly toxic in large doses (Laqueur, 1964). Following oral administration of cycasin to germfree rats, almost all of it was excreted unchanged in the feces, and neither neoplasia nor toxicity was observed (Spatz *et al.,* 1966). Conventional rats developed neoplasms of the kidney and liver. It is believed that germfree rats, in the absence of bacterial action, lack an enzyme that splits the glycoside into the more carcinogenic aglycine of cycasin and its sugar component.

2.3.3. Radiation-Induced Carcinogenesis

Pollard and Matsuzawa (1964) demonstrated that germfree mice developed lymphoid leukemias with localization in the thymus following four weekly doses of 150 rads of X irradiation. Viral particles of Bernhard C-type were found in the lesions. When Walburg *et al.* (1965) X-irradiated ICR germfree mice with a single high dose (600–900 rads), they obtained lymphoid leukemias, about half of which affected the thymus.

Pollard and Kajima (1966) X-irradiated germfree Sprague–Dawley, Wistar, and Fischer rats with a divided dose. The rats did not develop leukemia, but at 6 months postirradiation there was a generalized increase in thymic volume, and it became hard and fibrous; there was also atrophy of the lymph nodes. Six out of forty-nine Wistar and six out of forty-three Sprague–Dawley rats developed mammary adenofibromas at 9–12 months postirradiation. One Sprague–Dawley and one Fischer rat had mammary adenocarcinomas.

2.3.4. Viral Carcinogenesis

As mentioned in the previous sections on spontaneous leukemia in AKR mice and radiation-induced leukemia in all strains of germfree mice tested, germfree mice harbor type C viral particles that are identical to particles responsible for murine leukemias (DeHarven, 1964). Although no formal evidence for the oncogenic role of these viruses has been demonstrated, Salomon (1968) hypothesized that these C viral particles are responsible for the leukemias of germfree mice.

Mirand and Grace (1963) obtained typical Friend virus disease in ICR mice following i.p. inoculation of an acellular filtrate from the spleen of a conventional mouse with Friend virus leukemia. At the 14th and 21st days, germfree mice had higher splenic weights, greater leukocytosis and erythropoietic activity than conventional mice; i.e., the neoplasia produced by the Friend virus appeared sooner and was more severe in germfree than in conventional mice.

When Swiss mice with congenital lymphocytic choriomeningitis (LCM) virus were cesarian-derived and placed in isolators, they manifested no overt signs of disease up to 8 months of age and no early "runting" as seen in conventional LCM-infected mice (Pollard *et al.,* 1968). However, all the gnotobiotic LCM mice carried virulent LCM virus in their spleens, and the virus could also be detected in the embryos of their progeny. The virus was passed vertically to the offspring, and the resulting lesions were related to a hyperreactive immune system, which induced degenerative kidney changes, depletion of cortical thymocytes, lymphoid-cell infiltrations of many organs, and hypergammaglobulinemia.

Pollard (1964) and Pollard and Kajima (1966) produced fatal leukemia in both germfree mice and rats by inoculation with Gross virus strain A. Reyniers and Sacksteder (1959) induced fibrosarcomas in germfree Leghorn chickens using Rous sarcoma virus of the Bryan strain.

2.3.5. Conclusion

Although there is little or no essential difference between the response of germfree and conventional animals to some tumorigenic agents, significant data have been obtained from the germfree rat in the study of colon carcinogenesis with oral carcinogens. If oral carcinogens are metabolized by the microorganisms in the gut or their products, this alteration of the carcinogen may be either beneficial or harmful, depending upon its initial chemical nature. Germfree animals can also prove useful in additional studies on spontaneous carcinogenesis in mice and rats where an inapparent infection might otherwise cloud the interpretation of the data.

2.4. Gnotobiotic Animals and the Study of Disease

The gnotobiotic animal offers much promise for the study of disease, because it provides a model for studying the development of the particular syndrome in the absence of synergistic effects produced by the microflora and their products. This model has been used successfully in the study of shock, intestinal obstruction, thermal trauma, wound healing (Downey and Pisano, 1968), runting syndrome and other forms of GVHD, dental caries, periodontal disease, enteric infections, ulcerative colitis, uremia, mycoplasmosis, yeast infections, roundworm infections (Despommier *et al.,* 1967), and protozoal infections (Gordon and Pesti, 1971). These diseases are beyond the scope of this review, but they have been thoroughly reviewed by Gordon and Pesti (1971). The gnotobiotic animal has been especially helpful in studying host–parasite relationships, because in many instances host microflora influence the behavior of the parasite in the host as well as the response of the host to the parasite (Despommier *et al.,* 1967; Wescott, 1967, 1971).

2.5. Autoimmunity

There is only a suggestion that the germfree animal can give an autoimmune response (Gordon and Pesti, 1971). The consequences of necrotizing liver lesions in germfree and conventional mice were studied after i.p. injection of carbon tetrachloride (Arnason *et al.,* 1964). This led to the development of antiliver antibodies in the sera of both groups. Similar studies by Bauer *et al.* (1967) with 3′-methyl-4-dimethylamino-azobenzene demonstrated that both germfree and conventional rats developed liver injury, and became hypergammaglobulinemic following increased activity of the lymphatic tissue. No "antigen-free" diets were used in these studies, and thus dietary antigens might be responsible for these responses. Werder *et al.* (1978) also observed the development of a polycystic renal disease in germfree mice, which they believed to be autoimmune. Micklem and Asfi (1971) provided a direct demonstration of the existence in normal mice of cells carrying receptors for autologous erythrocyte antigens. Cell suspensions were prepared from the lymph nodes, thymus, bone marrow, spleen, and Peyer's patches. Mouse, sheep, and chicken erythrocytes were prepared by the usual washing procedures, and receptor-carrying cells were detected by a modification of the cluster (rosette) technique. Autologous and syngeneic erythrocytes were found to give similar results, and cells binding either were referred to as synclusterforming cells (SFC). Normal CBA mice had significantly more SFC in their thymuses than in their bone marrows or lymph nodes. The receptors for syngeneic erythrocytes are probably related antigenically to immunoglobulins. To test the specificity of cluster formation, lymphoid cells were incubated with a mixture of equal numbers of syngeneic mouse and chicken erythrocytes; less than 0.5% of the clusters were mixed. Micklem and Asfi (1971) concluded that it was impossible to attribute syncluster formation to any nonspecific adhesion between erythrocytes and mononuclear cells. When they tested nude mice for SFC, they were found to have significantly fewer SFC than their normal littermate controls. When germfree and conventional CBA mice were evaluated for SFC, no significant difference was found. The background SFC level, therefore, is not determined by previous infection with microorganisms carrying cross-reacting antigens unless these came from dead microorganisms in the diet.

It seems logical to conclude from these few experiments that germfree animals are capable of producing autoantibodies. Additional experiments with germfree mice maintained on the low-molecular-weight, "antigen-free" diet should help to clarify the status of autoantibody production in germfree animals.

3. The Colostrum-Deprived Piglet

3.1. Immunologically "Virgin" Animal

One of the most challenging and basic problems yet to be solved in modern immunology is the nature of the initiation of antibody synthesis. Despite the fact that many hypotheses have been proposed, the complete solution to this problem still escapes us. As mentioned in the Introduction, it has been demonstrated that animals and man have so-called "natural" antibodies and "background" antibody-forming cells to many antigens, e.g., SRBC. These "background" levels of antibody are even found in animals that are maintained germfree; they are believed to represent the response to ubiquitous antigens in the diet and environment (Thorbecke *et al.*, 1957; Springer *et al.*, 1959; Wostmann *et al.*, 1970). The question then arises whether the so-called "primary" immune responses are really primary or early secondary responses (Kim, 1975). A unique germfree animal has provided us with an animal model that has shed some light on this problem of antibody initiation; this animal model is the germfree, colostrum-deprived piglet obtained by aseptic hysterectomy from specific-pathogen-free (SPF) Minnesota miniature swine 3 to 5 days prior to term (Kim *et al.*, 1965, 1966a,b, 1967a,b, 1968a,b; Kim and Watson, 1968, 1971). This piglet has been proven to be an immunologically "virgin" animal as determined by the absence of immunoglobulin (Figs. 19, 20), background antibody-forming cells (Fig. 21), or immunoglobulin-synthesizing cells. Because of the absence of preexisting antibody-forming cells, plasma cells, and Ig in the immunologically "virgin" piglets (Kim *et al.*, 1965, 1966a,b, 1967a,b, 1968a,b), this animal is an excellent model for testing the presence or absence of Ig receptors on "virgin" uncommitted immunocompetent cells (Jarošková *et al.*, 1973). No classic Ig or Ig-chain receptors have been demonstrated on the "virgin" cells (uncommitted), while they are easily detected on the cells (committed) of the conventional animal by immunofluorescence techniques using fluorescent anti-Ig (Kim, 1975; Kim *et al.*, 1979).

These animals are immunologically competent (Watson *et al.*, 1968) as shown by

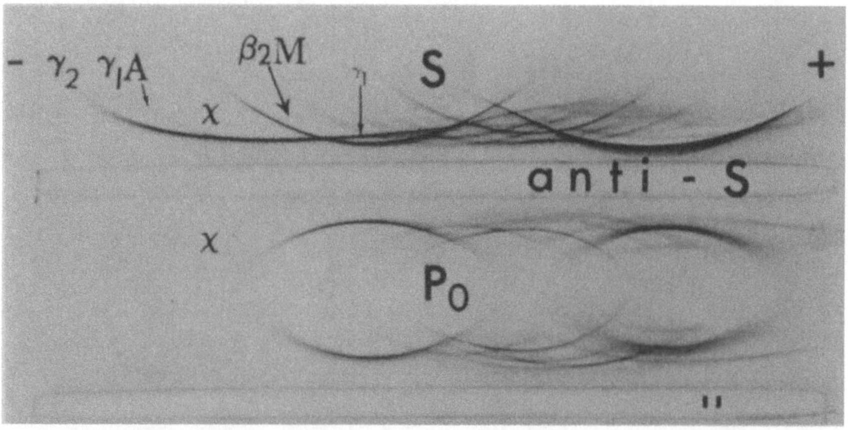

FIGURE 19. Immunoelectrophoretic analysis of sow serum and colostrum-deprived, germfree piglet serum. S is 0.2 ml of a 1:4 dilution of sow serum; P_0 is 0.2 ml of a 1:2 dilution of colostrum-deprived, germfree piglet serum obtained within 1 hr after hysterectomy. Anti-S is rabbit anti-sow serum D41, 1.5 ml in each trough. Lines of precipitation were developed at 30°C for 72 hr, washed, dried, and stained with azocarmine G. [From Kim *et al.* (1966a). Published with permission of Williams & Wilkins, Baltimore.]

their immune responses to various antigens; e.g., these immunologically "virgin" animals did not develop a single "background" hemolytic plaque-forming cell to SRBC over a 6-month period, but when injected with a specific antigen (SRBC) or actinophage (MSP-2), they responded well with antibody-forming cells within 48 hr, which peaked at 5 to 6 days after injection of antigen (Kim, 1979). When spleen cells from these "virgin" animals were cultured with SRBC, the *in vitro* immune response showed a definite lag period of 72 to 96 hr, followed by a rapid response. In contrast, "normal" cultures from conventional piglets had little or no lag period and gave a slower response (Tables III and IV). The "virgin" cell cultures without added SRBC did not produce anti-SRBC antibody, although the "normal" cultures did produce anti-SRBC antibody in the absence of antigen (Kim, 1975, 1979; Kim *et al.*, 1979). If extensive placental damage occurred during hysterectomy, contaminating immunoglobulin and "background" antibody could be detected (Setcavage and Kim, 1976a,b).

3.2. Complement Receptor-Bearing Lymphocytes

Germfree, colostrum-deprived piglets had a low number of complement receptor-bearing lymphocytes (CRL) on the day of hysterectomy (P_0); these CRL increased only slightly up to 3 weeks of age (Fig. 22). When these piglets were injected with antigen, CRL increased drastically to levels as high as those seen in conventional piglets of the same age. These investigators suggest that antigenic stimulation plays an essential role in the increase of CRL in the spleen and lymph nodes of germfree, colostrum-deprived piglets (Setcavage and Kim, 1976a,b; Kim, 1979).

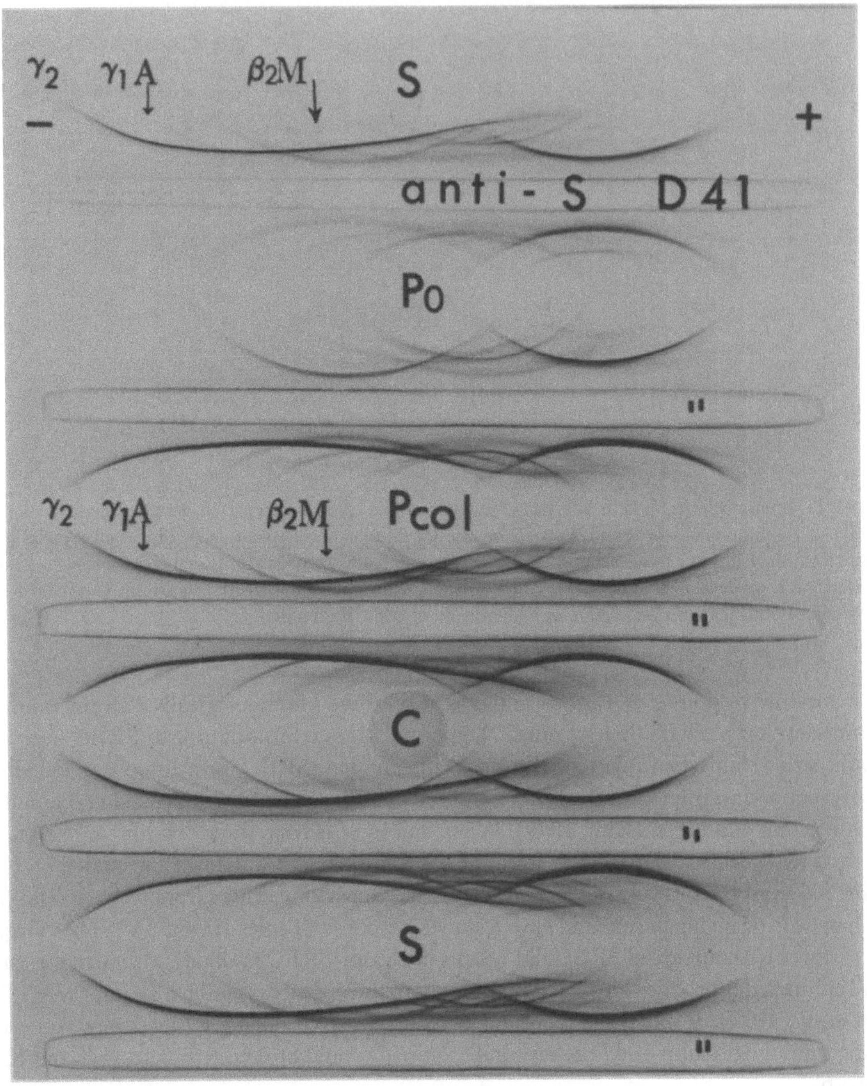

FIGURE 20. Immunoelectrophoretic analysis of sow serum and colostrum-deprived, germfree piglet serum. S is 0.2 ml of a 1:4 dilution of sow serum; P_0 is 0.2 ml of a 1:2 dilution of colostrum-deprived, germfree piglet serum obtained within 1 hr after hysterectomy; P_{col} is 0.2 ml of a 1:2 dilution of colostrum-fed, germfree piglet serum; C is conventional. Anti-S is rabbit anti-sow serum D41, 1.5 ml in each trough. Lines of precipitation were developed at 30°C for 72 hr, washed, dried, and stained with azocarmine G. [From Kim and Watson (1966). Published with permission.]

3.3. Polyclonal B-Cell Activators

Nonspecific, polyclonal B-cell activators (PBA), e.g., LPS, lipid A, dextran sulfate, pneumococcal polysaccharides (SSSIII), and purified protein derivative (PPD), could not activate B cells from immunologically "virgin" piglets, while they did activate B cells from "normal" animals (Table III). This suggests (Kim, 1975, 1979) that the observed polyclonal antibody response by nonspecific PBA (Andersson *et al.,* 1972; Coutinho and Möller, 1974; Coutinho *et al.,* 1974) may be due to nonspecific activation of already committed B cells (Y-Z cells) and not to responses by uncommitted "virgin" B cells (X cells); proper presentation of specific antigen and multicellular interactions are essential for the induction of the "virgin" B cells (X cells) to produce antibodies (Z cells). From these data, Kim and his associates (Kim, 1975; Kim *et al.,* 1979) have developed a conceptual model of ontogeny of the immune system (Fig. 23).

3.4. Membrane-Bound Enzymes of Lymphoid Tissues

Cells from the lymphoid organs of germfree colostrum-deprived piglets were also examined for their ability to bind *E. coli* β galactosidase (GZ), using a histochemical detection system (Decker *et al.,* 1974). From 0.3 to 1.4% of the cells in the various organs bound

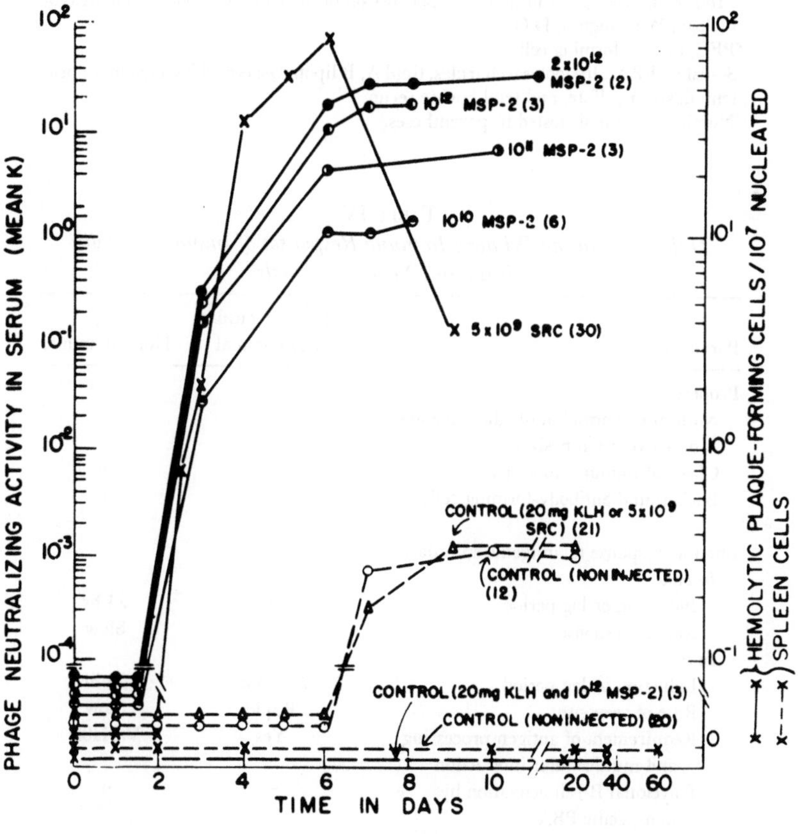

Figure 21. Primary immune response to actinophage (MSP-2) and sheep erythrocytes (SRC) in germfree colostrum-deprived piglets maintained on Mullsoy diet. Each curve represents the mean antibody response of the number of piglets indicated in parentheses. [From Kim (1975). Published with permission of the National Foundation–March of Dimes.]

TABLE III

Immunologic State of the Host in Response to Nonspecific Polyclonal B-Cell Activators and Specific Antigen[a]

	Anti-SRBC PFC[b] response *in vitro*		
	Background	PBA[c]	SRBC
Germfree, colostrum-deprived piglets			
0 day old (20)[d]	0	0	+++
14 days old (6)	0	0	+++
3 months old (6)	0	0	+++
Conventional, colostrum-fed piglets			
0 day old (10)	0	0	0
14 days old (6)	±	±	±
3 months old (6)	+	++	+++
Normal adults (20)	+	++	+++

[a]From Kim (1979). Published with permission of the American Society for Microbiology, Washington, D.C.
[b]PFC, plaque-forming cells.
[c]So-called PBAs: lipopolysaccharides, lipid A, B lipopolysaccharides, endotoxin protein, dextran sulfate, and fetal bovine serum.
[d]Number of animals tested in parentheses.

TABLE IV

Differences in the Primary Immune Responses of Immunologically Virgin and Normal Animals[a]

Parameter	Immunologically virgin animal	Normal animal
Property		
Natural or normal antibodies (passively or actively synthesized)	−	+
Classical immunoglobulins	−	+
Background antibody-forming cells	−	+
Immune response to antigenic stimulation		
In vivo		
Induction or lag period	36–48 hr	24 hr
Rate of response	Rapid	Slow
In vitro		
Induction or lag period	72–96 hr	24 hr
Rate of response	Rapid	Slow
Requirement of antigen processing and multicellular interaction	Yes	Variable
Polyclonal B-cell activation by nonspecific PBA	−	+
Requirement of specific antigen for antibody response	Yes (obligatory)	Variable

[a]From Kim (1979). Published with permission of the American Society for Microbiology, Washington, D.C.

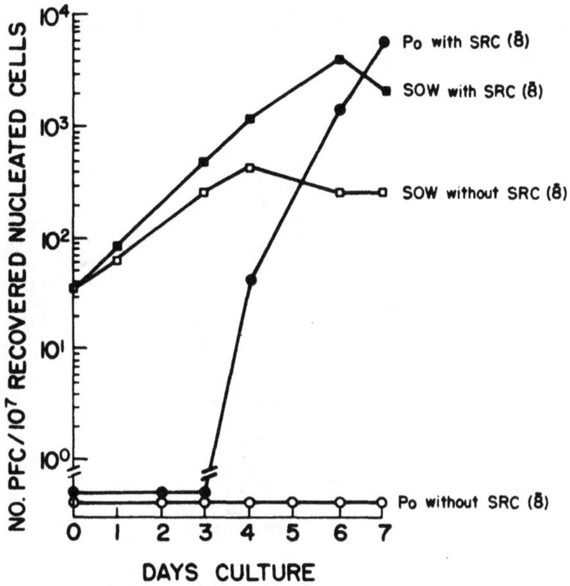

FIGURE 22. *In vitro* primary immune response to sheep erythrocytes (SRBC) with spleen cell cultures of immunologically "virgin" piglets (P_0) and "normal" pigs (sow). 1×10^7 spleen cells/ml per culture with or without 1% SRBC per culture were incubated in a 10% CO_2–air box on a rocking platform at 37°C and anti-SRBC PFC were assayed in triplicate at given time periods after incubation. Each curve represents the mean PFC response of eight different experiments. [From Kim *et al.,* 1979. Published with permission.]

GZ; the binding occurred rapidly and was specific as shown by competition experiments with high levels of GZ, hemocyanin, and ferritin. As there is no maternal antibody in the piglets and as they were kept free of antigens and of antibody contributed by colostrum, Decker *et al.* (1974) concluded that cells arise spontaneously before antigen contact that is capable of binding antigen. A proportion of the cells was shown to bind more than one antigen; this binding was noncompetitive, which suggested that these cells bear more than one receptor specificity.

Ferenčík *et al.* (1979) measured the activity of lysozyme (LYS), myeloperoxidase (MPO), elastase (ELA), and β-D-glucuronidase (BDG) in the leukocytes, sera, and spleens of conventional and germfree miniature piglets fed with a milk-type diet (GF-A) or with a nonantigenic diet (GF-NA) (Mandel *et al.,* 1979). In the leukocytes isolated by dextran sedimentation from conventional piglet peripheral blood, the specific activities of LYS, MPO, and ELA were higher than in the leukocytes from germfree piglets on GF-A and GF-NA diets. No differences were observed in the specific activity of BDG. Slightly higher activities were found in GF-A leukocytes than in GF-NA leukocytes. In the spleens of conventional piglets only the levels of LYS were higher than in the spleens of GF-A and GF-NA animals. The activities of other enzymes were unchanged. No differences in the activities of these enzymes were found in the sera. Ferenčík *et al.* (1979) concluded that enzyme and metabolic differences exist in the leukocytes from conventional and germfree piglets, and that these changes in the leukocytes may influence their antibacterial activity.

The proteolytic enzymes in leukocytes and spleens of conventional and germfree piglets were determined at various pH values (Bergendi *et al.,* 1979), and at all pH values tested, the spleen extracts from germfree piglets showed higher specific proteolytic activities than the extracts from conventional animals. Changes in proteolytic enzymes may influence the degradation activity of leukocytes.

Certain plasma membrane changes during mitogen action have been proposed to be the key events responsible for subsequent further changes in the plasma membrane and in

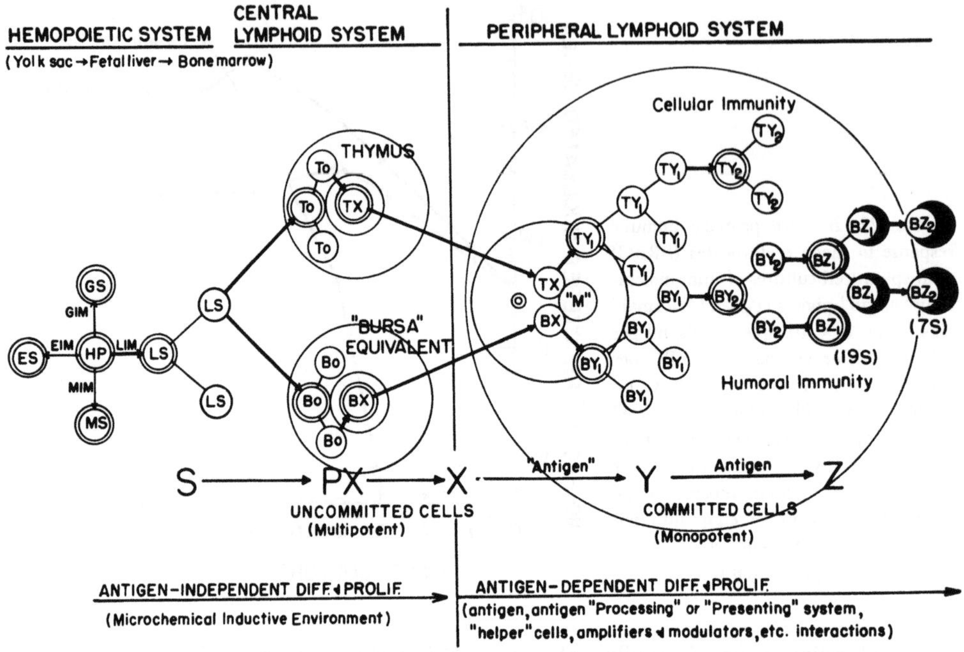

FIGURE 23. Conceptual model for ontogenetic development and differentiation of the immune system. *Stage I:* Antigen-independent physiologic differentiation and proliferation (up to X). Hemopoietic stem cells (HP) are omnipotent and may differentiate into lymphoid stem (LS) cells under the influence of a microchemical inductive environment (LIM) (yolk sac, fetal liver, bone marrow) and further differentiate into immunocompetent cells (PX→X), which are uncommitted multipotential cells. This step may be influenced by the thymic microchemical inductive environment to become T cells (To→Tx) and by the bursal or bursal-equivalent microchemical inductive environment to become B cells (Bo→Bx). Immunologically "virgin" piglets possess fully developed "virgin" T cells (Tx) and B cells (Bx), which are uncommitted multipotential cells, but they may not synthesize any Ig or antibodies (nonexpressive) until after antigenic exposure. *Stage II:* Antigen-dependent specific differentiation and proliferation (X→Y→Z). In the true primary immune response, the antigen initiates the differentiation of X cells to committed monopotent and monoexpressive Y cells (TX to TY_2; BX to BY_2). The primary role of antigen (properly processed by macrophages) in the humoral immune system (B cells) may be subcellular selection, derepression or induction of uncommitted multipotential cells to commit to the specific antigen and to turn on their synthesis of specific antibody (usually monoexpressive), but it is not selective for the cells. The committed Y cells carry specific antibody or Ig receptors. They can be selected and stimulated by the specific antigen to give clonal proliferation and differentiation and to become antibody-secreting Z cells (BY_1→BY_2 to BZ_1→BZ_2). It is also possible that immature plasma cells (BZ_1) produce 19 S antibody and that mature plasma cells (BZ_2) synthesize and secrete the 7 S IgG. The Y cells synthesize specific antibody Ig but do not actively secrete, particularly the Y_1 cells, which serve as memory cells. The X to Y step is, in a sense, inductive, and the Y to Z step could involve clonal selection as postulated by Burnet (1959). In conventional "normal" animals continually exposed to ubiquitous antigens, many of the immunocompetent cells are already committed Y cells. [From Kim (1975). Published with permission of the National Foundation–March of Dimes.]

the cells (Kovářů *et al.*, 1979a). It has been suggested that Na,K-ATPase participates in these initial membrane-mediated events (Lelievre *et al.*, 1977). Kovářů *et al.* (1979b) observed changes in Na,K-ATPase and γ-glutamyl transpeptidase (γ-GTPase) during the postnatal development of lymphoid tissues of germfree minipigs. A certain selective reactivity of thymus and lymph node Na,K-ATPase of minipigs during interaction with mitogen was found. Thymus and lymph node Na,K-ATPase were stimulated in the presence of PHA, in contrast to stimulation by PWM on lymph node Na,K-ATPase activity only. Con A had no effect on either thymus or lymph node ATPase activities. This selective interaction of ATPase with these mitogens might reflect the absence of specific carbohydrate moieties in enzyme molecules or tightly bound structures important for the enzyme function. In further experiments it was shown that thymus Na,K-ATPase was elevated in the presence of PHA only, in contrast to γ-GTPase, which was not stimulated by any of these mitogens during the 30- to 180-min period of observation. These results are consistent with the observation that γ-GTPase is not stimulated during the early incubation phase, but increases between 21 and 70 hr of incubation (Novogrodsky *et al.*, 1976). Kovářů *et al.* (1979a) concluded that: (1) Na,K-ATPase of pig lymphoid cells may be involved in mitogen-induced early events in cellular immunity; (2) γ-GTPase may participate in the mechanisms of antibody synthesis in humoral immunity; and (3) germfree animals are characterized by developmental and mitogen effects on Na,K-ATPase and γ-GTPase activities. These biochemical parameters may indicate certain prolonged maturation processes of lymphoid cells. The decreased γ-GTPase activity of germfree animals may reflect a lower function in membrane transport, changed cellular metabolic mechanisms, and also a decreased quantity of enzyme molecules per cell. All of these membrane-connected changes could result from different factors, e.g., physiological or immunologic, that influence animals reared under germfree conditions.

3.5. "Natural" Antibody

"Natural" antibody has been defined as antibody present in normal serum in the absence of specific antigenic contact (Scheffel and Kim, 1979). The sera of normal unimmunized animals contain agglutinins and hemolysins reactive with bacteria and various species of erythrocytes; the origin of this natural antibody is debatable, as antibodies are generally defined as products of antigenic stimulation. In order to examine the role of diet, normal flora, and environment in the ontogeny of natural agglutinins, serum hemagglutination (HA) titers against a panel of fixed erythrocyte antigens were determined for groups of colostrum-deprived piglets, SPF swine, and conventional swine. The resulting data supported the hypothesis that antigen is responsible for the development of natural hemagglutinins (Scheffel and Kim, 1979). Conventional young adult miniature swine assayed against the erythrocyte panel expressed the highest titers ($>$ 1:1000) against cat, dog, burro, and rabbit cells (Table V). Titers to ox and goat were low ($<$ 1:8) or absent, and titers to the remaining erythrocyte species were intermediate (1:64 to 1:1000). Sera from SPF young adult miniature swine contained readily detectable antibody to the panel of erythrocyte antigens, although the titers were significantly lower (2- to 16-fold) than those obtained with conventional animals because of the relatively restricted microbial and environmental antigenic contact within the SPF facility. Germfree colostrum-deprived miniature swine, 3 weeks to 1 year of age, demonstrated a virtual absence of hemagglutinins (except in two instances among the oldest group of animals), indicating a lack of sufficient dietary and microbial stimulation responsible for agglutinin production against the erythrocyte panel (Table VI). When these germfree piglets were introduced into the SPF facility, they developed significant levels of agglutinins, but if maintained germfree on Mullsoy diet they showed a continued absence of hemagglutinins, even up to 1 year of age. Scheffel and Kim (1979) concluded that in the absence of normal dietary and microbial antigenic stimulation,

TABLE V

Development of "Natural" Hemagglutinins against Erythrocytes from Various Animal Species in Naturally Farrowed Colostrum-Fed Conventional Miniature Swine[a]

Age	No. of animals	Log$_2$ of the reciprocal of the hemagglutinin titer (mean ± S.D.)											
		Ox	Goat	Pig	Human A	Human B	Human O	Sheep	Chicken	Cat	Dog	Burro	Rabbit
3–5 weeks	20	—[b]	—	7.2 ± 2.2	4.8 ± 2.9	4.6 ± 3.1	6.8 ± 2.8	4.7 ± 1.9	7.4 ± 1.8	8.0 ± 1.8	9.2 ± 1.3	7.4 ± 2.3	7.6 ± 1.5
6–10 weeks	13	—	—	7.0 ± 3.0	7.2 ± 1.7	7.1 ± 1.5	8.1 ± 1.7	4.4 ± 1.8	6.9 ± 2.3	7.9 ± 3.0	9.8 ± 1.3	7.5 ± 2.6	7.5 ± 2.2
3–5 months	18	—	—	8.7 ± 0.7	8.7 ± 0.6	8.7 ± 0.6	9.5 ± 0.5	6.6 ± 1.5	8.8 ± 0.8	10.3 ± 1.3	10.4 ± 1.1	9.7 ± 0.8	11.7 ± 0.7
6–11 months	15	—	2.5 ± 1.4	9.2 ± 1.2	10.1 ± 1.2	10.1 ± 1.5	10.8 ± 1.6	8.7 ± 1.8	9.7 ± 1.3	10.9 ± 1.7	11.3 ± 2.0	11.8 ± 1.7	11.7 ± 1.3

[a]From Scheffel and Kim (1979). Published with permission.
[b]Mean value 2.0 or less.

TABLE VI

Development of "Natural" Hemagglutinins against Erythrocytes from Various Animal Species in Germfree Miniature Swine[a]

Age	No. of animals	Log$_2$ of the reciprocal of the hemagglutinin titer (mean ± S.D.)											
		Ox	Goat	Pig	Human A	Human B	Human O	Sheep	Chicken	Cat	Dog	Burro	Rabbit
3–5 weeks	21	—[b]	—	—	—	—	—	—	—	—	—	—	—
6–10 weeks	11	—	—	—	—	—	—	—	—	—	—	—	—
3–5 months	6	—	—	—	—	—	—	—	—	—	—	—	—
6–12 months	3	—	—	—	—	2.0 ± 0.0	—	—	—	—	—	—	2.7 ± 1.7

[a]From Scheffel and Kim (1979). Published with permission.
[b]Mean value 2.0 or less.

no background hemagglutinins arise. Although the levels of serum immunoglobulins in germfree piglets are not comparable to those of conventional animals, the level of their serum immunoglobulin begins to rise shortly after Mullsoy feeding. Their Ig levels reach up to about 50% of the conventional level by about 1 year of age in some animals (Setcavage and Kim, 1979), but there are still virtually no detectable hemagglutinins at this time. Scheffel and Kim (1979) concluded that natural agglutinins result from dietary and microbial antigenic stimulation, not from the spontaneous appearance of specific serum globulins. This was consistent with their previous observation that animals reared under germfree conditions and maintained on a Mullsoy diet did not develop a single background antibody-forming cell reactive with sheep erythrocytes (Kim, 1975; Kim *et al.,* 1979).

3.6. Natural Killer Cells

A great deal of attention is currently focused on the natural killer (NK) cell because of its possible role in immune surveillance against tumors and in resistance to hematopoietic transplants. The NK cell has been described in mice, rats, and humans (Herberman and Holden, 1978). After extensive research, several questions still remain unanswered about the factors controlling the ontogeny of the NK cell and the influence of environmental antigens on its development. The precise relationships between the NK cell and the effector cell in antibody-dependent cell-mediated cytotoxicity (ADCC), the K cell, and the role of antibodies in natural killing (NK) are still not clear. The germfree colostrum-deprived piglet has provided a system for answering some of these questions (Koren *et al.,* 1978). Activities of NK and ADCC were tested by a short-term (2.5–4 hr) ^{51}Cr-release assay with the ^{51}Cr-labeled human myeloid-cell line K562 and the TNP-conjugated human B-cell line SB as target cells for NK and ADCC, respectively. Animals obtained by aseptic hysterectomy 3–5 days prior to term showed ADCC activities similar to adult levels, but lacked NK activity (Kim *et al.,* 1980). Hysterectomy-derived piglets, which were colostrum-deprived and maintained in germfree isolators, developed NK activity in 3–4 weeks. Naturally farrowed, colostrum-fed piglets developed NK activity 2–3 weeks after birth. There were no significant differences in the levels of either NK or ADCC between germfree and SPF animals. The differences in the ontogeny of the two activities suggested that they are distinct. Koren *et al.* (1978) concluded after preliminary characterization of the effector cells that: (1) NK and ADCC in the pig are physically not separable; (2) the majority of the cytotoxic activity on a cell-per-cell basis is mediated by the non-T-lymphocyte fraction; and (3) the rosetted T cells, which account for about 60% of the total pig peripheral blood lymphocytes, have low but demonstrable cytotoxic activity as well. The development of these two effector cells in piglets under environmentally controlled conditions requires further study to definitively answer these questions.

3.7. Radiobiology

The germfree animal has been an important tool in radiobiology and in the use of radiation to suppress the immune response, as has been discussed in greater detail in Section 2.2 on the germfree mouse. Because of the possibility of excluding the influence of the microflora on the radiation-induced changes in the organism, the germfree piglet was also used for studies on radiation pathology (Miler *et al.*, 1971; Mandel *et al.*, 1979). The $LD_{50(30)}$ following exposure to ^{60}Co γ radiation in 2-week-old conventional piglets was 73.9 mCi/kg (286 rads), and in 2-week-old germfree piglets, 87.2 mCi/kg (338 rads), thus making a difference of 13.3 mCi/kg (52 rads). The survival time in germfree piglets was about 6 days longer than that of their conventional counterparts. At autopsy, there was evidence of severe hemorrhaging in the skin, the submucosa, and the intestinal wall of both groups. After exposure to 206 mCi/kg (800 rads), a dose that causes death from hematopoietic failure, there was a significant difference in the peripheral blood picture of the germfree and conventional piglets; e.g., in the conventional piglets the total number of leukocytes decreased nearly to zero on day 3 postirradiation, but in the germfree piglets the leukocytes (mainly granulocytes) decreased to 2500 per mm^3 on day 3 and to approximately 100 per mm^3 on day 6. This was probably the result of the predominance of radioresistant granulocytes in the germfree piglets (Svoboda and Mandel, 1979).

3.8. Morphology of Lymphoid Organs

Kim and Watson (1971) examined the histologic changes in the lymphoid tissues of germfree immunologically "virgin" piglets in order to investigate the cellular origin of sequential synthesis of 19 to 7 S IgG-antibody formation in the "true" primary immune response.

3.8.1. Histology of Lymphoid Tissues of Immunologically "Virgin" Piglets

The histology of the spleen and lymph nodes of germfree colostrum-deprived piglets on the day of hysterectomy (109–112 days of gestation) is shown in Fig. 24 (Kim and Watson, 1971).

3.8.1a. Spleen. Discrete thin periarteriolar lymphatic sheaths composed of uniformly small nonpyroninophilic lymphoid cells with scanty cytoplasm loosely fill in the networks of reticulum cells surrounding the arterioles in the spleen (Fig. 24A,B). These sheaths comprise the only elements of white pulp in the spleens of these piglets. In contrast, the white pulp of the "normal" adult spleen is made up of thick periarteriolar lymphatic sheaths and lymphoid follicles with germinal centers. Germinal centers are not present in immunologically "virgin" piglets, but are present in conventional "normal" adult pigs. There are abundant erythroid cells (hematopoietic cells) in varying stages of differentiation; in prenatal and neonatal spleens these decrease rapidly after 3 days of life.

3.8.1b. Lymph Nodes. The histology of representative lymph nodes from immunologically "virgin" piglets on the day of hysterectomy is shown in Fig. 24C,D. The mesenteric lymph node has a well-developed cortex and medulla; in contrast to adult lymph nodes, there are neither germinal centers nor mature plasma cells. The cortex with primary follicles is loosely filled with small and medium-sized lymphoid cells. Occasional weakly pyroninophilic cells similar to those of the thymus, spleen, and bone marrow are observed

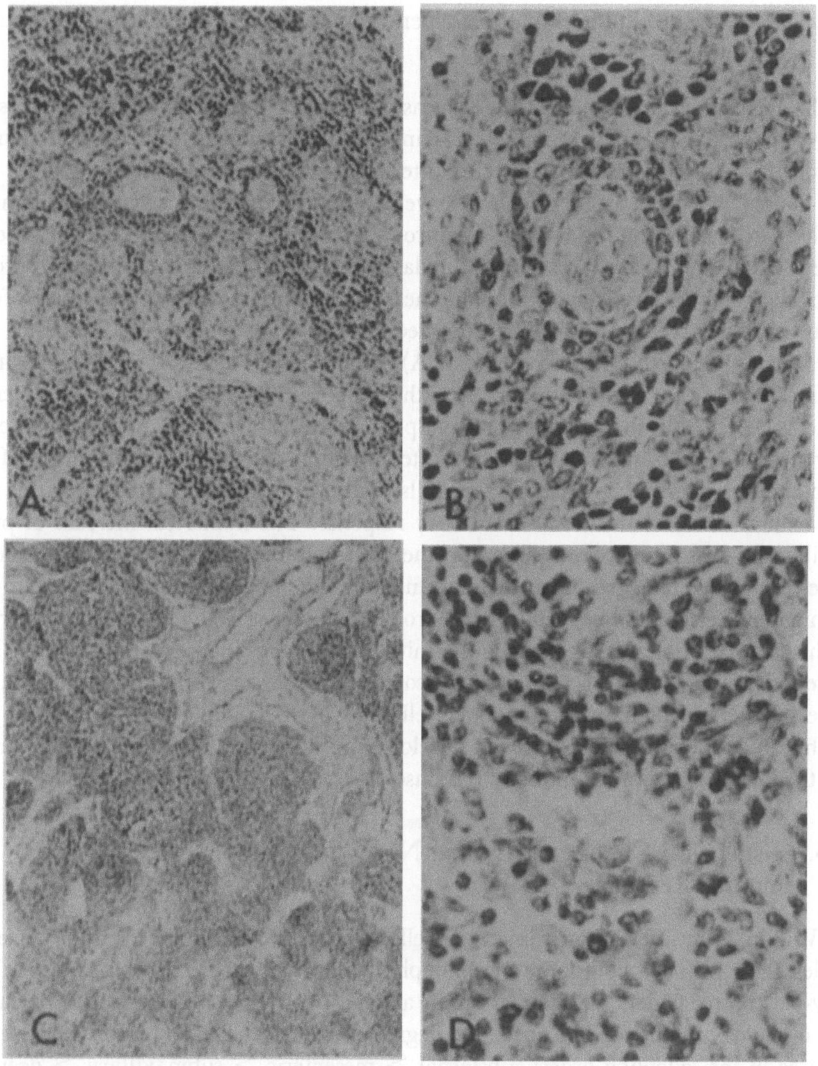

FIGURE 24. Spleen (A,B) and mesenteric lymph nodes (C,D) of a germfree colostrum-deprived piglet on the day of hysterectomy. Stained with methyl green–pyronin. (A) × 150; (B) × 600; (C) × 60; (D) × 600. [From Kim and Watson (1971). Published with permission.]

scattered in the cortex. These cells appear to be distinct from the large pyroninophilic cells arising in response to antigenic stimulation. The position of the cortex and medulla is inverted in contrast to most other species (Trautmann and Fiebiger, 1952; Binns and Hall, 1966); the cortex, including primary follicles, is located in the central position; it usually develops along the trabeculae, while the loosely arranged "medullary" area is located in the periphery of the node, the subcapsular area.

3.8.2. Histological Changes of the Spleen during the "True" Primary Immune Response

Histological examinations of the spleens were performed on the germfree colostrum-deprived piglets at 1, 1.5, 2, 3, 6, 8, 10, 14, and 16 days after injection of MSP-2 (Kim and Watson, 1971). Age-matched uninjected littermates were used as controls. Within 24 hr after antigen injection, cellular changes were observed in the periarteriolar lymphatic sheaths. The histology in the 1-day-old control piglets maintained on an "antigen-free" diet was essentially similar to that seen for the day 0 controls. The number of large pyroninophilic cells in the periarteriolar lymphatic sheaths increased on day 2 and increased further on day 3 in the spleens of the antigen-injected animals (Fig. 25B), while uninjected littermate controls remained unchanged (Fig. 25A). In contrast, in the early so-called "primary" immune response in 4-month-old piglets, the numbers of pyroninophilic cells increased markedly, not only in the periarteriolar lymphatic sheaths but also in the germinal centers and red pulp. On the 6th and 8th days after injection of antigen, the numbers of large pyroninophilic cells and small lymphoid cells increased further in the sheaths; this correlated with an increase in the 19 S antibody response. The numerical increase of pyroninophilic cells was even more striking on the 14th and 16th days after antigen injection. These cells tended to aggregate at the margin of the lymphatic sheaths; thus a group of the pyroninophilic cells could be seen separate from a group of small lymphoid cells surrrounding the same arterioles. This may be the initial site of germinal center formation. In the uninjected littermate controls fed with Mullsoy diet, there was a very slow, gradual increase in the numbers of both small lymphoid cells and pyroninophilic cells in the lymphatic sheaths; no germinal centers, however, developed in these animals. These results correlate with the low "background" antibody response observed in these control animals.

3.8.3. Histological Changes of Lymph Nodes during the "True" Primary Immune Response

Within 24 hr after antigen injection, cellular changes were detected in the cortex and follicles; these included both large pyroninophilic cells and small lymphoid cells, while the 1-day-old control piglets maintained on an antigen-free diet were essentially the same as those of the day 0 control. The onset and degree of the early cellular response in the nodes appeared in the following order: substernal > mesenteric > submaxillary > deep neck > axillary > superficial neck lymph nodes. On the second and third days after injection of antigen, the cellular response increased and included the large pyroninophilic cells in varying stages of maturation (Fig. 25D). The uninjected controls were unchanged due to the absence of antigenic stimulation (Fig. 25C). Within 48 hr after piglets were fed the Mullsoy diet, there were some large pyroninophilic and small lymphoid cell responses, especially in the mesenteric and submaxillary or pharyngeal lymph nodes. However, no changes were seen in the superficial neck and axillary lymph nodes until 72 hr. These results suggest that the Mullsoy diet contains antigens that are absorbed through the oro-gastrointestinal tract and stimulate the regional lymph nodes (Kim and Watson, 1971). The number of large pyroninophilic cells in the cortex of lymph nodes reached a maximum on day 4 after injection of antigen, and then gradually decreased. On day 6 there was a greater proportion of small lymphoid cells than large pyroninophilic cells. These changes

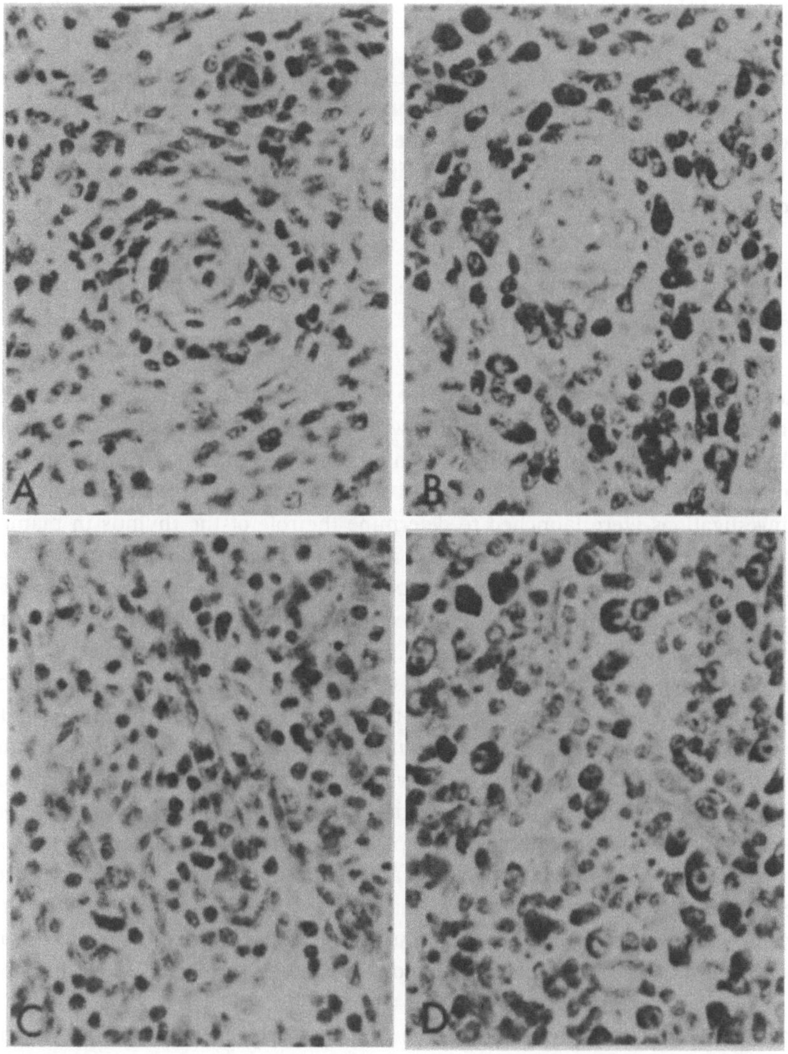

FIGURE 25. Spleen (A,B) and mesenteric lymph nodes (C,D) of a 3-day-old germfree colostrum-deprived piglet. (A,C) Unstimulated controls; (B,D) injected with 10^{12} MSP-2 i.p. on the day of hysterectomy. Note the large pyroninophilic cells in the periarteriolar lymphatic sheaths of the spleen (B) and in the cortex of the nodes (D) in the stimulated animal. Stained with methyl green–pyronin; × 600. [From Kim and Watson (1971). Published with permission.]

may be due to both cellular migrations and cellular differentiation, proliferation, and maturation of the nodes. On the 14th day after antigen injection, cortical areas were greatly increased including the development of germinal centers and immature and mature plasma cells in the corticomedullary junction. Uninjected controls showed less response than antigen-injected animals; however, significant cellular changes were observed in response to antigens in the diet.

3.9. Conclusion

The germfree colostrum-deprived piglet is a unique and valuable model for studying the role of antigen in the ontogeny of the immune system. Gnotobiotics will serve an important role in answering many fundamental questions in immunobiology, and especially those related to the role of microorganisms and their products on the development and regulation of the immune system.

4. Germfree Mouse

The germfree mouse made a significant contribution to the determination of the role of the thymus in the immune response following the renewed interest in the thymus after 1954. With the observation that patients with immune dysfunction, e.g., myasthenia gravis and agammaglobulinemia, had abnormalities of the thymus (Good, 1954; Good and Varco, 1955), investigations were launched to determine the role of the thymus in mammals by performing thymectomies in both adult and newborn rodents. Adult thymectomy resulted in no significant changes in the immune response, but after neonatal thymectomy the animals developed a wasting disease and died between 4 and 8 weeks of age (MacLean *et al.*, 1957; Fichtelius *et al.*, 1961; Jankovic *et al.*, 1962; Miller, 1963; Sherman *et al.*, 1963). These neonatally thymectomized animals also lost their ability to reject allogeneic transplants such as skin and tumor tissue, and they were incapable of producing antibodies against such standard antigens as *Salmonella typhimurium* H antigen, bovine serum albumin, and sheep erythrocytes. These effects were interpreted as evidence of an impairment of immune mechanisms. Thus, it seemed logical to conclude that the thymus was of prime importance in developing immunologic competence in the mammal. Because thymectomy after the first 2 days following birth usually failed to produce these effects, it was presumed that the thymus exerted its influence immediately after birth. All that seemed left to determine was whether the thymus produced a humoral factor that affected the development of other lymphoid organs or released immunologically potent cells that seeded lymph nodes, spleen, Peyer's patches, and other lymphoid tissues.

Complicating the interpretation of these laboratory experiments were postthymectomy infections. Presumably these infections occurred in the absence of immune processes that normally would have averted them. To what degree these infections affected the course of wasting disease and other postthymectomy phenomena was not known. At this point germfree technology was applied to the problem at the LOBUND Germfree Life Laboratory at the University of Notre Dame. By using laboratory animals reared in the complete absence of demonstrable microorganisms, a separation could be effected between infection and other pathological processes (Wilson *et al.*, 1964a,b; Bealmear, 1965).

Investigations into the role of the thymus in germfree mammals began with a histological comparison of the lymphoid organs of germfree and conventional thymectomized mice. The growth rates of these animals and of their various organs were also studied as were their respective hematological and immune responses. It was hoped that germfree thymectomized mice, free from the complications of infection and other manifestations of the presence of microorganisms in the host, might shed some light on the mechanism by which the thymus performed its function.

A correlation between immunologic activity and the development of the lymphoid tissues was sought in a comparison of germfree and conventional mice. The initial study was a comparison of the rate of growth and regression of the thymus in germfree and conventional mice (Bealmear, 1965; Wilson *et al.*, 1965). Qualitatively, the growth patterns of the thymus in germfree and conventional mice are very similar, and the gross appearance is the same. In both groups of animals the period of thymus growth extends from birth to about 50 days. There is accelerated growth during the first 2 weeks after birth with a peak thymus to body weight ratio at about 15 days (Fig. 26). Growth continues at a decelerated rate and regression of the thymus begins at about 50 days of age. A plateau is reached at about 200 days of age, but the thymus does not disappear completely. The thymus of conventional mice, however, grows more rapidly and attains greater size than that of the germfree mice.

A histological study of thymus tissue at various ages was also made in an attempt to correlate the microscopic anatomy with gross appearance and growth patterns of the thymus (Bealmear, 1965; Bealmear and Wilson, 1966). The histological picture correlated

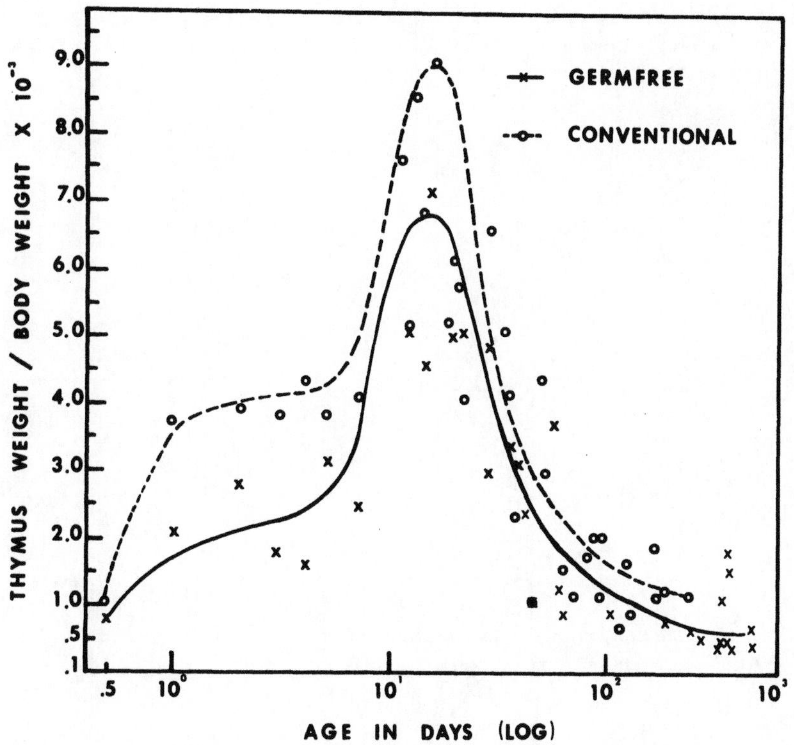

FIGURE 26. Ratio of thymus weight to body weight in germfree and conventional CFW mice from birth to 457 days of age. [From Wilson *et al.*, (1965). Published with permission.]

well with the growth pattern, and together they gave support to the hypothesis that the thymus initiates the production of lymphocytes in other tissues; the histology of the day 0 thymuses from germfree and conventional mice is shown in Fig. 27.

In subsequent studies a technique for germfree thymectomy was developed, and germfree neonatally thymectomized mice were observed for evidence of wasting syndrome, common to conventional neonatally thymectomized mice, and for alterations in granulocyte/agranulocyte ratios in peripheral blood (Wilson *et al.*, 1964b; Bealmear, 1965; Wilson *et al.*, 1966; Bealmear and Wilson, 1967b). It was in this series of experiments that germfree animals responded quite differently from conventional thymectomized mice. The germfree thymectomized CFW mice did not develop the wasting syndrome in 24 months, the maximum period of observation (the normal life span of this mouse strain). No significant differences in total leukocyte counts (Figs. 28, 29), differential leukocytes, or hematocrits were noted in germfree and conventional thymectomized mice after 14 days postthymectomy with the exception that conventional thymectomized mice showed increased total leukocyte, neutrophil, and lymphocyte counts at 60 days, when the onset of wasting disease was imminent (Bealmear, 1965; Bealmear and Wilson, 1967b).

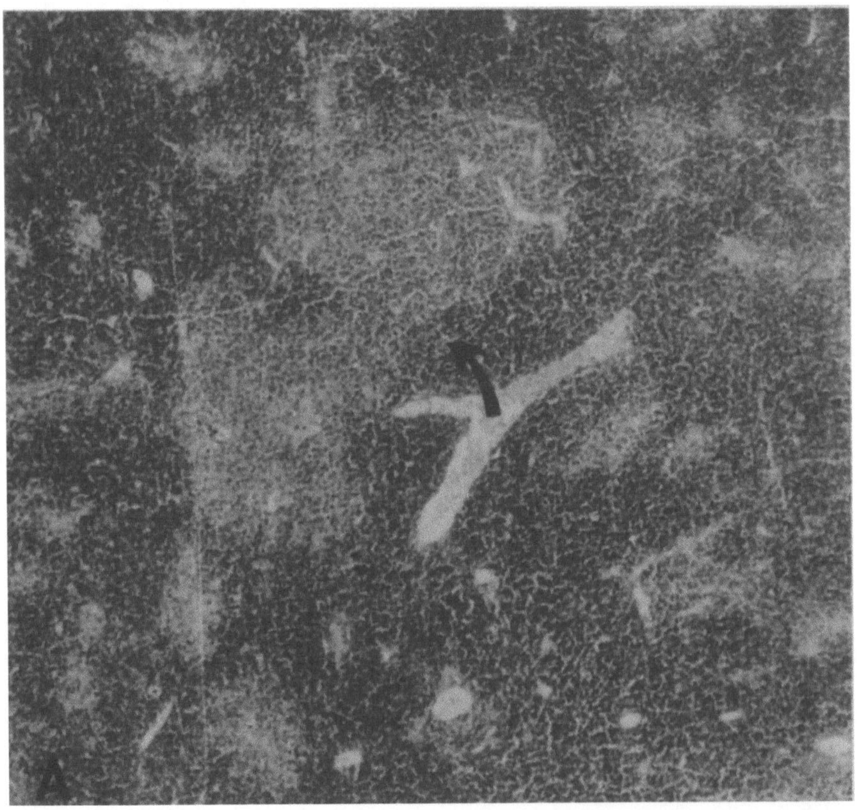

FIGURE 27. (A) Thymus of germfree CFW mouse at day 0 (approximately 9 hr old) (× 51). Note there are fewer small lymphocytes in the cortex than in (B), and there is a large distinct medulla (arrow). (B) Thymus

The ability of neonatally thymectomized mice to respond to antigenic stimulation was challenged by a variety of methods. *S. typhimurium* H antigen was inoculated i.p. into thymectomized animals to determine their response to a cellular antigen; response to a soluble antigen was measured with bovine serum albumin (Bealmear, 1965; Bealmear and Wilson, 1967d). Both germfree and conventional thymectomized mice produced detectable antibody titers against the cellular H antigen of *S. typhimurium* (Fig. 30) and the soluble antigen of bovine serum albumin. Antibody titers against both antigens were slightly lower in both germfree thymectomized and conventional thymectomized mice than in their respective controls; the latent period before antibody production against *S. typhimurium* H antigen was somewhat lengthened in both groups of thymectomized mice, but both were able to produce relatively high antibody titers against both antigens, with the titers in germfree thymectomized mice being significantly higher than in conventional thymectomized mice.

Finally, a response to tissue transplantation was elicited by challenge with homotransplantation of methylcholanthrene-induced tumor cells or neoplastic lymph node cells from animals inoculated with Gross leukemia virus strain A (Bealmear, 1965; Bealmear and Wilson, 1967a). Germfree thymectomized mice were able to reject both allogeneic meth-

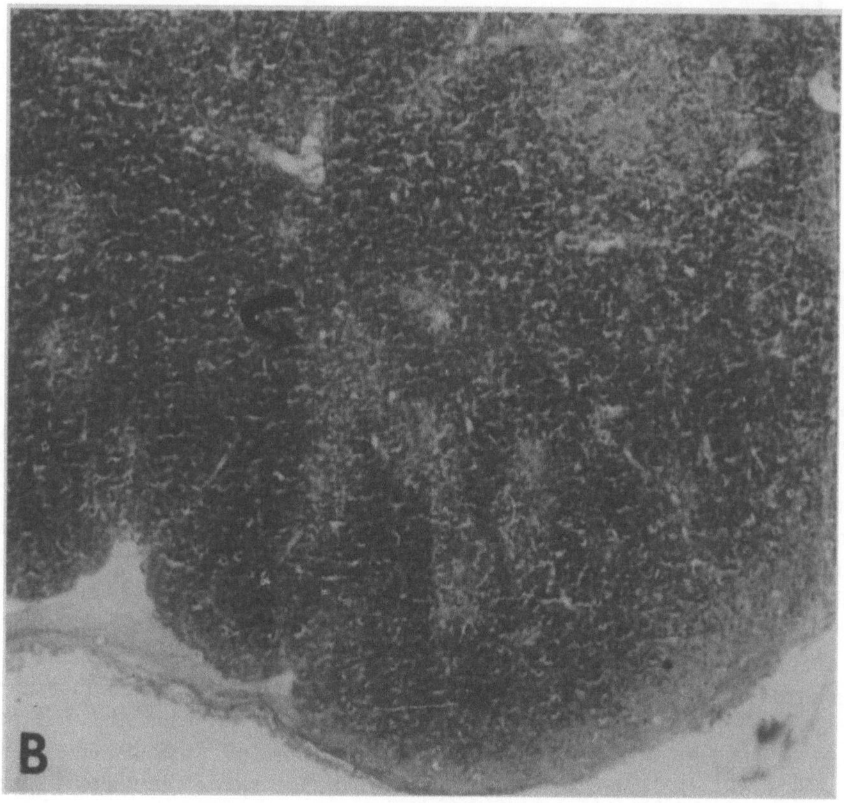

of conventional CFW mouse at day 0 (× 51). Note there are many small lymphocytes in the cortex, and the medulla is small and somewhat diffuse. [From Bealmear (1965), Published with permission.]

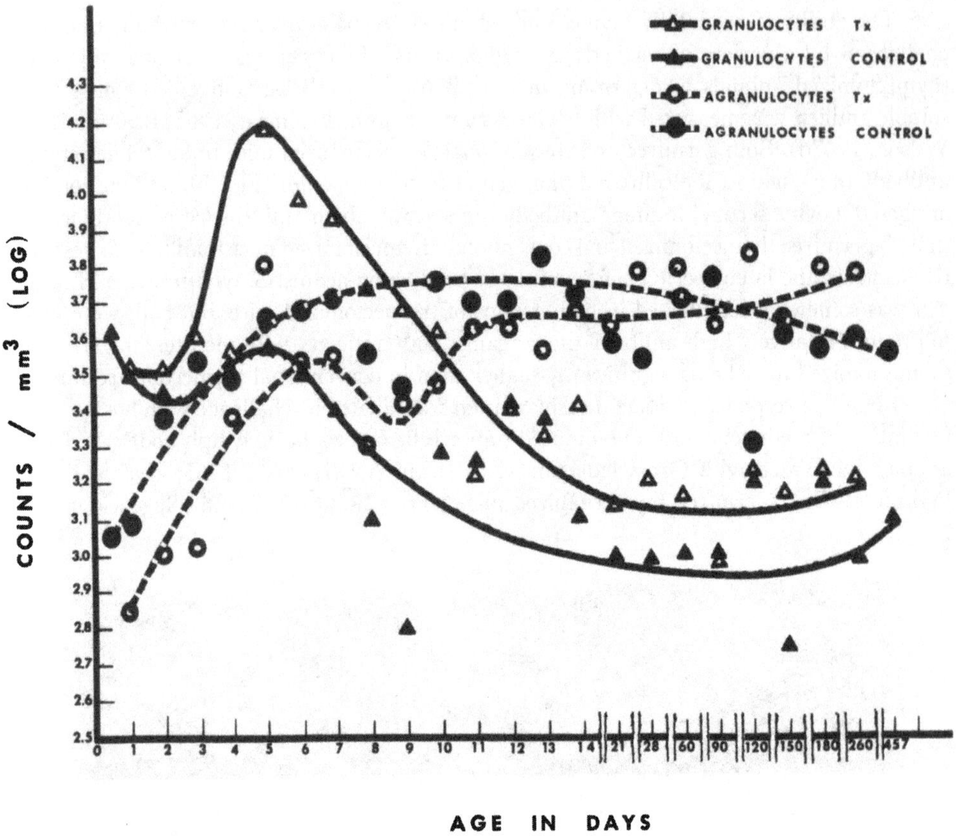

FIGURE 28. Granulocyte and agranulocyte counts per mm^3 in thymectomized and control germfree CFW mice. [From Bealmear and Wilson (1967d). Published with permission.]

ylcholanthrene-induced tumor cells and allogeneic neoplastic lymph node cells. Germfree thymectomized mice were not able to reject allogeneic methylcholanthrene-induced tumor cells and allogeneic neoplastic lymph node cells. Germfree thymectomized mice were not able to reject allogeneic methylcholanthrene-induced tumor cells after two doses of X irradiation (150 rads at 3-day intervals) when the transplant was made on the third day after the second exposure to X irradiation; however, 50% of the germfree thymectomized X-irradiated mice were able to reject the grafts when the tumor was transplanted 1 month after X irradiation and about 83% were capable of graft rejection at 2 months postirradiation. It appeared obvious then that the immunological competence of the neonatally thymectomized and X-irradiated mouse was capable of recovery if a longer time interval elapsed between X irradiation and tumor challenge. However, when the time period between X irradiation and challenge was further lengthened to 5 months, an interesting phenomenon occurred; only 63% of the experimental mice rejected the tumor despite the fact that they demonstrated normal peripheral blood counts and normal bone marrow, spleen, and lymph node histology. The remaining 37% did not reject their transplants; the transplants grew to large proportions, and the animals developed what appeared to be a

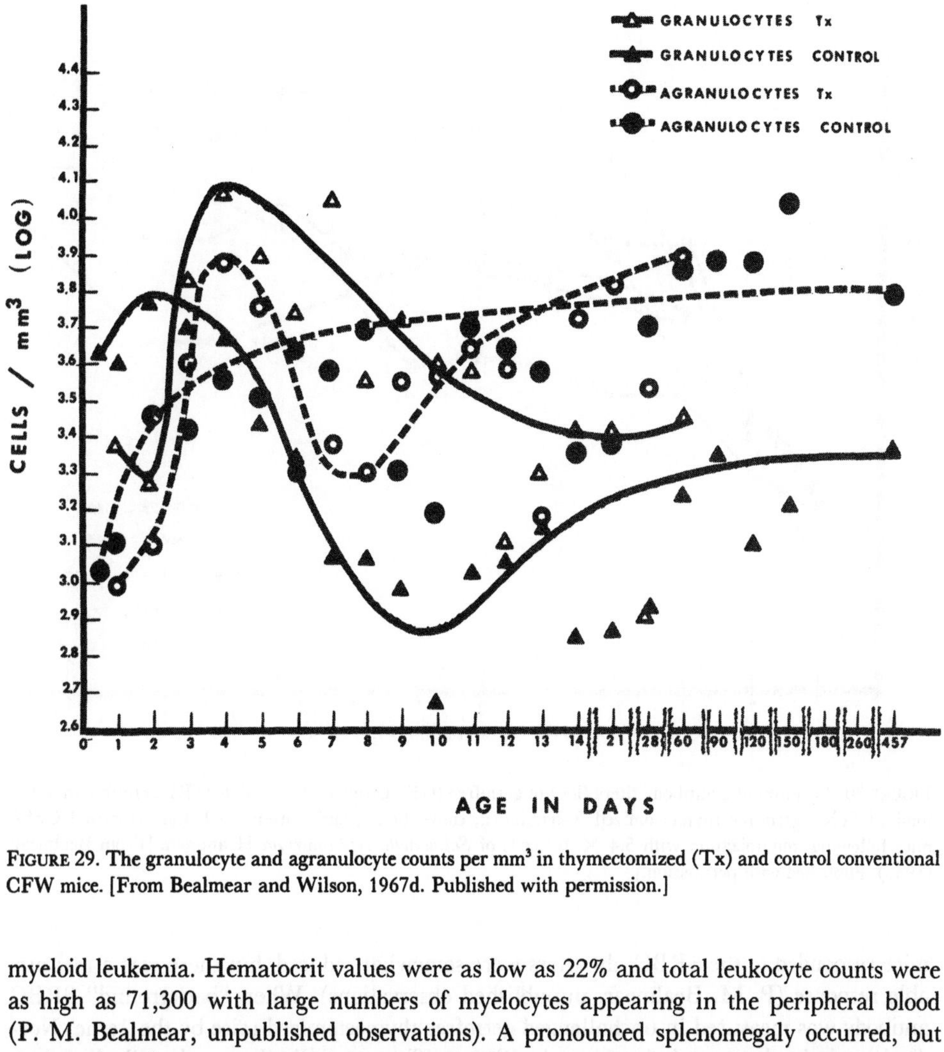

FIGURE 29. The granulocyte and agranulocyte counts per mm^3 in thymectomized (Tx) and control conventional CFW mice. [From Bealmear and Wilson, 1967d. Published with permission.]

myeloid leukemia. Hematocrit values were as low as 22% and total leukocyte counts were as high as 71,300 with large numbers of myelocytes appearing in the peripheral blood (P. M. Bealmear, unpublished observations). A pronounced splenomegaly occurred, but the lymph nodes were only slightly enlarged. The relationship between leukemogenesis and the combination of neonatal thymectomy, exposure to X irradiation, and tumor transplantation to older mice probably reflects a breakdown in the immune surveillance system following exposure to X irradiation.

From these observations it seemed logical to conclude that the reported complete immunological incompetence in thymectomized conventional mice was the result of their preoccupation with intercurrent infections and not a complete suppression of immunological response by thymectomy. By this period of the investigations "pathogen-free" conventional mice had been developed and thymectomized mice from this group performed in a manner similar to that of germfree mice studied earlier. Attempts to induce wasting disease in these "clean" mice also failed (P. M. Bealmear, unpublished observations).

When germfree neonatally thymectomized mice were immunized with SRBC and then tested for both agglutinating and hemolyzing antibody, all thymectomized germfree

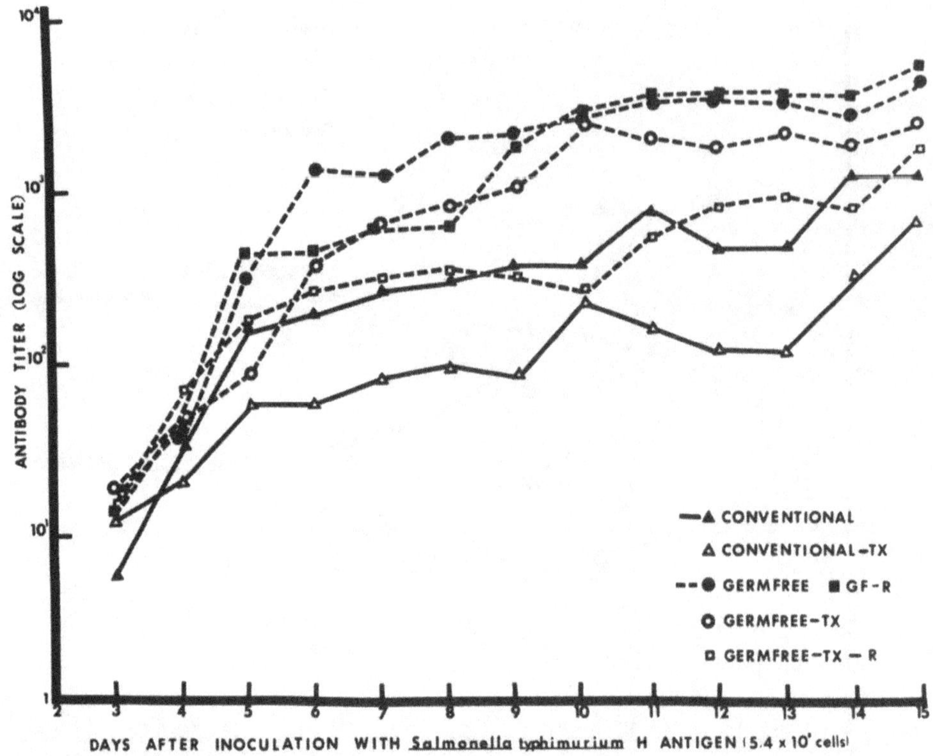

FIGURE 30. Agglutinating antibody titers (log) of germfree (GF); germfree X-irradiated (R); germfree thymectomized (TX); germfree thymectomized X-irradiated; conventional, and conventional thymectomized CFW mice following immunization with 5.4×10^8 cells of *Salmonella typhimurium* H antigen. [From Bealmear (1965). Published with permission.]

mice responded to the SRBC; the titers were somewhat reduced, but there was a measurable response (P. M. Bealmear, unpublished observations). When "background" SRBC antibody was measured in unchallenged germfree thymectomized mice by the Jerne–Nordin hemolytic plaque technique, no significant differences in background antibodies were found between the thymectomized germfree mice and their nonthymectomized germfree controls (A. A. Nordin and P. M. Bealmear, unpublished observations).

4.1. Serum Proteins of the Germfree Mouse

The electrophoretic patterns of the germfree Swiss mouse showed that all major protein fractions were present; numerical data indicated that β and γ fractions were definitely lower than in conventional mice (Wostmann, 1959). The electrophoretic patterns of the serum of a 60-day-old germfree mouse are shown in Fig. 31. Immunoelectrophoresis studies normally show that IgA is virtually absent from the sera of germfree mice. (See Section 2.1.3b.)

4.2. Germfree Nude Mouse

The germfree nude mouse has also provided a unique tool for the determination of thymic hormone levels in both mouse and man. A bioassay was developed that measures Thy-1.2 (θ) antigen in populations of prothymocytes or null lymphocytes taken from the spleens of germfree athymic (*nu/nu*) mice (Lewis *et al.*, 1976; Twomey *et al.*, 1977a,b). Spleen cells were gently teased from the organ and then incubated at 37°C for 45 min in syringes packed with nylon wool. About 5% of these cells were recovered by elution; this eluted cell population included about 80% null lymphocytes (Fig. 32), 3% B cells by immunofluorescence, with the remainder being hematopoietic cells. The eluted cells were washed, suspended in medium 199 and 5% BSA at a concentration of 5×10^6 cells/ml, and used as indicator cells in the bioassay. Induction incubations were for 18 hr at 37°C in a CO_2 incubator. They included 0.1 ml of the indicator cell suspension, 0.05 ml of the plasma to be tested after its passage through PM30 membranes or 1 μg thymopoietin in saline and 125 μg ubiquitin; incubations were brought to 0.2 ml with medium 199 and 5% BSA. Ubiquitin is a peptide extracted from many tissues that, in concentrations greater

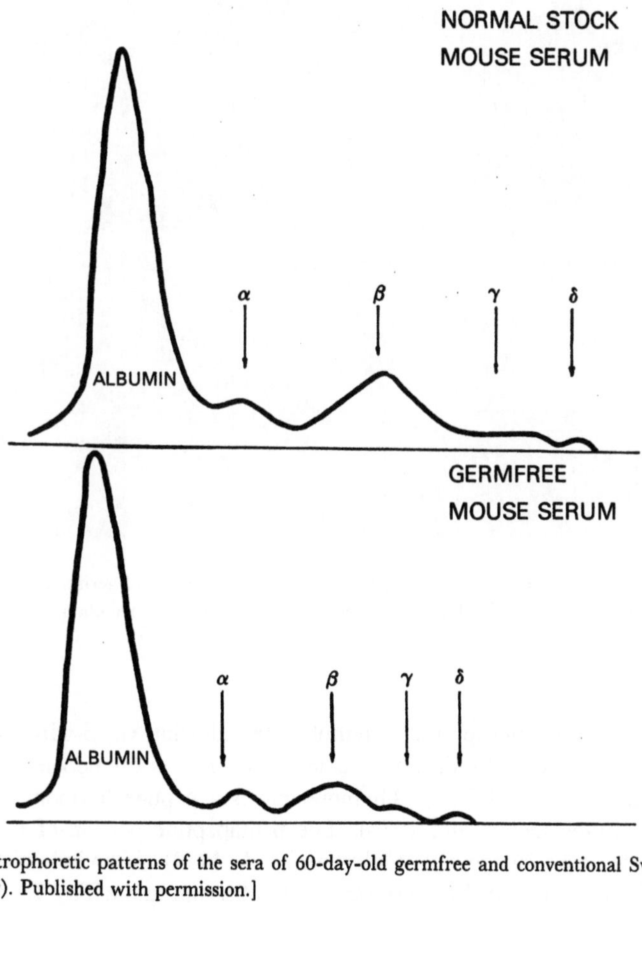

FIGURE 31. Electrophoretic patterns of the sera of 60-day-old germfree and conventional Swiss mice. [From Wostmann (1959). Published with permission.]

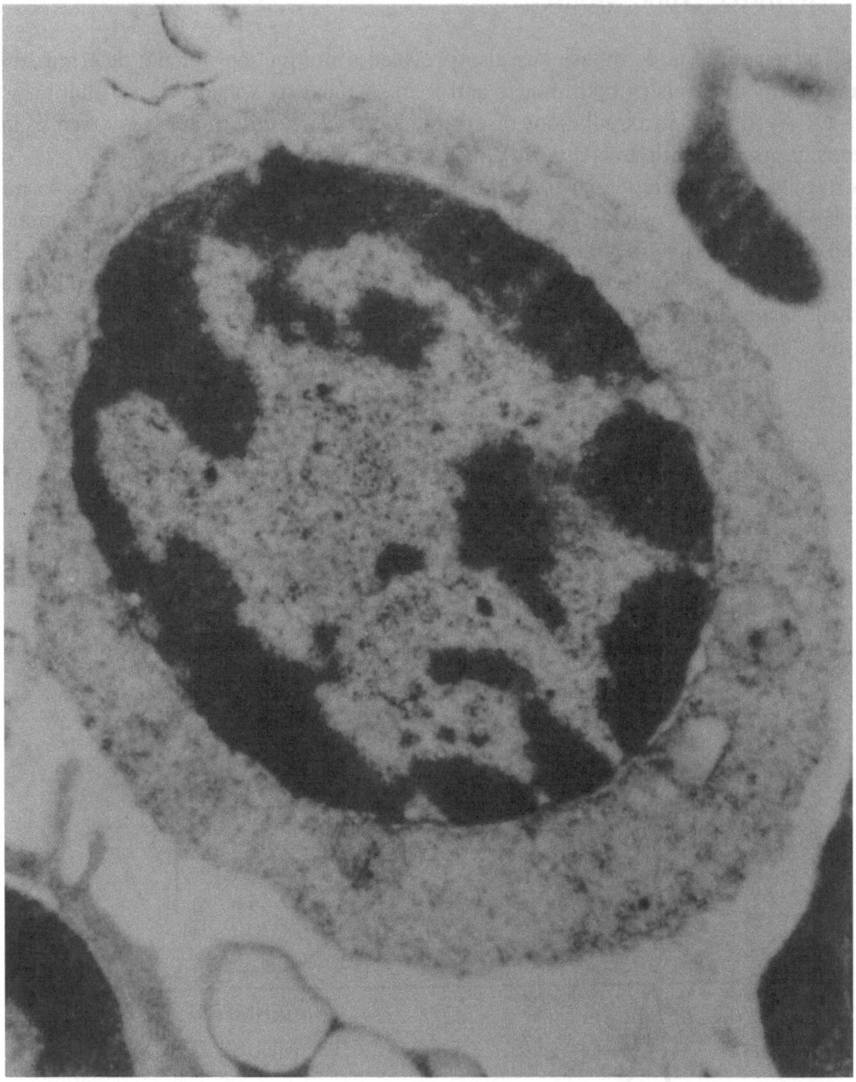

FIGURE 32. Ultrastructure of the null lymphocyte isolated from the spleens of germfree nude mice following nylon wool separation. [Beardsley, Bealmear, Twomey, and Lewis, unpublished observations. Published with permission.]

than 100 μg/ml, inhibits nonspecific T-lymphocyte induction via β-adrenergic receptors and increases the sensitivity of indicator cells to induction with thymopoietin (Goldstein *et al.,* 1975; Twomey *et al.,* 1977a,b). Thymopoietin is a peptide hormone of the thymus (Goldstein, 1974; Goldstein *et al.,* 1974). The pentapeptide portion of thymopoietin is believed to be the active component (Goldstein *et al.,* 1979). After induction incubation, 0.25×10^6 cells were incubated with 0.025 ml of antiserum to the Thy-1.2 antigen in the

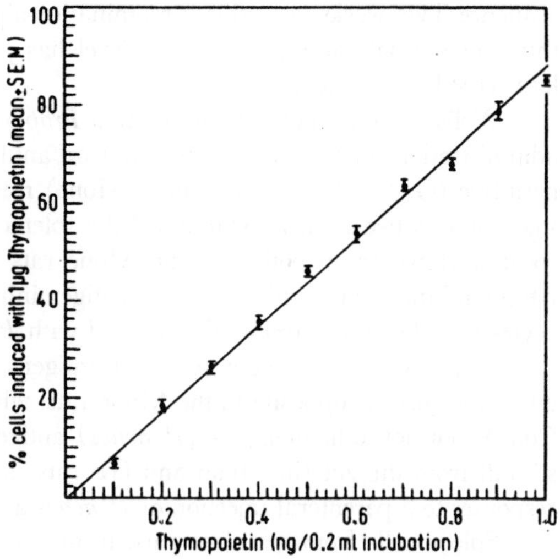

FIGURE 33. Relation between thymopoietin concentration and percentage of indicator cells induced to express Thy-1.2 antigen ($r = 0.997$). [From Twomey *et al.* (1977b). Published with permission.]

presence or absence of guinea pig complement for 1 hr. Then 0.625 mg of protease in 0.125 ml saline was added and incubation continued for an additional 30 min. Cell counts were made; injured cells were solubilized by incubation with protease, which was reflected in a lowering of cell counts (Stewart and Goldstein, 1974). Thy-1.2 induction was determined from the formula $(a - b)/a \times 100$, where a represents cell counts with heat-inactivated complement and b represents cell counts on paired incubations in the presence of complement. Optimal induction for individual indicator cell preparations was determined from incubations with 1 μg/ml of thymopoietin. Induction with test plasmas was first expressed as a percentage of optimal induction with 1 μg/ml of thymopoietin and finally as being equivalent to that amount of thymopoietin that induced a comparable percentage of Thy-1.2-antigen-positive cells (Fig. 33). In the range of activity studied, there was an excellent correlation between thymopoietin concentration and induction ($r = 0.997$). This assay has been used to evaluate thymic hormone activity in patients with immune deficiency disease (Lewis *et al.*, 1977), mycosis fungoides and Sézary syndrome (Safai *et al.*, 1979), and lymphoblastic leukemia (Twomey *et al.*, 1980), and to determine age-related changes in thymic hormone levels in normal individuals (Lewis *et al.*, 1978).

5. Germfree Rat

The serum protein patterns of 90-day-old germfree and conventional rats are shown in Fig. 34. Values of the serum protein fractions are given in Table VII. Significantly lower values in the alpha-2, beta, and gamma fractions are seen in germfree rats than in conventional rats (Wostmann, 1959). The deficit in these fractions is compensated for by a higher albumin level, which brings the total protein concentration of the serum (and thereby the colloid-osmotic value) to a value comparable to that of the conventional animals. If the germfree rat is removed from its germfree environment and associated with a "normal" cecal flora, the fractions alpha-2, beta, and gamma increase, one after the other, in this sequence. Two weeks after this contamination alpha-2 level has recovered to roughly two-thirds of normal values, the beta level has just started to increase, and the gamma level is still unchanged.

Wells and Balish (1979a) found that *Propionibacterium acnes* (oral and/or parenteral administration) had a modulating effect on antibody- and cell-mediated immunity in the germfree (GF) and monoassociated (Mono) rats. In conventional (CV) rats, parenteral injection of killed *P. acnes* stimulated the splenic hemolytic PFC response to sheep erythrocytes. However, in both GF and Mono rats (monoassociated with viable organisms), parenteral injection of killed *P. acnes* antigen inhibited the PFC response to SRBC. Splenocytes from GF rats parenterally injected with *P. acnes* had a decreased *in vitro* blastogenesis response to homologous bacterial antigens, when compared to GF control rats; the effect was just the opposite in the Mono rats, which had increased responses to PHA and Con A, but not to homologous *(P. acnes)* antigens. The presence or absence of antigenic stimuli from the gut (in Mono and GF rats) had a modulating effect on their immune responses to a parenteral injection of *P. acnes* antigen.

Spleen cells from germfree rats, from gnotobiotic rats associated with two *Pseudomonas* species (Mono), and from conventional rats gave no positive blastogenic responses when incubated with each of four LPS preparations from *E. coli,* with glycolipid extracted, from *Salmonella minnesota* R595, or with *S. minnesota* R595 lipid A (Wells and Balish,

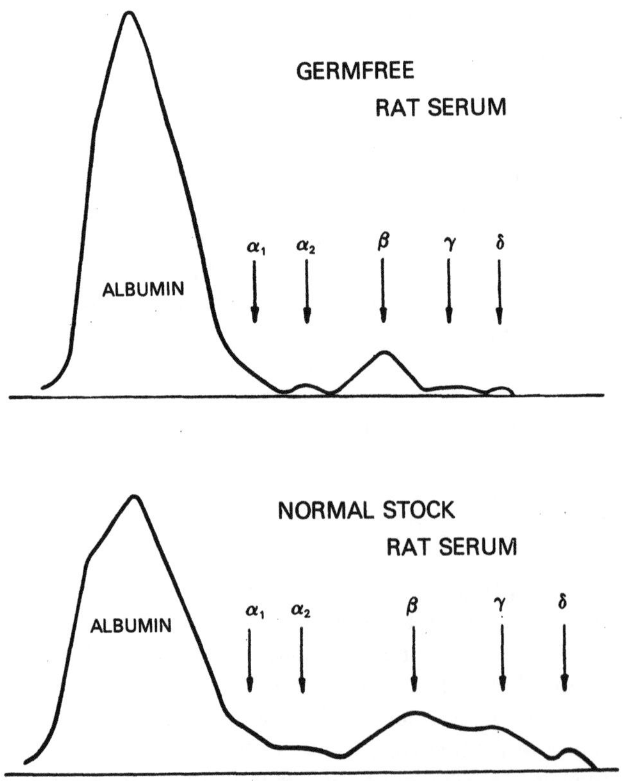

FIGURE 34. Electrophoretic patterns of the sera of 90-day-old germfree and conventional rats. [From Wostmann (1959). Published with permission.]

Table VII
Serum Protein Fractions of Germfree and Conventional 90-Day-Old Male Lobund Rats[a]

	Germfree (10)[b]	Conventional (9)	*p*
Total serum protein	5833 ± 163[c]	5961 ± 141	0.78
Albumin	5098 ± 179	4272 ± 136	0.01
Alpha-1 globulin	103 ± 14	135 ± 14	0.19
Alpha-2 globulin	69 ± 13	271 ± 41	0.01
Beta globulin	388 ± 48	804 ± 69	0.01
Gamma globulin	175 ± 28	479 ± 53	0.01

[a]From Wostmann (1959). Published with permission.
[b]Number of animals in parentheses.
[c]Values are mean (mg/100 ml) ± S.D.

1979b). Spleen preparations from nude mice, however, demonstrated a positive blastogenic response when incubated with *E. coli* LPS. Removal of adherent cells from GF and CV rat spleen cells did not increase the mitogenic activity of LPS for their nonadherent cells (0.5% esterase-positive cells). All rat spleen cells gave positive blastogenic responses to PHA and Con A. These data indicate that the rat may not be capable of responding to LPS as a mitogen, but perhaps this capacity is strain dependent as it is in the C3H/HeJ mice (Wells and Balish, 1979b) (see Chapter 10 of this volume).

McLeod and Balish (1978) assayed sera from Sprague–Dawley rats monoassociated with several different organisms, for homologous and cross-reacting antibodies against antigens on a variety of bacteria. A passive hemagglutination assay, with the bacterial antigens conjugated to tanned SRBC, was used to assess antibody titers. Both homologous and cross-reacting antibodies were detected in sera from the Mono animals. Sera from Mono rats had agglutinating antibodies to genera of bacteria that had not colonized their gastrointestinal tract. Although *S. aureus* shared some antigens with three species of *Clostridium,* the relationship was not reciprocal. Sera from rats monoassociated with *E. coli, P. mirabilis,* or *K. pneumoniae* had cross-reactive antibody against gram-positive and gram-negative bacteria, but did not have antibodies that could agglutinate *H. influenzae* type b. Monoassociation with *C. perfringens* resulted in the production of antibodies against a variety of gram-positive and gram-negative bacteria, but it did not induce cross-reactive antibodies against *Corynebacterium diphtheriae, streptococci* (groups A, C, and G), or *Staphylococcus epidermidis.*

Other characteristics of the germfree rat have been covered in earlier sections of this review on morphology (Section 2.1.1), cell-mediated immunity (Section 2.1.3a), and macrophages (Section 2.1.3c).

6. Germfree Guinea Pig

The failure of the germfree guinea pig to produce a delayed-type hypersensitivity (DTH) response has been discussed in Section 2.1.3a.

When the electrophoretic patterns of sera from adult germfree and conventional

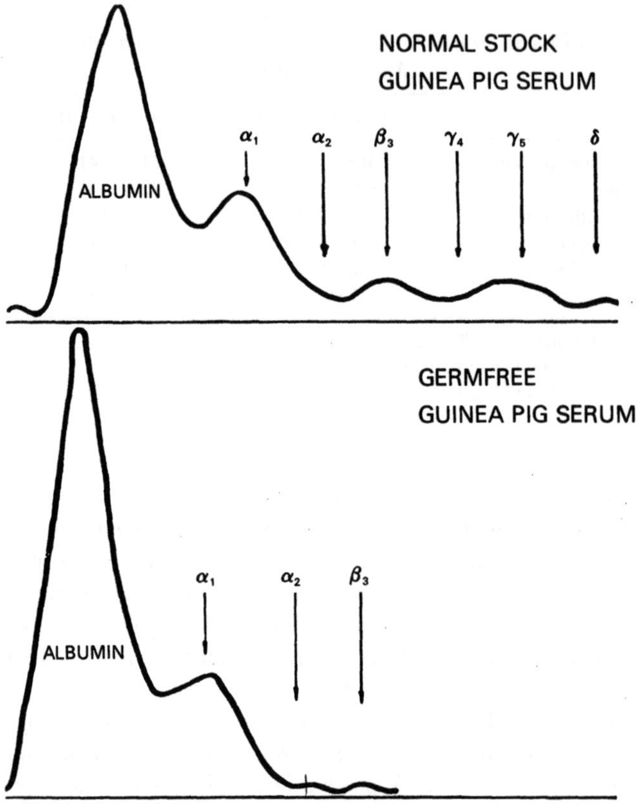

FIGURE 35. Electrophoretic patterns of the sera of adult germfree and conventional guinea pigs. [From Wostmann (1959). Published with permission.]

guinea pigs were determined by Wostmann (1959), he found that at least two of the slower moving (gamma) globulin fractions were missing. The beta fraction was present in most animals (Fig. 35), but it was present at lower levels than in conventional animals. This was the first species examined in which certain normally occurring serum protein fractions were completely missing in an animal with a germfree history.

In a subsequent age-related study of the serum proteins of germfree and conventional guinea pigs (Olson and Wostmann, 1964), it was determined that both microbial and dietary stimulation play an important role in determining the quantitative patterns of the serum globulins. The serum protein composition of germfree and conventional guinea pigs displayed significant differences during young adult life (2–4 months), at which time the alpha, beta, and gamma globulins were less than those of conventional animals. Immunoelectrophoretic analysis showed one component in the germfree gamma globulin range to be absent at this time. When animals were fed a diet containing only vegetable protein, there was little beta and no detectable gamma globulin in the sera of the adult germfree guinea pigs, but when the diet was composed of complex vegetable and animal proteins, larger amounts of globulin were formed. By 1 year of age, no significant differences were observed in germfree and conventional guinea pigs in their alpha, beta, or gamma fractions.

The germfree guinea pig has the unique advantage of being easily adapted to the germfree environment. As a newborn it does not require a period of suckling and it can be weaned at the time of cesarian derivation and fed a semisolid natural diet. This frees this species from the influence of maternal antigenic stimulation via the milk.

7. Germfree Rabbit

Germfree rabbits failed to reproduce in isolation until 1961, when the first litter of germfree rabbits was born at the LOBUND Laboratory at the University of Notre Dame. This first litter soon died, and it was not until 1963 that second-generation rabbits were weaned and reared to adulthood.

On the basis of free-boundary electrophoresis, rabbit serum can be separated into all the usual protein fractions (Wostmann, 1961). Only the gamma fraction is affected by the presence or absence of microflora. Most of the germfree rabbit sera tested have shown a trace of gamma globulin; it ranges from 0 to 2% of total serum proteins as compared with 10 to 12% in the sera of conventional animals. The cesarian-born germfree rabbit has an ample supply of maternal gamma globulin at birth, but there is no colostrum transfer and it rapidly disappears. In the presence of a microflora, it takes 4–5 weeks for the animal to produce an increased level of gamma globulin on its own. In the absence of a flora, a low minimum is reached and maintained.

Use of the germfree rabbit for immunologic studies is very limited, because of the expense involved in deriving and maintaining them. They are usually housed in steel isolators and must be housed singly; this presents obvious complications for any meaningful immunological work.

8. Germfree Primates

Although many species of laboratory animals have been successfully reared in a germfree environment, few nonhuman primates have been successfully reared in this system (Reyniers, 1952; Brant *et al.*, 1963; Wolfe *et al.*, 1966). A program to develop and maintain nonhuman primates has been developed at the Southwest Foundation for Research and Education in San Antonio, Texas (Kalter *et al.*, 1973).

In an early study, Berchelmann *et al.* (1975) observed changes in erythrocytes and leukocytes with age in baboons from birth through 11.5 months of age. The leukocyte alterations of these baboons are given in Table VIII. The total leukocyte counts were relatively stable, with slight fluctuations for the first 9 months, but changes in the differential were observed. The number of lymphocytes in cord blood at birth was greater than the number of segmented neutrophils; this condition persisted throughout the study. At 5 months, the number of lymphocytes doubled and remained at this level through the ninth month. Segmented neutrophils decreased in number after the second week, and at 5 months the number has greatly decreased. The number of monocytes was stable in the first 6 weeks but had diminished by the fifth month. At birth, the erythrocyte values of baboons main-

TABLE VIII

Leukocytic Changes (per mm^3) in Five Infant Baboons Maintained in a Germfree System from Birth through 11½ Months of Age[a]

Age	Number of observations	White blood cells	Segmented neutrophils	Lymphocytes	Monocytes	Eosinophils	Basophils	Stabs
Birth	5	5874[b]	2261.6	3359.9	211.4	23.4	11.7	0
		3680–9060[c]	662.4–5889.0	1104.0–6795.0	73.6–634.2	0–90.6	0–90.6	
2 weeks	5	7574	3044.7	3980.0	196.9	30.3	30.3	15.1
		4760–8580	999.6–4719.0	1761.2–6520.8	0–343.2	0–171.6	0–85.8	0–85.8
3 weeks	5	5140	1058.8	3680.2	272.4	82.2	20.4	15.4
		4040–6530	719.6–1828.4	2664.2–5028.1	121.2–522.4	0–315.9	0–65.3	0–65.3
4 weeks	5	6236	1883.2	4090.8	187.0	37.4	12.5	12.5
		4100–7990	574.0–3435.7	2173.0–6631.7	82.0–319.6	0–159.8	0–79.9	0–79.9
6 weeks	5	6220	2145.9	3700.9	217.7	31.1	0	0
		5530–6910	1880.2–2425.8	3152.1–4284.2	165.9–276.4	0–691.0		
5 months	5	6896	779.2	6067.0	34.0	50.0	68.9	34.5
		6170–7980	431.9–1117.2	5182.8–7102.2	0–79.8	0–167.6	0–79.8	0–79.8
7 months	2	7566	453.6	6900.1	37.8	75.6	37.8	37.8
		6540–8380	327.0–586.6	5820.6–7877.2	0–83.8	65.4–83.8	0–83.8	0–83.8
9 months	2	7395	369.7	6877.3	73.9	73.9	0	0
		7190–7600	71.9–684.0	6399.1–7372.0	0–152.0	0–152.0		
11½ months	1	3390	135.6	3152.7	67.8	33.9	0	0

[a]From Berchelmann *et al.* (1975). Published with permission.
[b]Average.
[c]Range.

TABLE IX
Erythrocytic Changes in Five Infant Baboons Maintained in a Germfree System from Birth through 11½ Months of Age[a]

Age	Number of observations	Hemoglobin (g/dl)	Packed cell volume (%)	RBC $\times 10^6/mm^3$	MCV (μm^3)	MCHC (%)
Birth	5	13.8[b]	40.3	4.21	95.7	31.7
		12.8–14.7[c]	40–47	3.53–5.95		
2 weeks	5	11.5	34.6	3.80	91.0	33.2
		10.8–12.1	32–38	3.51–4.19		
3 weeks	5	10.7	34.3	3.76	91.2	31.2
		10.2–11.2	33–37	3.32–4.12		
4 weeks	5	11.6	36.0	4.05	88.8	32.2
		11.4–12.1	34–38	3.94–4.19		
6 weeks	5	12.8	40.0	4.28	93.4	32.0
		12.5–13.2	38–42	4.32–4.65		
5 months	5	13.0	41.0	4.76	84.8	31.7
		12.1–13.5	39–42	4.48–5.08		
7 months	2	13.3	41.0	4.98	82.3	32.4
		13.2–13.5	40–42	4.88–5.08		
9 months	2	11.8	40.0	4.62	86.5	29.0
		11.2–12.5	37–43	4.33–4.92		
11½ months	1	13.2	41.0	4.59	89.1	32.1

[a]From Berchelmann *et al.* (1975). Published with permission.
[b]Average.
[c]Range.

tained in the germfree environment were very similar to those in the conventional environment (Table IX). At 2 weeks, the RBC, hemoglobin content, and packed cell volume (PCV) values of the germfree animals dropped below those of the conventional group. After 1 month the hemoglobin and PCV values of the germfree baboons increased and at 6 weeks surpassed the values of the conventional animals. They then remained at this level through the fifth month. The erythrocyte numbers of conventional animals increased and exceeded those of the germfree baboons shortly after birth. In the germfree animals the RBC values dropped in the first 2 weeks, then began to rise, and by the fifth month approached the levels seen in conventional animals. Erythrocytes of the conventional group were microcytic in comparison to those of the gnotobiotic group. The differences in erythrocyte values may possibly reflect the influence of nutritional as well as environmental factors (Berchelmann *et al.*, 1975). The lower numbers of leukocytes in the germfree infants with the subsequently decreased numbers of neutrophils may result from a lack of stimulation in their sterile environment. The sluggish increase in lymphocyte numbers in germfree infants could indicate a slower development of the immune system; this is in keeping with the observations on all other species of germfree animals.

As described in Section 2.1.3a, human dialyzable transfer factor was found to be capable of inducing *in vivo* and *in vitro* cellular immunity in one germfree chimpanzee and in two germfree baboons (Eichberg *et al.*, 1976a). These data supported the antigenic specificity of transfer factor, because the recipient primates developed cellular responses only against donor "positive" antigens but not against donor "negative" antigens.

The pathogenicity of a strain of simian herpesvirus, SA8, in 1-month-old conventional and germfree baboons was investigated by Eichberg *et al.* (1976b). Following intratracheal inoculation, there was mortality in one-fifth of the conventional baboons and in one-fourth of the germfree group. After 3 days, disease became apparent and was characterized by respiratory distress, reduced formula intake, weight loss, and fever in both groups. Isolation of herpesvirus from the respiratory tract, lymphoid organs, kidneys, adrenals, and central nervous system was more frequent by explant culturing than by routine procedures. There was a significant difference in total leukocyte counts, with higher values in the conventional than in the germfree animals, but the absolute number of lymphocytes was not different. The lower number of leukocytes was probably due to the low number of neutrophils in the germfree animals. Infection resulted in a leukopenia at 5 days postinfection and a leukocytosis at 10 days postinfection in both groups. The animals that died had acute necrotizing fibrinous pneumonia. Intranuclear inclusion bodies typical for herpesvirus were present. All the survivors had subacute interstitial pneumonia when sacrificed at 35 days postinfection.

9. Germfree Dogs

The first significant research with germfree dogs was performed by Griesemer (1963). Heneghan *et al.* (1968) reared germfree beagles to breeding age, and finally produced second-generation germfree beagles in April 1968 (Heneghan *et al.*, 1969).

The adult germfree dog is a valuable animal tool with great potential for immunologic investigations, particularly in the area of transplantation immunology. Little general information was known about this animal and no published data on their blood and serum chemistry values existed until the work of Yale and Balish in 1976. Blood values of germfree and conventionally raised beagle dogs are shown in Table X, and the blood serum chemistry values are shown in Table XI. The GF dogs had significantly lower erythrocyte, leukocyte, neutrophil, and band cell counts than either the confined conventional (CCV) or conventional dogs (CV). There were no significant differences in the monocyte counts among the three groups of animals. The lower leukocyte, neutrophil, and band cell counts in the GF dogs probably reflect the absence of subclinical bacterial infections which might frequently be in process in the CCV and CV dogs.

The GF dogs had significantly higher cholesterol and alkaline phosphatase values than the CCV or CV dogs; they had higher calcium and glucose, and lower serum glutamic oxalacetic acid levels than the CCV dogs (Table XI). All groups of dogs had similar blood urea nitrogen and creatinine values.

The GF dogs had significantly lower alpha-2 and gamma globulin values than the CCV or CV dogs, and they also had lower total protein and beta globulin levels than the CCV dogs. There were no statistically significant differences in the albumin determinations in any of the groups of dogs. Most of the deviations of the GF beagles from the control blood, serum chemistry, and serum protein values could be anticipated and/or explained by the absence of a microflora (Yale and Balish, 1976).

TABLE X
Blood Values of Germfree and Adult Conventionally Raised Beagle Dogs[a]

Parameter	Age (months)	Germfree			Confined conventional			Conventional			Significance[b]
		No.	Mean	S.D.	No.	Mean	S.D.	No.	Mean	S.D.	
Packed cell volume (%)	5	7	37.8	3.07	50	41.6	2.37	21	43.1	2.15	(1) $p < 0.001$; (2a) $p < 0.001$; (2b) $p < 0.001$; (3a) $p < 0.001$; (4) $p < 0.001$
	>5	18	44.4	2.59							
Hemoglobin (g/dl)	5	7	13.8	1.16	50	14.6	1.02	21	15.4	0.71	(1) $p < 0.001$; (2b) $p < 0.001$; (3a) $0.001 < p < 0.005$; (4) $p < 0.001$
	>5	18	16.2	0.88							
Red blood cells (10^6/mm^3)	5	7	5.52	0.50	50	6.81	0.49	20	6.70	0.42	(1) $p < 0.001$; (2a) $p < 0.001$; (2b) $0.001 < p < 0.005$; (3a) $p < 0.001$; (3b) $0.025 < p < 0.05$
	>5	18	6.43	0.37							
White blood cells (10^3/mm^3)		25	6.14	1.09	50	7.68	1.90	21	8.34	2.48	(2) $p < 0.001$; (3) $p < 0.001$
Neutrophils (10^3/mm^3)		25	3.61	0.88	49	5.24	1.84	19	5.37	2.42	(2) $p < 0.001$; (3) $p < 0.001$
Band cells (10^3/mm^3)		25	0.10	0.12	49	0.44	0.50	19	0.42	0.74	(2) $p < 0.001$; (3) $0.01 < p < 0.025$
Eosinophils (10^3/mm^3)		25	0.17	0.13	49	0.11	0.12	19	0.34	0.29	(2) $0.01 < p < 0.025$; (3) $0.005 < p < 0.01$; (4) $p < 0.001$
Lymphocytes (10^3/mm^3)	5	7	2.64	0.77	49	1.63	0.55	19	2.22	0.72	(1) $0.001 < p < 0.005$; (2a) $p < 0.001$; (3b) $0.025 < p < 0.05$; (4) $p < 0.001$
	>5	18	1.82	0.52							
Monocytes (10^3/mm^3)		25	0.19	0.11	49	0.23	0.17	19	0.25	0.17	

[a]From Yale and Balish (1976). Published with permission.
[b]Probability that compared items are the same is stated if $p < 0.05$. Data on germfree dogs 5 months and greater than 5 months of age are combined if $p > 0.05$; but if $p < 0.05$, then (a) denotes a comparison with germfree dogs 5 months old, and (b) denotes a comparison with germfree dogs over 5 months old; (1) compares germfree dogs 5 months and greater than 5 months of age; (2) compares germfree and confined conventional dogs; (3) compares germfree and conventional dogs; (4) compares confined conventional and conventional dogs.

TABLE XI

Blood Serum Chemistry Values of Germfree and Adult Conventionally Raised Beagle Dogs[a]

Parameter	Age (months)	Germfree			Confined conventional			Conventional			Significance[b]
		No.	Mean	S.D.	No.	Mean	S.D.	No.	Mean	S.D.	
Calcium (mg/dl)		32	10.15	0.57	81	9.65	0.88	19	9.76	0.78	(2) $p < 0.001$
Phosphorus, inorganic (mg/dl)	5	7	5.95	0.52	81	3.15	1.10	19	2.75	1.02	(1) $p < 0.001$; (2a) $p < 0.001$; (3a) $p < 0.001$
	>5	25	3.05	0.79							
Blood urea nitrogen (mg/dl)		32	15.1	4.1	80	15.6	5.2	19	19.4	10.8	
Creatinine (mg/dl)		14	0.67	0.20	39	0.72	0.16	9	0.62	0.34	
Uric acid (mg/dl)	5	7	0.24	0.07	89	0.55	0.26	22	0.57	0.21	(1) $p < 0.001$; (2a) $p < 0.001$; (3a) $p < 0.001$
	>5	32	0.50	0.15							
Glucose (mg/dl)		39	174	35.6	88	136	36.1	22	148	38.9	(2) $p < 0.001$
Cholesterol (mg/dl)		39	182	45.6	88	148	38.2	20	138	21.8	(2) $p < 0.001$; (3) $p < 0.001$
Bilirubin (mg/dl)		39	0.10	0.08	76	0.14	0.19	22	0.08	0.08	(4) $0.025 < p < 0.05$
Serum glutamic oxalacetic acid (mU/ml)		39	52.5	26.2	88	71.4	48.9	22	54.5	18.1	(2) $0.005 < p < 0.01$; (4) $0.01 < p < 0.02$
Lactic dehydrogenase (mU/ml)	>5	24	82.9	39.4	50	156	138	13	86.8	19.2	(2b) $0.005 < p < 0.01$; (4) $0.025 < p < 0.05$
Alkaline phosphatase (mU/ml)	5	7	120	16.6	82	38.1	15.0	19	40.7	11.8	(1) $p < 0.001$; (2a) $p < 0.001$; (2b) $p < 0.001$; (3a) $0.001 < p < 0.005$; (3b) $p < 0.001$; (4) $0.01 < p < 0.02$
	>5	25	69.0	17.1							

[a]From Yale and Balish (1976). Published with permission.
[b]See footnote[b] Table X.

10. Germfree Chickens

The first germfree chickens to be raised successfully were maintained by Cohendy (1912). Germfree chickens also provide a special type of germfree animal because they can be produced by passing the eggs through a germicidal bath prior to hatching in the isolator (Luckey, 1959).

The typical serum electrophoretic pattern of a 3-month-old germfree chicken (Fig. 36) demonstrates that the difference between germfree and conventional chickens is observed in the gamma globulin fraction, as has been noted in all species of germfree animals. Table XII shows that the gamma globulin values of germfree chickens remain low as the animal matures, while conventional chickens show a steady increase with time until about 30% of their total serum protein is gamma globulin. A rise in various antibody titers against coli-

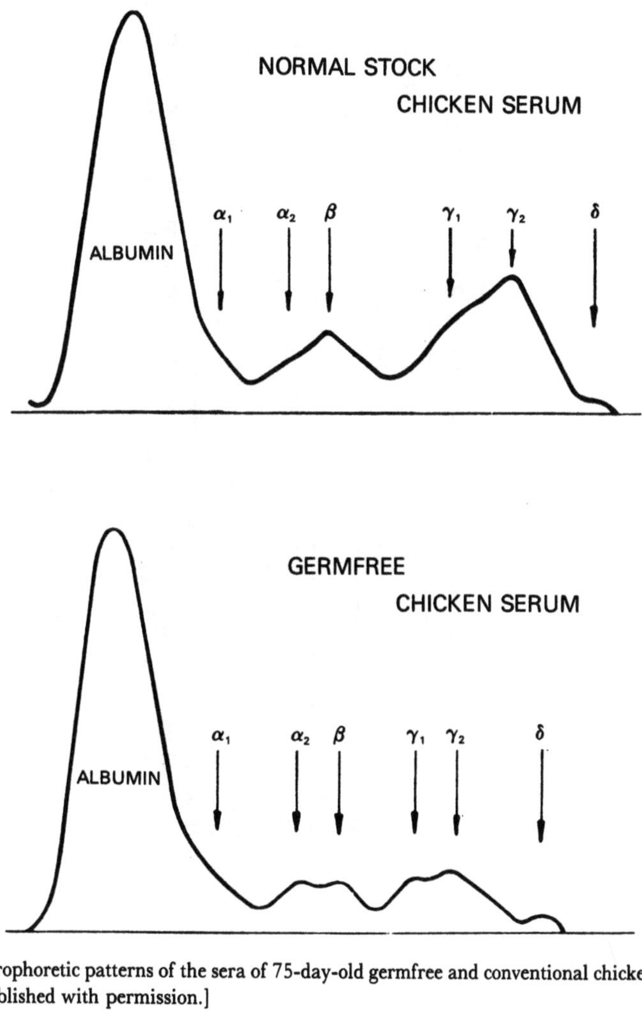

FIGURE 36. Electrophoretic patterns of the sera of 75-day-old germfree and conventional chickens. [From Wostmann (1959). Published with permission.]

TABLE XII

Serum Protein Fractions in Germfree (GF) and Conventional (CV) Chickens (Sexes Mixed)[a,b]

Age (days)	Albumin		Alpha-1 globulin		Alpha-2 plus beta globulin		Gamma globulin	
	GF	CV	GF	CV	GF	CV	GF	CV
35	2775 ± 94 (8)[c]	2315 ± 89 (9)	102 ± 8	95 ± 24	353 ± 31	332 ± 21	331 ± 21	443 ± 23
75	2673 ± 79 (5)	2623 ± 115 (11)	170 ± 31	108 ± 19	306 ± 27	341 ± 25	321 ± 43	800 ± 55
125	2753 ± 141 (4)	2450 ± 72 (10)	160 ± 50	106 ± 10	442 ± 31	506 ± 31	397 ± 20	1171 ± 77

[a]From Wostmann (1959). Published with permission.
[b]Chickens became contaminated with a pure culture of *Micrococcus* sp. 1–6 days before sacrifice.
[c]Values are mean (mg/100 ml) ± S.D. The number of experimental animals is indicated in parentheses.

forms and micrococci parallels the rise in gamma globulin with age (Wostmann, 1959). In a study on antibody production to BSA, Wostmann and Olson (1964) demonstrated that germfree chickens produced detectable precipitating antibody by the sixth day, but that peak titers were reached at approximately the same time as in the conventional chickens; the conventional chickens produced detectable antibody by the fifth day. The germfree chicken demonstrates an approximately 24-hr shift to the right in its immune response curve, as has been observed in most species of germfree animals. After peak titers had been reached, both gamma globulin content and antibody titers decreased, but the decline in antibody titer was much slower in the germfree than in the conventional animals. The half-life of the gamma globulin fraction showed no difference in metabolic degradation between the two groups. The magnitude of the response in the germfree chickens was less than that of the conventional chickens, with the peak titers for the germfree chickens being only half that for conventional chickens. Despite the fact that there are quantitative differences in their responses, the germfree chicken is still able to generate an adequate antibody response to this antigen.

Wostmann and Olson (1969) demonstrated that germfree chickens showed a prolonged formation of primary anti-BSA and anti-human gamma globulin (HGG) after peak concentrations had been reached. This was indicated by the half-value times of the antibody concentration, which in the germfree chickens showed 60–80% higher values than in conventional controls. Treatment of the BSA-treated germfree chickens with HGG suppressed the otherwise prolonged antibody production supposedly by competing for potential immunocompetent cells.

11. Conclusion

Germfree animals of all the species studied demonstrate a reduced development of the lymphatic system in the peripheral lymphoid organs, particularly when these organs are close to the gastrointestinal tract and would normally be influenced by the microflora or microbial products in the conventional state. The degree of developmental retardation is diet dependent and is maximal in the colostrum-deprived germfree piglet and the other species maintained on the low-molecular-weight, water-soluble, antigen-free diet. All species of germfree animals have reduced or absent immunoglobulins; this, too, is diet dependent. When immunized with specific antigens or stimulated with nonspecific mitogens, all species have been capable of an adequate response, with the exception of DTH in the germfree guinea pig. There is usually a shift to the right of approximately 24–48 hr in any response curve because of an increase in the lag phase. Once the germfree animal has responded, the response usually reaches a greater magnitude and is prolonged for a greater time period than that observed in the conventional counterpart of the species being tested.

Germfree animals have already made valuable contributions to basic immunology, especially in the area of the ontogeny of the immune response and in the demonstration of the role of the thymus in the immune response. It should be emphasized that when the thymus was removed from germfree mice during the first 18–24 hr of life, these animals lived out a normal life span of 24 months, demonstrated no diminution of reproductive capacity, and were capable of responding to antigenic challenge when given enough time to respond; this response was one of lower magnitude, however. These older germfree thymectomized mice were even capable of rejecting transplants of allogeneic neoplastic cells. These findings indicate that the thymus is not essential for life, unless the animal is maintained in an environment where it is subject to a constant onslaught of pathogenic microorganisms.

Much is still to be gained from the use of this unique animal tool. The studies in the future that can profit most from the use of an animal with little or no environmental antigenic stimulation are: (1) additional investigations of the ontogeny of the immune response with emphasis on the role of the thymus in the development of subpopulations of immunocompetent cells and their interactions; (2) identification and further characterization of NK and K cells; (3) bone marrow transplantation and amelioration or elimination of GVHD; (4) research on the influence of diet on immune mechanisms; (5) studies of host–parasite relationships and the regulatory influence of the microflora on these relationships; (6) evaluation of immune-deficient animal models, reconstitution of these deficient animals, the subsequent reevaluation of the reconstituted animal, and its return to a "normal" conventional environment; (7) carcinogenesis studies with oral carcinogens in such cases where an organism or organisms in the microflora may alter the carcinogen and render it harmless or may enhance its carcinogenicity; and (8) studies on dental caries and periodontal disease and the development of good therapeutic measures for controlling these diseases of bacterial origin.

ACKNOWLEDGMENTS

The author is grateful to the following members of the Association for Gnotobiotics, who have provided generous help in the preparation of this manuscript by helping to collect material and/or reading the manuscript: Drs. E. Balish, P. Carter, J. B. Heneghan, O. A. Holtermann, Y. B. Kim, J. Pleasants, M. Pollard, P. Trexler, M. Wagner, B. S. Wostmann, and C. E. Yale. Gratitude is also extended to staff members in the Medical Photography Department, Mail Room, and Print Shop at Roswell Park Memorial Institute, Buffalo, New York, for their generous help.

References

Abrams, G. D., Bauer, H., and Sprinz, H., 1963, Influence of the normal flora on mucosal morphology and cellular renewal in the ileum. A comparison of germfree and conventional mice, *Lab. Invest.* **12**:355.

Alexander, P., 1957, *Atomic Radiation and Life,* Penguin Press, London.

Andersson, J., Sjöberg, O., and Möller, G., 1972, Induction of immunoglobulin and antibody synthesis *in vitro* by lipopolysaccharides, *Eur. J. Immunol.* **2**:349.

Argyris, B. F., and Horowitz, M., 1976, Cell interaction in tumor allograft immunity, *Transplantation* **22**:112.

Arnason, B. G., Salomon, J. C., and Grabar, P., 1964, Anticorps antifoie et immunoglobulines seriques chez les souris classiques et axeniques; étude comparative après necrose hepatique aigue provoquée par le tetrachlorure de carbone, *C.R. Acad. Sci.* **259**:4882.

Asofsky, R., and Hylton, M. B., 1967, Some characteristics of a secretory immunoglobulin (IgA) in germfree and conventional mice, Symposium on Gnotobiotic Research, Madison, Wisconsin, p. 6.

Asofsky, R., Ikari, N. S., and Hylton, M. B., 1968, The relationship of specific antigenic stimulation to serum

IgM level in germfree mice, in: *Advances in Germfree Research and Gnotobiology* (M. Miyakawa and T. D. Luckey, eds.), pp. 219–224, Chemical Rubber Co., Cleveland.

Balish, E., Yale, C. E., and Hong, R., 1973, Influence of a defined flora on the serum proteins of gnotobiotic rats, in: *Germfree Research: Biological Effect of Gnotobiotic Environments* (J. B. Heneghan, ed.), pp. 485–492, Academic Press, New York.

Barclay, M., Calathes, D. N., Di Lorenzo, J. C., Helper, A., and Kaufman, R. J., 1959, The relation between plasma lipoproteins and breast carcinoma: Effect of degrees of breast disease on plasma lipoproteins and possible role of lipid metabolic aberration, *Cancer* **12:**1163.

Barry, T. A., 1967, Alterations in serum iron in germfree and conventional mice after whole-body X-irradiation, Doctoral dissertation, University of Notre Dame, Notre Dame, Indiana.

Barry, T. A., and Wilson, R., 1967, A semi-automated bioassay for the determination of bacterial endotoxin, in: *Automation in Analytical Chemistry, 1966,* pp. 246–249, Technicon, Inc.

Bauer, H., 1968, Part II: Cellular defense mechanisms, in: *The Germfree Animal in Research* (M. E. Coates, ed.), pp. 211–226, Academic Press, New York.

Bauer, H., Horowitz, R. E., Levenson, S. M., and Popper, H., 1963, The response of the lymphatic tissue to the microbial flora. Studies on germfree mice, *Am. J. Pathol.* **42:**471.

Bauer, H., Horowitz, R. E., Watkins, K. C., and Popper, H., 1964, Immunologic competence and phagocytosis in germfree animals with and without stress, *J. Am. Med. Assoc.* **187:**715.

Bauer, H., Paronetto, F., Porro, R. F., Einheber, A., and Popper, H., 1967, The influence of the microbial flora on liver injury and associated serum γ-globulin elevation. A study in germfree rats treated with 3′-methyl-4-dimethyl-amino-azobenzene, *Lab. Invest.* **16:**847.

Bealmear, P. M., 1965, Comparative study of the thymus of germfree and conventional CFW mice, Doctoral dissertation, University of Notre Dame, Notre Dame, Indiana.

Bealmear, P. M., 1975, Amelioration of graft-vs-host disease in germfree allogeneic chimeras by pre-treatment of the bone marrow with anti θ serum, Vth International Symposium on Gnotobiology, Stockholm, p. 18.

Bealmear, P. M., and Hale, L., 1977, Mitogenic response of germfree CFW and C3H/He mice, 15th Annual Meeting of the Association for Gnotobiotics, Saranac Lake, New York, No. 7.

Bealmear, P. M., and Wilson, R., 1966, Histological comparison of the thymus of germfree (axenic) and conventional CFW mice, *Anat. Rec.* **154:**261.

Bealmear, P. M., and Wilson, R., 1967a, Antibody responses of neonatally thymectomized germfree mice to *Salmonella typhimurium* H antigen, *Exp. Hematol.* **12:**37.

Bealmear, P. M., and Wilson, R., 1967b, Histological comparison of the thymus of germfree and conventional CFW mice and the lymphoid organs of thymectomized germfree and conventional CFW mice, Symposium on Gnotobiotic Research, Madison, Wisconsin, p. 27.

Bealmear, P. M., and Wilson, R., 1967c, Homograft rejection by neonatally thymectomized germfree mice, *Cancer Res.* **27:**358.

Bealmear, P. M., and Wilson, R., 1967d, The leukocyte counts of germfree neonatally thymectomized CFW mice, *Blood* **30:**112.

Bealmear, P. M., and Wilson, R., 1968, Changes in transferrin-[^{59}Fe] levels in the serum of germfree mice after whole-body X-irradiation, *Radiat. Res.* **35:**489.

Bealmear, P. M., Loughman, B. E., Nordin, A. A., and Wilson, R., 1973, Evidence for graft vs. host reaction in the germfree allogeneic radiation chimera, in: *Germfree Research: Biological Effect of Gnotobiotic Environments* (J. B. Heneghan, ed.), pp. 471–476, Academic Press, New York.

Bealmear, P. M., Mirand, E. A., and Holtermann, O. A., 1979, The influence of dietary lipids on immunocompetence, 17th Annual Meeting of the Association for Gnotobiotics, New York, New York, p. 40.

Beardsley, T., 1976, Factors affecting *in vitro* immune responsiveness of germfree allogeneic radiation chimeras, Doctoral dissertation, Baylor College of Medicine, Houston, Texas.

Berchelmann, M. L., Eichberg, J. W., and Kalter, S. S., 1975, Hemograms of gnotobiotic and conventional infant baboons *(Papio cynocephalus)*, *Lab. Anim. Sci.* **25:**621.

Bergendi, L., Ferenčík, M., Mandel, L., Kovářů, F., and Štefanovič, J., 1979, The proteolytic enzymes in leukocytes and spleens of conventional and germfree piglets, *Folia Microbiol. (Prague)* **24:**70.

Binns, R. M., and Hall, J. G., 1966, The paucity of lymphocytes in the lymph of unanesthetized pigs, *Br. J. Exp. Pathol.* **47:**275.

Bomford, R., and Christie, G. H., 1975, Mechanisms of macrophage activation by *Corynebacterium parvum.* II. *In vivo* experiments, *Cell. Immunol.* **17:**150.

Bradley, T. R., and Metcalf, D., 1966, The growth of mouse bone marrow cells *in vitro, Aust. J. Exp. Biol. Med.* **44**:287.

Brant, H. G., Kundzins, W., and Reese, W. H., 1963, The gnotobiotic primate. Procedure involving adaptations and maintenance in a germfree environment, *Lab. Anim. Sci.* **13**:557.

Burnet, F. M., 1959, *The Clonal Selection Theory of Acquired Immunity,* Vanderbilt University Press, Nashville, Tennessee.

Byrne, P., Heit, H., and Heit, W., 1979, Characterization of granulocyte macrophage progenitor cells of germfree mice by buoyant density analysis, in: *Clinical and Experimental Gnotobiotics* (T. M. Fliedner, H. Heit, D. Neithammer, and H. Pflieger, eds.), pp. 52–57, Gustav Fischer Verlag, Stuttgart.

Chakhava, O. V., 1973, Ascorbic acid and glucocorticoid levels in relation to underdeveloped lymphatic tissue in germfree guinea pigs, in: *Germfree Research: Biological Effect of Gnotobiotic Environments* (J. B. Heneghan, ed.), pp. 501–506, Academic Press, New York.

Chang, C. F., and Pollard, M., 1973, Effects of microbial flora on levels of colony stimulating factor in serums of irradiated CFW mice, *Proc. Soc. Exp. Biol. Med.* **144**:177.

Christie, G. H., and Bomford, R., 1975, Mechanisms of macrophage activation by *Corynebacterium parvum.* I. *In vitro* experiments, *Cell. Immunol.* **17**:141.

Cohendy, M., 1912, Expériences sur la vie sans microbes, *Ann. Inst. Pasteur* **26**:106.

Cohendy, M., and Wollman, E., 1914, Expériences sur la vie sans microbes. Élevage aseptique de cobayes, *Compt. Rend.* **158**:1283.

Connell, S. J., and Wilson, R., 1965, The treatment of x-irradiated germfree CFW and C3H mice with isologous and homologous bone marrow, *Life Sci.* **4**:721.

Considine, R. G., and Starr, T. J., 1967, Role of the reticuloendothelial system in interferon tolerance, *J. Reticuloendothelial Soc.* **4**:315.

Coutinho, A., and Möller, G., 1974, Evidence for "one nonspecific triggering signal" not delivered by the Ig receptors, *Scand. J. Immunol.* **3**:133.

Coutinho, A., Möller, G., and Richter, W., 1974, Molecular basis of B-cell activation. I. Mitogenicity of native and substituted dextrans, *Scand. J. Immunol.* **3**:321.

Curtiss, L. K., DeHeer, D. H., and Edgington, T. S., 1980, Influence of the immunoregulatory serum lipoprotein LDL-In on the *in vivo* proliferation and differentiation of antigen-binding and antibody-secreting lymphocytes during a primary immune response, *Cell. Immunol.* **49**:1.

Dauphinee, M., 1973, Cellular immune response in radiation induced chimeras, Doctoral dissertation, University of Notre Dame, Notre Dame, Indiana.

Decker, J. M., Kim, Y. B., and Miller, A., 1974, Binding of β-galactosidase by cells from germfree colostrum-deprived piglets, *J. Immunol.* **113**:1599.

DeHarven, E., 1964, Virus particles in the thymus of conventional and germfree mice, *J. Exp. Med.* **120**:857.

DeSomer, P., and Billiau, A., 1966, Interferon production by the spleen after intravenous injection of sindbis virus or heat-killed *Escherichia coli, Arch. Gesamte Virusforsch.* **19**:143.

Despommier, D. D., Kajima, M., and Wostmann, B. S., 1967, Ferritin-conjugated antibody studies on the larvae of *Trichinella spiralis, J. Parasitol.* **53**:618.

Dietschy, J. M., Solomon, H. S., and Siperstein, M. D., 1966, Bile acid metabolism. I. Studies on the mechanisms of intestinal transport, *J. Clin. Invest.* **45**:832.

Dixon, J. M. S., 1960, The fate of bacteria in the small intestine, *J. Pathol. Bacteriol.* **79**:131.

Dixon, J. M. S., and Paulley, J. W., 1963, Bacteriological and histological studies of the small intestine of rats treated with mecamylamine, *Gut* **4**:169.

Doll, J. P., 1962, Rate of carbon clearance in three strains of germfree mice, *Am. J. Physiol.* **203**:291.

Downey, R. J., and Pisano, J. C., 1968, A functional role for an alpha-l-glycoprotein from inflammatory exudate, in: *Advances in Germfree Research and Gnotobiology* (M. Miyakawa and T. D. Luckey, eds.), pp. 225–232, Chemical Rubber Co., Cleveland.

Eichberg, J. W., Steele, R. W., Kalter, S. S., Kniker, W. T., Heberling, R. L., Eller, J. J., and Rodriguez, A. R., 1976a, Cellular immunity in gnotobiotic primates induced by transfer factor, *Cell. Immunol.* **26**:114.

Eichberg, J. W., McCullough, B., Kalter, S. S., Thor, D. E., and Rodriguez, A. R., 1976b, Clinical, virological, and pathological features of herpesvirus SA 8 infection in conventional and gnotobiotic baboons *(Papio cynocephalus), Arch. Virol.* **50**:255.

Fahey, J. L., and Sell, S., 1965, The immunoglobulins of mice. V. The metabolic (catabolic) properties of five immunoglobulin classes, *J. Exp. Med.* **122**:41.

Ferenčík, M., Bergendi, L., Mandel, L., Kovářů, F., Štefanovič, J., and Ďuračková, Z., 1979, The lysosomal enzymes and antibacterial factors in leukocytes, serum and spleens of conventional and germfree piglets, *Folia Microbiol. (Prague)* **24**:70.

Fichtelius, K. E., Laurell, G., and Philipsson, L., 1961, The influence of thymectomy on antibody formation, *Acta Pathol. Microbiol. Scand.* **51**:81.

Foster, R., Metcalf, D., Robinson, W. A., and Bradley, T. R., 1968, Bone marrow colony stimulating activity in human sera. Results of two independent surveys in Buffalo and Melbourne, *Br. J. Haematol.* **15**:147.

Fray, A., Lorinet, A. M., and Halpern, B., 1978, *In vitro* antitumor properties of peritoneal exudate cells of conventional and germ-free mice. International Symposium on Biological Preparations in the Treatment of Cancer, London 1977, *Dev. Biol. Stand.* **38**:241.

Fry, R. J. M., and Staffeldt, E., 1964, Effect of a diet containing sodium deoxycholate on the intestinal mucosa of the mouse, *Nature (London)* **203**:1396.

Gershon, R. K., 1975, The importance of suppressor cells in immunoregulation, in: *Suppressor Cells in Immunity,* International Symposium, May, 1975 (S. K. Singhal and N. R. St. C. Sinclair, eds.), pp. 1–11, University of Western Ontario, London, Ontario.

Gilbert, R. P., 1960, Mechanism of the hemodynamic effects of endotoxin, *Physiol. Rev.* **40**:245.

Glimstedt, G., 1936, Bakterienfreie meerschweinchen, aufzucht, lebensfähigkeit und wachstum, nebst untersuchungen über das lymphytische gewebe, *Acta Pathol. Microbiol. Scand. Suppl.* **30**:1.

Goldstein, G., 1974, Isolation of bovine thymin: A polypeptide hormone of the thymus, *Nature (London)* **247**:11.

Goldstein, G. H., Scheid, M., Hammerling, U., Boyse, E. A., Schlesinger, D. H., and Niall, H. D., 1975, Isolation of a polypeptide that has lymphocyte-differentiating properties and is probably represented universally in living cells, *Proc. Natl. Acad. Sci. USA* **72**:11.

Goldstein, G., Scheid, M. P., Boyse, E. A., Schlesinger, D. H., and Wauwe, J. W., 1979, A synthetic pentapeptide with biological activity characteristic of the thymic hormone thymopoietin, *Science* **204**:1309.

Good, R. A., 1954, Agammaglobulinemia—A Provocative Experiment of Nature, Bull. Univ. Minnesota Hosp. and Minnesota Med. Found., p. 26.

Good, R. A., and Varco, R. L., 1955, A clinical and experimental study of agammaglobulinemia, *Lancet* **75**:245.

Gordon, H. A., 1959, Morphological and physiological characterization of germfree life, *Ann. N.Y. Acad. Sci.* **78**:208.

Gordon, H. A., and Pesti, L., 1971, The gnotobiotic animal as a tool in the study of host microbial relationships, *Bacteriol. Rev.* **35**:390.

Gordon, H. A., and Wostmann, B. S., 1960, Morphological studies on the germfree albino rat, *Anat. Rec.* **137**:65.

Gordon, H. A., Bruckner-Kardoss, E., and Wostmann, B. S., 1966, Aging in germ-free mice: Life tables and lesions observed at natural death, *J. Gerontol.* **21**:380.

Green, F., and Balish, E., 1977, Cell-mediated immunity in germfree guinea pigs infected with *Trichophyton mentagrophytes,* 15th Annual Meeting of the Association for Gnotobiotics, Saranac Lake, New York, No. 11.

Greenstein, J. P., Birnbaum, S. M., Winitz, M., and Otey, M. C., 1957, Quantitative nutritional studies with water soluble, chemically defined diets, *Arch. Biochem. Biophys.* **72**:396.

Griesemer, R. A., 1963, The gnotobiotic dog, *Lab. Anim. Care* **13**:643.

Griscelli, C., and Durandy, A., 1976, T and B cell chimerism in two patients with severe combined immunodeficiency (SCID) after transplantation, in: *Transplantation Abstracts* (M. Stephenson and A. Brzytwa, ed.), p. 18, Grune and Stratton, New York.

Grosch, D. S., 1965, *Biological Effects of Radiations,* Ginn (Blaisdell), Boston.

Gustafsson, B., 1948, Germfree rearing of rats: General technique, *Acta Pathol. Microbiol. Scand. Suppl.* **73**:1.

Gustafsson, B. E., and Laurell, C. B., 1960, Properdin titers in sera from germfree rats, *Proc. Soc. Exp. Biol. Med.* **105**:598.

Gustafsson, B. E., Bergstrom, S., Lindstedt, S., and Norman, A., 1957, Turnover and nature of fecal bile acids in germfree and infected rats fed cholic acid 24-^{14}C bile acid steroids, *Proc. Soc. Exp. Biol. Med.* **94**:467.

Gustafsson, B. E., Midtvedt, T., and Norman, A., 1966, Isolated fecal organisms capable of 7 d-dehydroxylating bile acids, *J. Exp. Med.* **123**:413.

Hashimoto, K., Onishi, N., Tajiri, T., Okada, Y., Maeda, R., Sugihara, Y., and Sasaki, S., 1973, Viral enteric infection and local IgA antibody: Studies in germfree mice thymectomized and/or infected with a mouse adenovirus, in: *Germfree Research: Biological Effect of Gnotobiotic Environments* (J. B. Heneghan, ed.), pp. 501–506, Academic Press, New York.

Hashimoto, M., and Hashimoto, N., 1968, Histological study of bone marrow in breeding rats, in: *Advances in Germfree Research and Gnotobiology* (M. Miyakawa and T. D. Luckey, eds.), pp. 149–161, Chemical Rubber Co., Cleveland.

Havell, E. A., Holtermann, O. A., and Starr, T. J., 1970, The influence of the indigenous bacterial flora upon the interferon response of mice, *J. Gen. Virol.* **9**:93.

Heise, E. R., and Myrvik, Q. N., 1966, Levels of lysosomal hydrolases in alveolar and peritoneal macrophages from conventional and germfree rats, *Fed. Proc.* **25**:439.

Heneghan, J. B., Longoria, S. G., and Cohn, I., Jr., 1968, Maintenance and growth of gnotobiotic beagle dogs, in: *Advances in Germfree Research and Gnotobiology* (M. Miyakawa and T. D. Luckey, eds.), pp. 63–67, Chemical Rubber Co., Cleveland.

Heneghan, J. B., Longoria, S. G., and Cohn, I., Jr., 1969, Reproduction in germfree beagles, in: *Advances in Experimental Biology and Medicine* (E. A. Mirand and N. Back, eds.), Vol. 3, pp. 367–371, Plenum, New York.

Herberman, R. B., and Holden, H. T., 1978, Natural cell-mediated immunity, in: *Advances in Cancer Research,* Vol. 27 (G. Klein and S. Weinhouse, eds.), pp. 305–377, Academic Press, New York.

Hill, M. J., and Drasar, B. S., 1967, Bacterial degradation of bile salts, *Biochem. J.* **104**:55.

Ho, M., 1964, Interferon-like viral inhibitor in rabbits after intravenous administration of endotoxin, *Science* **146**:1022.

Horowitz, R. E., and Bauer, H., 1967, Response of the lymphatic tissue to bacterial antigen, Symposium on Gnotobiotic Research, Madison, Wisconsin, p. 5.

Horowitz, R. E., and Bauer, H., 1968, Immunologic consequences of irradiation in the germfree mouse, in: *Advances in Germfree Research and Gnotobiology* (M. Miyakawa and T. D. Luckey, eds.), pp. 233–238, Chemical Rubber Co., Cleveland.

Hurlemann, J., Thorbecke, G. J., and Hochwald, G. M., 1966, The liver as the site of C-reactive protein formation, *J. Exp. Med.* **123**:365.

Iljima, S., and Yamane, T., 1968, The spleen of germfree animals, in: *Advances in Germfree Research and Gnotobiology* (M. Miyakawa and T. D. Luckey, eds.), pp. 139–148, Chemical Rubber Co., Cleveland.

Isa, A. M., Sanders, B. R., and Parham, C. A., 1976, T lymphocyte suppressor activity in the cytotoxic response to teratocarcinoma 402Ax in the mouse, in: *Transplantation Abstracts* (M. Stephenson and A. Brzytwa, eds.), p. 53, Grune & Stratton, New York.

Jankovic, B. D., Waksman, B. H., and Arnason, B. G., 1962, Role of the thymus in immune reactions in rats. I. The immunologic response to bovine serum albumin (antibody formation, Arthus reactivity, and delayed hypersensitivity) in rats thymectomized or splenectomized at various times after birth, *J. Exp. Med.* **116**:159.

Jarošková, L., Trebichavský, I., Říha, I., Kovářů, F., and Holub, M., 1973, Immunoglobulin determinants on lymphocytes in germfree piglets, *Eur. J. Immunol.* **1**:818.

Jones, J. M., 1970, A study of secondary disease in gnotobiotic mouse radiation chimeras, Doctoral dissertation, University of Notre Dame, Notre Dame, Indiana.

Jones, J. M., Wilson, R., and Bealmear, P. M., 1971, Mortality and gross pathology of secondary disease in germfree mouse radiation chimeras, *Radiat. Res.* **45**:577.

Joshi, J. H., Entringer, M. A., and Robinson, W. A., 1979, Bacterial stimulation of serum colony-stimulating activity and neutrophil production in germ-free mice, *Proc. Soc. Exp. Biol. Med.* **162**:44.

Kajima, M., and Pollard, M., 1965, Detection of virus-like particles in germfree mice, *J. Bacteriol.* **90**:1448.

Kalter, S. S., Eichberg, J., and Heberling, R. L., 1973, Raising germfree baboons *(Papio cynocephalus)*, in: *Germfree Research: Biological Effect of Gnotobiotic Environments* (J. B. Heneghan, ed.), pp. 611–613, Academic Press, New York.

Kellogg, T. F., and Wostmann, B. S., 1967, Rat and mouse stock diet L 485, *Gnotobiotics Newsletter* **3**:1.

Kim, Y. B., 1975, Developmental immunity in the piglet, in: *Immunodeficiency in Man and Animals* (D. Bergsma, R. A. Good, and J. Finstad, eds.), pp. 549–557, Sinauer Assoc., Sunderland, Mass.

Kim, Y. B., 1979, Role of antigen in ontogeny of the immune response, *Microbiology* **1979**:343.

Kim, Y. B., and Watson, D. W., 1966, Role of antibodies in reactions to gram-negative bacterial endotoxins, *Ann. N.Y. Acad. Sci.* **133**:733.

Kim, Y. B., and Watson, D. W., 1968, Antigenic competition in the true primary immune response in germfree colostrum-deprived piglets, *Bacteriol. Proc.* **68**:75.

Kim, Y. B., and Watson, D. W., 1971, Histological changes of lymphoid tissues in relation to the ontogeny of the immune response in germfree piglets, in: *Advances in Experimental Medicine and Biology* (K. Lindahl-Kiessling, G. Alm, and M. G. Hanna, Jr., eds.), Vol 12, pp. 169–177, Plenum, New York.

Kim, Y. B., Bradley, S. G., and Watson, D. W., 1965, Antibody synthesis in germfree colostrum-deprived miniature piglets, in: *Swine in Biomedical Research,* pp. 273–284, Battle Memorial Institute and U.S. Atomic Energy Commission, Richland, Wash.

Kim, Y. B., Bradley, S. G., and Watson, D. W., 1966a, Ontogeny of the immune response. I. Development of immunoglobulins in germfree and conventional colostrum-deprived piglets, *J. Immunol.* **97**:52.

Kim, Y. B., Bradley, S. G., and Watson, D. W., 1966b, Ontogeny of the immune response. II. Characterization of 19SγG- and 7SγG-immunoglobulins in the true and secondary responses in piglets, *J. Immunol.* **97**:189.

Kim, Y. B., Bradley, S. G., and Watson, D. W., 1967a, Ontogeny of the immune response. III. Characterization of x-component in germfree, colostrum-deprived piglets, *J. Immunol.* **98**:868.

Kim, Y. B., Bradley, S. G., and Watson, D. W., 1967b, Ontogeny of the immune response. IV. The role of antigen elimination in the true primary immune response in germfree, colostrum-deprived piglets, *J. Immunol.* **99**:320.

Kim, Y. B., Bradley, S. G., and Watson, D. W., 1968a, Ontogeny of the immune response. V. Further characterization of 19SγG- and 7SγG-immunoglobulins in the true primary immune response in germfree, colostrum-deprived piglets, *J. Immunol.* **101**:224.

Kim, Y. B., Bradley, S. G., and Watson, D. W., 1968b, 19SγG and 7SγG antibody synthesis in germfree colostrum-deprived piglets, in: *Advances in Germfree Research and Gnotobiology* (M. Miyakawa and T. D. Luckey, eds.), pp. 208–218, Chemical Rubber Co., Cleveland.

Kim, Y. B., Setcavage, T. M., Kim, D. J., Chun, H. G., and Scheffel, J. W., 1979, Ontogenic development and differentiation of the immune system in the gnotobiotic miniature swine, in: *Clinical and Experimental Gnotobiotics* (T. M. Fliedner, H. Heit, D. Neithammer, and H. Pflieger, eds.), pp. 203–213, Gustav Fischer Verlag, Stuttgart.

Kim, Y. B., Huh, N. D., Koren, H. S., and Amos, D. B., 1980, Natural killing (NK) and antibody-dependent cellular cytotoxicity (ADCC) in specific pathogen-free (SPF) miniature swine and germfree piglets. I. Comparison of NK and ADCC, *J. Immunol.* **125**:755.

Kiyono, H., McGhee, J. R., and Michalek, S. M., 1980, Lipopolysaccharide regulation of the immune response: Comparison of responses to LPS in germfree, *Escherichia coli*-monoassociated and conventional mice, *J. Immunol.* **124**:36.

Koren, H. S., Amos, D. B., and Kim, Y. B., 1978, Natural killing and antibody-dependent cellular cytotoxicity are independent immune functions in the Minnesota miniature swine, *Proc. Natl. Acad. Sci. USA* **75**:5127.

Kovářů, F., Štěpánková, R., Tlaskalová, H., and Kruml, J., 1975, The development and haemopoietic organs in germfree and conventional rabbits, Vth International Symposium on Gnotobiology, Stockholm, p. 20.

Kovářů, F., Štěpánková, R., Kruml, J., Mandel, L., and Kenig, E., 1979, Development of lymphatic and haemopoietic organs in germfree models, *Folia Microbiol. (Prague)* **24**:32.

Kovářů, H., Kovářů, F., and Mandel, L., 1979a, Membrane-bound enzyme of lymphoid tissues. I. ATPase and γ-glutamyl transpeptidase activities of intact cells during ontogenetic development of germfree and conventional pigs, *Folia Microbiol. (Prague)* **24**:51.

Kovářů, H., Kovářů, F., and Mandel, L., 1979b, Membrane-bound enzymes of lymphoid tissues. II. Mitogen and antigen effects on ATPase and γ-glutamyl transpeptidase activities of intact cells of germfree and conventional pigs, *Folia Microbiol. (Prague)* **24**:75.

Krumbhaar, E. B., 1948, A hematopoietic perifollicular envelope in the rat spleen, *Blood* **3**:953.

Kruml, J., Ludvík, J., Trebichavský, I., Mandel, L., and Kovářů, F., 1969, Morphology of germfree piglets, *Folia Microbiol. (Prague)* **14**:441.

Lapp, W. S., Elie, R. E., and Parthenais, E., 1975, Suppressor cells in graft-versus-host reactions, in: *Suppressor Cells in Immunity,* International Symposium, May, 1975 (S. K. Singhal and N. R. St. C. Sinclair, eds.), pp. 127–135, University of Western Ontario, London, Ontario.

Laqueur, G. L., 1964, Carcinogenic effects of cycadineal and cycasin, methylazoxymethanol glycoside in rats and effects of cycasin in germfree rats, *Fed. Proc.* **23**:1386.

Lelievre, L., Paraf, A., Charlemagne, D., and Sheppard, J. R., 1977, Plasma membrane studies on drug sensitive and resistant cell lines. I. Cross resistance and membrane enzyme coordination (ouabain/cAMP/Na^+K^+-ATPase/adenylate cyclase), *Exp. Cell Res.* **104**:191.

Lesher, S., Walburg, H. E., and Sacher, G. A., 1964, Generation cycle in the duodenal crypt cells of germfree and conventional mice, *Nature (London)* **202**:884.

Lewis, V., Twomey, J. J., and Bealmear, P. M., 1976, A new bioassay for thymic hormone, *Fed. Proc.* **35:**591.

Lewis, V., Twomey, J. J., Goldstein, G., O'Reilly, R., Smithwick, E., Pahwa, R., Pahwa, S., Good, R. A., Schultwisserman, H., Horowitz, S., Hong, R., Jones, J., Sieber, O., Kirkpatrick, C., Polmar, S., and Bealmear, P. M., 1977, Circulating thymic hormone activity in congenital immunodeficiency, *Lancet* **2:**471.

Lewis, V. M., Twomey, J. J., Bealmear, P. M., Goldstein, G., and Good, R. A., 1978, Age, thymic involution and circulating thymic hormone activity, *J. Clin. Endocrinol. Metab.* **47:**145.

Lindstedt, S., and Samuelsson, B., 1959, Bile acids and steroids. LXXXIII. On the interconversion of cholic and deoxycholic acid in the rat, *J. Biol. Chem.* **234:**2025.

Loughman, B. E., 1972, A study of the humoral immune response of mouse radiation chimeras: Thymus cell function in allogeneic radiation chimeras, Doctoral dissertation, University of Notre Dame, Notre Dame, Indiana.

Loughman, B. E., Nordin, A. A., and Bealmear, P. M., 1973, Studies of the immunological capacity of germfree mouse radiation chimeras. I. Chimerism and humoral immune response, *Cell. Immunol.* **9:**104.

Luckey, T. D., 1959, Nutrition and biochemistry of germfree chicks, *Ann. N.Y. Acad. Sci.* **78:**127.

Luckey, T. D., 1968, Gnotobiology and aerospace systems, in: *Advances in Germfree Research and Gnotobiology* (M. Miyakawa and T. D. Luckey, eds.), pp. 315–353, Chemical Rubber Co., Cleveland.

McCombs, C., Michalski, J. P., and Talal, N., 1976, Cellular interactions in lymphocyte proliferation: Effect of syngeneic and xenogeneic macrophages, *Cell. Immunol.* **23:**283.

MacDonald, T. T., and Carter, P. B., 1978, Contact sensitivity in the germ-free mouse, *J. Reticuloendothelial Soc.* **24:**287.

MacDonald, T. T., and Carter, P. B., 1979, Requirement for a bacterial flora before mice generate cells capable of mediating the delayed hypersensitivity reaction to sheep red blood cells, *J. Immunol.* **122:**2624.

MacDonald, T. T., Simkins, S. G., Spitalny, G. L., and Carter, P. B., 1978, Macrophage activation in the germ-free mouse, 16th Annual Meeting of the Association for Gnotobiotics, Madison, Wisconsin, p. 10.

McGhee, J. R., Kiyono, H., Michalek, S. M., Babb, J. L., Rosenstreich, D. L., and Mergenhagen, S. E., 1980, Lipopolysaccharide (LPS) regulation of the immune response: T lymphocytes from normal mice suppress mitogenic and immunogenic responses to LPS, *J. Immunol.* **124:**1603.

McLaughlin, M. M., Dacquisto, M. P., Jacobus, D. P., and Horowitz, R. E., 1964, Effects of the germfree state on responses of mice to wholebody irradiation, *Radiat. Res.* **23:**333.

MacLean, L. D., Zak, S. J., Varco, R. L., and Good, R. A., 1957, The role of the thymus and spleen in antibody production: An experimental study of the immune response in thymectomized rabbits, *Transplant. Bull.* **4:**21.

McLeod, J. C., and Balish, E., 1978, Homologous and cross-reacting antibodies in the sera of gnotobiotic rats, *Can. J. Microbiol.* **24:**365.

Mahmood, T., and Robinson, W. A., 1978, Granulocyte modulation of endotoxin-stimulated colony-stimulating (CSA) production, *Blood* **51:**879.

Mandel, L., Trebichavský, I., Kovářů, F., Trávníček, J., Talafantová, M., and Prokešová, L., 1979, The influence of a non-antigenic, synthetic diet on germfree piglets, *Folia Microbiol. (Prague)* **24:**27.

Matsuzawa, T., and Wilson, R., 1965, The intestinal mucosa of germfree mice after wholebody x-irradiation with 3 kiloroentgens, *Radiat. Res.* **25:**15.

Meltzer, M. S., 1976, Tumoricidal responses *in vitro* of peritoneal macrophages from conventionally housed and germ-free nude mice, *Cell. Immunol.* **22:**176.

Metcalf, D., and Foster, R., 1967, Bone marrow colony-stimulating activity of serum from mice with virus-induced leukemia, *J. Natl. Cancer Inst.* **39:**1235.

Metcalf, D., and Foster, R. S., 1969, In vitro colony-forming cells in the bone marrow of germfree mice, in: *Germ-Free Biology, Experimental and Clinical Aspects* (E. A. Mirand and N. Back, eds.), pp. 383–388, Plenum Press, New York.

Metcalf, D., and Wahren, B., 1968, Bone marrow colony stimulating activity of sera in mononucleosis, *Br. Med. J.* **3:**99.

Metcalf, D., Foster, R., and Pollard, M., 1967, Colony stimulating activity of serum from germfree normal and leukemic mice, *J. Cell. Physiol.* **70:**131.

Micklem, H. S., and Asfi, C., 1971, Lymphoid cells reacting against 'self' and their possible role in immune responses, in: *Morphological and Functional Aspects of Immunity* (K. Lindahl-Kiessling, G. Alm, and M. G. Hanna, Jr., eds.), pp. 57–61, Plenum Press, New York.

Miler, I., Mandel, L., and Morávek, F., 1971, The differences in the course of postirradiation gastrointestinal syndrome in newborn conventional and germfree piglets, *Folia Microbiol.* **16**:537.

Miller, J. F. A. P., 1963, Role of the thymus in immunity, *Br. Med. J.* **2**:459.

Miller, R. G., Clark, D. A., Harvey, E., and Phillips, R. A., 1975, Quantitation of killer T cell–target cell interaction: Suppressor cells at the effector stage, in: *Suppressor Cells in Immunity*, International Symposium, May, 1975 (S. K. Singhal and N. R. St. C. Sinclair, eds.), pp. 50–60, University of Western Ontario, London, Ontario.

Mirand, E. A., and Grace, J. T., Jr., 1963, Response of germfree mice to Friend virus, *Nature (London)* **200**:92.

Mirand, E. A., and Hoffman, J. G., 1973, The enhancement of bone marrow colony-stimulating factor produced by Friend virus (FV-P) infection in germfree mice, in: *Germfree Research: Biological Effect of Gnotobiotic Environments* (J. B. Heneghan, ed.), pp. 165–174, Academic Press, New York.

Mishell, R. I., and Dutton, R. W., 1967, Immunization of dissociated spleen cell cultures from normal mice, *J. Exp. Med.* **126**:423.

Miyakawa, M., 1959, The lymphatic system of germfree guinea pigs, *Ann. N.Y. Acad. Sci.* **78**:221.

Miyakawa, M., Iljima, S., Kobayashi, R., and Tosima, M., 1957, Observations on the lymphoid tissue of the germfree guinea pig, *Acta Pathol. Jpn.* **7**:183.

Miyakawa, M., Kishimoto, H., Itaya, J., Uei, Y., and Kashio, T., 1958, Homotransplantation experiment in germ-free animals, *Acta Pathol. Jpn.* **8**:177.

Miyakawa, M., Sumi, Y., and Uno, Y., 1973, Spontaneous tumors in Japan-born germfree rats, in: *Germfree Research: Biological Effect of Gnotobiotic Environments* (J. B. Heneghan, ed.), pp. 123–130, Academic Press, New York.

Moore, M. A. S., 1979, Hematopoietic stem cells, 17th Annual Meeting of the Association for Gnotobiotics, New York, New York, p. 10.

Moreau, C., Ducluzeau, R., and Guy-Grand, D., 1977, Effect of various bacterial strains in the stimulation of the development of IgA plasmocytes in the digestive mucosa of gnotobiotic mice, 15th Annual Meeting of the Association for Gnotobiotics, Saranac Lake, New York, No. 3.

Morley, A. P., Quesenberry, P., Bealmear, P. M., Stohlman, F., Jr., and Wilson, R., 1972, Serum colony stimulating factor levels in irradiated germfree and conventional CFW mice, *Proc. Soc. Exp. Biol. Med.* **140**:478.

Nair, P. P., Gordon, M., Gordon, S., Reback, J., and Mendeloff, A. I., 1965, The cleavage of bile acid conjugates by cell-free extracts from *Clostridium perfringens*, *Life Sci.* **4**:1887.

Newton, W. L., Pennington, R. M., and Lieberman, J. E., 1960, Comparative hemolytic complement activities of germfree and conventional guinea pig serum, *Proc. Soc. Exp. Biol. Med.* **104**:486.

Nielsen, A. H., Chapman, A. L., Cohen, H., and Werder, A. A., 1967, Long term oncogenic study in axenic and ex-axenic mice, Symposium on Gnotobiotic Research, Madison, Wisconsin, p. 26.

Nielsen, A. H., Amelunxen, R. E., Kornhaus, J., Sheek, M., and Werder, A. A., 1973, Natural transmission of leukemia and mammary tumors in conventional and germfree CFW_w mice, in: *Germfree Research: Biological Effect of Gnotobiotic Environments* (J. B. Heneghan, ed.), pp. 137–142, Academic Press, New York.

Norman, A., and Grubb, R., 1956, Hydrolysis of conjugated bile acids by *Clostridia* and *Enterococci*, *Acta Pathol. et Microbiol. Scand.* **35**:537.

Norman, A., and Sjovall, J., 1958, On the transformation and enterohepatic circulation of cholic acid in the rat, *J. Biol. Chem.* **233**:872.

Novogrodsky, A., Tate, S., and Meister, A., 1976, γ-Glutamyl transpeptidase, a lymphoid cell-surface marker: Relationship to blastogenesis, differentiation, and neoplasia, *Proc. Natl. Acad. Sci. USA* **73**:2414.

Ohwaki, M., Yasutake, N., Yasui, H., and Ogura, R., 1977, A comparative study on the humoral immune responses in germfree and conventional mice, *Immunology* **32**:43.

Olson, G. B., and Wostmann, B. S., 1964, Electrophoretic and immunoelectrophoretic studies of the serum of germfree and conventional guinea pigs, *Proc. Soc. Exp. Biol. Med.* **116**:914.

Olson, G. B., and Wostmann, B. S., 1966a, Cellular and humoral immune response of germfree mice stimulated with 7S HGG or *Salmonella typhimurium*, *J. Immunol.* **97**:275.

Olson, G. B., and Wostmann, B. S., 1966b, Lymphocytopoiesis, plasmacytopoiesis and cellular proliferation in nonantigenically stimulated germfree mice, *J. Immunol.* **97**:267.

Pasteur, L., 1885, Observations relatives á la note précédente de M. Declaux, *C.R. Acad. Sci.* **100**:68.

Perkins, E. H., Nettesheim, P., Morita, L., and Walburg, H. E., Jr., 1967, The engulfing potential of peritoneal phagocytes of conventional and germfree mice, in: *The Reticuloendothelial System and Atherosclerosis* (N. R. DiLuzio and R. Paoletti, eds.), pp. 175–187, Plenum Press, New York.

Phillips, B. P., Wolfe, P. A., and Gordon, H. A., 1959, Studies on rearing the guinea pig germfree, *Ann. N.Y. Acad. Sci.* **78:**183.

Pleasants, J. R., 1968, Characteristics of the germfree animal, in: *The Germfree Animal in Research* (M. E. Coates, ed.), pp. 113–125, Academic Press, New York.

Pleasants, J. R., Reddy, B. S., and Wostmann, B. S., 1964, Chemically defined, water soluble diets adequate for growth and reproduction of germfree mice, *Lab. Anim. Sci.* **14:**324.

Pleasants, J. R., Reddy, B. S., and Wostmann, B. S., 1968, Sudden death in germfree mice reared through successive generations on chemically defined liquid diet, in: *Germ-Free Biology: Experimental and Clinical Aspects* (E. A. Mirand and N. Back, eds.), pp. 307–315, Plenum Press, New York.

Podoprigora, G. I., and Zaitsev, T. I., 1979, Morphological and functional status of elements of the reticuloendothelial system of gnotobiotic animals, *Folia Microbiol. (Prague)* **24:**55.

Pollard, M., 1963, Chemical induction of mammary cancer in germfree rats, *Nature (London)* **200:**1289.

Pollard, M., 1964, Germfree animals and biological research, *Science* **145:**247.

Pollard, M., 1967a, Germinal centers in germfree animals, in: *Germinal Centers in Immune Responses* (H. Cottier, R. Odartchenko, C. Schindler, and C. Congdon, eds.), pp. 343–346, Springer-Verlag, New York.

Pollard, M., 1967b, Host responses to neoplastic diseases, *Proc. Inst. Med. Chicago* **26:**9.

Pollard, M., 1976, Lesions in aging germfree Wistar rats, 14th Annual Meeting of the Association for Gnotobiotics, Notre Dame, Indiana, p. 35.

Pollard, M., and Kajima, M., 1966, Leukemia in germfree rats, *Proc. Soc. Exp. Biol. Med.* **121:**585.

Pollard, M., and Kajima, M., 1967, Leukemia induced by 7,12-dimethylbenz (α)-anthracene in germfree rats, *J. Natl. Cancer Inst.* **39:**135.

Pollard, M., and Kajima, M., 1970, Lesions in aged germfree Wistar rats, *Am. J. Pathol.* **61:**25.

Pollard, M., and Luckert, P. J., 1977, Strain susceptibilities to radiation-induced mammary adenocarcinoma (CA) in germfree (GF) and conventional (Conv.) rats, 15th Annual Meeting of the Association for Gnotobiotics, Saranac Lake, New York, No. 22.

Pollard, M., and Matsuzawa, T., 1964, Radiation-induced leukemia in germfree mice, *Proc. Soc. Exp. Biol. Med.* **116:**967.

Pollard, M., and Matsuzawa, T., 1966, Chemical prevention of radiation-induced leukemia in mice, *Proc. Soc. Exp. Biol. Med.* **122:**539.

Pollard, M., and Nordin, A., 1971, Immune capabilities of germ-free animals, in: *Progress in Immunology* (B. Amos, ed.), pp. 1295–1297, Academic Press, New York.

Pollard, M., and Salomon, J. C., 1963, Oncogenic effect of methylcholanthrene in new-born germfree mice, *Proc. Soc. Exp. Biol. Med.* **112:**256.

Pollard, M., Matsuzawa, T., and Salomon, J. C., 1964, Inductions of neoplasms in germfree rodents by 3-methylcholanthrene, *J. Natl. Cancer Inst.* **33:**93.

Pollard, M., Sharon, N., and Teah, B. A., 1968, Congenital lymphocytic choriomeningitis virus infection in gnotobiotic mice, *Proc. Soc. Exp. Biol. Med.* **127:**755.

Quastler, H., 1956, The nature of intestinal radiation death, *Radiat. Res.* **4:**303.

Quesenberry, P. J., Bealmear, P., Ryan, M., and Stohlman, F., Jr., 1974a, Serum colony-stimulating and inhibiting factors in irradiated germfree and conventional mice, *Br. J. Haematol.* **28:**531.

Quesenberry, P., Morley, A., Bealmear, P., Niskanen, E., Halperin, J., Ryan, M., Howard, D., and Stohlman, F., Jr., 1974b, Studies on endotoxin-induced colony stimulating factor elevations, in: *Proceedings of the Second International Workshop on Hemopoiesis in Culture at Airlie House, 1973,* pp. 126–134.

Quesenberry, P., Cohen, H., Levin, J., Sullivan, R., Bealmear, P. M., and Ryan, M., 1978, The effects of bacterial infection and irradiation on serum colony-stimulating factor levels in tolerant and non-tolerant CF_1 mice, *Blood* **51:**229.

Ranken, R., Wilson, R., and Bealmear, P. M., 1969, The reduction in the life span of the intestinal mucosal cells of germfree mice by cholic acid, in: *The Germfree Animal as a Tool in Research,* p. 39, NATO Advanced Study Institute, Leuven, Belgium.

Ranken, R., Wilson, R., and Bealmear, P. M., 1971, Increased turnover of intestinal mucosal cells of germfree mice induced by cholic acid, *Proc. Soc. Exp. Biol. Med.* **138:**270.

Reddy, B. S., 1979, Colon carcinogenesis in germfree Fischer rats with 1,2-dimethylhydrazine (DMH), azoxymethane (AOM), and *N*-methyl-*N'*-nitro-*N'*-nitrosoguanidine (MNNG), 17th Annual Meeting of the Association for Gnotobiotics, New York, New York, p. 38.

Reddy, B. S., Watanabe, K., Louis, Y. M., Highland, K., and Choi, C., 1977, Colon cancer: Bile salts (BA) as tumor promoters in germfree (GF) and conventional (CONV) rats, 15th Annual Meeting of the Association for Gnotobiotics, Saranac Lake, New York, No. 23.

Reddy, B. S., Louis, Y., Haisley, A., and Choi, C., 1979, Effect of intestinal microflora and dietary fat on 3,2'-dimethyl-4-aminobiphenyl (DMAB)-induced colon and breast carcinogenesis in F344 rats, 17th Annual Meeting of the Association for Gnotobiotics, New York, New York, p. 21.

Reyniers, J. A., 1946, Germfree Life Applied to Nutrition Studies, Lobund Reports No. 1, p. 87, University of Notre Dame Press, Notre Dame, Ind.

Reyniers, J. A., 1952, Rearing of cesarean-born *M. rhesus* monkey under sterile conditions, *J. Bacteriol.* **43:**778.

Reyniers, J. A., 1957, Life in a germfree world, *Cincinnati J. Med.* **38:**417.

Reyniers, J. A., and Sacksteder, M. R., 1959, Tumorigenesis and the germfree chicken: A preliminary report, *Ann. N.Y. Acad. Sci.* **78:**328.

Rich, R. R., and Rich, S. S., 1976, Suppression of mixed lymphocyte reactions by alloantigen-activated spleen-localizing thymocytes, *Cell. Immunol.* **22:**358.

Richerson, H. B., Dvorak, H. F., and Leskowitz, S., 1969, Cutaneous basophil hypersensitivity: A new interpretation of the Jones–Mote reaction, *J. Immunol.* **103:**1431.

Robinson, W. A., Metcalf, D., and Bradley, T. R., 1967, Stimulation by normal and leukemic mouse sera of colony formation *in vitro* by mouse bone marrow cells, *J. Cell. Physiol.* **69:**83.

Robinson, W. A., Entringer, M. A., Bolin, R. W., and Stonington, O. G., 1977, Bacterial stimulation and granulocyte inhibition of granulopoietic factor production, *N. Engl. J. Med.* **297:**1129.

Rolls, B., Hedge, S. N., Turvey, A., and Coates, M. E., 1975, Some characteristics of germfree and conventional chickens, Vth International Symposium on Gnotobiology, Stockholm, p. 3.

Sacksteder, M. R., Kasza, L., Palmer, J. L., and Warren, J., 1973, Cell transformations in germfree Fischer rats, in: *Germfree Research: Biological Effect of Gnotobiotic Environments* (J. B. Heneghen, ed.), pp. 153–158, Academic Press, New York.

Safai, B., Good, R. A., Twomey, J. J., Lewis, V. M., and Goldstein, G., 1979, A novel lymphocyte differentiating factor in serum of patients with mycosis fungoides and Sézary syndrome, *Blood* **54:**837.

Salomon, J. C., 1968, Carcinogenesis in axenic animals, in: *The Germ-Free Animal in Research* (M. Coates, ed.), pp. 228–235, Academic Press, New York.

Scheffel, J. W., and Kim, Y. B., 1979, Role of environment in the development of "natural" hemagglutinins in Minnesota miniature swine, *Infect. Immun.* **26:**202.

Scott, M. T., MacDonald, T. T., and Carter, P. B., 1978, *C. parvum* in germfree mice, *Cancer Immunol. Immunother.* **4:**135.

Sell, S., 1964a, γ-Globulin metabolism in germfree guinea pigs, *J. Immunol.* **92:**559.

Sell, S., 1964b, Immunoglobulins of the germfree guinea pig, *J. Immunol.* **93:**122.

Sell, S., 1967, Immunization and the immunoglobulins of the germfree guinea pig, Symposium on Gnotobiotic Research, Madison, Wisconsin, p. 7.

Sell, S., and Fahey, J. L., 1964, Relationship between γ-globulin metabolism and low serum γ-globulin in germfree mice, *J. Immunol.* **93:**81.

Setcavage, T. M., and Kim, Y. B., 1976a, Characterization of porcine serum immunoglobulins IgG, IgM and IgA and the preparation of monospecific anti-chain sera, *Immunochemistry* **13:**643.

Setcavage, T. M., and Kim, Y. B., 1976b, Variability of the immunological state of germfree colostrum-deprived Minnesota miniature piglets, *Infect. Immun.* **13:**600.

Setcavage, T. M., and Kim, Y. B., 1979, Immunoglobulins of germfree colostrum-deprived and conventional colostrum-fed miniature piglets, in: *Clinical and Experimental Gnotobiotics* (T. M. Fliedner, H. Heit, D. Neithammer, and H. Pflieger, eds.), pp. 139–144, Gustav Fischer Verlag, Stuttgart.

Sharon, N., and Pollard, M., 1969, Spontaneous neoplastic transformation of germ-free rat embryo cell culture, *Cancer Res.* **29:**1523.

Sheid, M. P., Hoffman, M. K., Komuro, K., Hammering, V., Abbot, J., Boyse, E. A., Cohen, G. H., Hooper, J. A., Schulof, R. S., and Goldstein, A. L., 1973, Differentiation of T cells induced by preparations from thymus and nonthymic agents: The determined state of the precursor cell, *J. Exp. Med.* **138:**1027.

Sherman, J. D., Adner, M. M., and Dameshek, W., 1963, Effect of thymectomy on the golden hamster *(Mesocricetus auratus)*. I. Wasting disease, *Blood* **22:**252.

Shih, C. N., Balish, E., Lower, G. M., Jr., Yale, C. E., and Bryan, G. T., 1975, Induction of colonic tumors in germ-free and conventional rats, Vth International Symposium on Gnotobiology, Stockholm, p. 25.
Smith, C. S., Pilgrim, H. I., and Steinmuller, D., 1972, The survival of skin allografts and xenografts in germfree mice, *Transplantation* **13**:38.
Spatz, M., McDaniel, E. G., and Laqueur, G. L., 1966, Cycasin excretion in conventional and germfree rats, *Proc. Soc. Exp. Biol. Med.* **121**:417.
Springer, G. F., 1966, Relation of microbes to blood group-active substances, *Angew. Chem.* **5**:909.
Springer, G. F., Horton, R., and Forbes, M., 1959, Origin of antihuman blood group B agglutinins in germfree chicks, *Ann. N.Y. Acad. Sci.* **78**:272.
Stanley, E. R., Robinson, W. A., and Ada, G. L., 1968, Properties of the colony stimulating factor in leukemic and normal mouse serum, *Aust. J. Exp. Biol. Med. Sci.* **46**:715.
Starr, T. J., and Considine, R. G., 1968, Interferon production in gnotobiotic mice, in: *Advances in Germfree Research and Gnotobiology* (M. Miyakawa and T. D. Luckey, eds.), pp. 253–263, Chemical Rubber Co., Cleveland.
Steinman, R. M., Lustig, D. S., and Cohn, Z. A., 1974, Identification of a novel cell type in peripheral lymphoid organs of mice. III. Functional properties *in vivo, J. Exp. Med.* **139**:1431.
Štěpánková, R., Kovářů, F., and Kruml, J., 1979, The development of intestinal lymphatic tissues in germfree and conventional rabbits, *Folia Microbiol. (Prague)* **24**:48.
Šterzl, J., 1963, The opsonic activity of complement in sera without antibody, *Folia Microbiol.* **8**:240.
Šterzl, J., and Kovářů, F., 1978, Development of lymphatic tissue and immunocompetency in pig foetuses and germfree piglets, *Acta Vet. (Brno) Suppl.*
Šterzl, J., and Silverstein, A. M., 1967, Developmental aspects of immunity, *Adv. Immunol.* **6**:337.
Stewart, C. C., and Goldstein, S., 1974, The direct determination of cytolytic antibody titer and specificity, *J. Lab. Clin. Med.* **84**:425.
Stinebring, W. R., and Youngner, J. S., 1964, Patterns of interferon appearance in mice injected with bacteria or bacterial endotoxin, *Nature (London)* **204**:712.
Svoboda, V., and Mandel, L., 1979, Influence of the microbial factor on the physiological condition of hemopoiesis and its response to partial body irradiation in piglets, *Folia Microbiol.* **24**:108.
Tajima, M., 1955, Fundamental studies on the blood of germfree guinea pigs, *Nagoya Igakkai Zasshi* **70**: 766.
Thonard, J. C., Dalbow, M. H., and Crosby, R. G., 1968, Immune response in germfree mice after intragingival antigenic stimulation, in: *Advances in Germfree Research and Gnotobiology* (M. Miyakawa and T. D. Luckey, eds.), pp. 233–238, Chemical Rubber Co., Cleveland.
Thorbecke, G. J., 1959, Some histological and functional aspects of lymphoid tissue in germfree animals. I. Morphological studies, *Ann. N.Y. Acad. Sci.* **78**:237.
Thorbecke, G. J., and Benacerraf, B., 1959, Some histological and functional aspects of lymphoid tissue in germfree animals. II. Studies on phagocytosis *in vivo, Ann. N.Y. Acad. Sci.* **78**:247.
Thorbecke, G. J., Gordon, H. A., Wostmann, B. S., Wagner, M., and Reyniers, J. A., 1957, Lymphoid tissue and serum gamma globulin in young germfree chickens, *J. Infect. Dis.* **101**:237.
Till, J. E., and McCulloch, E. A., 1961, A direct measurement of the radiation sensitivity of normal mouse bone marrow cells, *Radiat. Res.* **14**:213.
Trautman, A., and Fiebiger, J., 1952, in: *Fundamentals of the Histology of Domestic Animals* (R. E. Habel and E. L. Biberstein, eds.) (translated, revised, and edited from the 1949 German edition), p. 126, Comstock Publishing Associates, Ithaca, New York. Cited in Kim and Watson (1971).
Trávníček, J., Kostka, J., Lanc, A., Rejnek, J., and Šterzl, J., 1968, The effect of specific antibodies in the intestine of germfree piglets monoinoculated with the pathogenic strain of *E. coli* 055, in: *Advances in Germfree Research and Gnotobiology* (M. Miyakawa and T. D. Luckey, eds.), pp. 247–252, Chemical Rubber Co., Cleveland.
Trexler, P. C., 1963, An isolator system for control of contamination, *Lab. Anim. Sci. Suppl.* **13**:572.
Trexler, P. C., 1979, Transfer isolator to protect personnel from exposure to peracetic acid, *Lab Animals* **13**:163.
Trexler, P. C., and Reynolds, L. I., 1957, Flexible film apparatus for the rearing and use of germfree animals, *Appl. Microbiol.* **5**:406.
Tuttle, R. L., and North, R. J., 1975, Mechanisms of antitumor action of *Corynebacterium parvum:* Nonspecific tumor cell destruction at site of an immunologically mediated sensitivity reaction to *C. parvum, J. Natl. Cancer Inst.* **55**:1403.

Twomey, J. J., Bealmear, P. M., and Lewis, V., 1977a, Assay for thymosin reactivity in germfree nude mice, 15th Annual Meeting of the Association for Gnotobiotics, Saranac Lake, New York, No. 13.
Twomey, J. J., Lewis, V. M., Bealmear, P. M., Goldstein, G., and Good, R. A., 1977b, Bioassay determination of thymopoietin levels in human plasma, *Proc. Natl. Acad. Sci. USA* **74**:2541.
Twomey, J. J., Lewis, V. M., Ford, R., and Goldstein, G., 1980, An inhibitor of thymic hormone activity in serum from patients with lymphoblastic leukemia, *Am. J. Med.* **68**:377.
Ueda, U., Yamazaki, S., and Someya, S., 1975, Hyporeactivity to tuberculin in germ-free mice, *J. Reticuloendothel. Soc.* **18**:107.
Uno, Y., Sumi, U., and Sakurai, K., 1968, Studies on the thymus of germfree rats, in: *Advances in Germfree Research and Gnotobiology* (M. Miyakawa and T. D. Luckey, eds.), pp. 129–138, Chemical Rubber Co., Cleveland.
Veit, B. C., and Feldman, J. D., 1976, Suppressor macrophages in spleens of tumor bearing animals, *J. Immunol.* **117**:655.
Wagner, M., 1955, Germfree research: A basic study in host contaminant relationship. II. Serologic observations in germfree animals, *Bull. N.Y. Acad. Med.* **31**:236.
Wagner, M., 1959, Serologic aspects of germfree life, *Ann. N.Y. Acad. Sci.* **78**:261.
Wagner, M., and Wostmann, B. S., 1959, Studies on monocontaminated chickens (*Clostridium perfringens* or *Streptococcus faecalis*) fed penicillin, in: *Antibiotics Annual, 1958–1959,* Medical Encyclopedia, New York.
Wagner, M., and Wostmann, B. S., 1961, Serum protein fractions and antibody studies in gnotobiotic animals reared germfree or monocontaminated, *Ann. N.Y. Acad. Sci.* **94**:210.
Wagner, R. R., 1961, Biological studies of interferon. I. Suppression of cellular infection with Eastern equine encephalitis virus, *Virology* **13**:323.
Walburg, H. E., Jr., 1973, Carcinogenesis in gnotobiotic rodents, in: *Germfree Research: Biological Effect of Gnotobiotic Environments* (J. B. Heneghan, ed.), pp. 115–122, Academic Press, New York.
Walburg, H. E., Jr., and Cosgrove, G. E., 1967, Aging in irradiated and unirradiated germfree ICR mice, *Exp. Gerontol.* **2**:143.
Walburg, H. E., Jr., Mynatt, E. I., and Robie, D. M., 1964, Survival of x-irradiated germfree mice, *Radiat. Res.* **22**:247.
Walburg, H. E., Jr., Upton, A. C., Tyndall, R. L., Harris, W. W., and Cosgrove, G. E., 1965, Preliminary observations on spontaneous and radiation-induced leukemia in germfree mice, *Proc. Soc. Exp. Biol. Med.* **118**:11.
Watson, D. W., Kim, Y. B., and Bradley, S. G., 1968, Immune response in germfree colostrum-deprived piglets, in: *Advances in Germfree Research and Gnotobiology* (M. Miyakawa and T. D. Luckey, eds.), pp. 199–207, Chemical Rubber Co., Cleveland.
Wells, C. L., and Balish, E., 1979a, Immune response modulation by colonization of germfree rats with *Propionibacterium acnes*, *Infect. Immun.* **26**:473.
Wells, C., and Balish, E., 1979b, The mitogenic activity of lipopolysaccharide for spleen cells from germfree, conventional, and gnotobiotic rats, *Can. J. Microbiol.* **25**:1087.
Werder, A. A., Amos, M. A., and Nielsen, A. H., 1978, Studies on polycystic renal disease (PCRD) in germfree and ex-germfree CFW_{wd} mice, 16th Annual Meeting of the Association for Gnotobiotics, Madison, Wisconsin, p. 20.
Wescott, R. B., 1967, A comparative study of *Nematospiroides dubius* infection in germfree and conventional mice, Symposium on Gnotobiotic Research, Madison, Wisconsin, p. 17.
Wescott, R. B., 1971, Present status and future prospects of the use of germfree host animals for research in parasitology, Symposium on Gnotobiotic Research, Columbus, Ohio, p. 14.
Wheater, D. W. F., and Hurst, E. W., 1962, The effect of sex on bacterial infections in mice and on chemotherapy on one of them, *J. Pathol. Bacteriol.* **82**:117.
Wilson, R., 1963, Survival studies of whole-body x-irradiated germfree (axenic) mice, *Radiat. Res.* **20**:477.
Wilson, R., and Bealmear, P. M., 1967, Radiation pathology in germfree mice and bone marrow transplantation in irradiated germfree mice, Proceedings of the International Conference on Radiation Biology and Cancer, Kyoto, Japan, pp. 107–115.
Wilson, R., Sjodin, K., and Bealmear, P. M., 1964a, Thymus studies in germfree mice, in: *The Thymus* (V. Defendi and D. Metcalf, eds.), pp. 89–93, Wistar Institute Press, Philadelphia.
Wilson, R., Sjodin, K., and Bealmear, P. M., 1964b, The absence of wasting in thymectomized germfree (axenic) mice, *Proc. Soc. Exp. Biol. Med.* **117**:237.

Wilson, R., Matsuzawa, T., and Connell, S. J., 1964c, Hematological changes in germfree mice following wholebody x-irradiation, *Radiat. Res.* **22**:249.

Wilson, R., Bealmear, M., and Sobonya, R., 1965, Growth and regression of the germfree (axenic) thymus, *Proc. Soc. Exp. Biol. Med.* **118**:97.

Wilson, R., Bealmear, M., and Sjodin, K., 1966, A technique for thymectomizing germfree mice, *J. Appl. Physiol.* **21**:279.

Wilson, R., Bealmear, P. M., and Matsuzawa, T., 1968, Acute intestinal radiation death in germfree and conventional mice, in: *Gastrointestinal Radiation Injury* (M. F. Sullivan, ed.), pp. 148–158, Excerpta Medica, Amsterdam.

Wilson, R., Barry, T. A., and Bealmear, P. M., 1969, Alterations in serum iron in germfree and conventional mice after wholebody x-irradiation, *Life Sci.* **8**:401.

Wilson, R., Barry, T. A., and Bealmear, P. M., 1970, Evidence for a toxic substance of bacterial origin in the blood of irradiated mice, *Radiat. Res.* **41**:89.

Wolberg, G., Duncan, G. S., Adlam, C., and Whisnant, J. K., 1977, Antibody to *Corynebacterium parvum* in normal human and animal sera, *Infect. Immun.* **15**:1004.

Wolfe, L. G., Griesemer, R. A., and Rohovsky, M. W., 1966, Germfree cynomologus monkeys, *Lab. Anim. Sci.* **16**:364.

Woodward, B., 1978, The peritoneal leukocytes of the germfree mouse, *Experientia* **34**:1639.

Wostmann, B. S., 1959, Serum proteins in germfree vertebrates, *Ann. N.Y. Acad. Sci.* **78**:255.

Wostmann, B. S., 1961, Recent studies on the serum proteins of germfree animals, *Ann. N.Y. Acad. Sci.* **94**:272.

Wostmann, B. S., 1968, Defence mechanisms in germ-free animals. Part I. Humoral defence mechanisms, in: *The Germ-Free Animal in Research* (M. Coates, ed.), pp. 197–209, Academic Press, New York.

Wostmann, B. S., and Gordon, H. A., 1960, Electrophoretic studies on the serum protein pattern of the germfree rat and its changes upon exposure to a conventional bacterial flora, *J. Immunol.* **84**:27.

Wostmann, B. S., and Kellogg, T. F., 1967, Purified starch casein diet for nutritional research with germfree rats, *Lab. Anim. Sci.* **17**:580.

Wostmann, B. S., and Olson, G. B., 1964, Precipitating antibody production in germfree chickens, *J. Immunol.* **92**:41.

Wostmann, B. S., and Olson, G. B., 1969, Persistence of primary antibody formation caused by the absence of antigen of microbial origin in the germfree chicken, *Immunology* **17**:199.

Wostmann, B. S., Pleasants, J. R., and Bealmear, P. M., 1969, Diet and immune mechanisms, in: *Germ-Free Biology: Experimental and Clinical Aspects* (E. A. Mirand and N. Back, eds.), pp. 287–292, Plenum Press, New York.

Wostmann, B. S., Pleasants, J. R., Bealmear, P. M., and Kincade, P. W., 1970, Serum proteins and lymphoid tissues in germfree mice fed a chemically defined, water soluble, low molecular weight diet, *Immunology* **19**:443.

Wostmann, B. S., Pleasants, J. R., and Bealmear, P. M., 1971, Dietary stimulation of immune mechanisms, *Fed. Proc.* **30**:1779.

Yale, C. E., and Balish, E., 1976, Blood and serum chemistry values of gnotobiotic beagles, *Lab. Anim. Sci.* **26**:633.

Zacharia, T. P., and Pollard, M., 1969, Elevated levels of α-globulins in sera from germfree rats with 3-methylcholanthrene-induced tumors, *J. Natl. Cancer Inst.* **42**:35.

Part X
Cryopreservation of Embryos for Strain Preservation

12

Physiological Basis of the Freezing of Mammalian Embryos

S. P. Leibo

1. Introduction

The ability to store animal cells in the frozen state for extended times offers a number of advantages to the experimental biologist as well as to the clinician. Because it offers the unique option of stopping biological time, cryobiology permits the accumulation of cells at different times but the assay of their activity to be performed at the same time. It also ensures against the irrevocable loss of cell lines caused by contamination or genetic alteration, e.g., the spontaneous loss of chromosomes as occasionally occurs in hybridoma cells. If frozen under optimum conditions and stored at a sufficiently low temperature, the thawed cell is identical to itself before freezing. As applied to most biological systems, cryobiology seeks to devise procedures by which the majority of a large population of cells, usually numbering in the tens or hundreds of millions, can be preserved by freezing. Sometimes the desired endpoint is simply the preservation of an intact cell, although usually it is the preservation of cellular function as well. To the noncryobiologist, it is frequently sufficient to preserve the ability of a frozen-thawed cell population to reproduce itself in culture. The preservation of mammalian embryos by freezing, however, is a unique application of cryobiology. In this case, very small numbers of cells, usually 10–20 embryos, are frozen with the ultimate goal of generating an entire living organism that is itself capable of reproducing yet another organism. In other words, because so few embryos are usually frozen, their freezing must be performed reliably enough for most of the embryos to survive; this restriction frequently does not apply to the freezing of large populations of cells. The freezing of embryos was first reliably demonstrated in 1972 when Whittingham, Leibo, and Mazur and, independently, Wilmut reported that preimplantation-stage mouse embryos could be successfully frozen and stored in liquid nitrogen at −196°C. The frozen-thawed embryos not only exhibited cleavage in culture, but also developed *in utero* into

S. P. Leibo • Biology Division, Oak Ridge National Laboratory, Oak Ridge, Tennessee 37830.

living mouse pups when transferred into suitably primed foster mothers (Whittingham *et al.*, 1972). Since that time, the freezing of mouse embryos has been repeated in at least two dozen laboratories, and has been demonstrated with the embryos of the rat, rabbit, hamster, sheep, goat, and cattle. Although the overall survival of embryos of species other than the mouse is still relatively low, it has been reported that living young have developed from frozen-thawed embryos of all of the above-named species, with the exception of the hamster (Whittingham, 1975; Bank and Maurer, 1974; Tsunoda *et al.*, 1976; Willadsen *et al.*, 1976; Bilton and Moore, 1976; Wilmut and Rowson, 1973).

Most studies of embryo freezing have essentially followed the procedure first described by Whittingham *et al.* and Wilmut:

1. Suspend the embryos in a multimolar solution of a cryoprotective agent, usually dimethylsulfoxide (Me_2SO).
2. Cool the embryos to a temperature somewhat below the freezing point of the suspending solution, i.e., -5 to $-10°C$, and induce ice formation by "seeding" the sample.
3. Cool the partially frozen suspension to $-80°C$ or below at a low rate of ≈ 0.2 to $0.8°C/min$.
4. Transfer the frozen embryos into liquid nitrogen at $-196°C$. They may be stored at that temperature for months or years with no loss of developmental capability.
5. Warm and thaw the frozen embryos at a moderate rate of $\sim$ 5 to $25°C/min$.
6. Dilute the Me_2SO in several steps, and wash the frozen-thawed embryos in a balanced salt solution.
7. Culture the embryos to assess developmental capacity, and transfer them into pseudopregnant foster mothers to develop to term.

Recently, it has been found that embryos can also be successfully frozen by modifications of this basic procedure. These modifications are based on the following observations.

1. Embryos slowly cooled to ~ -35 to $-40°C$ may be transferred directly into liquid nitrogen; they survive if subsequently warmed rapidly at $\sim$ $300°C/min$, but not if warmed slowly at $\sim 25°C/min$ or less (Willadsen *et al.*, 1976, 1978; Whittingham *et al.*, 1979). In other words, slow cooling of embryos need be continued only down to $\sim -40°C$, thus reducing the total time required for cooling.
2. The cooling procedure apparently can be shortened even further. Wood and Farrant (1980) have reported very recently that embryos survive if cooled directly to $-20°C$, held for $\sim$ 15 min, plunged directly into liquid nitrogen, and then warmed rapidly at $\sim$ $500°C/min$.
3. It has also been found that various glycols will protect embryos against freezing damage to the same extent as dimethylsulfoxide or glycerol (Miyamoto and Ishibashi, 1977, 1978).
4. Embryos may be successfully frozen while still contained in the oviduct itself; they do not need to be flushed out of the tube to be frozen (Zeilmaker and Verhamme, 1979).

The basic procedure of embryo freezing itself, and the recent modifications of it, can be viewed as a simple application of the fundamental principles of cryobiology. More importantly, it is beginning to appear that the response of cells exposed to subzero temperatures can be viewed and understood as an extension of fundamental cellular physiology.

The purpose of this review, then, is to consider the freezing of mammalian embryos from this viewpoint.

2. Cryobiology of Mammalian Embryos

2.1. Introduction

It has been known for many years that the survival of frozen-thawed cells is highly dependent on the rate at which they have been cooled (see Mazur, 1970, 1977a; Leibo and Mazur, 1971, for review). Different types of cells exhibit maximum survival when cooled at different optimum rates. In 1963, Mazur had theorized that these differences can be explained by a quantitative analysis of the kinetics of water loss from cells at subzero temperatures. He derived mathematical expressions to calculate this water loss and, thereby, to predict the likelihood of intracellular ice formation, a major cause of freezing injury. In their first study of embryo freezing, Whittingham *et al.* (1972) used these equations to estimate the cooling rate necessary to avoid intracellular ice formation in cells the size of embryos (~ 70-μm diameter). They estimated this rate to be 1°C/min or less. In other words, they predicted that embryos would have to be cooled very slowly if they were to survive freezing.

As they expected, Whittingham *et al.* (1972) actually observed that the survival of frozen-thawed embryos was highly dependent on the rate at which they had been cooled from −5 to −100°C. Survival was high over the range of ~ 0.2 to 2°C/min, but dropped precipitously if cooled at rates of ~ 5°C/min (Fig. 1). This was true both for cleavage-stage embryos, e.g., two- and eight-cell stages, and for unfertilized ova as well. In this respect, embryos seemed to respond to cooling and freezing much as do many other types of cells. What seemed unique about the response of embryos was that their survival required that the slowly frozen embryos had to be warmed relatively slowly as well (Whittingham *et al.*, 1972; Leibo *et al.*, 1974). It is now known that this is not necessarily true, because it has been observed that embryos will survive rapid warming if their slow cooling has been interrupted by rapid cooling from about −40°C into liquid nitrogen at −196°C (Whittingham *et al.*, 1979). To understand these rate-dependent phenomena requires appreciation both of the changes produced in aqueous solutions by freezing and of the time-dependent responses of cells exposed to such altered solutions.

2.2. Solutions at Subzero Temperatures

Mammalian cells are normally suspended in aqueous solutions of primarily NaCl at a concentration of ~ 0.16 M (0.9%). When such a solution is cooled a few degrees below 0°C, it freezes either spontaneously or by having been induced to freeze by "seeding" it with an ice crystal. Freezing removes some of the liquid water and converts it to ice. This removal of water concentrates the remaining unfrozen solution. As the partially frozen solution is cooled to lower temperatures, more water is converted to ice, resulting in a concomitant increase in the concentration of dissolved solutes. This behavior of aqueous solutions can be described quantitatively by phase diagrams such as those shown in Fig. 2. The curve for NaCl in that figure shows the increase in salt concentration that results from the removal of water as ice with decreasing temperature. The curve for Me_2SO in that figure

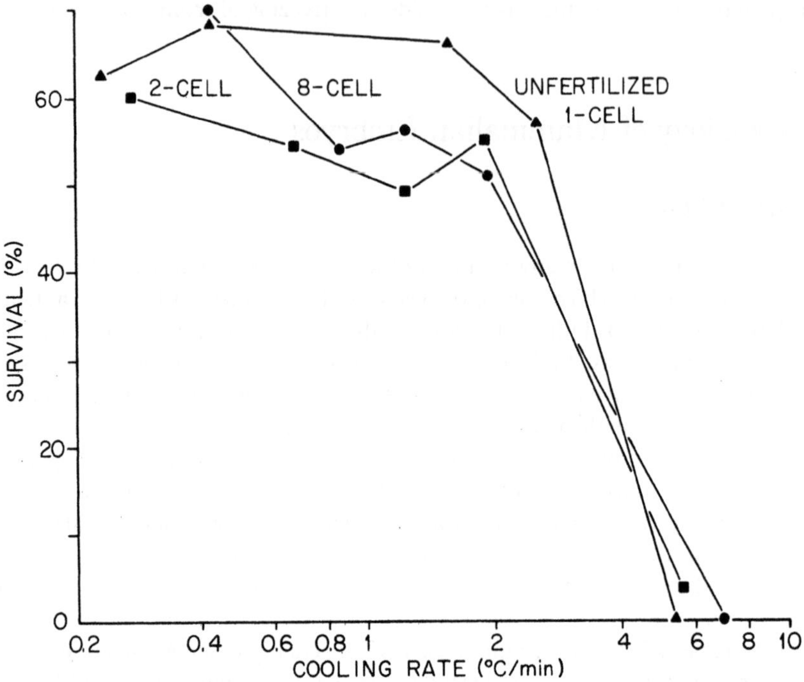

FIGURE 1. Survival of preimplantation-stage mouse ova and embryos as a function of the rate at which they were cooled to −196°C. Survival of the two- and eight-cell-stage embryos was measured by their ability to develop in culture (data of Whittingham *et al.*, 1972); survival of the unfertilized ova was measured by their ability to fluoresce in the presence of fluorescein diacetate, an excellent indicator of embryo survival (data of Leibo *et al.*, 1978).

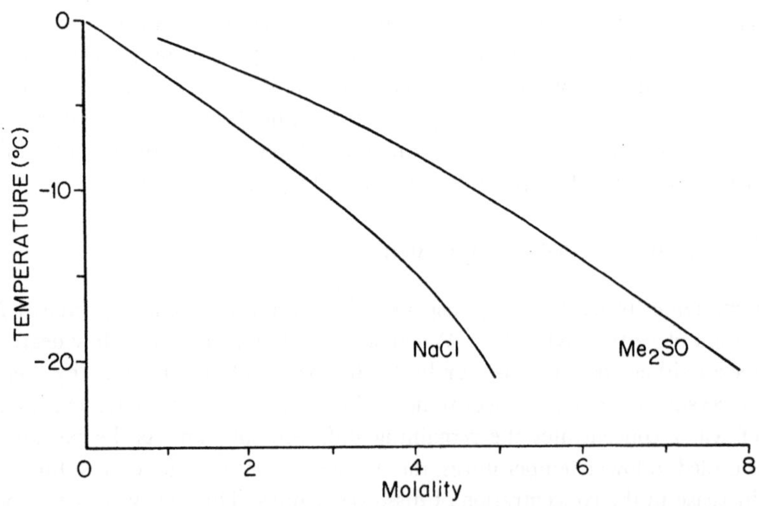

FIGURE 2. Partial phase diagrams of aqueous solutions of NaCl and of dimethylsulfoxide (Me_2SO). The curves were redrawn from original data of Landolt-Börnstein (NaCl) and of Rasmussen and MacKenzie (Me_2SO) as given by Cocks and Brower (1974).

TABLE I
Eutectic Points of Various Aqueous Solutions

Solute	Eutectic point (°C)	Approximate eutectic concn. (molal)
KCl	−10.7	3.0
NaCl	−21.1	5.2
$MgCl_2$	−33.5	2.9
$CaCl_2$	−55.0	3.8
Sucrose	−10.8	4.0
Glycerol	−46.2	21.8
Dimethylsulfoxide	−62.0	15.6

shows that aqueous solutions of nonelectrolytes respond similarly, although the solute itself determines the quantitative relationship. The concentration of solutes by freezing is not a rate-dependent process. That is, once ice forms, the concentration of solute in the unfrozen portion of the solution is fixed by temperature alone, and not by the rate at which the solution is cooled to that temperature. There is a temperature, called the eutectic point, below which the entire solution has become completely solid; all the solute has either solidified or crystallized, and all the water has either formed ice or become bound to solid solute in the form of hydrates. Eutectic points are characteristic of given solutes, although mixtures of solutes may exhibit eutectic zones rather than discrete eutectic points. In cryobiology, one is primarily concerned with the response of cells at subzero temperatures above eutectic points of solutions. Eutectic points of different solutes occur over a wide range of temperatures (Table I). When more than one solute is present in a solution, liquid solution will exist at least down to the lowest eutectic point of all the solutes present. Cells frozen in solutions of mixtures of solutes will be exposed to extremely concentrated solutions at low subzero temperatures. Cryobiology is concerned with the response of cells in concentrated solutions and at low temperatures.

2.3. Cell Responses to Altered Solutions and Temperatures

When mouse ova are cooled slightly below 0°C and the suspending solution freezes, they respond as shown in Fig. 3. As described above, the formation of ice produces an increase in solute concentration. When frozen in a solution of NaCl alone to ~ −2°C, an ovum is exposed to ~ 0.5 M NaCl. Over a period of ~ 6 min, the ovum gradually shrinks by losing water (Fig. 3A). When frozen in a Me_2SO–NaCl solution to a slighly lower temperature, ova are exposed to an even more concentrated solution (see Fig. 2). Although the concentrations of the NaCl and Me_2SO–NaCl solutions differ, the osmolalities of both solutions are approximately the same at the same temperature. That is, ova frozen to ~ −2 to −3°C in these two solutions are exposed to equal osmotic pressures. The ova frozen in the dimethylsulfoxide solution also shrink, but at a lower rate (Fig. 3B). Such time-dependent water loss from cells in the presence of ice, with its concomitant increase in solute concentration, provides the beginnings of an explanation of the rate dependency of survival of frozen-thawed cells.

All mammalian cells respond osmotically when transferred from an isotonic solution into a hypertonic one. An example of such an osmotic response of a mouse ovum transferred from an isotonic balanced salt solution into 0.5 M NaCl is shown in Fig. 4. Over a period

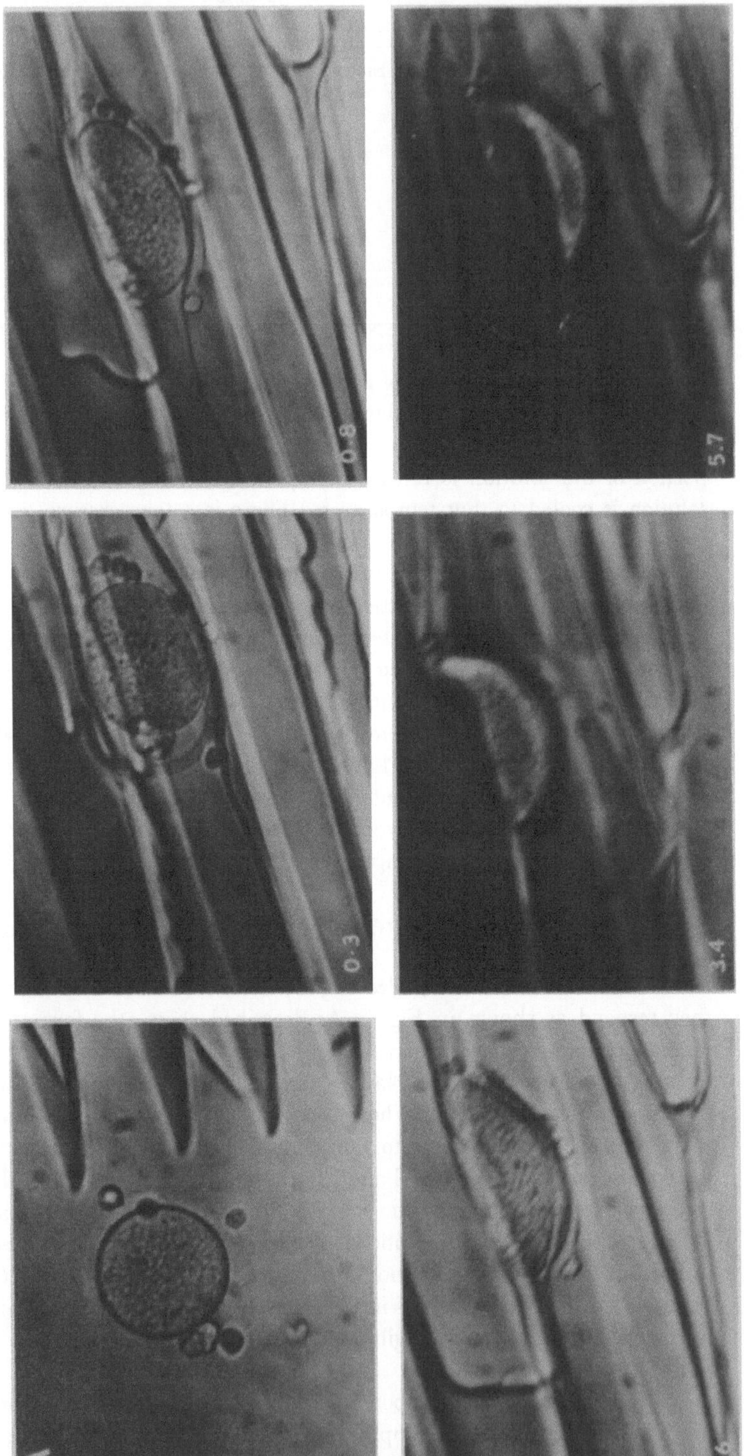
A
0
0.3
0.8
1.6
3.4
5.7

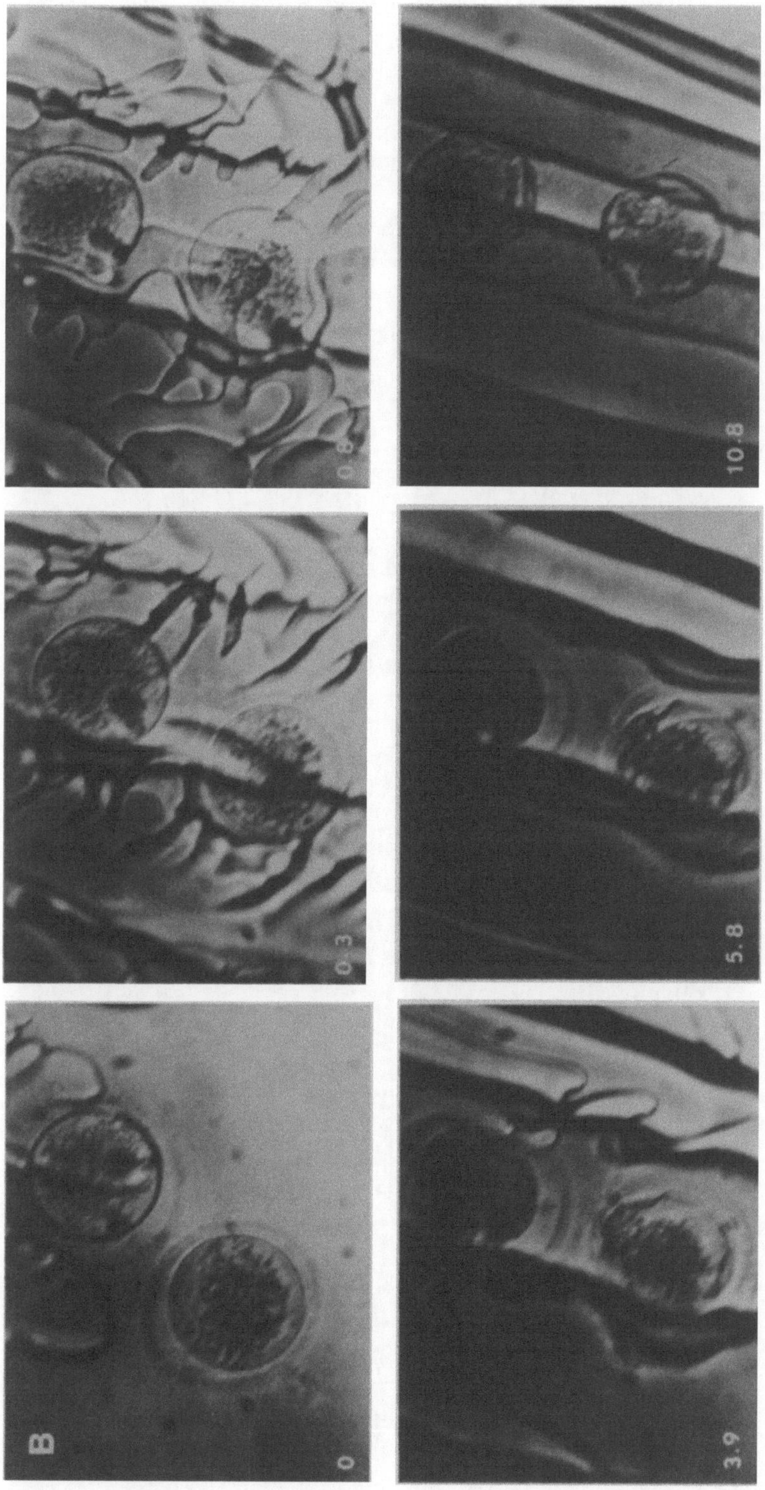

FIGURE 3. Low-temperature microscopy of mouse ova. The observations were made using the cryomicroscope of E. G. Cravalho (Massachusetts Institute of Technology), as described in Leibo *et al.* (1978). (A) An ovum suspended in isotonic NaCl was cooled and held at $-2°C$. (B) Two ova suspended in 1 M Me_2SO in isotonic balanced salt solution were cooled and held at $\sim -3°C$.

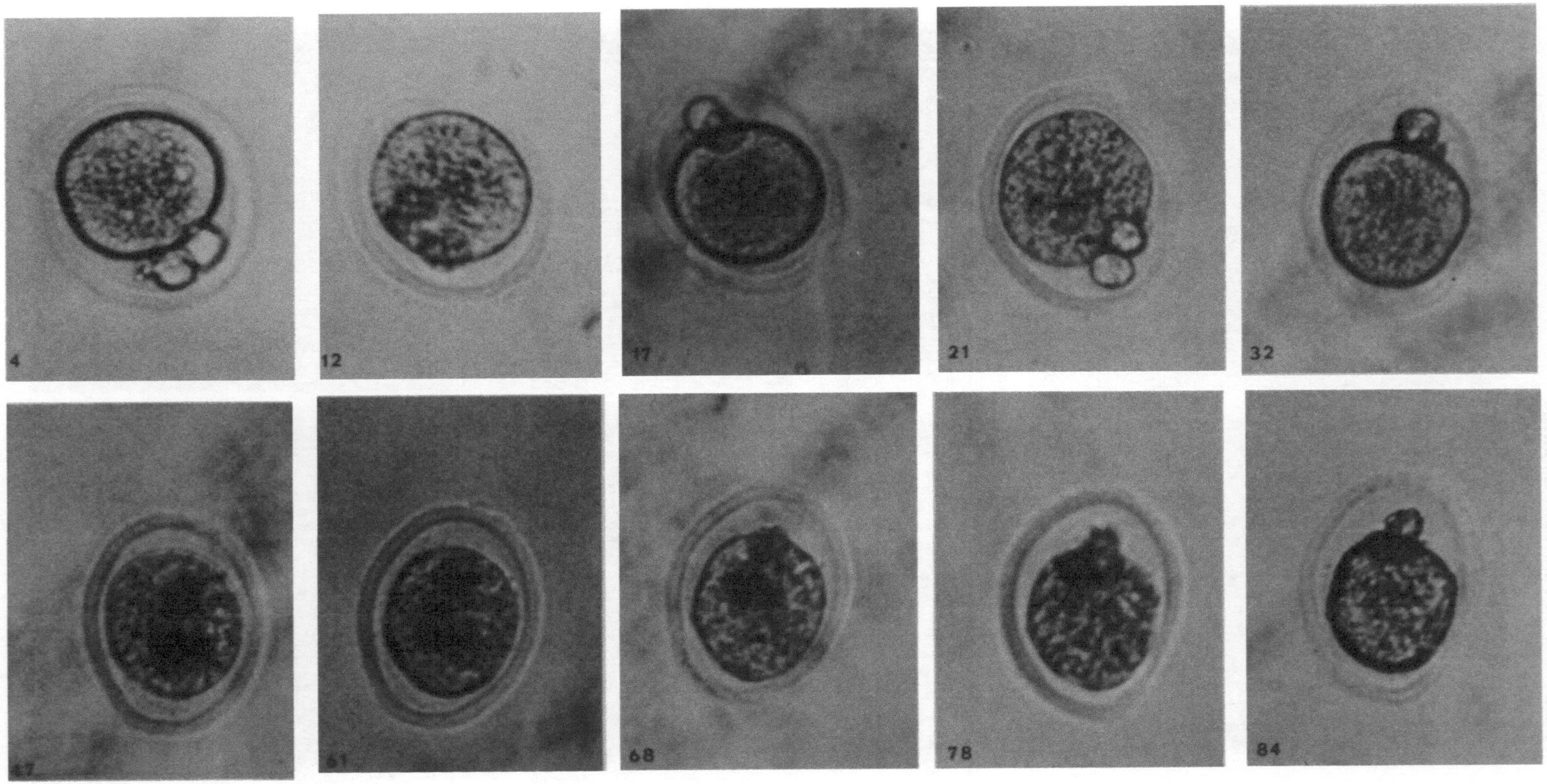
4
12
17
21
32
61
68
78
84

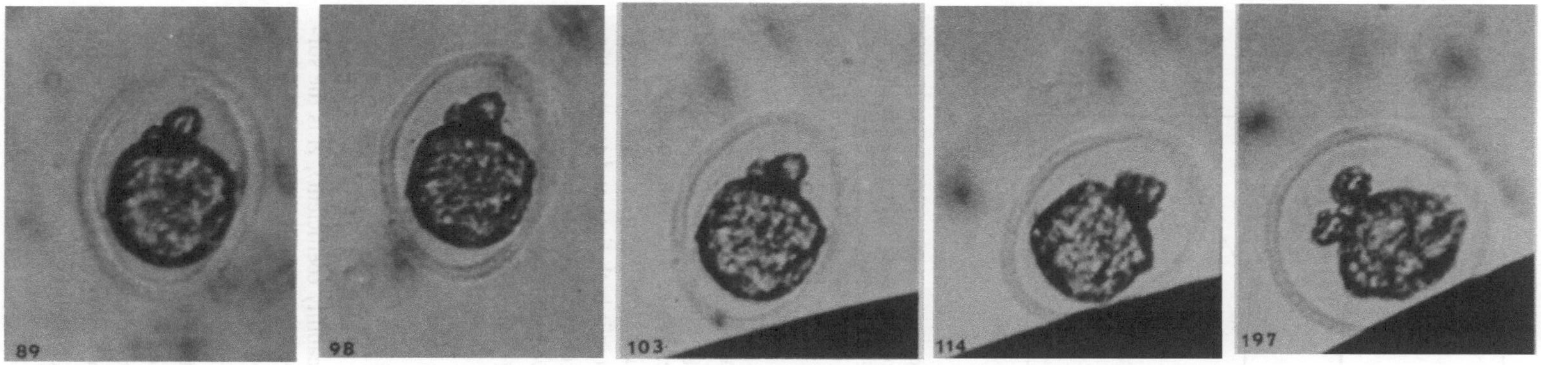

FIGURE 4. The osmotic response of a fertilized mouse ovum transferred from an isotonic salt solution into 0.5 M NaCl at 20°C. The number at the lower left of each photograph is the time (in seconds) after transfer. The actual method used is described in Leibo (1980), which is the original source of the figure. (Published with permission.)

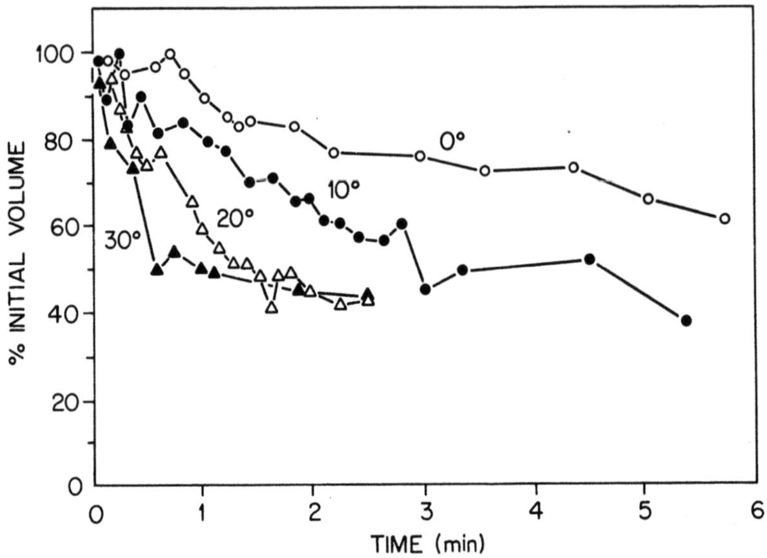

FIGURE 5. The change in volume of mouse ova transferred from an isotonic salt solution into 0.5 M NaCl at each of four constant temperatures. The observations were made as shown in Fig. 4. The cross-sectional areas of the ova were measured, and their volumes calculated as a percentage of the initial volume of the same ovum in isotonic saline. The data are those of Leibo (1980), who describes the method used and the details of the calculation. (Published with permission.)

of a few minutes, the ovum shrinks by losing water. By measuring the cross-sectional area of the ovum, and assuming it to be a sphere, the changing volume of the ovum relative to its initial volume can be calculated as a function of time. Water loss from mouse ova at different temperatures has been measured in this way (Leibo, 1980). An example is shown in Fig. 5. The results in that figure show that the water loss is both time and temperature dependent; the lower the temperature, the slower the water loss. Numerous mathematical expressions have been written to describe the osmotic movement of water across cell membranes. In general terms, all these mathematical equations express the relationship between total cell volume, V, and time, t, as

$$\frac{dV}{dt} = L_pA(\pi_i - \pi_e)$$

in which L_p is the hydraulic conductivity (or water permeability coefficient), A is the surface area of the cell, and π_i and π_e are the osmotic pressures of the intracellular and extracellular solutions, respectively. By fitting the observations of dV/dt to those calculated using an integrated form of this equation, one can determine the L_p that gives the best fit for a given set of conditions. In other words, one can calculate the theoretical loss of water from a cell having the characteristics of a mouse ovum exposed to a hypertonic solution at a constant temperature (Fig. 6). Furthermore, the rate of water loss, expressed as L_p in various units of (volume of H_2O)/(area of cell surface)/(unit time)/(unit of osmotic pressure difference), can be related to temperature in an Arrhenius plot, as shown in Fig. 7. The

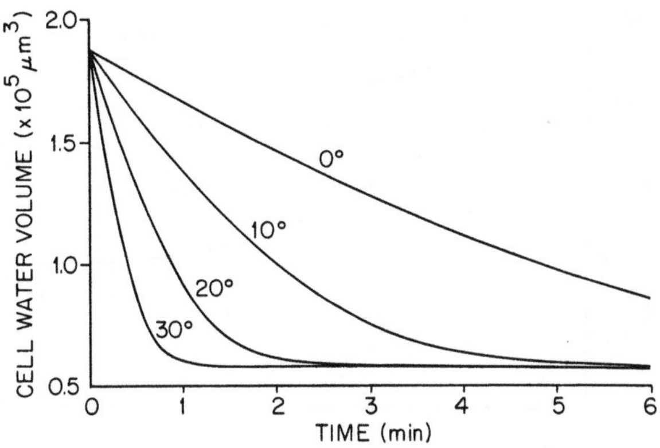

FIGURE 6. The calculated change in cell water volume of mouse ova at each of the indicated temperatures as a function of time. The method of calculation and the equations used are those of Leibo (1980).

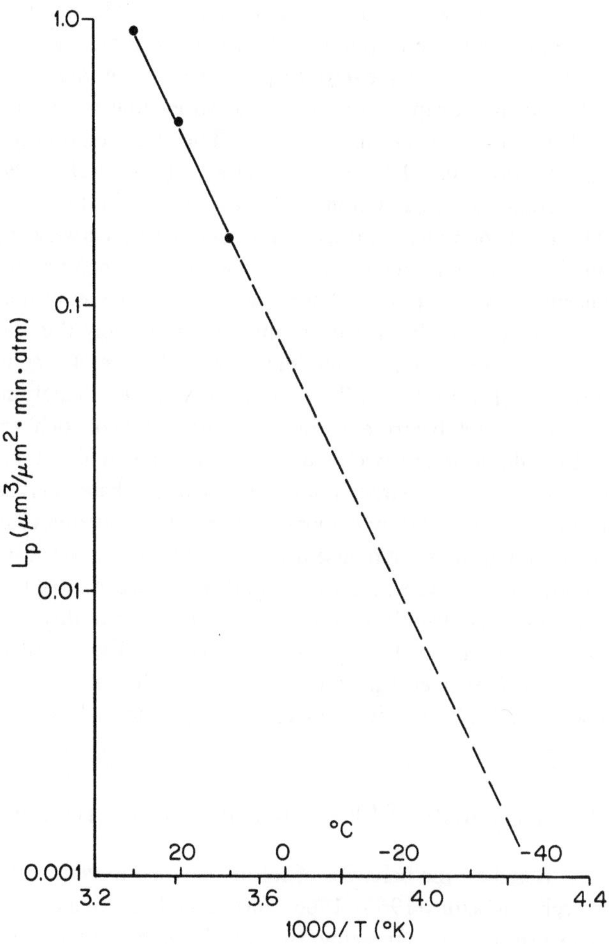

FIGURE 7. Arrhenius plot of the hydraulic conductivity (L_p) of unfertilized mouse ova. The measured points and the solid line are the data of Leibo (1980). The dashed line is the simple extrapolation of the measured values.

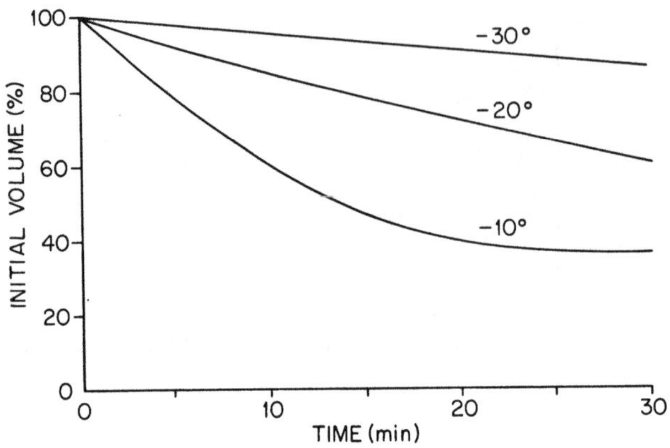

FIGURE 8. The calculated change in the total cell volume, expressed as a percentage of the initial volume, as a function of time at three subzero temperatures. The calculations were performed with the values for L_p read from the extrapolated line shown in Fig. 7.

solid portion of the line shows the actual measured relationship between L_p and temperature ($1/T$ in °K) for unfertilized mouse ova (Leibo, 1980). But here we are primarily concerned with the response of mouse ova and embryos at subzero temperatures. An estimate of the effect of subzero temperatures on the rate of water loss from mouse ova can be obtained by extrapolation of the measured line to lower temperatures, as shown by the dashed portion of the line in Fig. 7. The extrapolation in Fig. 7 shows that lowering the temperature from 0 to -20°C reduces the rate of water loss by almost a full order of magnitude, i.e., L_p at 0 and -20°C is $\sim$ 0.07 and 0.009 $\mu m^3/\mu m^2$ min atm, respectively. This extrapolation assumes that the relationship between L_p and T at subzero temperatures and in partially frozen solutions is governed by the same mechanisms that determine water permeability above 0°C. These two assumptions may not be completely valid. However, it will be extremely difficult to measure accurately the water permeability of cells in partially frozen solutions. Therefore, in the absence of actual measurements, these estimates of L_p as a function of $1/T$ permit approximate calculations to be made of water loss from mouse ova at subzero temperatures. The results of such approximations are shown in Fig. 8. In addition to the two assumptions mentioned above, it was also assumed for these calculations that the extracellular solution would have increased to $\gg$ 2 M (see Fig. 2). The results in Fig. 8 clearly demonstrate that the reduction of temperature drastically increases the time required for mouse ova to equilibrate osmotically with their suspending solution. In other words, when mouse ova and embryos are cooled to subzero temperatures, they are exposed to hypertonic solutions. However, their ability to respond osmotically will depend on the rate at which they have been cooled. Expressed formally, cooling rate is dT/dt, where T is temperature and t is time. As the osmotic response of a cell is dependent on both T and t, herein lies the explanation of the cooling rate dependence of embryo survival.

2.4. Responses of Ova and Embryos to Freezing

Based on the behavior of solutions when frozen, and on the known osmotic responses of cells, Mazur (1963, 1966) formulated mathematical expressions for describing the response of a cell to freezing. Briefly, his analysis predicted that for a given cell with defined characteristics, there is a cooling rate that will permit the cell to dehydrate during freezing. That is, if a cell is cooled sufficiently slowly, it can lose water so as to remain in osmotic equilibrium with its extracellular solution as the solution itself freezes. This analysis for cells in general (Mazur, 1977b) and for embryos in particular (Leibo, 1977b) has recently been reviewed. One corollary of Mazur's analysis is that a cell cooled at rates higher than the equilibrium rate will become increasingly supercooled with decreasing temperature. That is, the cell will contain unfrozen water because there has been insufficient time for the cell to equilibrate by losing water as the extracellular solution becomes more concentrated. Mazur (1963) argued, then, that at some subzero temperature, the metastable supercooled state of this cell water would be "broken" by the formation of intracellular ice; the cell itself would freeze.

Recently, a semiquantitative test of Mazur's hypothesis has been conducted for mouse ova (Leibo *et al.*, 1978). The test consisted of observing the response of mouse ova as they were cooled at various rates on the low-temperature stage of a specially constructed cryomicroscope. The observations showed that the fraction of ova that froze intracellularly was directly proportional to the rate at which they were cooled. Few ova cooled at ~ 1°C/min froze intracellularly, but virtually all ova cooled at ~ 5°C/min or faster froze intracellularly. Moreover, those ova that did freeze did so, on the average, at ~ −45°C, regardless of the cooling rate. There were other features of those observations that are also relevant to this discussion. The photographs in Fig. 9 illustrate the microscopic appearance of unfertilized mouse ova suspended in Me_2SO as they were cooled at each of three rates. An ovum cooled at ~ 30°C/min underwent essentially no change in volume as it was cooled; it froze intracellularly at −40°C. An ovum cooled at ~ 3°C/min exhibited some shrinkage as it was cooled; nevertheless, it also froze intracellularly at −50°C. Once intracellular ice formed, obviously there could be no further change in cell volume. An ovum cooled at ~ 1°C/min, however, exhibited substantial shrinkage as it was cooled, and it did not freeze intracellularly, even when cooled to −135°C. Qualitatively, these observations demonstrate the point made above that cells cooled at high rates freeze intracellularly, whereas cells cooled at low rates do not. Moreover, they illustrate directly the osmotic response of mouse ova during cooling and freezing. It is difficult to measure the perimeter of the cells in the presence of ice, and the ova clearly did not shrink as spheres during freezing. Nevertheless, an estimate of the water volume lost during cooling can be made from such micrographs by measurement of the cross-sectional areas of ova cooled at various rates. Volumes were estimated for ova at the temperature at which they froze, and could lose no more water, or of ova at a temperature below which they could no longer shrink, e.g., −135°C. These volumes were then subtracted from the initial volumes of the ova before freezing. The results in Fig. 10 compare the volumes lost by ova during cooling at various rates with the measured survival of ova cooled over the same range of rates. This comparison clearly indicates that cooling rates that permit water loss during freezing are coincident with those rates that yield substantial survival. Conversely, cooling rates that are too rapid to permit cellular dehydration are the same as those that yield no survival.

As mentioned initially, the "standard" procedure for successful embryo freezing has been to cool the embryos rather slowly, and then to warm the frozen embryos slowly as well. The very first observations on embryo freezing demonstrated that few if any embryos survived cooling to −80°C or below when cooled at rates of ~ 7°C/min or higher (Whittingham *et al.*, 1972). The micrographs described above provide part of the explanation;

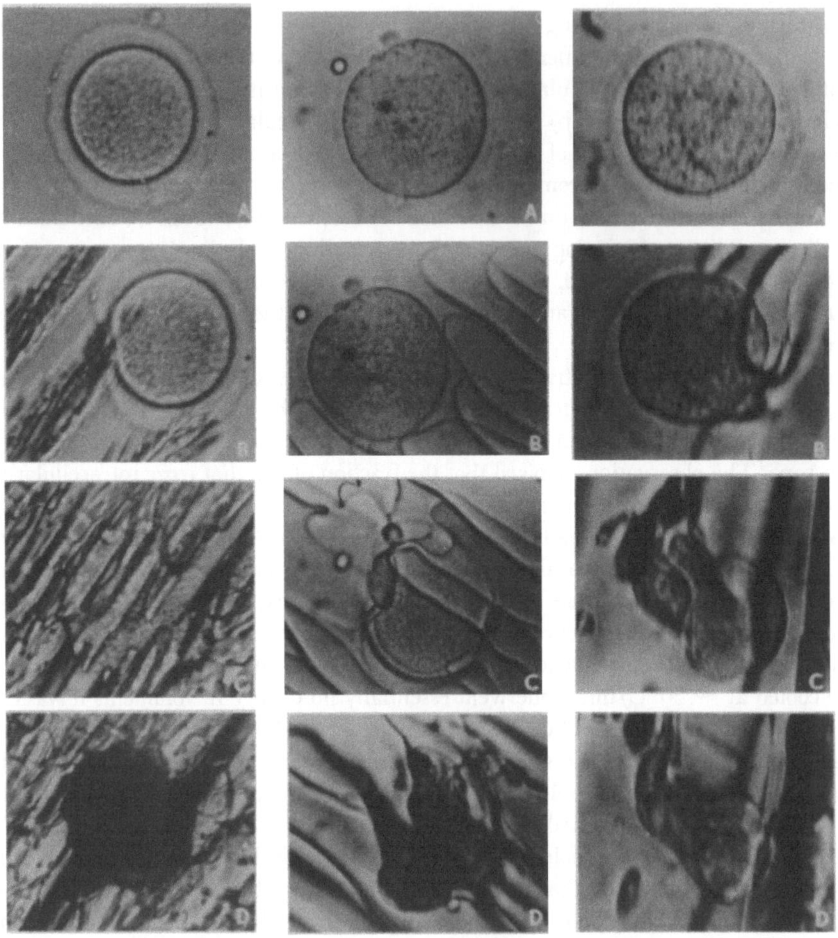

FIGURE 9. Low-temperature microscopy of mouse ova suspended in 1 M Me_2SO. The ovum in the left column was cooled at 30°C/min from 0°C (A) to −3°C (B), −20°C (C), and −40°C (D). The ovum in the middle column was cooled at 3°C/min from 0°C (A) to −3°C (B), −20°C (C), and −50°C (D). The one in the right column was cooled at ~ 1°C/min from 0°C (A) to −6°C (B), −20°C (C), and −135°C (D). The observations are those of Leibo *et al.* (1978), and the figure itself is from Leibo (1977a). (Published with permission.)

embryos cooled at rates greater than 5°C/min freeze intracellularly. It has been well documented that cells that have frozen intracellularly are extremely sensitive to warming rate. They survive if warmed rapidly but not if warmed slowly (Mazur, 1970, 1977a; Mazur *et al.*, 1972; Leibo, 1977a). Furthermore, it has been directly demonstrated by freeze-etch electron microscopy that intracellular ice undergoes recrystallization when frozen cells are warmed from −196°C to rather high subzero temperatures for a brief time before being cooled again to below −125°C to be replicated (Bank and Mazur, 1973; Bank, 1974). One interpretation of these observations is that recrystallization of ice formed intracellularly during rapid cooling disrupts the fine organization of organelles of cells. Such recrystallization can occur either during slow warming or during a brief holding time at a rather high subzero temperature. In both cases, recrystallization would result in cell death. Rapid warming is presumed to cause melting of intracellular ice, thus preventing or, at least, reducing the extent of damage (Mazur, 1977a). It has been suggested on circumstantial grounds that rapid warming can "rescue" cells frozen in such a way as to contain some intracellular ice (Leibo, 1977a).

There are survival data for mouse embryos that are fully consistent with these interpretations. Consider the data shown in Fig. 11. When embryos are slowly cooled and then rapidly warmed, they survive if the slow cooling was terminated at −30°C or above (curve labeled SC-RW); few survive if slowly cooled to −40°C or below and then rapidly warmed. When embryos are slowly cooled, then rapidly cooled to −196°C, and then slowly warmed, they survive if the slow cooling was terminated at −50°C or below (curve labeled SC, RC-SW); none survive if slowly cooled only to −30°C or above. When, however, embryos are slowly cooled, then rapidly cooled to −196°C, and then rapidly warmed, they survive only if the slow cooling was terminated between −30 and −50°C (curve labeled SC, RC-RW; data of Whittingham *et al.*, 1979). All of these rather complicated observations can be explained by the osmotic response of ova and embryos described above.

During slow cooling, the embryos must begin to lose water to equilibrate with the increasingly concentrated extracellular solution. If slowly cooled to ~ −50°C or below, the embryos will have dehydrated almost to the maximum possible extent. When they are

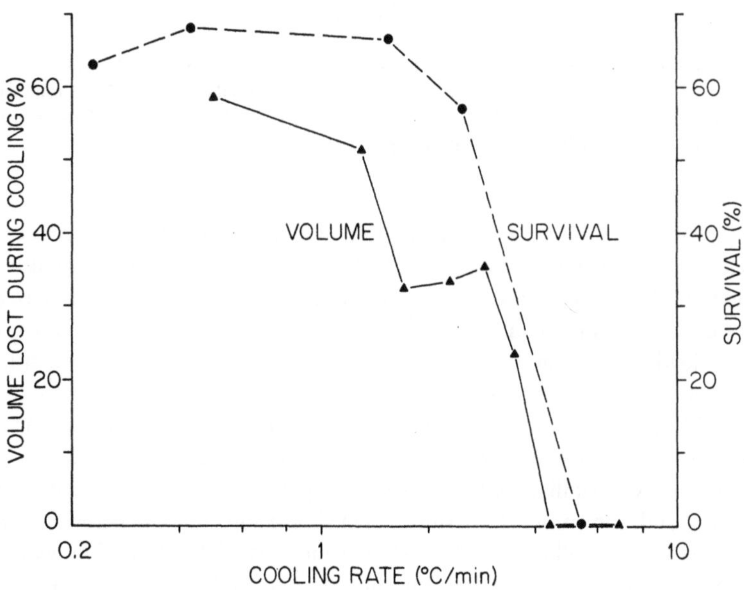

FIGURE 10. Comparison of the volume lost from mouse ova with survival of ova, both as functions of cooling rate. The survival measurements and the microscopic observations from which the volumes were calculated are from Leibo *et al.* (1978). The volume calculation is described in more detail in Leibo (1977b).

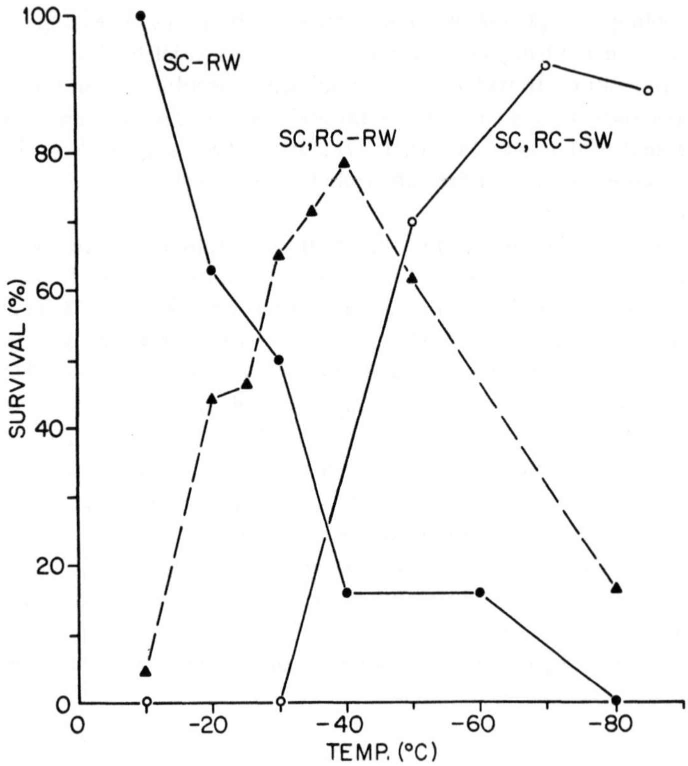

FIGURE 11. Survival of eight-cell mouse embryos suspended in Me_2SO and subjected to various cooling and warming treatments. The details are described in the text. The curve labeled SC-RW is from unpublished observations of Whittingham, Leibo, and Mazur, that labeled SC, RC-SW is from data of Leibo *et al.* (1974), and that labeled SC, RC-RW is from data of Whittingham *et al.* (1979).

then rapidly warmed, the extracellular solution will melt rapidly and produce a large osmotic pressure difference between the intracellular and the extracellular solutions. This rapid dilution will be equivalent to a large osmotic shock. The embryos will be required to rehydrate rapidly while still at subzero temperatures. Apparently, this is a damaging event (SC-RW). If slowly cooled only to about −30 to −40°C, the embryos will still contain water capable of freezing when they are rapidly cooled to −196°C. Under these conditions, the embryos will survive if rapidly warmed (SC, RC-RW), but not if slowly warmed (SC, RC-SW). Presumably, rapid warming causes the intracellular ice to melt before it can undergo damaging recrystallization; slow warming permits this recrystallization to occur. If, on the other hand, the embryos are slowly cooled to below −50°C before being rapidly cooled to −196°C, they will survive if slowly warmed (SC, RC-SW). Slow warming will be equivalent to a slow dilution of the concentrated extracellular solution. They will not survive, however, if rapidly warmed (SC, RC-RW). Again, rapid warming will be equivalent to a large and rapid dilution of the extracellular solution. If the embryos are slowly cooled just to −50°C before being rapidly cooled to −196°C, they are rather insensitive to warming rate (cf. SC, RC-RW and SC, RC-SW at −50°C). This may reflect a balance between the potentially damaging consequences of rapid dilution and of recrystallization of some intracellular ice. In any case, the response of embryos subjected to very complicated cooling and warming treatments can be qualitatively explained in terms of (1) the concentration and dilution of solutions produced by freezing and thawing, respectively, and (2) by the osmotic responses of embryos exposed to such concentrated and diluted solutions. More importantly, however, all of these results taken together indicate that the freezing of mammalian ova and embryos is beginning to rest on a rather firm analytical base. The freezing of embryos has not been accomplished by simple empiricism.

2.5. Other Aspects of Embryo Freezing

2.5.1. Suspending Media

A wide variety of compounds has been shown to protect mammalian cells against freezing damage. Generally, these so-called protective compounds have been divided into two broad categories: (1) low-molecular-weight permeating solutes and (2) high- and low-molecular-weight nonpermeating solutes. The more commonly used ones in the latter category are sucrose, hydroxyethyl starch, and polyvinylpyrrolidone (PVP). The former category includes glycerol, dimethylsulfoxide, and various glycols. Thus far, no nonpermeating compounds have been demonstrated to protect embryos against freezing to $-196°C$. Whittingham (1971) had reported that PVP would protect embryos against freezing to $\sim -75°C$ for 30 min or less, although a number of other studies, including that by Whittingham *et al.* (1972), were unable to confirm this observation (Wilmut, 1972; Maurer *et al.,* 1977). Most studies of embryo freezing have used Me_2SO as the protective medium, although glycerol also protects both mouse ova and embryos against freezing damage (Jackowski, 1977; Whittingham *et al.,* 1972). Several hypotheses have been advanced to explain the mechanism(s) by which protective compounds act (e.g., Meryman *et al.,* 1977; Farrant and Morris, 1973), although none is completely satisfactory. The most generally accepted hypothesis is that of Lovelock (1953), who argued that protective compounds act colligatively to reduce the electrolyte concentration at any given temperature. Recently, however, this hypothesis has also been found to be flawed (Mazur, 1977b).

Despite the lack of an adequate explanation of their effect, such compounds do, in fact, protect ova and embryos against freezing damage. It also appears that, at least for some of these compounds, their protective action is related to their ability to permeate the embryos. The situation is complex, however, in that the permeability of ova and embryos of a given species changes during development. For example, Jackowski *et al.* (1980) have recently shown that the fertilized mouse ovum is significantly more permeable to glycerol than is the unfertilized egg. Comparison of the change in volume (a corollary of solute permeability) of unfertilized and fertilized ova as a function of time in glycerol at various constant temperatures clearly shows that the fertilized egg takes up glycerol faster than the unfertilized egg (Fig. 12). Moreover, these workers have also shown that the permeability of the fertilized egg itself changes with time after fertilization (Fig. 13). A further complication, shared with many other types of cells (Davson and Danielli, 1952), is that embryos of a given stage differ in their permeability to different compounds. Circumstantial evidence for this is found by comparing the survival of eight-cell mouse embryos frozen in solutions

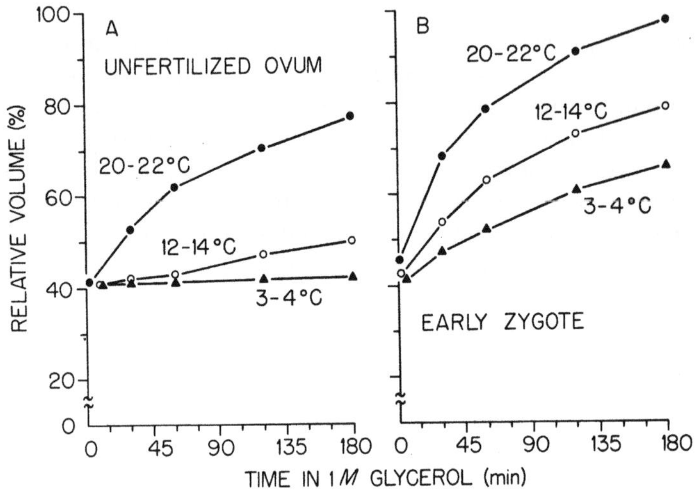

FIGURE 12. Comparison of the volume changes of fertilized and unfertilized mouse ova, relative to their volumes in isotonic solution, as a function of time in 1 M glycerol at the indicated temperatures. The figure is from Jackowski *et al.* (1980), who describe the methods and the calculations. (Published with permission.)

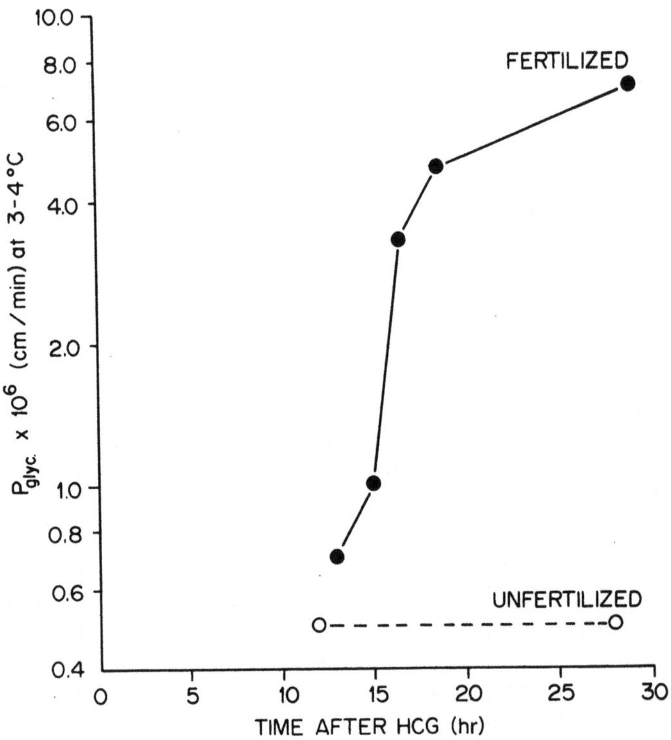

FIGURE 13. The glycerol permeability coefficient, $P_{glyc.}$, of fertilized and unfertilized mouse ova as a function of time after the injection of human chorionic gonadotropin (HCG) at which the ova were collected. Fertilization is estimated to occur ~ 10–12 hr after injection of HCG. The figure is from Jackowski *et al.* (1980). (Published with permission.)

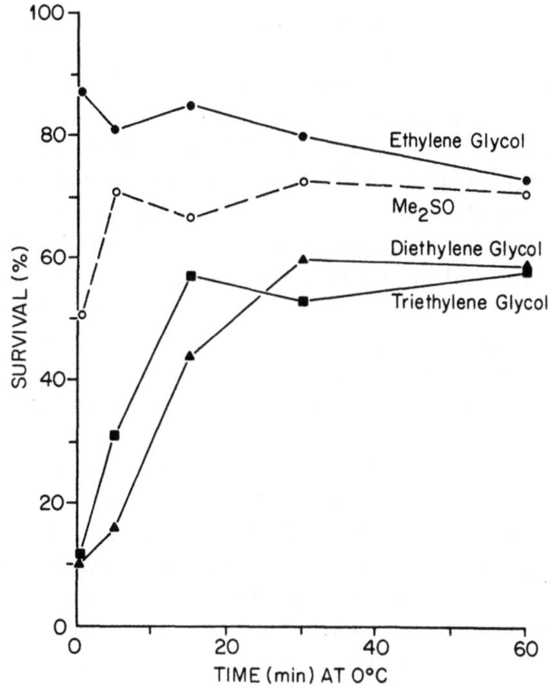

FIGURE 14. Survival of eight-cell mouse embryos frozen to −79°C or below in 1.2 M solutions of each of the indicated solutions, as a function of the time the embryos were exposed to the solutions at 0°C prior to freezing. The data for the glycols are those of Miyamoto and Ishibashi (1978), and for Me_2SO are those of Leibo *et al.* (1974).

of Me_2SO (Leibo *et al.,* 1974) and of various glycols (Miyamoto and Ishibashi, 1978). The results in Fig. 14 indicate that the length of time that the embryos are exposed to solute prior to freezing has no effect with ethylene glycol (mol. wt. 62), little effect with Me_2SO (mol. wt. 78), but a rather significant effect with both di- and triethylene glycol (mol. wt. 106 and 150, respectively). Presumably, the effect of time prior to freezing reflects the permeation of the embryos by the solutes. The problem, of course, is that it is impossible to analyze such data fully in the absence of actual permeability measurements. An additional complication is that embryos of different species undoubtedly differ in their permeability to different compounds. An example of this for a cell type other than embryo is that the bovine erythrocyte is 20-fold more permeable to triethylene glycol than to glycerol, whereas the rabbit erythrocyte is somewhat more permeable to glycerol than to triethylene glycol (Davson and Danielli, 1952). Finally, there is the recent significant observation of Miyamoto and Ishibashi (1978) that polyethylene glycol (mol. wt. 380–420) also protects mouse embryos against freezing damage. This compound may not permeate eight-cell embryos at all at 0°C prior to freezing. True understanding of the role of protective compounds in embryo freezing will come only with a more complete analysis of the permeability properties of different embryo stages of different species to different solutes.

2.5.2. Storage Time of Frozen Embryos

A common variable in many cryobiological studies is the length of time that the frozen cells are stored in liquid nitrogen at −196°C before being thawed. Although there have been occasional reports of the deterioration of cells stored at −196°C, in none of these cases can the simplest explanation be excluded, namely, that all of the liquid nitrogen simply evaporated and the cells thawed. In the case of frozen embryos, a number of studies have demonstrated that the survival after storage at −196°C did not decrease for periods of ~ 1 year (Leibo *et al.*, 1974; Whittingham and Whitten, 1974) and $2\frac{1}{4}$ years (Whittingham *et al.*, 1977). This latter observation has now been extended. Very recently, M. Lyon, P. Glenister, and D. G. Whittingham (1980, personal communication) completed a study in which embryos were stored at −196°C for 24 hr, ~ 1, ~ 3, and ~5 years. Based on both *in vitro* and *in vivo* development, these workers have found no effect whatsoever on the survival of embryos frozen for extended times. They have produced living mice from embryos stored for > 5 years and then transferred into pseudopregnant foster mothers. There now seems little reason to doubt that frozen embryos can be stored at −196°C for extremely long periods of time. It has been estimated, for example, that embryos in the frozen state can probably be stored for periods of 200 to 3000 years before accumulating sufficient radiation damage to adversely affect the mice to be derived from those embryos (Mazur, 1976; DuFrain, 1976). Whittingham *e al.* (1977) have attempted to assess this possibility directly by exposing frozen embryos at −196°C to γ radiation for a period of 29 months at a dose level equivlent to 84 times the background level. They reported little if any effect of this treatment on the number of living young born after the frozen-thawed, irradiated embryos were transferred into foster mothers. This observation has also been extended by Lyon *et al.* (1980, as cited above) to exposure times of 5 years with very similar results.

3. Embryological Aspects of Embryo Freezing

3.1. Introduction

By now, it should be clear that mammalian embryos can indeed be successfully stored in the frozen state. There are other aspects of embryo freezing and storage, however, that are not directly related to cryobiology but rather to embryology and/or whole animal physiology. These include (1) the question of differences between the response to freezing of embryos of different species, (2) differences between embryos of different strains, especially mouse strains, (3) the possibility of *in vitro* fertilization of frozen-thawed ova, (4) the ability to culture frozen-thawed embryos to assess survival, and (5), most importantly, the transfer of frozen-thawed embryos to foster mothers to generate living young. Although the latter three items are somewhat out of the purview of this review, they will be discussed very briefly below.

3.2. Species Differences

As mentioned in Section 1, successful embryo preservation has been reported for seven mammalian species. The procedures used for embryos of different species are roughly the same, although there are some minor differences. The first successful report of freezing an embryo from a species other than the mouse was that of Wilmut and Rowson (1973) for a bovine embryo, although only 1 of 21 frozen-thawed blastocysts developed into a living calf when transferred into a synchronized recipient. Since that first meager success, extensive experiments with embryos of cattle, sheep, and goats have indicated that higher survival is obtained with 1.5 M Me_2SO, slow cooling to $\sim -60°C$, transfer into liquid nitrogen, followed by rapid warming to $-50°C$, and finally by slow warming (Bilton and Moore, 1979; Willadsen *et al.,* 1976, 1978; Trounson *et al.,* 1978). It may be significant that no successful freezing of bovine embryos other than the late morula or blastocyst stages has yet been reported. Of course, this may only reflect the fact that blastocyst-stage embryos are required for the nonsurgical transfer of bovine embryos; for obvious reasons, this procedure is preferred to that of surgical embryo transfer.

The report of the first successful rabbit embryo freezing (Bank and Maurer, 1974) emphasized differences between the optimum procedures for mouse versus rabbit embryos. The differences were primarily the Me_2SO concentration (1.2 M for the mouse, 1.6 M for the rabbit) and the method of dilution of the Me_2SO after thawing. These authors were apparently unable to freeze rabbit embryos earlier than the eight-cell stage. Later work by Whittingham and Adams (1976), however, showed that four-cell as well as eight-cell rabbit embryos can be frozen using procedures very similar to those used for mouse embryos. Similar results with rabbit embryos of the four- to eight-cell stages, including some developed from *in vitro*-fertilized eggs (Schneider *et al.,* 1974), have also been reported (Tsunoda and Sugie, 1977). An extensive study by Maurer and Haseman (1976) demonstrated that maximum survival of morula-stage rabbit embryos was obtained using 2 M Me_2SO. With that exception, then, the freezing procedure that yields good survival of mouse embryos also yields good survival of rabbit embryos. Examining yet another species, Whittingham (1975) reported that two- to eight-cell-stage embryos of the rat can also be successfully frozen by the procedure that yields survival of frozen-thawed mouse embryos. In summary, preimplantation-stage embryos, especially those of late developmental stages, of a wide variety of animal species can all be frozen essentially by the methods first described for mouse embryos. The only variables that may have a significant effect on embryo survival appear to be the Me_2SO concentration (1.5–2 M appears better than 1 M) and the dilution temperature after thawing.

3.3. Mouse Strain Differences

A reasonable estimate is that more than 1000 different strains of mice are currently used for biological and medical research. Institutions such as the National Institutes of Health, The Jackson Laboratory, and the Oak Ridge National Laboratory in the United States, the Radiobiology Unit at Harwell and the Laboratory Animals Centre in Carshalton in the United Kingdom, and the Netherlands Cancer Institute in Amsterdam normally maintain dozens if not hundreds of different inbred and congenic lines of mice. In addition, there are mouse colonies maintained in universities, commercial breeding laboratories, and private pharmaceutical and testing companies. If embryo freezing is to have wide applicability for the preservation of this enormous genetic pool of mouse strains, it is important

TABLE II
Strains of Successfully Frozen Mouse Embryos

Reference	Institution[a]	Strain Frozen	Maximum survival (%)
Leibo *et al.* (1974)	ORNL, U.S.	(C57BL × C3H/AN)F_1	90
Leibo *et al.* (1978)	MIT, U.S.	CD1	65
Maurer *et al.* (1977)	NIEHS, U.S.	CD1	55
Miyamoto and Ishibashi (1978)	Kyoto Univ., Japan	ICR	86
Parkening *et al.* (1976)	Worcester Found., U.S.	CD1	65
Schneider *et al.* (1974)	TH Hannover, W. Germany	NMR1/Hann.	84
Whittingham *et al.* (1972)	ORNL, U.S.	(C57BL × C3H/AN)F_1	70
		Swiss-Webster albino	70
Whittingham (1977)	Univ. London, U.K.	MF1	55
Whittingham and Whitten (1974)	Jackson Lab., U.S.	C57BL/6J	95
		BALB/c-*wt*	58
Whittingham *et al.* (1977)	London, Harwell, U.K.	(C3H/HeH × 101/H)F_1	54
Whittingham *et al.* (1979)	London, LAC, CRC, U.K.	(C57BL/6J-*Lac*	88
		MF1	88
Wilmut (1972)	ARC Cambridge, U.K.	CFLP	65
Zeilmaker and Verhamme (1979)	Erasmus Univ., Netherlands	(C57BL × CBA)F_1	67

[a]Abbreviations: ORNL, Oak Ridge National Laboratory; MIT, Massachusetts Institute of Technology; NIEHS, National Institute of Environmental Health Sciences; TH, Tierärztlichen Hochschule; LAC, Laboratory Animals Centre; CRC, Clinical Research Centre; ARC, Agricultural Research Centre.

to know whether the procedure for freezing need be modified for each new strain to be preserved. Although it may be premature to draw a definitive conclusion, as no prospective studies of this question have yet been published, the available data suggest that the same freezing methods can be applied to the embryos of all mouse strains.

Documentary evidence of the successful freezing of mouse embryos has been reported from at least 15 different laboratories in the United States, Great Britain, The Netherlands, Germany, and Japan (Table II). In addition, as a result of a Training Course held at Erasmus University in The Netherlands and "Roving Seminars" sponsored by UNESCO, the International Cell Research Organization, and the International Committee on Laboratory Animals, mouse embryos have been successfully frozen in Denmark, Czechoslovakia, Hungary, Italy, Poland, and Yugoslavia. A variety of mouse strains were successfully used for demonstration purposes. More importantly, as shown in Table II, successful embryo freezing has been formally documented for at least 13 different strains. Most of the mice used for these studies have been random-bred or F1 animals, as many of the reports published thus far have been primarily concerned with rather fundamental aspects of the subject. But enough inbred lines of mice have been used to demonstrate that their embryos can also be successfully frozen. Regardless of the source of embryos, all have exhibited good to excellent survival after freezing. Apparent differences between strains suggested by the survival values in the table are misleading. In those cases in which survival was somewhat lower (~ 55%), experimental variables other than mouse strain were clearly responsible.

Embryos of virtually all of the strains listed have developed to living fetuses* or live-born mouse pups upon transfer into pseudopregnant recipients. Because most of the embryos listed in Table II were frozen by the "standard" procedure, it seems reasonable to conclude that embryos of all mouse strains can be frozen by the same procedure.

Only a few data have been published regarding the reproductive performance of mice that developed from frozen-thawed embryos. Although the observations were based on limited numbers of mated pairs of animals, Leibo (1976) reported that "frozen-thawed" mice exhibited essentially the same fertility as control animals. For example, 7 females of 10 mated pairs (in which both the males and the females developed from frozen-thawed embryos) littered an average of 8.6 mouse pups. The corresponding figures for control animals were that 75% of the females of 121 mated pairs littered an average of 7.6 mouse pups. Subsequent generations of the offspring of "frozen-thawed" mice also exhibited completely normal fertility. Maurer *et al.* (1977) have reported similar observations. They found that the reproductive performance at 65 days of age of mice that had been frozen as eight-cell embryos was virtually the same as that of control animals. In other words, animals that develop to term from frozen-thawed embryos are as capable of reproducing themselves as animals obtained by natural mating.

3.4. *In Vitro* Fertilization of Frozen-Thawed Ova

The usual goal of embryo freezing is the preservation of the entire genome of an animal. In this respect, embryo freezing is clearly superior to sperm freezing. However, the ability to store unfertilized oocyes opens the possibility of establishing banks from which specific genetic combinations of male and female genomes could be originated (Whittingham, 1977).

As a curious aside, it is interesting that the entire discipline of cryobiology has as its origin an observation by Polge *et al.* (1949) that glycerol protects turkey spermatozoa against freezing damage. It was quickly established that glycerol would also protect bovine and human sperm as well. It has even been reported (Seager, 1977, Society for Cryobiology Annual Meeting) that the sperm of approximately 100 different species of wild animals have been successfully frozen, yielding living young when used for artificial insemination. Nevertheless, there are no documented reports of the successful preservation of the sperm of laboratory animals, such as the mouse, rat, and rabbit. Despite that present flaw in the

*Throughout this review, I have referred in passing to "living fetuses" developed from frozen-thawed embryos transferred into pseudopregnant recipients. In many studies involving embryo transfer, the foster mothers are sacrificed a few days prior to their expected delivery of living young. The reason is that additional information about the transfer can be obtained by examination of the reproductive tract prior to birth. For example, some transferred embryos may implant into the uterine wall, begin fetal development, but then degenerate. These appear as "implantation sites" in the uterus, demonstrating that the embryo was alive at transfer, although its subsequent development was abnormal. Normal living fetuses can easily be identified while still in the uterus several days before birth. The use of eye-color markers, e.g., embryos obtained from albino, pink-eyed strains transferred into dark-eyed recipients, provides assurance that the living fetuses did in fact develop from the transferred embryos and not from some "accidental" natural pregnancy.

system, banks of frozen sperm and of frozen ova of laboratory species seem to open new possibilities for research.

Since the first study of embryo freezing (Whittingham *et al.,* 1972) showed that fertilized ova could be successfully frozen, it was a relatively small step to apply these procedures to unfertilized ova as well. Three such studies have been published (Parkening *et al.,* 1976; Tsunoda *et al.,* 1976; Whittingham, 1977). Briefly, these workers have shown that freshly ovulated oocytes of the mouse can be frozen and stored at −196°C. After thawing, the ova were fertilized *in vitro,* and then transferred directly into pseudopregnant recipients or cultured *in vitro* before transfer. Although the efficiency of *in vitro* fertilization of frozen-thawed ova was diminished, live mice pups have been born. Hamster oocytes have apparently also been frozen successfully, although no transfers have been attempted. Future study should undoubtedly yield the successful preservation of oocytes of a variety of mammalian species, as well as the preservation of the sperm of these species. When applied to the ova of commercially important animals, these procedures should permit even greater control and improvement of the reproductive process of these animals.

3.5. Culture and Transfer of Frozen-Thawed Embryos

As mentioned previously, the usual application of embryo preservation is simply to interrupt the normal development of laboratory or domestic animals. In most cases, living young are ultimately to be produced from the frozen embryos. Two approaches have been adopted with approximately equal success. In the first, the thawed embryos are cultured *in vitro* to assess their survival. They are then transferred into suitably primed recipients to develop to term. The second approach is to transfer the embryos immediately after they have been thawed. The real success of an embryo-freezing experiment may depend as much on the culture and transfer procedures as on the freezing itself. Many excellent reviews of both subjects have been published (see, for example, Daniel, 1971, 1978).

Briefly, *in vitro* culture can be reliably performed with all preimplantation stages of the mouse (e.g., Brinster, 1972) and of the rabbit (e.g., Kane and Foote, 1970). Rat embryos, however, can usually only be cultured from the eight-cell stage to the blastocyst (Mayer and Fritz, 1974). The advantage of a culture system, when available, is that embryo survival can usually be estimated within 12–24 hr after the embryos have been thawed. In general, frozen-thawed embryos that begin to develop normally very shortly after being placed in culture will usually develop to the blastocyst stage, the final stage of preimplantation development. Shortly after reaching the expanded blastocyst stage, embryos of laboratory animals implant into the uterine wall of the mother to begin true fetal development.

The usual final step in embryo freezing is the transfer of the frozen-thawed embryos into primed recipient females. For laboratory animals, this normally involves a rather simple surgical procedure. The reproductive tract of the recipient is exposed, a puncture made through to the lumen of the uterus, and the embryos gently blown into the uterus with a fine pipet. Most transfers are of eight-cell- to blastocyst-stage embryos, although earlier stages, especially of the mouse, can also be transferred. The entire complicated subject of recipient preparation, surgical procedures, and embryo manipulation has been discussed in many articles (again, see Daniel, 1971). Recently, a simple nonsurgical transfer procedure for the mouse, similar in principle to that used with large domestic animals, has been described (Moler *et al.,* 1979).

4. Summary

There now exist reliable procedures by which early embryos of laboratory animals can be preserved for extended times in the frozen state. When thawed, these embryos develop normally in culture and yield living young when transferred into suitably primed recipient foster mothers. The availability of this procedure means that valuable, unusual, or rarely used strains of mice can be protected against perhaps irrevocable loss by storing their embryos in the frozen state. It also means that laboratory animals, in the form of frozen embryos, can be moved from place to place or held in a state of suspended animation.

At a more fundamental level, the freezing of mammalian embryos can be explained in terms of cellular physiology and the physical chemistry of solutions. The freezing of embryos has prompted investigation of other fundamental properties of these unusual cells, such as their permeability to water and to various solutes. Furthermore, although embryo freezing was achieved by the application of basic principles of cryobiology, the study of their responses to freezing has led, in turn, to a greater understanding of cryobiology itself.

ACKNOWLEDGMENTS

Oak Ridge National Laboratory is operated by Union Carbide Corporation under Contract W-7405-eng-26 with the U.S. Department of Energy.

By acceptance of this article, the publisher or recipient acknowledges the U.S. Government's right to retain a nonexclusive, royalty-free license in and to any copyright covering the article.

References

Bank, H., 1974, Freezing injury in tissue cultured cells as visualized by freeze-etching, *Exp. Cell Res.* **85**:367.

Bank, H., and Maurer, R. R., 1974, Survival of frozen rabbit embryos, *Exp. Cell Res.* **89**:188.

Bank, H., and Mazur, P., 1973, Visualization of freezing damage, *J. Cell Biol.* **57**:729.

Bilton, R. J., and Moore, N. W., 1976, *In vitro* culture, storage, and transfer of goat embryos, *Aust. J. Biol. Sci.* **29**:125.

Bilton, R. J., and Moore, N. W., 1979, Factors affecting the viability of frozen stored cattle embryos, *Aust. J. Biol. Sci.* **32**:101.

Brinster, R. L., 1972, Cultivation of the mammalian embryo, in: *Growth, Nutrition, and Metabolism of Cells in Culture* (G. H. Rothblat and V. J. Cristofalo, eds.), pp. 251–286, Academic Press, New York.

Cocks, F. H., and Brower, W. E., 1974, Phase diagram relationships in cryobiology, *Cryobiology* **11**:340.

Daniel, J. C., Jr., (ed.), 1971, *Methods in Mammalian Embryology,* Freeman, San Francisco.

Daniel, J. C., Jr., (ed.), 1978, *Methods in Mammalian Reproduction,* Academic Press, New York.

Davson, H., and Danielli, J. F., 1952, *The Permeability of Natural Membranes,* Cambridge University Press, Cambridge.

DuFrain, R. J., 1976, The effects of ionizing radiation on preimplantation mouse embryos developing *in vitro,* in: *Basic Aspects of Freeze Preservation of Mouse Strains* (O. Mühlbock, ed.), pp. 73–84, Gustav Fischer Verlag, Stuttgart.

Farrant, J., and Morris, G. J., 1973, Thermal shock and dilution shock as the causes of freezing injury, *Cryobiology* **10:**134.

Jackowski, S. C., 1977, Physiological differences between fertilized and unfertilized mouse ova: Glycerol permeability and freezing sensitivity, Ph.D. dissertation, The University of Tennessee.

Jackowski, S. C., Leibo, S. P., and Mazur, P., 1980, Glycerol permeabilities of fertilized and unfertilized mouse ova, *J. Exp. Zool.* **212:**329.

Kane, M. T., and Foote, R. H., 1970, Culture of two- and four-cell rabbit embryos to the expanding blastocyst stage in synthetic media, *Proc. Soc. Exp. Biol. Med.* **133:**921.

Leibo, S. P., 1976, Sensitivity of mouse embryos to freezing and thawing, in: *Basic Aspects of Freeze Preservation of Mouse Strains* (O. Mühlbock, ed.), pp. 13–33, Gustav Fischer Verlag, Stuttgart.

Leibo, S. P., 1977a, Preservation of mammalian cells and embryos by freezing, in: *Cryoimmunologie* (D. Simatos, D. M. Strong, and J. M. Turc, eds.), pp. 311–334, INSERM, Paris.

Leibo, S. P., 1977b, Fundamental cryobiology of mouse ova and embryos, *Ciba Found. Symp.* **52:**69.

Leibo, S. P., 1980, Water permeability and its activation energy of fertilized and unfertilized mouse ova, *J. Membrane Biol.* **53:**179.

Leibo, S. P., and Mazur, P., 1971, The role of cooling rates in low temperature preservation, *Cryobiology* **8:**447.

Leibo, S. P., Mazur, P., and Jackowski, S. C., 1974, Factors affecting survival of mouse embryos during freezing and thawing, *Exp. Cell Res.* **89:**79.

Leibo, S. P., McGrath, J. J., and Cravalho, E. G., 1978, Microscopic observation of intracellular ice formation in unfertilized mouse ova as a function of cooling rate, *Cryobiology* **15:**257.

Lovelock, J. E., 1953, The mechanism of the protective action of glycerol against haemolysis by freezing and thawing, *Biochim. Biophys. Acta* **11:**28.

Maurer, R. R., and Haseman, J. K., 1976, Freezing morula stage rabbit embryos, *Biol. Reprod.* **14:**256.

Maurer, R. R., Bank, H., and Staples, R. E., 1977, Pre- and postnatal development of mouse embryos after storage for different periods at cryogenic temperatures, *Biol. Reprod.* **16:**139.

Mayer, J. F., Jr., and Fritz, H. I., 1974, The culture of preimplantation rat embryos and the production of allophenic rats, *J. Reprod. Fertil.* **39:**1.

Mazur, P., 1963, Kinetics of water loss from cells at subzero temperatures and the likelihood of intracellular freezing, *J. Gen. Physiol.* **47:**347.

Mazur, P., 1966, Physical and chemical basis of injury in single-celled microorganisms subjected to freezing and thawing, in: *Cryobiology* (H. T. Meryman, ed.), pp. 213–315, Academic Press, New York.

Mazur, P., 1970, Cryobiology: The freezing of biological systems, *Science* **168:**939.

Mazur, P., 1976, Freezing and low-temperature storage of living cells, in: *Basic Aspects of Freeze Preservation of Mouse Strains* (O. Mühlbock, ed.), pp. 1–12, Gustav Fischer Verlag, Stuttgart.

Mazur, P., 1977a, The role of intracellular freezing in the death of cells cooled at supraoptimal rates, *Cryobiology* **14:**251.

Mazur, P., 1977b, Slow freezing injury in mammalian cells, *Ciba Found. Symp.* **52:**19.

Mazur, P., Leibo, S. P., and Chu, E. H. Y., 1972, A two-factor hypothesis of freezing injury, *Exp. Cell Res.* **71:**345.

Meryman, H. T., Williams, R. J., and Douglas, M. St. J., 1977, Freezing injury from "solution effects" and its prevention by natural or artificial cryoprotection, *Cryobiology* **14:**287.

Miyamoto, H., and Ishibashi, T., 1977, Survival of frozen-thawed mouse and rat embryos in the presence of ethylene glycol, *J. Reprod. Fertil.* **50:**373.

Miyamoto, H., and Ishibashi, T., 1978, The protective action of glycols against freezing damage of mouse and rat embryos, *J. Reprod. Fertil.* **54:**427.

Moler, T. L., Donahue, S. E., and Anderson, G. B., 1979, A simple technique for nonsurgical embryo transfer in mice. *Lab. Anim. Sci.* **29:**353.

Parkening, T. A., Tsunoda, Y., and Chang, M. C., 1976, Effects of various low temperatures, cryoprotective agents, and cooling rates on the survival, fertilizability and development of frozen-thawed mouse eggs, *J. Exp. Zool.* **197:**369.

Polge, C., Smith, A. U., and Parkes, A. S., 1949, Revival of spermatozoa after vitrification and dehydration at low temperatures, *Nature (London)* **164:**666.

Schneider, U., Hahn, J., and Sulzer, H., 1974, Erste Ergebnisse der Tiefgefrierkonservierung von Mäuse- und Kanincheneizellen, *D. tsch. Tieraerztl. Wochenschr.* **81**:445.

Trounson, A. O., Shea, B. F., Ollis, G. W., and Jacobson, M. E., 1978, Frozen storage and transfer of bovine embryos, *J. Anim. Sci.* **47**:677.

Tsunoda, Y., and Sugie, T., 1977, Survival of rabbit eggs preserved in plastic straws in liquid nitrogen, *J. Reprod. Fertil.* **49**:173.

Tsunoda, Y., Parkening, T. A., and Chang, M. C., 1976, *In vitro* fertilization of mouse and hamster eggs after freezing and thawing, *Experientia* **32**:223.

Whittingham, D. G., 1971, Survival of mouse embryos after freezing and thawing, *Nature (London)* **233**:125.

Whittingham, D. G., 1975, Survival of rat embryos after freezing and thawing, *J. Reprod. Fertil.* **43**:575.

Whittingham, D. G., 1977, Fertilization *in vitro* and development to term of unfertilized mouse oocytes previously stored at −196°C, *J. Reprod. Fertil.* **49**:89.

Whittingham, D. G., and Adams, C. E., 1976, Low temperature preservation of rabbit embryos, *J. Reprod. Fertil.* **47**:269.

Whittingham, D. G., and Whitten, W. K., 1974, Long-term storage and aerial transport of frozen mouse embryos, *J. Reprod. Fertil.* **36**:433.

Whittingham, D. G., Leibo, S. P., and Mazur, P., 1972, Survival of mouse embryos frozen to −196° and −269°C, *Science* **178**:411.

Whittingham, D. G., Lyon, M. F., and Glenister, P. H., 1977, Long-term storage of mouse embryos at −196°C: The effect of background radiation, *Genet. Res.* **29**:171.

Whittingham, D. G., Wood, M., Farrant, J., Lee, H., and Halsey, J. A., 1979, Survival of frozen mouse embryos after rapid thawing from −196°C, *J. Reprod. Fertil.* **56**:11.

Willadsen, S. M., Polge, C., Rowson, L. E. A., and Moor, R. M., 1976, Deep freezing of sheep embryos, *J. Reprod. Fertil.* **46**:151.

Willadsen, S., Polge, C., and Rowson, L. E. A., 1978, The viability of deep-frozen cow embryos, *J. Reprod. Fertil.* **52**:391.

Wilmut, I., 1972, The effect of cooling rate, warming rate, cryoprotective agent and stage of development on survival of mouse embryos during freezing and thawing, *Life Sci.* **11**:1071.

Wilmut, I., and Rowson, L. E. A., 1973, Experiments on the low-temperature preservation of cow embryos, *Vet. Rec.* **92**:686.

Wood, M. J., and Farrant, J., 1980, Preservation of mouse embryos by two-step freezing, *Cryobiology* **17**:178.

Zeilmaker, G. H., and Verhamme, C. M. P. M., 1979, A simplified method for freezing mouse embryos, *Cryobiology* **16**:6.

Index

If you have any concerns about our products,
you can contact us on
ProductSafety@springernature.com

In case Publisher is established outside the EU,
the EU authorized representative is:
Springer Nature Customer Service Center GmbH
Europaplatz 3, 69115 Heidelberg, Germany

Printed by Libri Plureos GmbH
in Hamburg, Germany

MIX
Papier aus verantwortungsvollen Quellen
Paper from responsible sources
FSC® C105338

If you have any concerns about our products,
you can contact us on
ProductSafety@springernature.com

In case Publisher is established outside the EU,
the EU authorized representative is:
Springer Nature Customer Service Center GmbH
Europaplatz 3, 69115 Heidelberg, Germany

Printed by Libri Plureos GmbH
in Hamburg, Germany